Advances in Neurobiology

Volume 26

Series Editor
Arne Schousboe, Department of Drug Design & Pharmacology,
University of Copenhagen, Copenhagen, Denmark

More information about this series at http://www.springer.com/series/8787

Baoman Li • Vladimir Parpura
Alexei Verkhratsky • Caterina Scuderi
Editors

Astrocytes in Psychiatric Disorders

Springer

Editors
Baoman Li
Department of Forensic Analytical
Toxicology, School of Forensic Medicine
China Medical University
Shenyang, China

Vladimir Parpura
Department of Neurobiology
The University of Alabama at Birmingham
Birmingham, AL, USA

Alexei Verkhratsky
Faculty of Biology, Medicine and Health
University of Manchester
Manchester, UK

Caterina Scuderi
Department of Physiology and
Pharmacology "Vittorio Erspamer"
Sapienza University of Rome
Rome, Italy

ISSN 2190-5215 ISSN 2190-5223 (electronic)
Advances in Neurobiology
ISBN 978-3-030-77374-8 ISBN 978-3-030-77375-5 (eBook)
https://doi.org/10.1007/978-3-030-77375-5

This Springer imprint is published by the registered company Springer Nature Switzerland AG
The registered company address is: Gewerbestrasse 11, 6330 Cham, Switzerland

Contents

Part I
General Aspects of Astrogliopathology with Emphasis on Neurocognitive Disorders

Neuroglia in Psychiatric Disorders

Caterina Scuderi, Alexei Verkhratsky, Vladimir Parpura, and Baoman Li

Introduction: Definition, Classification, and Main Functions of Neuroglia

The human brain has a considerable complexity. In a rather limited volume, it contains a population of more than 200 billion neural cells, including neurones and neuroglia. Altogether, these neural cells form intricate networks connecting the various parts that make up this organ through trillions of chemical and electrical synapses. The concept of neuroglia was initially formalized by Rudolf Virchow who introduced it in the mid-1800s. According to Virchow, neuroglia was a "substance also which lies between the proper nervous parts, holds them together and gives the whole its form in a greater or lesser degree" (Virchow 1860). The neuroglia is present in both the peripheral nervous system (PNS) and the central nervous system (CNS) (Fig. 1). The PNS neuroglia arises from the neural crest, similarly to peripheral neurones, and is classified into Schwann cells (Kidd et al. 2013), satellite

C. Scuderi (✉)
Department of Physiology and Pharmacology "Vittorio Erspamer",
SAPIENZA University of Rome, Rome, Italy
e-mail: caterina.scuderi@uniroma1.it

A. Verkhratsky
Faculty of Biology, Medicine and Health, The University of Manchester, Manchester, UK

Achucarro Center for Neuroscience, IKERBASQUE, Basque Foundation for Science,
Bilbao, Spain

V. Parpura
Department of Neurobiology, The University of Alabama at Birmingham,
Birmingham, AL, USA

B. Li
Department of Forensic Analytical Toxicology, School of Forensic Medicine,
China Medical University, Shenyang, China

B. Li et al. (eds.), *Astrocytes in Psychiatric Disorders*, Advances in
Neurobiology 26, https://doi.org/10.1007/978-3-030-77375-5_1

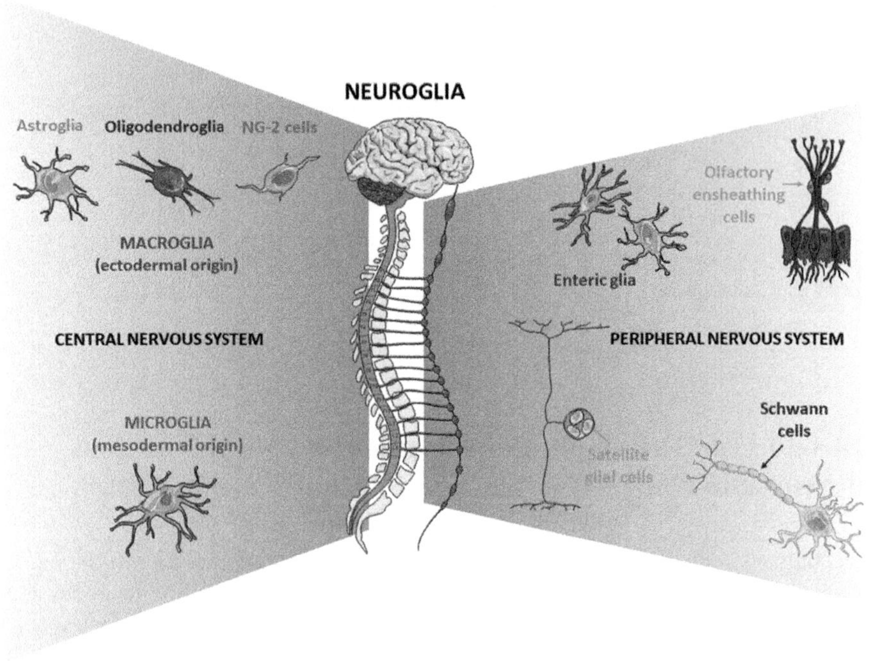

Fig. 1 Neuroglia classification

glial cells (Hanani and Verkhratsky 2021), olfactory ensheathing cells (Ruitenberg et al. 2006), and enteric glia (Grubisic et al. 2018). The neuroglia cells of the CNS are divided into macroglia cells (ectodermal, neuroepithelial origin) and microglia (mesodermal, myeloid origin) (Verkhratsky and Butt 2013). Macroglia is further classified into astroglia, oligodendroglia, and NG-2 glia, the latter also known as oligodendrocyte progenitor cells, or synantocytes, or polydendrocytes (Verkhratsky and Butt 2013). Each of these populations listed above can, in turn, be divided into further subtypes, making the complexity that these cells possess to parallel the multitude of functions they govern. The large number of subtypes of glial cells fueled for years the belief that in the human brain glial cells outnumber neurones by a factor of 10 up to 50 (Bear et al. 2007; Kandel et al. 2000). However, the views of numerical preponderance of glial cells in the brain and spinal cord with respect to the number of neurones have been proven erroneous, because none of the concepts that had been adopted as a demonstration of big glial numbers has been corroborated experimentally (Hilgetag and Barbas 2009; von Bartheld et al. 2016). It is generally agreed upon that the total number of neuronal and non-neuronal cells in the human brain is almost on par. Nonetheless, even if not in a linear manner, the evolution of the nervous system paralleled with trend of an increase in glia to neurone ratio (Verkhratsky et al. 2019), suggesting glial involvement in cerebral superior functions, although the largest numbers of glial cells are observed in the largest brains of whales and elephants.

The unifying fundamental function of all types of glial cells, regardless of their origin, structure, morphological appearance, and function, is the maintenance of the homeostasis of the nervous system (Verkhratsky and Butt 2013). This function is of crucial importance in the healthy brain, when neuroglia perform the normal housekeeping duties, as well as in pathology, when glial cells react to unusual stimuli and undergo morphofunctional modifications aimed at restoring brain homeostasis. Any deviation from this delicate equilibrium may have serious consequences in the correct development or functioning of the brain. The homeostatic support of neuroglia takes place at all levels of brain organization, thus allowing the brain to function properly.

Microglia

Microglia are the main type of immune cells that permanently reside in the CNS. Unlike all other brain parenchymal cells that have multiple neuroectodermal lineages, microglia originate from a mesodermal source. Microglia are of the myeloid origin, colonize the CNS very early in evolution, and are conserved across species (Ginhoux et al. 2010; Monier et al. 2007; Schlegelmilch et al. 2011; Swinnen et al. 2013; Verney et al. 2010). For a long time studies on microglia have been focused on their function as resident macrophages and their role in the immune response (Cartier et al. 2014; Prinz et al. 2011; Prinz and Priller 2014). Being the main immunocompetent cells of the nervous system, microglia fulfill fundamental defensive function by the virtue of their phagocytic capacity and ability to secrete numerous pro- and anti-inflammatory factors. Through phagocytosis, microglia can incorporate waste products, cellular debris, and pathogens (Nayak et al. 2014). Advances in the available technologies have enabled a better understanding of the microglial functions across different conditions (Tay et al. 2019). Microglia is fundamental for brain development, activity, and plasticity (Tay et al. 2017), including the creation and remodeling of synapses. Through the modulation of synapse number and synaptic activity, microglia can regulate the processes of learning, memory, and cognition (Weinhard et al. 2018). Microglia also regulate neurogenesis, neuronal density and connectivity, as well as neuronal survival and turnover (Shigemoto-Mogami et al. 2014). Most of these processes begun during the period of perinatal development and persist through to the late adolescence/adulthood (Sellner et al. 2016; Sierra et al. 2010; Ueno et al. 2013). Microglial functions are based on the scavenging of cellular debris as well as the intense exchange of communication between microglia and neurones, achieved through the production and release of numerous neurotrophic mediators (Tay et al. 2017). This former property explains the reason why microglial numerical preponderance occurs in areas containing debris or apoptotic neurones as well as in regions with high density of neural precursor cells where microglia can drive neuronal turnover during development (Ayata et al. 2018; Cunningham et al. 2013; Swinnen et al. 2013). Microglia can be considered as key contributors to normal brain functioning, mainly because these cells regularly scan the surrounding environment and adapt their morphology and functions to restore homeostasis. Therefore, dysfunctions of microglial cells could have deleterious consequences at any stage of human life. During the pre- and perinatal brain development, the modification of microglial functions could impair essential processes such as neural connectivity and synaptic plasticity (Kettenmann et al. 2013). In the same way during adult life, changes to microglial functions could cause a remodeling of the neuronal circuits with serious consequences on learning and memory (Weinhard et al. 2018). In conclusion, dysfunctional microglia play a fundamental role in the onset, evolution, and outcome of neurological diseases throughout life span (Scuderi and Verkhratsky 2020; Tay et al. 2018).

Oligodendrocyte and NG-2 Glia

Oligodendrocytes are cells of fundamental importance for the CNS because they form the myelin sheath necessary for a fast and efficient transmission of the nervous impulse. Oligodendrocytes originate from the oligodendrocyte precursor cells (OPCs) that arise from multipotent neural stem cells (NSCs), mainly localized in the ventricular zones of the brain from which they migrate to the developing CNS where they become active oligodendrocytes. This process starts shortly before birth and continues throughout life (Bergles and Richardson 2015) as a significant amount of OPCs persists in the adult brain. These OPCs have been also identified as NG-2 glia because they express CSPG4, the NG2 chondroitin sulfate proteoglycan (Almeida and Lyons 2017). The differentiation of NG-2 glia into oligodendrocytes is essential for myelin repair in the adult brain (Ortiz et al. 2019), and for ensheathing new neuronal connections with myelin in response to new experiences (McKenzie et al. 2014; Xiao et al. 2016). These observations suggest that neurotransmission drives the differentiation of NG-2 glia and are consistent with the evidence that NG-2 glia exhibit a wide range of ion channels and neurotransmitter receptors (Larson et al. 2016), and respond to synaptic transmission (Bergles et al. 2000). Despite these findings, further studies are required to decipher how neuronal activity drives NG-2 glia conversion in oligodendrocytes. Indeed, data acquired so far demonstrated that blocking, or stimulating, synaptic signaling has only weak effects on NG-2 glia, suggesting that neurotransmitters alone are not sufficient to start oligodendrogenesis (Butt et al. 2019).

Several factors modulate OPC migration, proliferation, differentiation, and myelination (Elbaz and Popko 2019). These factors include extrinsic as well as intrinsic transcription factors, epigenetic modulators, and signaling pathways (Elbaz and Popko 2019). Oligodendrocytes express many receptors belonging to different classes suggesting that these cells receive impulses from different signaling pathways indispensable for their development and functions, mainly the formation of myelin (Butt et al. 2019; Habermacher et al. 2019; Kiray et al. 2016; Patel and Klein 2011). For instance, it has been demonstrated that the Wnt signaling controls OPC expansion throughout life (Azim et al. 2017) and that estrogen favors oligodendrocyte differentiation and myelination by regulating cholesterol homeostasis (Voskuhl et al. 2019).

Myelin is mostly composed of lipids (about 70%, of which the primary component is cholesterol) and proteins (about 30%, of which the main components are the myelin basic protein and proteolipid protein) (Muller et al. 2013; Saher and Stumpf 2015). Although oligodendrocytes seem capable of de novo synthesis of cholesterol, it has been suggested that the lipid used to form the myelin sheath comes from astrocytes, as the blood-brain barrier (BBB) does not allow dietary cholesterol to enter the CNS (Kirayet al. 2016). Myelin also contains gap junctions formed by connexins, which are fundamental for ion homeostasis and axonal metabolism and integrity (Vejar et al. 2019). Besides the establishment of the optimal conditions for rapid electrical conduction, myelin is also required for axonal integrity (Alexandra et al. 2018). The underlying mechanisms are not fully clarified, but recent evidence indicates oligodendrocytes as essential for fulfilling axonal metabolic needs. They indeed provide glucose (Meyer et al. 2018) and lactate to axons (Funfschilling et al. 2012; Lee et al. 2012) depending on the axonal activity requirements (Micu et al. 2018; Saab et al. 2016).

Given the above, dysfunction or loss of oligodendroglia or of their ability to make the myelin sheath can cause devastating effects on CNS function and eventually lead to neuronal death. Moreover, the pleiotropism of factors involved in oligodendrocyte development and myelination helps to ensure that the disruption of any single factor does not result in their loss of function. On the other side, they represent multiple targets that could be involved in oligodendrocyte pathologies offering exciting new perspectives of research.

Astrocytes

Astroglia (to which astrocytes belong) are a class of highly heterogeneous in form and function neural cells of the ectodermal, neuroepithelial origin; these cells maintain homeostasis and defence of the CNS (Verkhratsky and Nedergaard 2018). Astrocytes reside in the white and gray matter of the brain and the spinal cord (Verkhratsky and Butt 2013). Numerous distinct morphological and functional subtypes of astrocytes have been identified, including (i) *protoplasmic astrocytes* of the gray matter; (ii) *fibrous astrocytes* of the white matter; (iii) *velate astrocytes*, localized in brain regions where neurones are small and densely packed (e.g., the olfactory bulb or the granular layer of the cerebellar cortex); (iv) *radial glia*, which are the pluripotent neural cell precursors that mostly disappear at birth; (v) *radial astrocytes*, which comprise the cerebellar Bergmann glia, the retinal Müller glia, radial glia-like neural stem cells of the neurogenic niches, and tanycytes; (vi) *pituicytes*, localized in the neurohypophysis; (vii) the iron-enriched astrocytes, named *Gomori astrocytes*, localized in the hypothalamus and the hippocampus; (viii) *perivascular astrocytes*, whose endfeet connect with blood vessels and are fundamental for the establishment of the glia limitans barriers; (ix) juxtavascular astrocytes somata of which are in close apposition with blood vessels; (x) *ependymocytes*, which are choroid plexus cells, lining the ventricles and producing the cerebrospinal fluid, and retinal pigment epithelial cells, which line up the retinal space; and specialized astrocytes observed only in the brain of higher primates (including humans) which include (xi) *interlaminar astrocytes*, (xii) *polarized astrocytes*, and (xiii) *varicose projection astrocyte;* functions of all these types are still unclear (Colombo 2018; Verkhratsky et al. 2018, 2019).

The heterogeneity of astroglia correlates with the multiplicity of functions that they perform. For instance, astroglia (i) control the levels of neurotransmitters, ions, reactive oxygen species, and metabolites (Deitmer and Rose 1996; Hertz et al. 1999; Kirischuk et al. 2007; Kofuji and Newman 2004); (ii) drive neurogenesis (Verkhratsky and Nedergaard 2018); (iii) regulate synapse formation, pruning, and elimination (Kettenmann et al. 2011, 2013; Pfrieger 2010); (iv) form and maintain the myelin sheath (Butt et al. 2019; Kuhn et al. 2019); and (v) control the BBB and the blood flow (Abbott et al. 2010; Attwell et al. 2010) (Fig. 2). The type and relative number of astrocytes vary among brain regions (Verkhratsky et al. 2019). Despite their great variety of morphology and functions, all astrocytes are best at performing their homeostatic function. To this end, astrocytes cooperate to form, through gap junctions, cellular networks called syncytia, comprised of apposing membranes of adjacent astrocytes pierced by hundreds of intercellular channels or connexons (Giaume et al. 2010). Gap junctions represent highly specialized areas for the transport of second messengers, ions, and bioactive molecules (Houades et al. 2008; Roux et al. 2011). Networks between astrocytes and oligodendrocytes have been identified in both the hippocampus and the neocortex (Butt and Ransom 1989; Griemsmann et al. 2015; Pastor et al. 1998) and named "panglial syncytia."

Astrocytic membrane carries a multitude of receptors for neurotransmitters and neurohormones, ion channels, and membrane transporter systems. Astrocytes

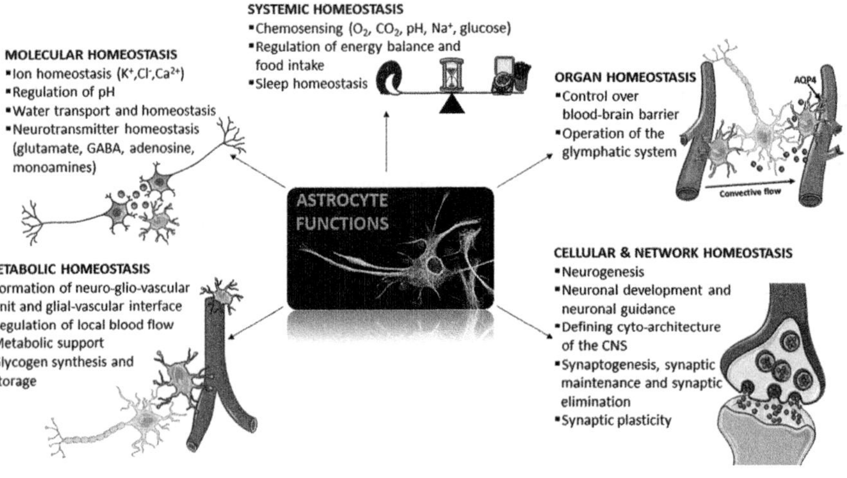

Fig. 2 Astrocytes as homeostatic cells of the CNS

integrate the signals from all other cells to operate their homeostatic function and foster neuronal activity. Channels for K^+ (voltage-independent, voltage-gated and Ca^{2+}-dependent K^+ channels), Na^+ (voltage-gated, specific type of Na^+ channels regulated by extracellular Na^+ concentration and epithelial Na^+ channels), and Ca^{2+} (voltage-gated, Orai, and Ca^{2+} release channels), as well as for many other ions, have been registered (for a comprehensive review, refer to Verkhratsky and Nedergaard 2018). Also, astrocytes express receptors for almost all neuroactive agents (Kettenmann and Zorec 2013; Verkhratsky 2010), including adenosine receptors (Dare et al. 2007; Pilitsis and Kimelberg 1998), purinoreceptors (Franke et al. 2001; Fumagalli et al. 2003; Verkhratsky et al. 2009), GABA receptors (MacVicar et al. 1989; Nilsson et al. 1993), glycine receptors (Kirchhoff et al. 1996; Pastor et al. 1995), acetylcholine receptors (Graham et al. 2003; Sharma and Vijayaraghavan 2001), monoamines receptors (Hertz et al. 2010; Miyazaki et al. 2004; Shelton and McCarthy 2000), cannabinoid receptors (Navarrete and Araque 2008; Navarrete and Araque 2010), and both ionotropic and metabotropic glutamate receptors (Lalo et al. 2006; Sun et al. 2013; Verkhratsky and Butt 2013; Verkhratsky and Chvatal 2020). Lastly, numerous membrane transporter systems for different ions and neuroactive substances complete the complex astrocytic machinery required to exert their homeostatic function, such as the Na^+-K^+ ATPase (Hertz et al. 2015), Ca^{2+}-ATPases (Verkhratsky and Nedergaard 2018), as well as plasmalemmal transporters for GABA (Ribak et al. 1996), glycine (Zafra et al. 1995), glutamate (Verkhratsky and Rose 2020), glutamine (Scalise et al. 2016), and monocarboxylates (Halestrap 2012). In this way, astrocytes control the CNS microenvironment by adjusting extracellular neurotransmitters, ions, and pH, regulating blood flow through the release of vasoactive molecules, and buffering reactive oxygen species (Parpura and Verkhratsky, 2012). It has been demonstrated the ability of a single astrocyte to be in contact with several neurones. In this way, they finely regulate synaptic transmission by tuning neurotransmitter levels in the synaptic cleft (Verkhratsky and Nedergaard 2018). Astrocytes are fundamental components of the BBB where their presence is essential for a protective function and the control of cerebral flow, thus regulating the communication between the CNS and the periphery (Verkhratsky and Parpura 2015). Astrocytes are also a part of the so-called gliocrine system, releasing around 200 molecules, mainly neurotrophic factors, and energy substrates, fundamental for the maintenance of CNS functions (Verkhratsky et al. 2016).

Given the above, all types of glial cells contribute to neuropathological developments. As astrocytes are a part of neural networks, interacting with neurones, with other glial cells, and with blood vessels, they are the key players in maintaining the structural and functional integrity of the brain tissue. The role of astrocytes in driving neuronal function and survival both in physiology and pathology has been widely documented (Verkhratsky and Nedergaard 2018). The hypothesis that astrocyte dysfunctions allow the creation of a disease-permissive context, which may favor neuronal deficits and death, has gained great attention in the recent years. Here, we provide a brief recap of the evidence accumulated so far on the active role of astrocytes in neuropsychiatric disorders, which are discussed in detail in the following chapters.

Astrogliopathology in Neuropsychiatric Disorders

Considering the above-mentioned multiple homeostatic and supportive functions that astrocytes perform, it becomes clear that any changes in the physiological performance of these cells are having a role in the etiology or progression of neuropsychiatric pathologies. Astrocyte impairments can be generic or disease-specific, and they often differ depending on the stage of the disease (Pekny et al. 2016). To complicate this scenario, human diseases are frequently modified both by age and by the presence of other comorbidities. Schematically, we can divide astrogliopathies into three main categories: (i) reactive astrogliosis; (ii) astroglial atrophy, characterized by degeneration and loss of function; and (iii) pathological remodeling of astrocytes (Verkhratsky et al. 2017a, 2019) (Fig. 3). It should be remembered that all three of these reactions are considered pathological, as well as that they can occur simultaneously or singly.

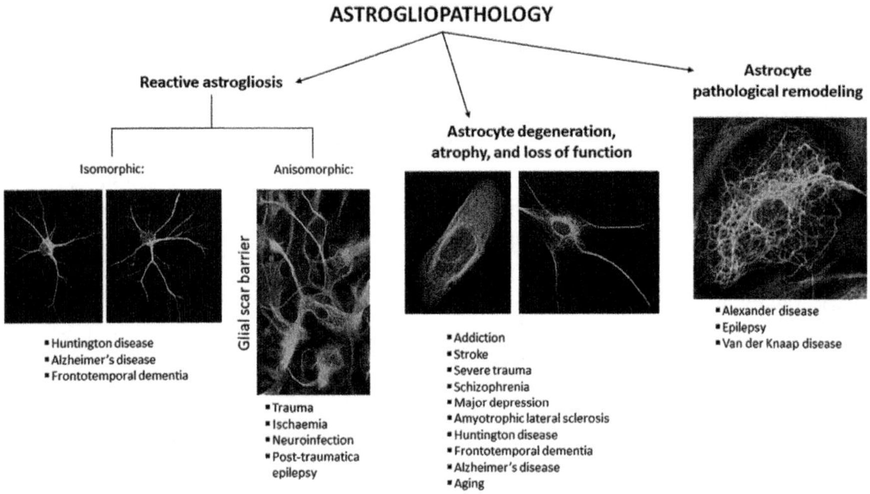

Fig. 3 Astrocyte contribution to neuropsychiatric disorders

Reactive Astrogliosis in Neuropsychiatric Disorders

Reactive astrogliosis represents the most studied type of astrocytic response (Escartin et al. 2021), which has been considered for a long time the stereotypic and universal response to pathology. According to the severity, reactive astrogliosis can be classified as mild to moderate astrogliosis, diffuse severe astrogliosis, and severe astrogliosis with scar formation (Sofroniew 2009, 2014). According to cellular morphology, two types of reactive astrogliosis have been identified: the isomorphic one, which is reversible and characterized by the preservation of the territorial astroglial domains, and the anisomorphic astrogliosis, characterized by a lack of maintenance of the territorial domains, presence of cell migration, territorial overlap, and ultimately scar formation (Pekny et al. 2016). Histopathologically, reactive astrocytes display hypertrophic extensions due to the upregulation of vimentin and glial fibrillary acidic protein (GFAP), two cytoskeletal intermediate filaments/proteins (Hol and Pekny 2015; Sofroniew 2014). Broadly speaking, reactive astrocytes undergo numerous morphological and functional modifications, acquiring different phenotypes that are believed to be disease specific (Pekny et al. 2016; Escartin et al. 2021).

Reactive astrogliosis is an evolutionary-conserved defensive program aimed at isolating the damaged region, increasing neuroprotection, and starting the reparation of the damaged nervous tissue, as well as the BBB. Growing experimental evidence supports this notion, demonstrating that the suppression of reactive astrogliosis often increases the extent of the traumatic brain injury, exacerbates post-traumatic synaptic loss, and aggravates disease progression (Li et al. 2008; Okada et al. 2006; Pekny et al. 1999, 2014, 2016). Thus, astrocytic reactivity is broadly considered neuroprotective, albeit in some circumstances, especially if sustained for a too long time, it can become a maladaptive process, the consequences of which may override the initial benefits.

Reactive astrogliosis has been widely documented in numerous neurological diseases, including multiple sclerosis, Alzheimer's disease, and autism spectrum disorders (Bronzuoli et al. 2018a, b; das Neves et al. 2020; Scuderi et al. 2018; Scuderi and Verkhratsky 2020; Tang et al. 2006; Zeidan-Chulia et al. 2014). Accumulating evidence indicates the presence of reactive astrocytes even in the course of some neuropsychiatric disorders. For instance, the astrocytic responses to chronic alcohol use may also lead to secondary activation of gliosis-like astrocyte responses (Miguel-Hidalgo 2009; Miguel-Hidalgo and Rajkowska 2003).

Astrocyte Degeneration, Atrophy, and Loss of Function in Neuropsychiatric Disorders

Astrodegeneration is characterized by morphological atrophy and functional asthenia of astrocytes. Thickness and extension of astrocyte branches appear reduced, while some of their homeostatic functions are compromised. This astrocytic response has been detected in several neurodegenerative and neuropsychiatric disorders (Heneka et al. 2010; Verkhratsky et al. 2014; Verkhratsky et al. 2017b). Schizophrenia, major depressive disorder, alcohol abuse disorder, and obsessive-compulsive disorders are all characterized by a reduction in the number or the packing density of astrocytes, accompanied by failure of their homeostatic function, especially in glutamate homeostasis (Aida et al. 2015; Czeh and Nagy 2018; Korbo 1999; Rajkowska et al. 2002; Rajkowska and Stockmeier 2013). Aberrant glutamate metabolism and transport, as well as the subsequent alteration in Ca^{2+} homeostasis, likely provoke an alteration in neurotransmission and excitotoxic neuronal death, both resulting in psychotic symptoms (Verkhratsky et al. 2014). Of note, Miguel-Hidalgo in his chapter published in this book offers evidence in the field of alcohol abuse disorder suggesting that the reduced number of astrocytes and the shrinkage of their processes may impair some of their critical functions. For instance, it has been shown that alcohol inhibits astrocyte proliferation, as well as DNA and protein synthesis, in cultured neonatal astrocytes (Davies and Cox 1991; Guerri and Renau-Piqueras 1997). Similar findings have also been achieved analyzing postmortem human brain tissue (Kane et al. 1996). In the chapter by Kruyer and Scofield, the emerging research highlighting the critical contribution of astrocytes to the encoding and expression of motivated behaviors relevant to drug addiction is extensively discussed. The chapter by Tanaka discusses the role of astrocytic control of the synaptic efficacy and its dysfunction in the pathophysiology of obsessive-compulsive and related disorders.

Astrocyte Pathological Remodeling in Cognitive Disorders

Astrocytes can undergo modifications in their intracellular cascade signaling or in their functional properties acquiring a pathological phenotype. This process is called pathological remodeling, and it has been implicated in the progression of several neurological diseases (Ferrer 2018; Pekny et al. 2016). Astrocytic pathological remodeling has been documented in diseases with severe damage to the developing white matter, mainly leukodystrophies. These are a group of hereditary diseases characterized by the accumulation of substances in the myelin that, therefore, gradually undergoes destruction. Alexander's disease is a rare neurodegenerative disease of astrocytes that display sporadically mutated GFAP gene that causes early and severe leukomalacia (Messing et al. 2012). Pathological remodeling in astrocytes has also been observed in other pathologies, that is, mesial temporal lobe epilepsy and Van der Knaap disease (Bedner et al. 2015; Lanciotti et al. 2013; Verkhratsky et al. 2019).

Envoi

We concisely examined the neuroglia, the origin of these cells, their classification, and some of their functions. We have deepened the aspects connected to the homeostatic function of astrocytes, and then we tersely reviewed the modification that astrocytes undergo in neuropsychiatric diseases. In the follow-up chapters collected for this book, we explore the role of astrocytes in the progression of neuropathological diseases, particularly in neuropsychiatric disorders.

Acknowledgments We are grateful to Giorgia Menegoni for her help in preparing the figures.

BL's work is supported by the National Natural Science Foundation of China (grant number 8187185), LiaoNing Revitalization Talents Program (grant number XLYC1807137), the Scientific Research Foundation for Returned Scholars of Education Ministry of China (grant number 20151098), LiaoNing Thousand Talents Program (grant number 202078), and "ChunHui" Program of Education Ministry of China (grant number 2020703). CS's work is supported by a grant from the Italian Ministry of Education, University and Research (2015KP7T2Y_002) and a grant from Sapienza University of Rome (RM11916B7A8D0225). VP's work is supported by a grant from the National Institute of General Medical Sciences of the National Institutes of Health (R01GM123971). VP is an Honorary Professor at University of Rijeka, Croatia.

References

Abbott NJ, Patabendige AA, Dolman DE, Yusof SR, Begley DJ (2010) Structure and function of the blood-brain barrier. Neurobiol Dis 37:13–25

Aida T, Yoshida J, Nomura M, Tanimura A, Iino Y, Soma M, Bai N, Ito Y, Cui W, Aizawa H, Yanagisawa M, Nagai T, Takata N, Tanaka KF, Takayanagi R, Kano M, Götz M, Hirase H, Tanaka K (2015) Astroglial glutamate transporter deficiency increases synaptic excitability and leads to pathological repetitive behaviors in mice. Neuropsychopharmacology 2015(40):1569–7159

Alexandra IM, Constanze D, Klaus-Armin N (2018) An emerging role of dysfunctional axon-oligodendrocyte coupling in neurodegenerative diseases. Dialogues Clin Neurosci 20:283–292

Almeida RG, Lyons DA (2017) On myelinated axon plasticity and neuronal circuit formation and function. J Neurosci 37:10023–10034

Attwell D, Buchan AM, Charpak S, Lauritzen M, Macvicar BA, Newman EA (2010) Glial and neuronal control of brain blood flow. Nature 468:232–243

Ayata P, Badimon A, Strasburger HJ et al (2018) Epigenetic regulation of brain region-specific microglia clearance activity. Nat Neurosci 21:1049–1060. https://doi.org/10.1038/s41593-018-0192-3

Azim K, Angonin D, Marcy G, Pieropan F, Rivera A, Donega V, Cantu C, Williams G, Berninger B, Butt AM, Raineteau O (2017) Pharmacogenomic identification of small molecules for lineage specific manipulation of subventricular zone germinal activity. PLoS Biol 15:e2000698

Bear MF, Connors BW, Paradiso MA (2007) Exploring the brain. Lippincott Williams & Wilkins, Philadelphia

Bedner P, Dupper A, Huttmann K, Muller J, Herde MK, Dublin P, Deshpande T, Schramm J, Haussler U, Haas CA, Henneberger C, Theis M, Steinhauser C (2015) Astrocyte uncoupling as a cause of human temporal lobe epilepsy. Brain 138:1208–1222

Bergles DE, Roberts JD, Somogyi P, Jahr CE (2000) Glutamatergic synapses on oligodendrocyte precursor cells in the hippocampus. Nature 405:187–191

Bergles DE, Richardson WD (2015) Oligodendrocyte development and plasticity. Cold Spring Harb Perspect Biol 8(2):a020453. https://doi.org/10.1101/cshperspect.a020453

Bronzuoli MR, Facchinetti R, Ingrassia D, Sarvadio M, Schiavi S, Steardo L, Verkhratsky A, Trezza V, Scuderi C (2018a) Neuroglia in the autistic brain: evidence from a preclinical model. Mol Autism 9:66. https://doi.org/10.1186/s13229-018-0254-0

Bronzuoli MR, Facchinetti R, Steardo L Jr, Romano A, Stecca C, Passarella S, Steardo L, Cassano T, Scuderi C (2018b) Palmitoylethanolamide dampens reactive astrogliosis and improves neuronal trophic support in a triple transgenic model of Alzheimer's disease: in vitro and in vivo evidence. Oxid Med Cell Longev 16:4720532. https://doi.org/10.1155/2018/4720532

Butt AM, Ransom BR (1989) Visualization of oligodendrocytes and astrocytes in the intact rat optic nerve by intracellular injection of lucifer yellow and horseradish peroxidase. Glia 2:470–475

Butt AM, Papanikolaou M, Rivera A (2019) Physiology of oligodendroglia. In: Verkhratsky A, Ho M, Zorec R, Parpura V (eds) Neuroglia in neurodegenerative diseases. Advances in experimental medicine and biology, vol 1175. Springer, Singapore. https://doi.org/10.1007/978-981-13-9913-8_5

Cartier N, Lewis C-A, Zhang R, Rossi FMV (2014) The role of microglia in human disease: therapeutic tool or target? Acta Neuropathol 128:363–380. https://doi.org/10.1007/s00401-014-1330-y

Colombo JA (2018) Interlaminar glia and other glial themes revisited: pending answers following three decades of glial research. Neuroglia 1:7–20

Cunningham CL, Martínez-Cerdeño V, Noctor SC (2013) Microglia regulate the number of neural precursor cells in the developing cerebral cortex. J Neurosci 33:4216–4233. https://doi.org/10.1523/JNEUROSCI.3441-12.2013

Czeh B, Nagy SA (2018) Clinical findings documenting cellular and molecular abnormalities of glia in depressive disorders. Front Mol Neurosci 11:56

Dare E, Schulte G, Karovic O, Hammarberg C, Fredholm BB (2007) Modulation of glial cell functions by adenosine receptors. Physiol Behav 92:15–20

das Neves SP, Sousa JC, Sousa N, Cerqueira JJ, Marques F (2020) Altered astrocytic function in experimental neuroinflammation and multiple sclerosis. Glia 69(6):1341–1368. https://doi.org/10.1002/glia.23940

Davies DL, Cox WE (1991) Delayed growth and maturation of astrocytic cultures following exposure to ethanol: electron microscopic observations. Brain Res 547:53–61

Deitmer JW, Rose CR (1996) pH regulation and proton signalling by glial cells. Prog Neurobiol 48:73–103

Elbaz B, Popko B (2019) Molecular control of oligodendrocyte development. Trends Neurosci 42:263–277

Escartin C, Galea E, Lakatos A, O'Callaghan JP, Petzold GC, Serrano-Pozo A, Steinhauser C, Volterra A, Carmignoto G, Agarwal A, Allen NJ, Araque A, Luis Barbeito L, Barzilai A, Bergles DE, Bonvento G, Butt AM, Chen W-T, Cohen-Salmon M, Cunningham C, Deneen B, De Strooper B, Díaz-Castro B, Farina C, Freeman M, Gallo V, Goldman JE, Goldman SA, Götz M, Gutiérrez A, Haydon PG, Heiland DH, Hol EM, Holt MG, Iino M, Kastanenka KV, Kettenmann H, Khakh BS, Koizumi S, Lee CJ, Liddelow SA, MacVicar BA, Magistretti P, Messing A, Mishra A, Molofsky AV, Murai K, Norris CM, Okada S, SHR O, Oliveira JF, Panatier A, Parpura V, Pekna M, Pekny M, Pellerin L, Perea G, Pérez-Nievas BG, Pfrieger FW, Poskanzer KE, Quintana FJ, Ransohoff RM, Riquelme-Perez M, Robel S, Rose CR, Rothstein J, Rouach N, Rowitch DH, Semyanov A, Sirko S, Sontheimer H, Swanson RA, Vitorica J, Wanner I-B, Wood LB, Wu J, Zheng B, Zimmer ER, Zorec R, Sofroniew MV, Verkhratsky A

(2021) Working consensus on reactive astrocyte nomenclature, definitions and future directions. Nat Neurosci 24(3):312–325. https://doi.org/10.1038/s41593-020-00783-4
Ferrer I (2018) Astrogliopathy in tauopathies. Neuroglia 1:126–150
Franke H, Grosche J, Schadlich H, Krugel U, Allgaier C, Illes P (2001) P2X receptor expression on astrocytes in the nucleus accumbens of rats. Neuroscience 108:421–429
Fumagalli M, Brambilla R, D'Ambrosi N, Volonte C, Matteoli M, Verderio C, Abbracchio MP (2003) Nucleotide-mediated calcium signaling in rat cortical astrocytes: role of P2X and P2Y receptors. Glia 43:218–230
Funfschilling U, Supplie LM, Mahad D, Boretius S, Saab AS, Edgar J, Brinkmann BG, Kassmann CM, Tzvetanova ID, Mobius W, Diaz F, Meijer D, Suter U, Hamprecht B, Sereda MW, Moraes CT, Frahm J, Goebbels S, Nave KA (2012) Glycolytic oligodendrocytes maintain myelin and long-term axonal integrity. Nature 485:517–521
Giaume C, Koulakoff A, Roux L, Holcman D, Rouach N (2010) Astroglial networks: a step further in neuroglial and gliovascular interactions. Nat Rev Neurosci 11:87–99
Ginhoux F, Greter M, Leboeuf M et al (2010) Fate mapping analysis reveals that adult microglia derive from primitive macrophages. Science 330:841–845. https://doi.org/10.1126/science.1194637
Graham AJ, Ray MA, Perry EK, Jaros E, Perry RH, Volsen SG, Bose S, Evans N, Lindstrom J, Court JA (2003) Differential nicotinic acetylcholine receptor subunit expression in the human hippocampus. J Chem Neuroanat 25:97–113
Griemsmann S, Hoft SP, Bedner P, Zhang J, von Staden E, Beinhauer A, Degen J, Dublin P, Cope DW, Richter N, Crunelli V, Jabs R, Willecke K, Theis M, Seifert G, Kettenmann H, Steinhauser C (2015) Characterization of panglial gap junction networks in the thalamus, neocortex, and hippocampus reveals a unique population of glial cells. Cereb Cortex 25:3420–3433
Grubisic V, Verkhratsky A, Zorec R, Parpura V (2018) Enteric glia regulate gut motility in health and disease. Brain Res Bull 136:109–117
Guerri C, Renau-Piqueras J (1997) Alcohol, astroglia, and brain development. Mol Neurobiol 15:65–81
Habermacher C, Angulo MC, Benamer N (2019) Glutamate versus GABA in neuron-oligodendroglia communication. Glia 67(11):2092–2106. https://doi.org/10.1002/glia.23618
Halestrap AP (2012) The monocarboxylate transporter family—structure and functional characterization. IUBMB Life 64:1–9
Hanani M, Verkhratsky A (2021) Satellite glial cells and astrocytes, a comparative review. Neurochem Res:1–13. https://doi.org/10.1007/s11064-021-03255-8
Heneka MT, Rodriguez JJ, Verkhratsky A (2010) Neuroglia in neurodegeneration. Brain Res Rev 63:189–211
Hertz L, Dringen R, Schousboe A, Robinson SR (1999) Astrocytes: glutamate producers for neurons. J Neurosci Res 57:417–428
Hertz L, Lovatt D, Goldman SA, Nedergaard M (2010) Adrenoceptors in brain: cellular gene expression and effects on astrocytic metabolism and $[Ca^{2+}]_i$. Neurochem Int 57:411–420
Hertz L, Song D, Xu J, Peng L, Gibbs ME (2015) Role of the astrocytic Na+, K+-ATPase in K+ homeostasis in brain: K^+ uptake, signaling pathways and substrate utilization. Neurochem Res 40:2505–2516
Hilgetag CC, Barbas H (2009) Are there ten times more glia than neurons in the brain? Brain Struct Funct 213:365–366
Hol EM, Pekny M (2015) Glial fibrillary acidic protein (GFAP) and the astrocyte intermediate filament system in diseases of the central nervous system. Curr Opin Cell Biol 32:121–130
Houades V, Koulakoff A, Ezan P, Seif I, Giaume C (2008) Gap junction-mediated astrocytic networks in the mouse barrel cortex. J Neurosci 28:5207–5217
Kandel ER, Schwartz JH, Jessell TM (2000) Principles of neural science. McGrawhill, New York
Kane CJ, Berry A, Boop FA, Davies DL (1996) Proliferation of astroglia from the adult human cerebrum is inhibited by ethanol in vitro. Brain Res 731:39–44

Kettenmann H, Hanisch UK, Noda M, Verkhratsky A (2011) Physiology of microglia. Physiol Rev 91:461–553

Kettenmann H, Kirchhoff F, Verkhratsky A (2013) Microglia: new roles for the synaptic stripper. Neuron 77:10–18

Kettenmann H, Zorec R (2013) Release of gliotransmitters and transmitter receptors in astrocytes. In: Kettenmann H, Ransom BR (eds) Neuroglia. Oxford University Press, New York, pp 197–211

Kidd GJ, Ohno N, Trapp BD (2013) Biology of Schwann cells. Handb Clin Neurol 115:55–79

Kiray H, Lindsay SL, Hosseinzadeh S, Barnett SC (2016) The multifaceted role of astrocytes in regulating myelination. Exp Neurol 283:541–549

Kirchhoff F, Mulhardt C, Pastor A, Becker CM, Kettenmann H (1996) Expression of glycine receptor subunits in glial cells of the rat spinal cord. J Neurochem 66:1383–1390

Kirischuk S, Kettenmann H, Verkhratsky A (2007) Membrane currents and cytoplasmic sodium transients generated by glutamate transport in Bergmann glial cells. Pflugers Arch 454:245–252

Kofuji P, Newman EA (2004) Potassium buffering in the central nervous system. Neuroscience 129:1045–1056

Korbo L (1999) Glial cell loss in the hippocampus of alcoholics. Alcohol Clin Exp Res 23:164–168

Kuhn S, Gritti L, Crooks D, Dombrowski Y (2019) Oligodendrocytes in development, myelin generation and beyond. Cells 8(11):1424

Lalo U, Pankratov Y, Kirchhoff F, North RA, Verkhratsky A (2006) NMDA receptors mediate neuron-to-glia signaling in mouse cortical astrocytes. J Neurosci 26:2673–2683

Lanciotti A, Brignone MS, Bertini E, Petrucci TC, Aloisi F, Ambrosini E (2013) Astrocytes: emerging stars in leukodystrophy pathogenesis. Transl Neurosci 4(2):144–164

Larson VA, Zhang Y, Bergles DE (2016) Electrophysiological properties of NG2 + cells: matching physiological studies with gene expression profiles. Brain Res 1638:138–160

Lee Y, Morrison BM, Li Y, Lengacher S, Farah MH, Hoffman PN, Liu Y, Tsingalia A, Jin L, Zhang PW, Pellerin L, Magistretti PJ, Rothstein JD (2012) Oligodendroglia metabolically support axons and contribute to neurodegeneration. Nature 487:443–448

Li L, Lundkvist A, Andersson D, Wilhelmsson U, Nagai N, Pardo AC, Nodin C, Stahlberg A, Aprico K, Larsson K, Yabe T, Moons L, Fotheringham A, Davies I, Carmeliet P, Schwartz JP, Pekna M, Kubista M, Blomstrand F, Maragakis N, Nilsson M, Pekny M (2008) Protective role of reactive astrocytes in brain ischemia. J Cereb Blood Flow Metab 28:468–481

MacVicar BA, Tse FW, Crichton SA, Kettenmann H (1989) GABA-activated Cl^- channels in astrocytes of hippocampal slices. J Neurosci 9:3577–3583

McKenzie IA, Ohayon D, Li H, de Faria JP, Emery B, Tohyama K, Richardson WD (2014) Motor skill learning requires active central myelination. Science 346:318–322

Messing A, Brenner M, Feany MB, Nedergaard M, Goldman JE (2012) Alexander disease. J Neurosci 32:5017–5023

Meyer N, Richter N, Fan Z, Siemonsmeier G, Pivneva T, Jordan P, Steinhauser C, Semtner M, Nolte C, Kettenmann H (2018) Oligodendrocytes in the mouse corpus callosum maintain axonal function by delivery of glucose. Cell Rep 22:2383–2394

Micu I, Plemel JR, Caprariello AV, Nave KA, Stys PK (2018) Axo-myelinic neurotransmission: a novel mode of cell signalling in the central nervous system. Nat Rev Neurosci 19:49–58

Miguel-Hidalgo JJ (2009) The role of glial cells in drug abuse. Curr Drug Abuse Rev 2:76–82

Miguel-Hidalgo JJ, Rajkowska G (2003) Comparison of prefrontal cell pathology between depression and alcohol dependence. J Psychiatr Res 37:411–420

Miyazaki I, Asanuma M, Diaz-Corrales FJ, Miyoshi K, Ogawa N (2004) Direct evidence for expression of dopamine receptors in astrocytes from basal ganglia. Brain Res 1029:120–123

Monier A, Adle-Biassette H, Delezoide A-L et al (2007) Entry and distribution of microglial cells in human embryonic and fetal cerebral cortex. J Neuropathol Exp Neurol 66:372–382. https://doi.org/10.1097/nen.0b013e3180517b46

Muller C, Bauer NM, Schafer I, White R (2013) Making myelin basic protein -from mRNA transport to localized translation. Front Cell Neurosci 7:169

Navarrete M, Araque A (2008) Endocannabinoids mediate neuron-astrocyte communication. Neuron 57:883–893
Navarrete M, Araque A (2010) Endocannabinoids potentiate synaptic transmission through stimulation of astrocytes. Neuron 68:113–126
Nayak D, Roth TL, McGavern DB (2014) Microglia development and function. Annu Rev Immunol 32:367–402. https://doi.org/10.1146/annurev-immunol-032713-120240
Nilsson M, Eriksson PS, Ronnback L, Hansson E (1993) GABA induces Ca2+ transients in astrocytes. Neuroscience 54:605–614
Okada S, Nakamura M, Katoh H, Miyao T, Shimazaki T, Ishii K, Yamane J, Yoshimura A, Iwamoto Y, Toyama Y, Okano H (2006) Conditional ablation of Stat3 or Socs3 discloses a dual role for reactive astrocytes after spinal cord injury. Nat Med 12:829–834
Ortiz FC, Habermacher C, Graciarena M, Houry PY, Nishiyama A, Oumesmar BN, Angulo MC (2019) Neuronal activity in vivo enhances functional myelin repair. JCI Insight 5(9):e123434. https://doi.org/10.1172/jci.insight.123434
Parpura V, Verkhratsky A (2012) Neuroglia at the crossroads of homoeostasis, metabolism and signalling: evolution of the concept. ASN Neuro4:201–205. https://doi.org/10.1042/AN20120019
Pastor A, Chvatal A, Sykova E, Kettenmann H (1995) Glycine- and GABA-activated currents in identified glial cells of the developing rat spinal cord slice. Eur J Neurosci 7:1188–1198
Pastor A, Kremer M, Moller T, Kettenmann H, Dermietzel R (1998) Dye coupling between spinal cord oligodendrocytes: differences in coupling efficiency between gray and white matter. Glia 24:108–120
Patel JR, Klein RS (2011) Mediators of oligodendrocyte differentiation during remyelination. FEBS Lett 585:3730–3737
Pekny M, Johansson CB, Eliasson C, Stakeberg J, Wallen A, Perlmann T, Lendahl U, Betsholtz C, Berthold CH, Frisen J (1999) Abnormal reaction to central nervous system injury in mice lacking glial fibrillary acidic protein and vimentin. J Cell Biol 145:503–514
Pekny M, Wilhelmsson U, Pekna M (2014) The dual role of astrocyte activation and reactive gliosis. Neurosci Lett 565:30–38
Pekny M, Pekna M, Messing A, Steinhauser C, Lee JM, Parpura V, Hol EM, Sofroniew MV, Verkhratsky A (2016) Astrocytes: a central element in neurological diseases. Acta Neuropathol 131:323–345
Pfrieger FW (2010) Role of glial cells in the formation and maintenance of synapses. Brain Res Rev 63:39–46
Pilitsis JG, Kimelberg HK (1998) Adenosine receptor mediated stimulation of intracellular calcium in acutely isolated astrocytes. Brain Res 798:294–303
Prinz M, Priller J, Sisodia SS, Ransohoff RM (2011) Heterogeneity of CNS myeloid cells and their roles in neurodegeneration. Nat Neurosci 14:1227–1235. https://doi.org/10.1038/nn.2923
Prinz M, Priller J (2014) Microglia and brain macrophages in the molecular age: from origin to neuropsychiatric disease. Nat Rev Neurosci 15:300–312. https://doi.org/10.1038/nrn3722
Rajkowska G, Miguel-Hidalgo JJ, Makkos Z, Meltzer H, Overholser J, Stockmeier C (2002) Layer-specific reductions in GFAP-reactive astroglia in the dorsolateral prefrontal cortex in schizophrenia. Schizophr Res 57:127–138
Rajkowska G, Stockmeier CA (2013) Astrocyte pathology in major depressive disorder: insights from human postmortem brain tissue. Curr Drug Targets 14:1225–1236
Ribak CE, Tong WM, Brecha NC (1996) GABA plasma membrane transporters, GAT-1 and GAT-3, display different distributions in the rat hippocampus. J Comp Neurol 367:595–606
Roux L, Benchenane K, Rothstein JD, Bonvento G, Giaume C (2011) Plasticity of astroglial networks in olfactory glomeruli. Proc Natl Acad Sci U S A 108:18442–18446
Ruitenberg MJ, Vukovic J, Sarich J, Busfield SJ, Plant GW (2006) Olfactory ensheathing cells: characteristics, genetic engineering, and therapeutic potential. J Neurotrauma 23:468–478
Saab AS, Tzvetavona ID, Trevisiol A, Baltan S, Dibaj P, Kusch K, Mobius W, Goetze B, Jahn HM, Huang W, Steffens H, Schomburg ED, Perez-Samartin A, Perez-Cerda F, Bakhtiari D, Matute

C, Lowel S, Griesinger C, Hirrlinger J, Kirchhoff F, Nave KA (2016) Oligodendroglial NMDA receptors regulate glucose import and axonal energy metabolism. Neuron 91:119–132

Saher G, Stumpf SK (2015) Cholesterol in myelin biogenesis and hypomyelinating disorders. Biochim Biophys Acta 1851:1083–1094

Scalise M, Pochini L, Galluccio M, Indiveri C (2016) Glutamine transport. From energy supply to sensing and beyond. Biochim Biophys Acta 1857(8):1147–1157

Schlegelmilch T, Henke K, Peri F (2011) Microglia in the developing brain: from immunity to behaviour. Curr Opin Neurobiol 21:5–10. https://doi.org/10.1016/j.conb.2010.08.004

Scuderi C, Bronzuoli MR, Facchinetti R, Pace L, Ferraro L, Broad KD, Serviddio G, Bellanti F, Palombelli G, Carpinelli G, Canese R, Gaetani S, Steardo L Jr, Steardo L, Cassano T (2018) Ultramicronized palmitoylethanolamide rescues learning and memory impairments in a triple transgenic mouse model of Alzheimer's disease by exerting anti-inflammatory and neuroprotective effects. Transl Psychiatry 8(1):32. https://doi.org/10.1038/s41398-017-0076-4

Scuderi C, Verkhratsky A (2020) The role of neuroglia in autism spectrum disorders. Prog Mol Biol Transl Sci 173:301–330. https://doi.org/10.1016/bs.pmbts.2020.04.011

Sellner S, Paricio-Montesinos R, Spieß A et al (2016) Microglial CX3CR128 promotes adult neurogenesis by inhibiting Sirt 1/p65 signaling independent of CX3CL1. Acta Neuropathol Commun 4:102. https://doi.org/10.1186/s40478-016-0374-8

Sharma G, Vijayaraghavan S (2001) Nicotinic cholinergic signaling in hippocampal astrocytes involves calcium-induced calcium release from intracellular stores. Proc Natl Acad Sci U S A 98:4148–4153

Shelton MK, McCarthy KD (2000) Hippocampal astrocytes exhibit Ca^{2+}-elevating muscarinic cholinergic and histaminergic receptors in situ. J Neurochem 74:555–563

Shigemoto-Mogami Y, Hoshikawa K, Goldman JE et al (2014) Microglia enhance neurogenesis and oligodendrogenesis in the early postnatal subventricular zone. J Neurosci 34:2231–2243. https://doi.org/10.1523/JNEUROSCI.1619-13.2014

Sierra A, Encinas JM, Deudero JJP et al (2010) Microglia shape adult hippocampal neurogenesis through apoptosis-coupled phagocytosis. Cell Stem Cell 7:483–495. https://doi.org/10.1016/j.stem.2010.08.014

Sofroniew MV (2009) Molecular dissection of reactive astrogliosis and glial scar formation. Trends Neurosci 32:638–647

Sofroniew MV (2014) Astrogliosis. Cold Spring Harb Perspect Biol 7:a020420

Sun W, McConnell E, Pare JF, Xu Q, Chen M, Peng W, Lovatt D, Han X, Smith Y, Nedergaard M (2013) Glutamate-dependent neuroglial calcium signaling differs between young and adult brain. Science 339:197–200

Swinnen N, Smolders S, Avila A et al (2013) Complex invasion pattern of the cerebral cortex by microglial cells during development of the mouse embryo. Glia 61:150–163. https://doi.org/10.1002/glia.22421

Tang G, Xu Z, Goldman JE (2006) Synergistic effects of the SAPK/JNK and the proteasome pathway on glial fibrillary acidic protein (GFAP) accumulation in Alexander disease. J Biol Chem 281:38634–38643

Tay TL, Savage JC, Hui CW et al (2017) Microglia across the lifespan: from origin to function in brain development, plasticity and cognition. J Physiol (Lond) 595:1929–1945. https://doi.org/10.1113/JP272134

Tay TL, Béchade C, D'Andrea I et al (2018) Microglia gone rogue: impacts on psychiatric disorders across the lifespan. Front Mol Neurosci 10:421. https://doi.org/10.3389/fnmol.2017.00421

Tay TL, Carrier M, Tremblay MÈ (2019) Physiology of microglia. In: Verkhratsky A, Ho M, Zorec R, Parpura V (eds) Neuroglia in neurodegenerative diseases. Advances in experimental medicine and biology, vol 1175. Springer, Singapore. https://doi.org/10.1007/978-981-13-9913-8_6

Ueno M, Fujita Y, Tanaka T et al (2013) Layer V cortical neurons require microglial support for survival during postnatal development. Nat Neurosci 16:543–551. https://doi.org/10.1038/nn.3358

Vejar S, Oyarzun JE, Retamal MA, Ortiz FC, Orellana JA (2019) Connexin and pannexin-based channels in oligodendrocytes: implications in brain health and disease. Front Cell Neurosci 13:3
Verkhratsky A (2010) Physiology of neuronal-glial networking. Neurochem Int 57:332–343
Verkhratsky A, Butt AM (2013) Glial physiology and pathophysiology. Wiley-Blackwell, Chichester, p 560
Verkhratsky A, Chvatal A (2020) NMDA receptors in astrocytes. Neurochem Res 45:122–133
Verkhratsky A, Krishtal OA, Burnstock G (2009) Purinoceptors on neuroglia. Mol Neurobiol 39:190–208
Verkhratsky A, Rodriguez JJ, Steardo L (2014) Astrogliopathology: a central element of neuropsychiatric diseases? Neuroscientist 20:576–588
Verkhratsky A, Parpura V (2015) Physiology of astroglia: channels, receptors, transporters, ion signaling and gliotransmission. Morgan & Claypool Publishers, p 172
Verkhratsky A, Matteoli M, Parpura V, Mothet JP, Zorec R (2016) Astrocytes as secretory cells of the central nervous system: idiosyncrasies of vesicular secretion. EMBO J 35(3):239–257. https://doi.org/10.15252/embj.201592705
Verkhratsky A, Zorec R, Parpura V (2017a) Stratification of astrocytes in healthy and diseased brain. Brain Pathol 27:629–644
Verkhratsky A, Zorec R, Rodriguez JJ, Parpura V (2017b) Neuroglia: functional paralysis and reactivity in Alzheimer's disease and other neurodegenerative pathologies. Adv Neurobiol 15:427–449
Verkhratsky A, Nedergaard M (2018) Physiology of astroglia. Physiol Rev 98:239–389
Verkhratsky A, Rose CR (2020) Na^+-dependent transporters: the backbone of astroglial homeostatic function. Cell Calcium 85:102136
Verkhratsky A, Oberheim Bush NA, Nedergaard M, Butt AM (2018) The special case of human astrocytes. Neuroglia 1:21–29
Verkhratsky A, Ho MS, Zorec R, Parpura V (2019) The concept of neuroglia. In: Verkhratsky A, Ho M, Zorec R, Parpura V (eds) Neuroglia in neurodegenerative diseases. Advances in experimental medicine and biology, vol 1175. Springer, Singapore
Verney C, Monier A, Fallet-Bianco C, Gressens P (2010) Early microglial colonization of the human forebrain and possible involvement in periventricular white-matter injury of preterm infants. J Anat 217:436–448. https://doi.org/10.1111/j.1469-7580.2010.01245.x
Virchow R (1860) Cellular pathology. Robert M de Witt, New York
von Bartheld CS, Bahney J, Herculano-Houzel S (2016) The search for true numbers of neurons and glial cells in the human brain: a review of 150 years of cell counting. J Comp Neurol 524:3865–3895
Voskuhl RR, Itoh N, Tassoni A, Matsukawa MA, Ren E, Tse V, Jang E, Suen TT, Itoh Y (2019) Gene expression in oligodendrocytes during remyelination reveals cholesterol homeostasis as a therapeutic target in multiple sclerosis. Proc Natl Acad Sci U S A 116:10130–10139
Weinhard L, d'Errico P, Tay TL (2018) Headmasters: microglial regulation of learning and memory in health and disease. Molecular 5:63–89. https://doi.org/10.3934/molsci.2018.1.63
Xiao L, Ohayon D, McKenzie IA, Sinclair-Wilson A, Wright JL, Fudge AD, Emery B, Li H, Richardson WD (2016) Rapid production of new oligodendrocytes is required in the earliest stages of motor-skill learning. Nat Neurosci 19:1210–1217
Zafra F, Aragon C, Olivares L, Danbolt NC, Gimenez C, Storm-Mathisen J (1995) Glycine transporters are differentially expressed among CNS cells. J Neurosci 15:3952–3969
Zeidan-Chulia F, Salmina AB, Malinovskaya NA, Noda M, Verkhratsky A, Moreira JC (2014) The glial perspective of autism spectrum disorders. Neurosci Biobehav Rev 38:160–172

Astrocytes: The Housekeepers and Guardians of the CNS

Alexei Verkhratsky, Vladimir Parpura, Baoman Li, and Caterina Scuderi

Astroglia: Definition and Evolutionary Origins

Astroglia represents a class of neural cells of the ectodermal, neuroepithelial origin which are the principal homeostatic cells of the central nervous system (CNS) (Verkhratsky and Nedergaard 2016, 2018). The class of astroglia includes several cellular subtypes with distinct morphology and function (Fig. 1); astroglia includes parenchymal astrocytes of several types, radial astrocytes, tanycytes, pituicytes, ependymocytes, choroid plexus cells and retinal pigment epithelial cells. The term astrocyte (αστρον κψτοσ; *astron, star and kytos, a hollow vessel,* later *cell* that is a star-like cell) was introduced by Michael von Lenhossék in 1895 (Lenhossék 1895); for the history of glial research see also (Kettenmann and Verkhratsky 2008; Chvatal

---

A. Verkhratsky (✉)
Faculty of Biology, Medicine and Health, The University of Manchester, Manchester, UK

Achucarro Center for Neuroscience, IKERBASQUE, Basque Foundation for Science, Bilbao, Spain
e-mail: alexej.verkhratsky@manchester.ac.uk

V. Parpura
Department of Neurobiology, The University of Alabama at Birmingham, Birmingham, AL, USA
e-mail: vlad@uab.edu

B. Li
Department of Forensic Analytical Toxicology, School of Forensic Medicine, China Medical University, Shenyang, China
e-mail: bmli@cmu.edu.cn

C. Scuderi
Department of Physiology and Pharmacology "Vittorio Erspamer", SAPIENZA University of Rome, Rome, Italy
e-mail: caterina.scuderi@uniroma1.it

B. Li et al. (eds.), *Astrocytes in Psychiatric Disorders*, Advances in Neurobiology 26, https://doi.org/10.1007/978-3-030-77375-5_2

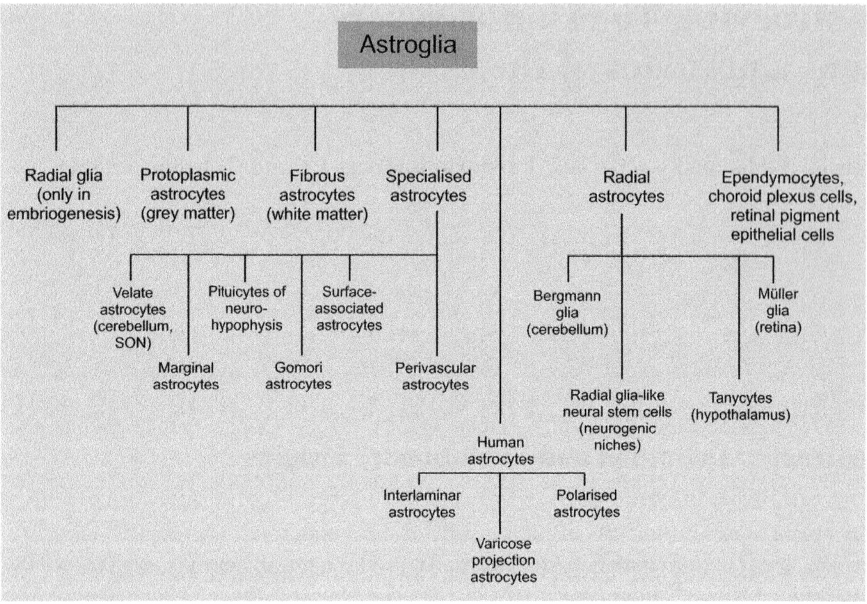

Fig. 1 Classification of astroglia

and Verkhratsky 2018). Lenhossék suggested naming all parenchymal neuroglia 'spongiocytes,' while the term astrocyte he reserved to a sub-population of cells with star-like appearance when stained with the Golgi black reaction. Parenchymal (or protoplasmic) astrocytes in the nervous tissue indeed have complex spongioform morphology defined by extensive arborisation.

Astroglial cells originate from radial glia that produce, through asymmetric division, astroglial progenitors. These progenitors populate the brain and, in the early postnatal period, propagate through symmetric division, thus producing the bulk of parenchymal astrocytes. After birth radial glial cells directly transform (transdifferentiate) into astrocytes; in neurogenic niches, these astrocytes can act as stem cells underlying adult neurogenesis.

Types of Astroglia

Protoplasmic Astrocytes

The terms of protoplasmic and fibrous glia, populating grey and white matter, respectively, have been introduced by Albert von Kölliker and William Lloyd Andriezen (Andriezen 1893; Kölliker 1896). Protoplasmic astrocytes are located in the grey matter of the brain and of the spinal cord; they have round somata of ~10µm in diameter and several primary processes. The primary processes divide into processes of higher order; these processes host tiny terminal processes, frequently referred to as peripheral astrocytic processes (PAPs); they are presented in a form of membranous leaflets, which give protoplasmic astrocyte distinctive spongioform appearance (Bushong et al. 2002; Popov et al. 2020). Processes of protoplasmic astrocytes can also be classified into (i) branches that represent processes containing organelles and (ii) leaflets which are thin membranous structures devoid of organelles (Fig. 2); protoplasmic astrocytes extend at least one process to the nearby

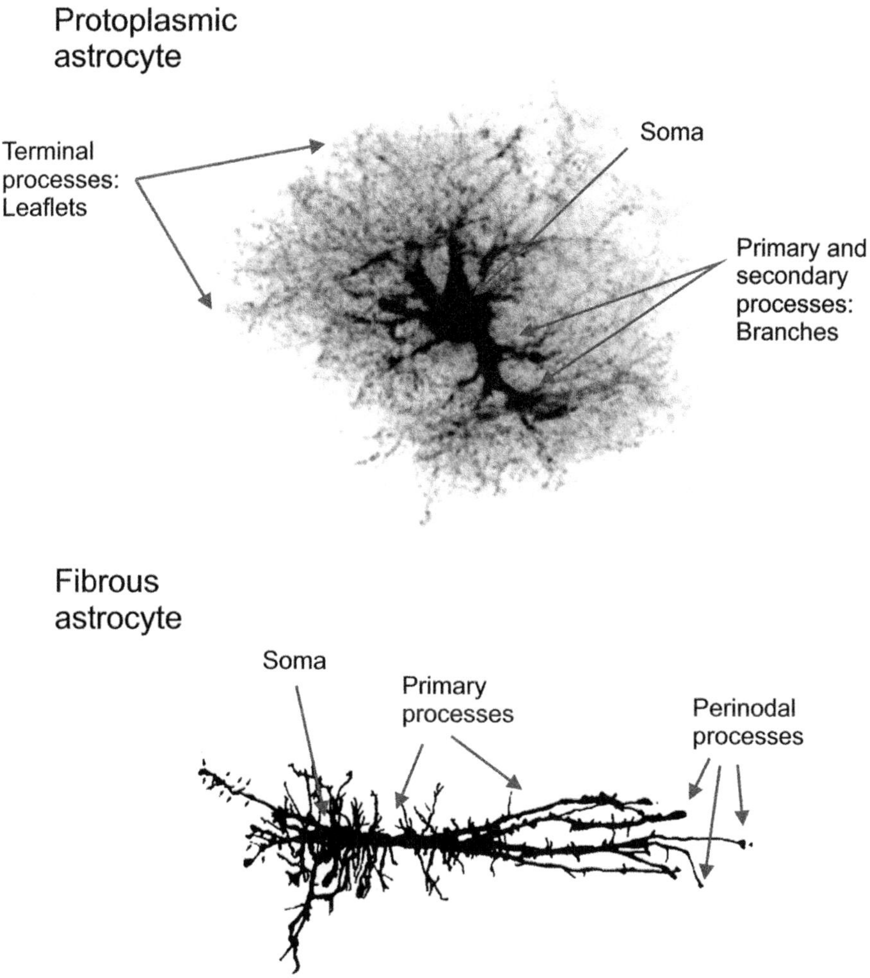

Fig. 2 Protoplasmic and fibrous astrocytes. Protoplasmic astrocytes of the grey matter are composed of soma, primary processes or branches and organelle-free thin leaflets. Fibrous astrocytes of the white matter contain perinodal processes by which they contact oligodendrocytes and nodes of Ranvier. The protoplasmic astrocyte was traced using image provided by Milos Pekny (Gothenburg University); the camera lucida image for fibrous astrocyte was provided by Arthur Butt from University of Portsmouth

blood vessel. These vascular processes end up with endfeet that cover blood vessels and form *glia limitans vascularis* (Khakh and Sofroniew 2015). Protoplasmic astrocytes occupy independent territorial domains; the neighbouring astrocytes overlap only with their distal processes and the degree of overlap does not exceed 5% (Bushong et al. 2002); this however may not occur in all species. Territories occupied by protoplasmic astrocytes define the confines of neurogliovascular units, which integrate parenchymal cellular elements with blood vessels (Iadecola 2017). Gap junctions, made from connexons and located at distal processes, integrate astrocytes into anatomically segregated syncytia (Giaume et al. 2021). The leaflets and peripheral branches contact synapses, creating a multipartite synapse or astroglial cradle that supports many aspects of synaptic function (Verkhratsky and Nedergaard 2014). Human protoplasmic astrocytes are substantially (10–20 times) larger and more complex than protoplasmic astrocytes in rodents (Oberheim et al. 2009).

Fibrous Astrocytes

Fibrous astrocytes dwell in the white matter, in the optic nerve and in the nerve fibre layer of the retina. Fibrous astrocytes (Fig. 2) have small somata and straight nonbranched processes (up to 100μm in length) oriented in parallel to axonal bundles. These processes end up with multiple finger-like cytoplasmic protrusions (perinodal processes) that are establishing contacts with axonal perinodal spaces of the surrounding axons (Lundgaard et al. 2014). Fibrous astrocytes also contact blood vessels and create perivascular or subpial endfeet. Morphology of fibrous astrocytes is heterogeneous. In rodent optic nerve, fibrous astrocytes are classified (according to morphology) into transverse, random and longitudinal (Butt et al. 1994). Human fibrous astrocytes are approximately two times larger than the same cells in rodents (Oberheim et al. 2009).

Juxtavascular Astrocytes

A sub-population of protoplasmic astrocytes located in close proximity to the blood vessels is known as juxtavascular astrocytes. These cells have some distinct physiology and demonstrate proliferative potential in response to traumatic brain injury (Bardehle et al. 2013; Gotz et al. 2021).

Surface-Associated Astrocytes

Surface-associated astrocytes are the main cellular elements of the glia limitans externa in the posterior prefrontal and amygdaloid cortex. The cell bodies of surface-associated astrocytes are located at the cortical surface, and there are two types of processes: long parallel superficial process running beneath the pia mater vessels and shorter processes extending in all directions, with some of them projecting into the cortical layer 1 (Feig and Haberly 2011).

Velate Astrocytes

Velate astrocytes represent a special population of parenchymal astrocytes covering somata of densely packed neurones in the olfactory bulb or in the granular layer of the cerebellar cortex. Velate astrocytes are characterised by a small soma with several primary processes, which form large leaflets (with high surface-to-volume ratio of 20–30μm^{-1}) enwrapping neuronal cell bodies. These perineuronal processes look like extended leaves, which defined their name derived from *vellum* (parchment in Latin) or *velatus* (which is Latin word for covered, wrapped, veiled). In the cerebellum these veil-like processes mainly cover the somata of granule cells/neurones (with a single astrocyte covering several neurones) as well as somata of Purkinje cells/neurones (Chan-Palay and Palay 1972; Buffo and Rossi 2013).

Radial Astrocytes

Radial astrocytes, named so because of their radially extended processes, are represented by (i) cerebellar Bergmann glial cells; (ii) retinal Müller glia; (iii) radial astrocytes of the supraoptic nucleus; (iv) radial glia-like neural stem cells, localised in neurogenic niches, and (v) tanycytes present in the periventricular organs, in the hypothalamus, in the hypophysis/pituitary gland and in the raphe part of the spinal cord (Reichenbach and Bringmann 2017; Verkhratsky and Nedergaard 2018).

Interlaminar and Varicose Projection Astrocytes in the Primate Brain

The brain of higher primates, and most notably of humans, contains several types of astrocytes which cannot be found in other species. These include (i) interlaminar astrocytes, with small somata located in the upper cortical layers and long processes penetrating through cortical layers; (ii) polarised astrocytes, with somata positioned in deep cortical layers, with several long processes penetrating into superficial cortical layers, and (iii) varicose projection astrocytes characterised by several very long (up to 1 mm) unbranched processes bearing varicosities and extending in all directions through the deep cortical layers (Oberheim et al. 2009; Sosunov et al. 2014; Colombo 2017).

Physiology of Astrocytes

Astrocytes control the homeostasis of the CNS at all levels of organisation. This includes homeostasis of ions, pH and neurotransmitters; supplying neurones with metabolic substrates; supporting oligodendrocytes and axons; regulating synaptogenesis, neurogenesis, and formation and maintenance of the blood-brain barrier (BBB); contributing to operation of the glymphatic system and regulating systemic homeostasis with some astrocytes being central chemosensors for oxygen, CO_2 and Na^+ (Verkhratsky and Nedergaard 2018).

Despite substantial morphological and functional heterogeneity, all astrocytes have common basic physiological features. First and foremost, all astrocytes are electrically non-excitable cells maintaining negative resting membrane potential (V_m) of about −80 mV. This is set up by transmembrane ionic gradients and very high membrane K^+ permeability. Intracellular concentrations of major ions in astrocytes are somewhat different from neurones: in particular astrocytic cytoplasmic Na^+ concentration is around 15–20 mM (which is twice as high as in neurones), and cytoplasmic concentration of Cl^- is maintained at 30–50 mM (compare to ~5–10 mM in neurones). Intracellular concentration of Ca^{2+} in astrocytes is again higher than in the majority of neurones (~ 120–200 nM vs. 50–100 nM, respectively). Astrocytic intracellular concentration of K^+ is ~120–140 mM, which is quite similar to neuronal cytosolic K^+ (Verkhratsky and Nedergaard 2018). High membrane K^+ permeability stipulates low input resistance (5–20 MΩ) of astrocytes (Fig. 3).

Ion Channels

As mentioned above astroglial membrane permeability is dominated by K^+ channels. Astrocytes express several types of K^+ channels with distinct voltage dependence, which covers the whole range of physiological membrane potentials. As a result astrocytes demonstrate linear current-voltage relationship, which is their most characteristic electrophysiological signature maintained in vitro, in situ and in vivo (Fig. 4). This diverse complement of K^+ channels allows maintenance of hyperpolarised membrane potential, which in turn provides the electrical gradient governing numerous membrane transporters and is thus critical for astroglial homeostatic function.

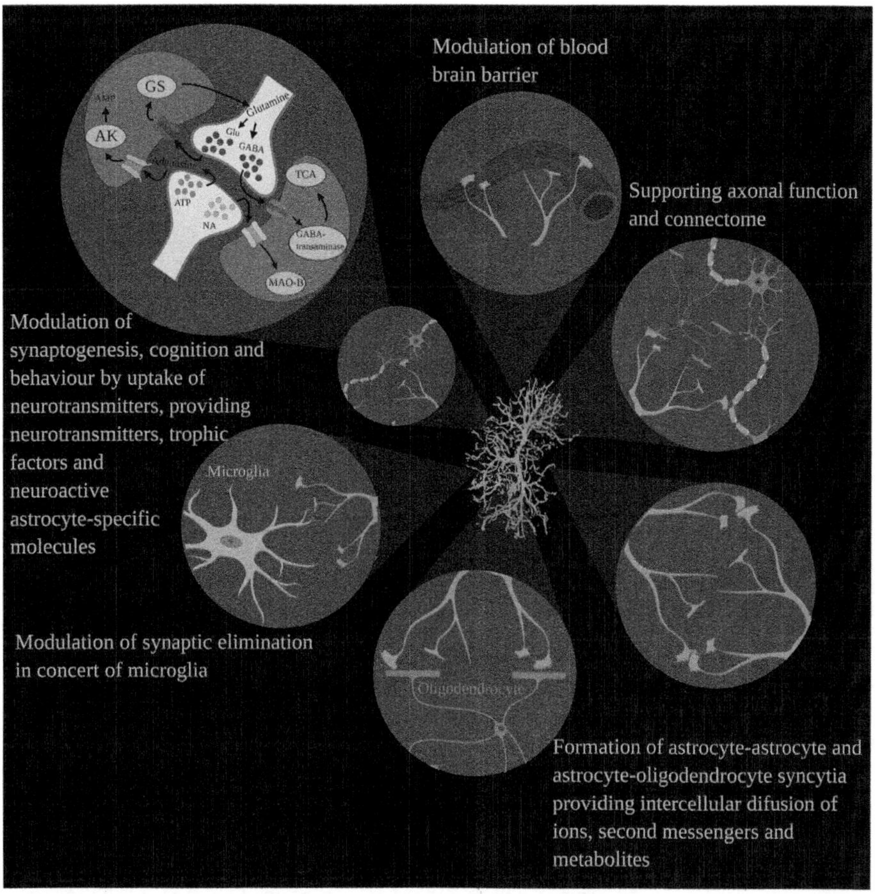

Fig. 3 Functions of astrocytes. Astrocytes interact with diverse cellular structures including blood vessels, oligodendrocytes, neurones, microglia and other astrocytes, contributing to various functions (clockwise from the top) such as regulation of the blood-brain barrier, supporting axons, myelination and connectome, forming astrocyte-astrocyte and astrocyte-oligodendrocyte syncytia, controlling (in concert with microglia) synaptic elimination and modulating synaptogenesis, cognition and behaviour by the neurochemical dialogue with synapses. (Reproduced from Augusto-Oliveira et al. (2020))

Inward rectifier potassium channels are most abundantly expressed in astrocytes. The main type of astrocytic inward rectifier channels is the $K_{ir}4.1$ subtype (encoded by *KCNJ10* gene). These $K_{ir}4.1$ channels are expressed in astrocytes throughout the brain, in the spinal cord and in the retina (Kalsi et al. 2004; Butt and Kalsi 2006). K^+ flux through $K_{ir}4.1$ channels refines the resting membrane potential of astrocytes; pharmacological suppression or genetic deletion of these channels depolarises astrocytes by ~20 mV (Olsen et al. 2007; Seifert et al. 2009). Astrocytes also express $K_{ir}5.1$ channels which may co-assemble with $K_{ir}4.1$. channels; the resulting heteromeric channels are particularly densely expressed in perisynaptic

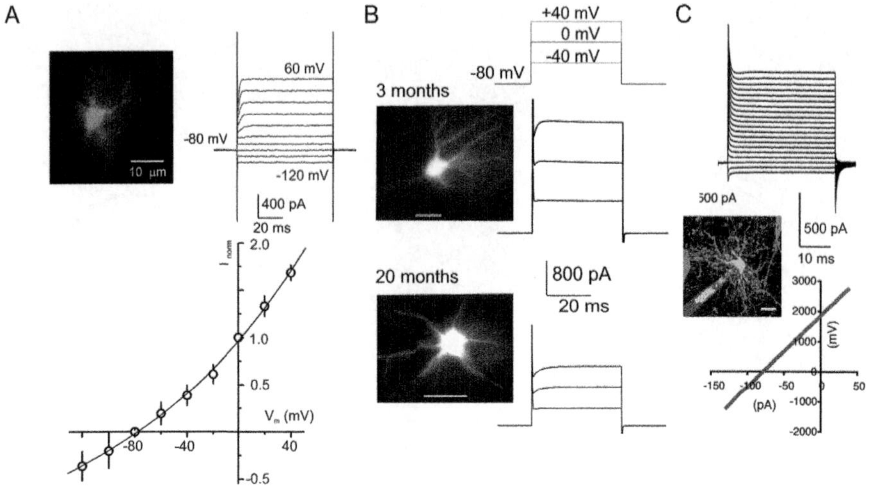

Fig. 4 Passive membrane properties of astrocytes. (**a**) Voltage-clamp recordings from astrocytes freshly isolated from the cortex of transgenic mice expressing EGFP under control of the GFAP promoter. Astrocytes were identified by specific EGFP fluorescence; whole-cell currents were recorded in response to hyper- and depolarising steps from −120 to +60 mV (step interval 20 mV). To construct the I–V relationship, amplitudes of currents were normalised to the value measured at 0 mV; every point is mean ± SD for 20 cells. (**b**) Voltage-clamp recordings from astrocytes in acute slices obtained from 3-month-old and 20-month-old *gfa*::EGFP mice; astrocytes were identified by fluorescence. (**c**) Voltage-clamp recordings from human astrocytes grafted into mouse brain. (Reproduced with permission from Verkhratsky and Nedergaard (2018))

and perivascular astroglial processes (Hibino et al. 2004; Brasko et al. 2017). The ATP-sensitive inward rectifying K^+ channels (assembled from $K_{ir}6.x$ subunit and SUR1/2) have been also detected in astrocytes; these channels are activated in conditions of intracellular ATP depletion (Eaton et al. 2002; Skatchkov et al. 2002). Resting astroglial K^+ permeability is also mediated by several voltage-independent channels of TREK and TWIK families including TREK1/$K_{2P}2.1$, TREK2/$K_{2P}10.1$ and TWIK1/$K_{2P}1.1$ channels (Seifert et al. 2009; Zhou et al. 2009).

In addition to inward rectifying channels, astrocytes express voltage-dependent K^+ channels, which include $K_v1.5$ (KCNA5), $K_v1.4$ (KCNA4) and $K_v11.1$/ERG1 (Bordey and Sontheimer 2000; Verkhratsky and Steinhauser 2000; Edwards et al. 2002) and K_v1, K_v3 and K_v4 mediating fast A-type K^+ currents (Bekar et al. 2005). Finally, cortical astrocytes express SK (small conductance K_{Ca} 2.3/KCNN3) and IK (intermediate conductance $K_{Ca}3.1$/KCNN4) Ca^{2+}-dependent K^+ channels in their somata (Armstrong et al. 2005; Longden et al. 2011) and large-conductance (225 pS) BK (big conductance) channels in their endfeet (Filosa et al. 2006).

Several types of other voltage-dependent channels have been identified in astrocytes. These include tetrodotoxin (TTX)-sensitive and TTX-resistant Na^+ channels which were mainly characterised in cultured astrocytes (Sontheimer et al. 1991; Sontheimer and Waxman 1992; Sontheimer et al. 1994). At transcriptional and protein levels, astrocytes were found to express mainly $Na_v1.5$ subunit, with lower levels of $Na_v1.2$, $Na_v1.3$ and $Na_v1.6$ channels (Pappalardo et al. 2014a, 2016; Zhu et al. 2016). Similarly, astrocytes were reported (at mRNA and sometimes at protein levels) to possess voltage-gated Ca^{2+} channels mainly of $Ca_v1.2$ and $Ca_v1.3$ types (Latour et al. 2003; Cahoy et al. 2008; Zhang et al. 2014). There are only few reports for functionally active Ca^{2+} channels in astrocytes in slice preparations (Parri et al. 2001; Letellier et al. 2016) with no evidence for these channels being operative in vivo.

Astrocytes in the subfornical organ express a specific type of Na^+ channel known as Na_x (encoded by SCN7A gene). This channel is activated by an increase in extracellular Na^+ concentration above 140 mM and contribute to astroglia-dependent chemosensing of Na^+ fluctuations in circulation (Hiyama et al. 2013; Noda and Sakuta 2013). Another channel type that may be involved in Na^+ chemosensing is the epithelial Na^+ channel (ENaC). Immunoreactivity for ENaC has been found in astrocytes in circumventricular organs, white matter and pia mater (Miller and Loewy 2013).

Astrocytes are in possession of several types of transient receptor potential channels (TRP). The TRPA1 (ankyrin-type) channel is operational in the somata and processes of astrocytes in the brain stem and hippocampus; these channels provide for background Ca^{2+} entry thus regulating resting Ca^{2+} concentration in the cytoplasm (Shigetomi et al. 2012, 2013). The TRPC1 (or canonical-type)-containing channels mediate the store-operated Ca^{2+} entry following depletion of Ca^{2+} stores (Malarkey et al. 2008; Reyes et al. 2013; Verkhratsky and Parpura 2014). The TRPV1 and TRPV4 (vanilloid-type) are identified in cortical, hippocampal and spinal cord astrocytes, where they are activated in response to hypo-osmotic stress and cell swelling (Benfenati et al. 2007; Butenko et al. 2012). The store-operated Ca^{2+} entry in astrocytes is also mediated by ORAI 1 and ORAI3 channels activated by the endoplasmic reticulum (ER) Ca^{2+} sensor STIM1 (Kwon et al. 2017; Toth et al. 2019).

Astrocytes are the main type of neural cells that express water channels or aquaporins (AQP). The most abundant is AQP4, with much lesser expression of AQP1 and AQP9. In healthy conditions, the AQP4 channels are predominantly concentrated in astrocytic perivascular and pial endfeet (Nagelhus and Ottersen 2013). These channels contribute to many functions from olfaction and K^+ buffering to synaptic plasticity and memory (Lu et al. 2008; Skucas et al. 2011; Scharfman and Binder 2013; Lisjak et al. 2017). The endfeet AQP4 channels are particularly important for the function of the glymphatic system (Mestre et al. 2020).

Astrocytes are functionally integrated into syncytia by gap junctional plaques which are composed of several hundred of intercellular channels known as connexons, with each connexon being assembled from 12 subunits known as connexins (Giaume et al. 2021). Astrocytes express three types of connexins, namely, Cx26, Cx30 and Cx43, of which Cx43 is the most abundant and most widespread being expressed in all CNS regions (Nagy et al. 2004). Connexons assembled from Cx30 are mainly present in the thalamus and leptomeninges (Nagy et al. 2004; Sohl et al. 2004), whereas Cx26 is confined to the hypothalamus, reticular thalamic and subtalamic nuclei (Nagy et al. 2011). Astrocytes also establish syncytial connections with oligodendrocytes; the astrocyte-oligodendrocyte connexons are assembled from heterotypic channels composed of Cx47/Cx43, Cx32/Cx30, Cx47/Cx30 or Cx32/Cx26, of which the first two are predominant types in vivo (Orthmann-Murphy et al. 2007; Magnotti et al. 2011). Unpaired connexons, also known as hemichannels, have been identified in astrocytes in various brain regions, and they may mediate (together with Pannexin channels) diffusional secretion of various neuroactive substances (Orellana et al. 2009; Giaume et al. 2013; Verkhratsky et al. 2016).

Receptors of Neurotransmitters

Fundamentally, astrocytes are capable of expressing virtually every type of neurotransmitter/neuromodulator receptor existing in the CNS. At the same time this potential promiscuity is very much restricted by the neurochemical environment in different parts of the CNS. Generally, astrocytes express receptors of the same modality as their neuronal neighbours; these modalities are congruent to the neurotransmitters released in a given brain region (Verkhratsky et al. 1998; Verkhratsky 2010). For example, in the spinal cord, where glycine is the main inhibitory neurotransmitter, astrocytes specifically express glycine receptors; in the basal ganglia, which utilise dopamine, astrocytes are endowed with dopamine receptors. Thus, the expression of astroglial receptors in vivo is regulated by neurochemical input, which makes astrocytes perceptive to regional neurochemical environment.

Probably the most abundant astrocytic receptors are receptors of purines, as the purinergic signalling system is omnipresent in the CNS (Burnstock and Verkhratsky 2012). Metabotropic purinoceptors, represented by ATP/ADP $P2Y_{1,2,4,6,12,13}$ and UDP-glucose $P2Y_{14}$ receptor, predominate (Abbracchio and Ceruti 2006; Verkhratsky et al. 2009). Activation of metabotropic P2Y receptors is linked, through phospholipase C (PLC) and inositol-1,4,5-trisphophate ($InsP_3$) receptors to Ca^{2+} signalling. Astrocytes also possess ionotropic purinoceptors, mainly of $P2X_7$ type; murine cortical astrocytes also have $P2X_{1/5}$ receptors (Lalo et al. 2008, 2011; Illes et al. 2012). Activation of ionotropic purinoceptors contributes to both Ca^{2+} and Na^+ signalling in astrocytes.

The second class of astrocytic neurotransmitter receptors is represented by receptors of amino acid glutamate, GABA and glycine. Again astrocytes express several types of glutamate receptors. Ionotropic glutamate receptors functionally expressed in astrocytes are represented by AMPA receptors (with all four subunits being expressed, albeit in brain region-dependent fashion) and NMDA receptors. The latter are composed of two GluN1 subunits assembled with GluN2C or D and GluN3 subunits. Such composition infers weak Mg^{2+} block (which develops at ~ −120 mV) and relatively low Ca^{2+} permeability ($P_{Ca}/P_{monovalent}$ ~ 3), as well as sensitivity to memantine and GluN2C/D subunit-selective antagonist UBP141 (Lalo et al. 2006; Palygin et al. 2010; Verkhratsky and Chvatal 2020). Astrocytes also express metabotropic glutamate receptors, of which mGluR3 (connected to cAMP signalling cascade) and $mGuR_{1/5}$ (connected to PLC-$InsP_3$-Ca^{2+} signalling cascade) predominate. Astrocytes express ionotropic $GABA_A$ receptors, which mediate Cl^- currents and, because of high $[Cl^-]i$, depolarise the membrane (Kettenmann et al. 1987); astrocytic metabotropic $GABA_B$ receptors are linked to PLC-$InsP_3$-Ca^{2+} signalling cascade (Nilsson et al. 1993). Spinal cord astrocytes express glycine receptors which mediate Cl^- efflux and astrocytic depolarisation (Pastor et al. 1995).

The third class of receptors abundantly present in astrocytes is represented by receptors for monoamines. In particular, most of astrocytes express α- and β-adrenoceptors that were detected in culture, in slices and in vivo at transcriptional, protein and functional levels (Verkhratsky and Nedergaard 2018). Both α_1 and α_2 adrenoceptors are linked to Ca^{2+} signalling (Kirischuk et al. 1996; Ding et al. 2013); β_1-adrenoceptors regulate glycogen synthesis, β_2-adrenoceptors are linked to cAMP signalling and β_3-adrenoceptors regulate glucose uptake by modulating GLUT1 plasmalemmal glucose transporter (Hutchinson et al. 2007; Dong et al. 2012). In addition, astrocytes express $5\text{-}HT_{1A}$, $5\text{-}HT_{2A}$, $5\text{-}HT_{2B}$ and $5\text{-}HT_{5A}$ metabotropic serotonin receptors, as well as D_1, D_2, D_4 and D_5 dopamine receptors, and H_1, H_2 and H_3 histamine receptors (Verkhratsky and Nedergaard 2018).

Other receptors expressed in astrocytes include B_2 bradykinin receptors, cannabinoid CB_1 receptors, V_1 vasopressin receptors, oxytocin receptors, ET_A and ET_B endothelin receptors, atrial natriuretic peptide receptors, receptors for leptin and insulin, platelet-activating factor receptors and protease-activated receptors, which fulfil various functions and are linked to different signalling cascades (Verkhratsky and Nedergaard 2018).

Astroglial Plasma Membrane Transporters

Plasmalemmal transporters are fundamentally important for astroglial homeostatic function because they accomplish transport of ions, neurotransmitters, scavengers of reactive oxygen species (ROS), metabolic substrates and various neuromodulators and hormones. All plasmalemmal transporters are classified into pumps (which utilise energy of ATP) and secondary plasmalemmal transporters or solute carriers (SLC) which use concentration gradients.

Astroglial Na^+-K^+ pump (NKA) has the idiosyncratic α2 catalytic subunit, which distinguishes it from the neuronal NKA, which contains α1 or α3 subunits. The affinity of the astroglial α2 subunit to extracellular K^+ is much lower as compared to neuronal α1 and α3 subunits: $[K^+]_{0.5}$ for NKA composed from α2/β1 subunits is ~3.6 mM, whereas $[K^+]_{0.5}$ for neuronal NKAs assembled from α1/β1, α1/β2, α3/β1 or α3/β2 subunits lies between 0.25 and 0.65 mM (Larsen et al. 2014). As consequence, the activity of astroglial NKA is stimulated by physiological increases in the extracellular K^+ concentration. To the contrary neuronal NKA is fully saturated and hence cannot respond to increases in extracellular K^+ (Hertz et al. 2015). This stipulates the leading role of NKA in astrocytic K^+ buffering (Verkhratsky and Nedergaard 2018).

Astrocytic SLC transporters are many and they perform numerous homeostatic functions. The largest group of homeostatic SLC transporters is represented by transporters for neurotransmitters; the majority of these transporters utilise transmembrane Na^+ gradient to move neurotransmitters across the plasma membrane. Astrocytic transporters for glutamate are known as excitatory amino acid transporters 1 and 2 (EAAT1/SLC1A6 and EAAT2/SLC1A2 (Zhou and Danbolt 2013)), which in rodent experiments are referred to as glutamate-aspartate transporter (GLAST (Storck et al. 1992)) and glutamate transporter 1 (GLT-1 (Pines et al. 1992)). EAAT1 is mainly present in the cerebellum, in the retina and in circumventricular organs, whereas EAAT2 is expressed in all other parts of the brain; at the cellular level both transporters localise to perisynaptic astroglial processes (Zhou and Danbolt 2013; Verkhratsky and Nedergaard 2018). Glutamate transporters are Na^+-dependent with stoichiometry 3 Na^+, 1 H^+ and 1 glutamate (in): 1 K^+ (out) (Zerangue and Kavanaugh 1996). This transporter is electrogenic and generates transmembrane current carried mainly by Na^+ ions (Kirischuk et al. 2007). In addition to EAATs, astrocytes express the cystine/glutamate antiporter Sxc^- (SLC7A11) which is critical for cystine accumulation needed for the production of glutathione (Bridges et al. 2012). Glutamate homeostasis also depends on astrocytic glutamine transporters SNAT3/SLC38A3 and SNAT5/SLC38A5 which couple transport of glutamine with co-transport of 1 Na^+ and counter-transport of 1 H^+ (Todd et al. 2017). Astroglial GABA transporters are mainly represented by GAT3/SLC6A11 with stoichiometry of 1 GABA: 2 Na^+: 1 Cl^- (Kavanaugh et al. 1992); moderate increases in cytoplasmic Na^+ may reverse this transporter (Unichenko et al. 2012). Astrocytes in the spinal cord possess glycine transporter GlyT1/SLC6A9 with stoichiometry of 1 glycine: 2 Na^+: 1 Cl^-; this transporter also can reverse in physiological conditions (Shibasaki et al. 2017). Monoamine accumulation into astrocytes is mediated by norepinephrine transporter NET/SLC6A2 that co-transports monoamines with 2 Na^+ and 1 Cl^- (Takeda et al. 2002). Astrocytes transport adenosine by equilibrative (i.e. controlled by adenosine transmembrane gradient (King et al. 2006)) plasmalemmal transporters ENT-1/SLC29A1, ENT-2/SLC29A2, ENT-3/SLC29A3 and ENT-4/SLC29A4, and Na^+-dependent concentrative nucleoside plasmalemmal transporters CNT2/SLC28A2 and CNT3/SLC28A3 (King et al. 2006; Li et al. 2013).

The next group of SLC transporters is responsible for ion exchange. The most prominent member of this group is plasmalemmal sodium-calcium exchanger present in astrocytes in three isoforms NCX1/SLC8A1, NCX2/SLC8A2 and NCX3/SLC8A3 (Pappalardo et al. 2014b). Stoichiometry of astrocytic NCX is 3 Na^+: 1 Ca^{2+} which sets the equilibrium potential at ~ −85 to −90 mV; this is very close to the resting membrane potential of astrocytes, and hence minor depolarisations or moderate increases in cytosolic Na^+ readily reverse NCX operation from the forward mode (extrusion of Ca^{2+} in exchange for Na^+) to the reverse mode (Ca^{2+} influx in exchange for Na^+ exit) (Verkhratsky et al. 2018). Another transporter expressed in astrocytes is the NHE1/SLC9A1 Na^+-H^+ exchanger (Deitmer and Rose 1996;

Chesler 2003) with electroneutral stoichiometry of 1 Na^+ (in): 1 H^+ (out) (Orlowski and Grinstein 1997); this transporter mediates efflux of protons from astrocytes (Rose and Ransom 1996; Chesler 2003). Additional transporter involved in pH homeostasis is the sodium-bicarbonate transporter NBCe1/SLC4A4 with stoichiometry of 1 Na^+: 2 HCO_3^- or 1 Na^+: 3 HCO_3^-, which allows its operation in both forward and reverse modes (Theparambil et al. 2015). The Na^+-K^+-Cl^- co-transporter NKCC1/SLC12A2 with electroneutral stoichiometry of 1 Na^+: 1 K^+: 2 Cl^- contributes to K^+ and Cl^- homeostasis especially in pathology (Macaulay and Zeuthen 2012).

Finally, astrocytes express several metabolic transporters responsible for transmembrane movements of energy substrates. In particular, astrocytes possess glucose transporter GLUT1/SLC2A1 predominantly localised in endfeet (Allen and Messier 2013). Monocarboxylate transporters 1 and 4 (MCT1/SLC16A1, MCT4/SLC16A3) are responsible for the export of lactate from astrocytes for local energy support of neurones; in certain conditions, however, these transporters can mediate lactate accumulation (Halestrap 2012).

Intracellular Excitability of Astrocytes

Astrocytes (similarly to other neuroglia) are electrically non-excitable cells of the CNS; in contrast to neurones astrocytes do not generate regenerative and propagating action potentials. At the same time astrocytes do generate active responses to all changes in nervous tissue environment; these responses are linked to transient fluctuations in cellular ionic content or generation/degradation of intracellular messengers. These molecular changes are the substrate for astrocytic excitability.

Astrocyte Ca^{2+} Excitability

Calcium ions are universal and evolutionary conserved intracellular messengers, involved in the regulation of numerous cellular processes, such as excitation-contraction coupling, secretion, gene expression, metabolism and cell death (Berridge et al. 2000; Carafoli 2002; Petersen et al. 2005; Plattner and Verkhratsky 2015). Cellular Ca^{2+} signalling and homeostatic control over cytosolic Ca^{2+} concentration ($[Ca^{2+}]_i$) rely on dedicated molecules, which include membrane channels and transporters, intracellular Ca^{2+}-binding proteins and numerous Ca^{2+} sensors represented by Ca^{2+}-regulated proteins.

Astrocytic Ca^{2+} excitability was discovered in experiments in vitro which demonstrated that challenging cultured astrocytes with neurotransmitters and neurohormones triggered $[Ca^{2+}]_i$ transients and propagating Ca^{2+} waves (Enkvist et al. 1989; Cornell-Bell et al. 1990; McCarthy and Salm 1991; Verkhratsky and Kettenmann 1996; Petersen et al. 2005). Molecular pathways underlying astrocytic

Ca^{2+} excitability rely on both intracellular and extracellular Ca^{2+} sources. Generation of Ca^{2+} signals in the soma and primary processes mainly depends on metabotropic receptors, which activate PLC that produces InsP*3*, which in turn activates $InsP_3$ receptors residing in the membrane of the ER. This organelle is the main Ca^{2+} storage organelle in living cells; it is capable of accumulating, storing and releasing Ca^{2+} in physiological context. Concentration of ionised Ca^{2+} in the lumen of the ER is rather high, ranging between 200μM and 1 mM (Alonso et al. 1999; Alvarez and Montero 2002; Solovyova and Verkhratsky 2002; Solovyova et al. 2002). High intraluminal $[Ca^{2+}]$ creates a gradient aimed at the cellular cytosol; hence opening of Ca^{2+} release channels dwelling in the endomembrane generates Ca^{2+} efflux and thus generates cytosolic Ca^{2+} signals; in addition, intraluminal Ca^{2+} regulates both Ca^{2+} pump (SERCA, SarcoEndoplasmic Reticulum Ca^{2+}-ATPase) and Ca^{2+} release channels (Burdakov et al. 2005). Besides being a dynamic Ca^{2+} store, the ER is responsible for protein synthesis, folding and haulage to their respective destinations, with all these processes being regulated by luminal $[Ca^{2+}]$ (Verkhratsky and Petersen 2002). Finally, the ER creates intracellular Ca^{2+} tunnels which allow long-distance diffusion of Ca^{2+}, which is impossible in heavy Ca^{2+}-buffered cytosol (Petersen and Verkhratsky 2007).

Both classes of intracellular Ca^{2+} release channels, the $InsP_3$ receptors and ryanodine receptors (which are Ca^{2+}-gated Ca^{2+}channels) are expressed in glial cells (Verkhratsky et al. 1998). In astrocytes the type II $InsP_3$ receptors predominate; these receptors are activated by InsP*3* and positively modulated by $[Ca^{2+}]_i$; local increases in $[Ca^{2+}]_i$ therefore can generate propagating wave of opening of $InsP_3$ receptors along the ER membrane, making the latter an excitable media sustaining propagating Ca^{2+} waves (Verkhratsky et al. 2012).

As alluded before, release of Ca^{2+} from the ER, mediated by $InsP_3$ receptors, represents the leading mechanism for Ca^{2+} signal generation in the soma and organelle-bearing processes; depletion of ER Ca^{2+} store activates secondary Ca^{2+} influx through plasmalemmal store-operated pathway mediated by TRP and ORAI channels (Burnstock and Verkhratsky 2012). lGeneration of Ca^{2+} signals in the distal leaflets is different as it mainly relies on plasmalemmal Ca^{2+} entry (Fig. 5). The principal route for Ca^{2+} entry in response to neuronal activity is associated with the reversal of Na^+-Ca^{2+} exchanger (NCX). The latter is controlled by transmembrane Na^+ gradient; increase in $[Na^+]_i$ associated with operation of glutamate transporters reverses the NCX (Reyes et al. 2012; Rose et al. 2020). In addition, Ca^{2+} entry into the leaflets is mediated by ionotropic receptors and possibly by TRP channels (Verkhratsky et al. 2012). Such a dichotomy is characteristic for protoplasmic astrocytes; in radial Bergmann glial cells, the main mechanism for Ca^{2+} signal generation is represented by PLC-$InsP_3$ pathway (Kirischuk et al. 1999). The complex molecular machinery responsible for Ca^{2+} signalling in astrocytes results in emergence of spatially and temporally distinct Ca^{2+} signals in the form of microdomains or propagating Ca^{2+} waves of global Ca^{2+} signals (Grosche et al. 1999; Shigetomi et al. 2013; Khakh and Sofroniew 2015; Arizono et al. 2020).

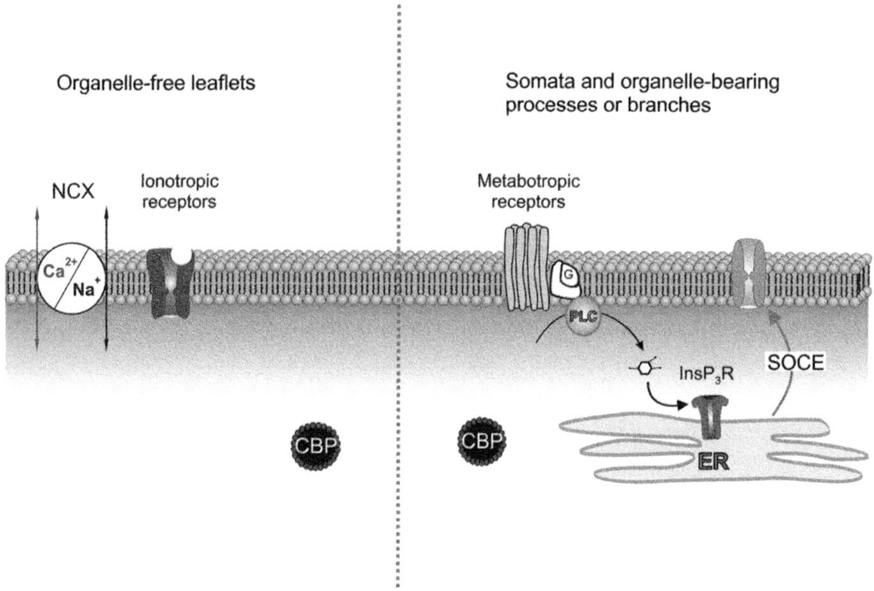

Fig. 5 Ca^{2+} signalling in different compartments in protoplasmic astrocyte. Calcium signalling in the organelle-free leaflets is associated with Ca^{2+} entry through ionotropic receptors (NMDA glutamate receptors or P2X purinoceptors) or Ca^{2+}-permeable channels (e.g. TRPA1 channels). Plasmalemmal Ca^{2+} influx can also be mediated by the Na^+/Ca^{2+} exchanger (NCX) operating in the reverse mode. Calcium signalling in soma and branches is mainly associated with Ca^{2+} release from the endoplasmic reticulum (ER) with subsequent store-operated Ca^{2+} entry (SOCE). This Ca^{2+} release is mediated by $InsP_3$ receptors ($InsP_3R$); $InsP_3$ is synthesised by phospholipase C (PLC) linked to G-protein metabotropic receptors. CBP, calcium binding proteins. (Reproduced with permission from Verkhratsky et al. (2020))

Sodium Signalling

The foundations for astrocytic Na^+ signalling are associated solely with Na^+ entry mediated by various proteins in the plasma membrane (Fig. 6). Cells have neither intracellular Na^+ stores nor cytosolic Na^+ buffers (Kirischuk et al. 2012; Rose and Verkhratsky 2016). Physiological stimulation triggers $[Na^+]_i$ transients in astrocytes in vitro and in situ (Rose and Ransom 1996; Kirischuk et al. 1997; Langer and Rose 2009). The main source for generation of Na^+ signals is mediated by Na^+ entry through Na^+-coupled SLC transporters, whereas extrusion of Na^+ is primarily mediated by NKA (Rose and Verkhratsky 2016). In addition, Na^+ influx can be mediated by ionotropic receptors and plasmalemmal channels, in particular TRPC channels (Reyes et al. 2013). Among Na^+-dependent SLC transporters, glutamate and GABA transporters play the leading role in generation of Na^+ signals. Increase in extracellular glutamate associated with neurotransmission is followed by rapid accumulation of glutamate into astrocytes; translocation of a single glutamate molecule is coupled with entry of 3 Na^+. Increase in cytosolic Na^+ concentration

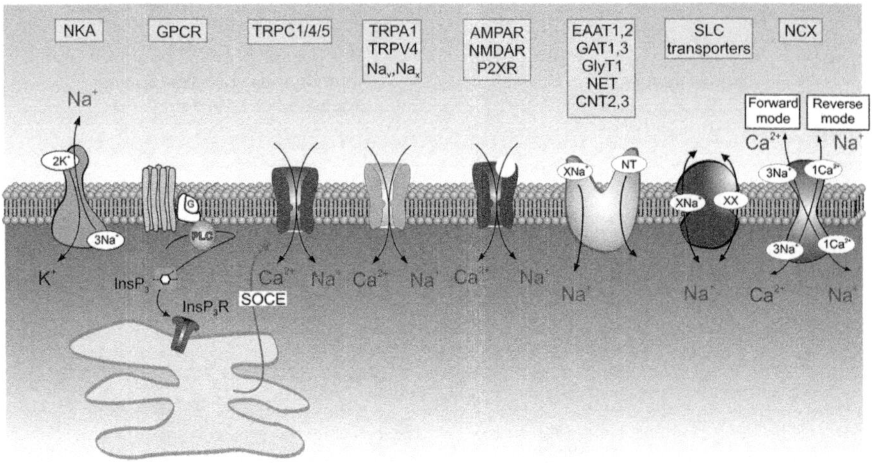

Fig. 6 Membrane molecular pathways of Na^+ signalling in astrocytes. Influx of Na^+ occurs though (i) Na^+-permeable channels which include ionotropic receptors (AMPAR, NMDAR, P2XR: AMPA, NMDA glutamate receptors and ionotropic purinoceptors, respectively); channels of the transient receptor potential (TRP) family (TRPC1/4/5 channels that operate as a part of store-operated Ca^{2+} entry and hence generate Na^+ influx in response to the depletion of endoplasmic reticulum Ca^{2+} stores; as well as TRPA1 and TRPV4 channels); voltage-dependent Na_v channels and $[Na^+]_o$-activated Na_x channels; (ii) through Na^+-dependent SLC transporters that include excitatory amino acid transporters EAAT1,2, GABA transporters GAT 1,3, glycine transporters GlyT, noradrenaline transporters NET and concentrative adenosine transporters CNT2,3. The main pathway for Na^+ exit is provided by Na^+-K^+ pump, NKA. The Na^+-Ca^{2+} exchanger NCX fluctuates between forward and reverse modes and couples Na^+ and Ca^{2+} signalling. Other abbreivations as in Fig. 5. (Reproduced with permission from Verkhratsky et al. (2020))

($[Na^+]_i$) in response to glutamate may reach 10–20 mM (Kirischuk et al. 2007; Bennay et al. 2008). The main physiological target of astroglial Na^+ signals is represented by SLC transporters sensitive to transmembrane Na^+ gradients (Fig. 7); in addition $[Na^+]_i$ regulates glutamine-glutamate (GABA) shuttle through direct action on glutamine synthetase and regulation of glutamine transporters (Benjamin 1987; Todd et al. 2017).

Astrocyte Functions

Astrocytes perform every known housekeeping and homeostatic function in the CNS from structural support and control over molecular homeostasis to regulation of blood flow, synaptogenesis, neurogenesis and development of the nervous system.

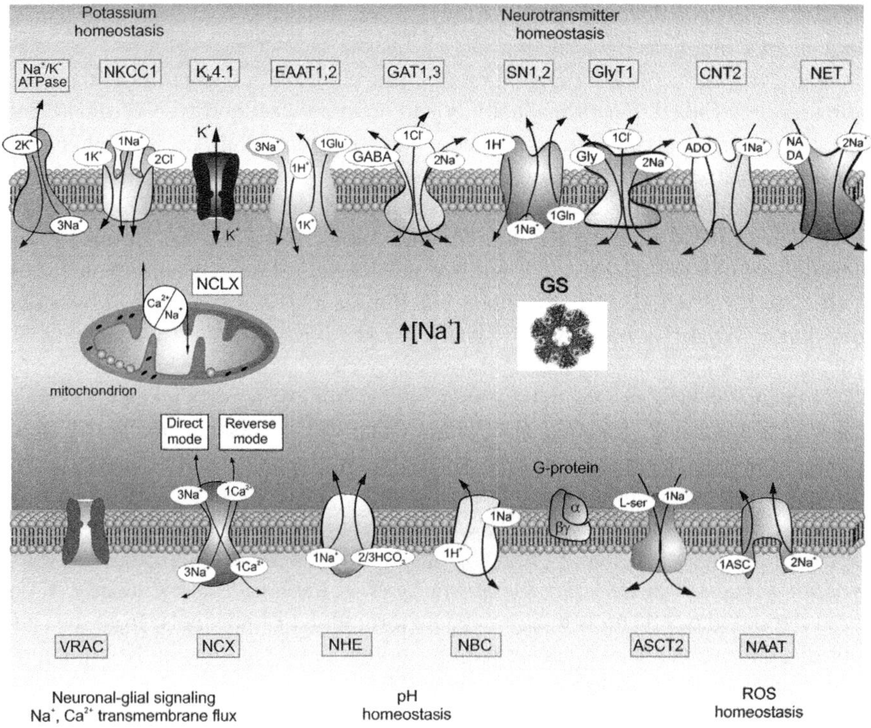

Fig. 7 Molecular targets of Na^+ signalling in astroglia. Abbreviations: ASCT2 alanine-serine-cysteine transporter 2, ASIC acid sensing ion channels, CNT2 concentrative nucleoside transporters, EAAT excitatory amino acid transporters, ENaC epithelial sodium channels, GAT GABA transporters, GS glutamine synthetase, GlyT1 glycine transporter, iGluRs ionotropic glutamate receptors, Na_x Na^+ channels activated by extracellular Na^+, NAAT Na^+-dependent ascorbic acid transporter, NBC Na^+/HCO_3^- (sodium-bicarbonate) co-transporter, NCX Na^+/Ca^{2+} exchanger, NCLX mitochondrial Na^+/Ca^{2+} exchanger, NHE Na^+/H^+ exchanger, NKCC1 $Na^+/K^+/Cl^-$ co-transporter, NET norepinephrine transporter, MCT1 monocarboxylase transporter 1, P2XRs ionotropic purinoceptors, SN1/2 sodium-coupled neutral amino acid transporters which underlie exit of glutamine, TRP transient receptor potential channels, ROS reactive oxygen species, VRAC volume-regulated anion channels. Other abbreviations as in Fig. 5. See text for further explanation. (Modified and reproduced from Verkhratsky and Nedergaard (2018))

Ion Homeostasis in the CNS, or Ionostasis

Controlling ion concentrations in the interstitial fluids is of paramount importance for the functional activity of the nervous tissue, as even minor fluctuation in ionic composition may have profound influences on neuronal excitability and information processing in the neuronal networks. In particular, changes in ion concentrations are linked to changes in the brain state, such as sleep and arousal (Ding et al. 2016), while fluctuations in ion composition of interstitial fluids affect learning and memory (Hertz and Chen 2016). Control over ionostasis is one of the most fundamental functions of astrocytes.

Astrocytes are central for regulation of extracellular K^+ concentration, the latter being directly linked to neuronal activity and neuronal excitability. Neuronal firing and synaptic transmission are associated with substantial K^+ efflux, which may significantly change K^+ concentration in the narrow and diffusion-restricted extracellular compartments. The leading role of astrocytes in the regulation of extracellular K^+ homeostasis has been proposed in the mid-1960s by Leif Hertz, Steven Kuffler and Richard Orkand (Hertz 1965; Kuffler and Nicholls 1966; Orkand et al. 1966). Astrocytic K^+ clearance from the interstitial fluid is mainly accomplished by active transport by NKA (Larsen et al. 2014), as astrocytic NKA is directly activated by K^+ rise (see above). After being taken up by astrocytes, K^+ is subsequently released by $K_{ir}4.1$ channels back into the extracellular space, from which it is taken up by neurones to restore ionic gradients (Hertz and Chen 2016; Larsen et al. 2016). Such scenario implies the generation of K^+ microdomains, which might be supported by physical properties of bilipid membranes trapping ions in narrow compartments of astrocytic leaflets (Breslin et al. 2018). In Müller glial cells in the retina, K^+ buffering is mainly accomplished by K^+ entry through $K_{ir}4.1$ channels localised in perisynaptic processes in the inner plexiform layer of the retina. Subsequently, K^+ is spred through the cell and is released (again through $K_{ir}4.1$ channels) from the endfoot distantly to the site of accumulation; this process is known as K^+ siphoning (Newman et al. 1984; Kofuji and Newman 2004).

Another important extracellular ion that may fluctuate during neuronal activity and synaptic transmission is Ca^{2+}. Opening of plasmallemal Ca^{2+} channels with subsequent Ca^{2+} influx into presynaptic boutons may result in substantial drop in Ca^{2+} concentration in the narrow synaptic cleft, thus affecting neurotransmitter release (Rusakov and Fine 2003). Astrocytes seem to provide a pathway for replenishment of extracellular Ca^{2+} as decrease in $[Ca^{2+}]_o$ below ~0.5 mM induces $InsP_3$-induced Ca^{2+} release from the ER in astrocytes (Zanotti and Charles 1997). This Ca^{2+} then can be transported to extracellular space through plasmalemmal Ca^{2+} pumps or NCX.

A similar situation may also happen with extracellular Cl^-. Overstimulation of neuronal ionotropic $GABA_A$ receptors may result in substantial Cl^- accumulation into neurones; reduction in extracellular Cl^- concentration may subsequently impair inhibitory transmission. As astrocytes have high cytoplasmic Cl^- concentration, activation of astrocytic ionotropic $GABA_A$ receptors may initiate substantial Cl^- efflux, thus restoring extracellular Cl^- concentration (Kettenmann and Verkhratsky 2008). Indeed, manipulations with astrocytic intracellular Cl^- were demonstrated to affect inhibitory transmission in hippocampal slices (Egawa et al. 2013). Astrocytes are also principal players in regulation of extracellular pH through transport of H^+ and HCO_3^- (Deitmer and Rose 2010).

Neurotransmitter Homeostasis

Astrocytes control neurotransmitter homeostasis in the brain through removal (by dedicated transporters) and inactivation (by enzymatic conversion) of all four major neurotransmitters, namely, glutamate, GABA, adenosine and noradrenaline (Fig. 8).

Astrocytic glutamate homeostatic system is an indispensable element for glutamatergic and GABAergic transmission because it provides for (i) clearance of glutamate and GABA from the synaptic cleft, (ii) prevention of glutamate spillover and (iii) replenishment of the glutamate and GABA releasable pools of vesicles in the neuronal terminals by providing the obligatory precursor glutamine (Hertz et al. 1999; Verkhratsky and Nedergaard 2018). Astrocytic glutamate clearance also protects nervous tissue against glutamate excitotoxicity.

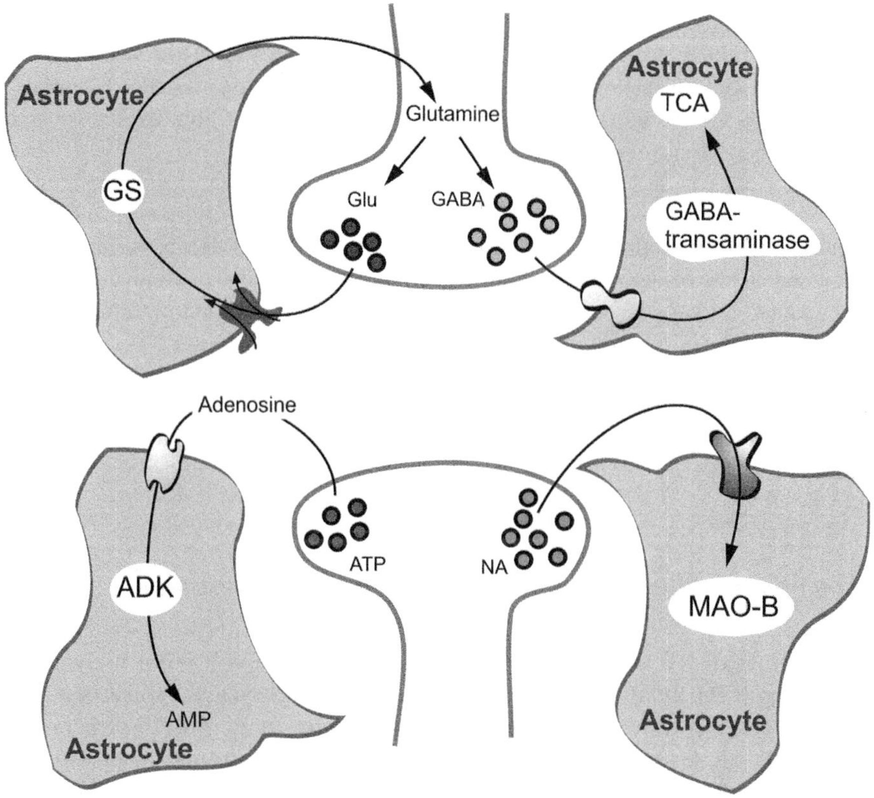

Fig. 8 Astrocytes and neurotransmitter homeostasis. Astrocytes take up glutamate, GABA, adenosine and monoamines. Glutamate (Glu) is converted to glutamine by glutamine synthetase (GS) in astrocytes which also synthesise this transmitter de novo. In turn, glutamine is shuttled to neurones for subsequent conversion into glutamate and GABA. Astroglial accumulated GABA is mainly transaminated and consumed in tricarboxylic acid cycle (TCA). Adenosine is converted to AMP by adenosine kinase (ADK), while monoamines are degraded by astroglial monoamine oxidase type B (MAO-B). (Modified and reproduced from Verkhratsky and Nedergaard (2018))

Neurones do not have enzymes for de novo synthesis of glutamate from pyruvate (generated from glucose or lactate); this synthesis occurs solely in astrocytes (Hertz et al. 1999). Glutamate produced in astrocytes is converted, with another astrocyte-specific enzyme, glutamine synthetase, into biologically inert glutamine, which is subsequently transported to neurones; when in neurones glutamine is converted into glutamate by glutaminase. In inhibitory terminals, glutamate is further converted into GABA.

Astrocytic glutamate transporters EAAT1/2 are responsible for the clearance of ~80% of all glutamate released during synaptic transmission (Danbolt 2001). Removal of glutamate by astrocytes dynamically modulates glutamate concentration in the synaptic cleft thus shaping kinetics of neuronal postsynaptic excitatory potentials (Marcaggi and Attwell 2004; Tzingounis and Wadiche 2007). Glutamate entering into astrocytes can undergo conversion to glutamine, which is subsequently shuttled to neurones through Na^+-dependent glutamine transporters (described in a previous section). Increase in $[Na^+]_i$ resulting from EAAT1/2-mediated uptake of glutamate stimulates glutamine efflux (Todd et al. 2017). This sequence of plasmalemmal transporting and enzymatic conversion is known as the glutamine-glutamate (GABA) shuttle that is fundamental for sustaining both excitatory and inhibitory neurotransmission in the CNS (Hertz 2013).

Astrocytes provide substantial contribution to the catabolism of monoamines in the CNS. The main converting enzyme that catabolises noradrenalin (norepinephrine), dopamine and serotonin, the monoamine oxidase B (MAO A/B), is preferentially expressed in astrocytes (Saura et al. 1992; Hertz et al. 2004). Monoamines are accumulated into astrocytes through plasmalemmal Na^+-dependent norepinephrine transporter NET/SLC6A2 and possibly by dopamine transporter DAT/SLC6A3 (Verkhratsky and Nedergaard 2018). Finally, astrocytes express high levels of adenosine kinase which catalyses the bulk of adenosine conversion into AMP (Boison 2008); genetic deletion of adenosine kinase is incompatible with life (Boison et al. 2010).

Astroglial Cradle in Regulation of Synaptic Transmission

Astrocytic processes, including branches and leaflets, establish intimate dynamic contacts with synaptic structures; at least 50–60% of all synaptic contacts in the CNS are covered with astrocytic membranes (Witcher et al. 2007; Reichenbach et al. 2010). These astrocytic perisynaptic structures play most fundamental role in synaptic connectivity being involved in the regulation of synaptogenesis, synaptic maturation, synaptic maintenance, synaptic isolation and synaptic extinction; this multitude of functions stipulated the emergence of the concept of astroglial cradle (Nedergaard and Verkhratsky 2012; Verkhratsky and Nedergaard 2014) (Fig. 9). Astrocytes express numerous specific pathways through which they can regulate various aspects of synaptic transmission (Augusto-Oliveira et al. 2020).

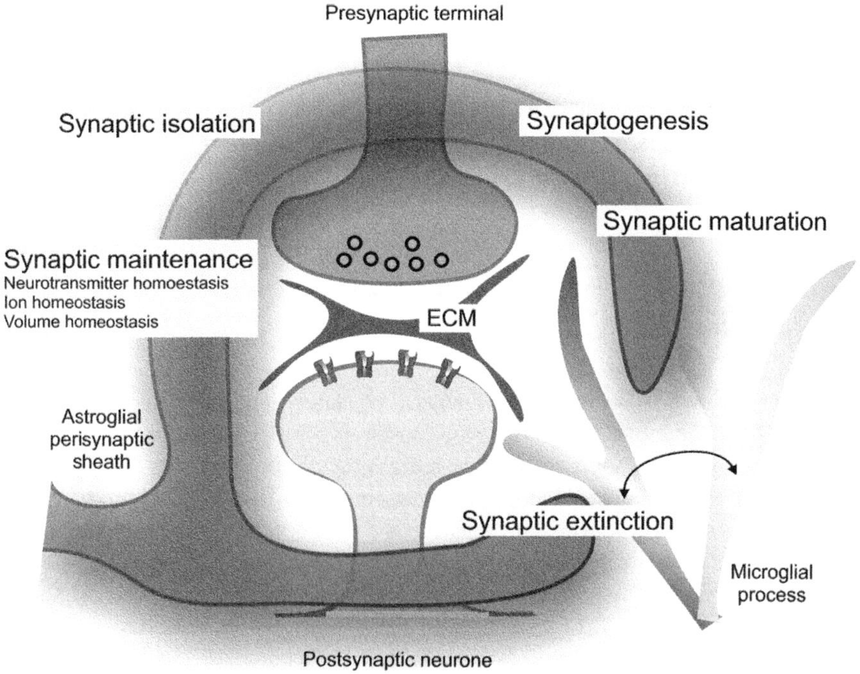

Fig. 9 Astrocytic synaptic cradle. Astroglial cradle embraces and fosters multipartite synapse in the CNS. The majority of synapses in the brain and in the spinal cord is composed of several components that include the presynaptic terminal. These components are: the postsynaptic part, the perisynaptic process of the astrocyte, the process of neighbouring microglial cell that periodically contacts the synaptic structure and the extracellular matrix (ECM) present in the synaptic cleft and also extending extra-synaptically. Astroglial perisynaptic sheath enwraps synaptic structures; regulates, influences and assists synaptogenesis, synaptic maturation, synaptic maintenance and synaptic extinction; and also modulates synaptic transmission and plasticity. (Reproduced from Verkhratsky and Nedergaard (2014))

Synaptogenesis in particular requires astrocytes and astrocytic factors. The massive emergence of excitatory glutamatergic synapses occurs in the early postnatal period and coincides with massive wave of astrocytogenesis. Without astrocytes, synapses do not form in vitro, and similarly astrocytes are necessary for synapse formation in vivo (Pfrieger and Barres 1997; Eroglu and Barres 2010). Astrocytic support on synaptogenesis is mediated by the release of several factors, such as cholesterol, needed for membrane formation (Mauch et al. 2001), thrombospondins (Eroglu et al. 2009) estradiol, protocadherins or integrins (Pfrieger 2010). Astrocytes also produce and release hevin, which promotes the formation of excitatory synapses (Kucukdereli et al. 2011). Synaptic maturation also depends on astrocytic factors such as activity-dependent neurotrophic factor and tumour necrosis factor α, which regulate the trafficking of glutamate receptors to postsynaptic membranes (Eroglu and Barres 2010) and glypicans 4 and 6 that up-regulate the density of AMPA-type glutamate receptors at postsynaptic sites (Allen et al. 2012). Astrocytes maintain synaptic transmission through numerous homeostatic cascades and membrane transporters that control clearance of neurotransmitters and supply neuronal terminals with neurotransmitter precursors (such as glutamine or L-serine). Finally, astrocytes contribute to synaptic elimination by labelling terminals with complement factor C1q, which is recognised by microglia as an 'eat-me' signal that initiates synaptic pruning (Kettenmann et al. 2013).

Astrocytes Protect Nervous Tissue Against Reactive Oxygen Species

Nervous tissue has exceptional energy demands. The human brain, while representing only 2% of the body mass, consumes >20% of organism energy; most of this energy fuels NKA-dependent redressing of ionic gradients (Magistretti 2009). This high-energy consumption is associated with excessive production of ROS that need to be scavenged and neutralised. The anti-oxidant system of the nervous tissue mostly relies on glutathione and ascorbic acid (Makar et al. 1994). In the brain the bulk of glutathione is concentrated in astrocytes (Dringen et al. 2000), while neuronal synthesis of glutathione relies on astrocytes providing obligatory precursors cysteine or glutamylcysteine. Astrocytes accumulate cystine through the Sxc^- glutamate/cystine exchanger, which is not expressed in neurones. Subsequently cystine is reduced to cysteine, which is further converted to glutamylcysteine (CysGlu) and glutathione. Astrocytes release cysteine which is shuttled to neurones (Chen and Swanson 2003). In the presence of astrocytes, in vitro neurones sustain high levels of glutathione, whereas removal of astrocytes instigates ROS neuronal damage (Dringen et al. 1999). Astrocytes also act as a reservoir for another anti-oxidant, ascorbic acid, which similarly protects neurones against oxidative stress (Wilson et al. 2000).

Neurogliovascular Unit and the Blood-Brain Barrier

The brain is one of the most vascularised organs in the human body. In addition, the brain vasculature possesses the BBB, which controls the nature of molecules entering the nervous tissue. Neuronal activity is functionally linked to the local circulation, and an increase in neuronal firing initiates focal vasodilatation of arterioles and capillaries, a phenomenon known as functional hyperemia (Mosso 1880; Roy and Sherrington 1890). Astrocytes are active contributors to both the barrier function and regulation of the local blood flow.

Protoplasmic astrocytes divide the grey matter into spatially segregated and relatively independent domains, within which astrocytes integrate neurones, synapses, microglial cells and neighbouring capillaries into the neurogliovascular (also known as neurovascular) unit (Iadecola 2017). The *glia limitans perivascularis* formed by astroglial endfeet cover the parenchymal part of brain vessels almost entirely (Mathiisen et al. 2010). The endfeet are functionally coupled to capillaries through the release of vasoconstrictors and vasodilatators, which include derivatives of arachidonic acid. In addition, the vascular tone can be regulated by local release of K^+ (Zonta et al. 2003; Mulligan and MacVicar 2004; Filosa et al. 2006). Astrocytes can also function as intracranial baroreceptors contributing to the regulation of arterial blood pressure (Marina et al. 2020).

Astrocytes also form the parenchymal part of BBB. The barrier as such is created by the specialised brain capillary endothelial cells which are (together with pericytes and vascular smooth muscle cells) integrated into neurogliovascular unit. The contacts between brain endothelial cells are sealed with tight and adherent junctions, while several transporting systems ensure BBB permeability for selected molecules, which are allowed to enter the brain parenchyma (Sweeney et al. 2019). At the capillary level, astroglial endfeet, endothelial cells and pericytes share a common basement membrane. At the level of arteriolae, two basement membranes, the parenchymal (in contact with astroglial endfeet) and vascular (in contact with endothelial cells), create the perivascular space. Astrocytes are secreting numerous factors that control formation and maintenance of the BBB (Sweeney et al. 2019); in pathology, astroglia support dwindles which impairs the integrity of the barrier (Kriauciunaite et al. 2021).

Astrocytes and Gliotransmission

Astrocyte-neurone signalling, a.k.a. 'gliotransmission' (Santello et al. 2012; Verkhratsky et al. 2016), modulates synaptic transmission/plasticity at tripartite synapses as a part of the synaptic cradle (Perez-Alvarez and Araque 2013). Among the processes regulated by gliotransmission are sleep regulation (Haydon 2017), respiration (Sheikhbahaei et al. 2018) and learning/memory (Gibbs et al. 2011). Gliotransmission is a result of the astrocytic capacity to release several neurotransmitters and neuromodulators, which include ATP, glutamate, GABA, taurine and kynurenic acid. This release occurs through several pathways, which have been characterised in situ and in vitro astroglial preparations and include exocytotic (vesicular) release, diffusion through plasmalemmal channels (e.g. through unpaired connexons, i.e. hemichannels, or several types of large anion/cation channels), via reversed neurotransmitter transporters or previously mentioned cystine-glutamate exchangers (Verkhratsky et al. 2016).

Astrocytes and the Glymphatic System

The brain has a specialised system for the removal of soluble waste products known as a glymphatic system (Iliff et al. 2012; Nedergaard 2013). This system utilises perivascular space between parenchymal and vascular basement membranes and astrocytic AQP4 channels concentrated at the perivascular endfeet to create fluid flow through the brain parenchyma. This fluid flow is instrumental for interstitial solute clearance and is critical for the normal function of the brain. In pathology, under stress or in ageing, the glymphatic subsystem is often impaired; deficient glymphatic clearance contributes to the pathophysiology of several brain disorders including neurodegenerative and psychiatric diseases (Liang et al. 2020; Nedergaard and Goldman 2020).

Envoi

We have tersely summarised the morphology and function of astroglia. We presented them as a group of ectodermal cells with diverse morphology and specific functions, superordinate to the principle function of maintaining homeostasis of the central nervous system. We consider this summary as a necessity in order to better understand a possible role of these cells in pathophysiological condition and diseases.

Acknowledgements BL's work is supported by the National Natural Science Foundation of China (grant number 8187185), LiaoNing Revitalization Talents Program (grant number XLYC1807137), the Scientific Research Foundation for Returned Scholars of Education Ministry of China (grant number 20151098), LiaoNing Thousands Talents Program (grant number 202078) and 'ChunHui' Program of Education Ministry of China (grant number 2020703). CS's work is supported by a grant from the Italian Ministry of Education, University and Research (2015KP7T2Y_002) and a grant from Sapienza University of Rome (RM11916B7A8D0225).

VP's work is supported by a grant from the National Institute of General Medical Sciences of the National Institutes of Health (R01GM123971). VP is an Honorary Professor at University of Rijeka, Croatia.

References

Abbracchio MP, Ceruti S (2006) Roles of P2 receptors in glial cells: focus on astrocytes. Purinergic Signal 2:595–604

Allen A, Messier C (2013) Plastic changes in the astrocyte GLUT1 glucose transporter and beta-tubulin microtubule protein following voluntary exercise in mice. Behav Brain Res 240:95–102

Allen NJ, Bennett ML, Foo LC, Wang GX, Chakraborty C, Smith SJ, Barres BA (2012) Astrocyte glypicans 4 and 6 promote formation of excitatory synapses via GluA1 AMPA receptors. Nature 486:410–414

Alonso MT, Barrero MJ, Michelena P, Carnicero E, Cuchillo I, Garcia AG, Garcia-Sancho J, Montero M, Alvarez J (1999) Ca^{2+}-induced Ca^{2+} release in chromaffin cells seen from inside the ER with targeted aequorin. J Cell Biol 144:241–254

Alvarez J, Montero M (2002) Measuring $[Ca^{2+}]$ in the endoplasmic reticulum with aequorin. Cell Calcium 32:251–260

Andriezen WL (1893) The neuroglia elements of the brain. Br Med J 2:227–230

Arizono M, Inavalli V, Panatier A, Pfeiffer T, Angibaud J, Levet F, Ter Veer MJT, Stobart J, Bellocchio L, Mikoshiba K, Marsicano G, Weber B, Oliet SHR, Nagerl UV (2020) Structural basis of astrocytic Ca(2+) signals at tripartite synapses. Nat Commun 11:1906

Armstrong WE, Rubrum A, Teruyama R, Bond CT, Adelman JP (2005) Immunocytochemical localization of small-conductance, calcium-dependent potassium channels in astrocytes of the rat supraoptic nucleus. J Comp Neurol 491:175–185

Augusto-Oliveira M, Arrifano GP, Takeda PY, Lopes-Araujo A, Santos-Sacramento L, Anthony DC, Verkhratsky A, Crespo-Lopez ME (2020) Astroglia-specific contributions to the regulation of synapses, cognition and behaviour. Neurosci Biobehav Rev 118:331–357

Bardehle S, Kruger M, Buggenthin F, Schwausch J, Ninkovic J, Clevers H, Snippert HJ, Theis FJ, Meyer-Luehmann M, Bechmann I, Dimou L, Gotz M (2013) Live imaging of astrocyte responses to acute injury reveals selective juxtavascular proliferation. Nat Neurosci 16:580–586

Bekar LK, Loewen ME, Cao K, Sun X, Leis J, Wang R, Forsyth GW, Walz W (2005) Complex expression and localization of inactivating K_v channels in cultured hippocampal astrocytes. J Neurophysiol 93:1699–1709

Benfenati V, Amiry-Moghaddam M, Caprini M, Mylonakou MN, Rapisarda C, Ottersen OP, Ferroni S (2007) Expression and functional characterization of transient receptor potential vanilloid-related channel 4 (TRPV4) in rat cortical astrocytes. Neuroscience 148:876–892

Benjamin AM (1987) Influence of Na^+, K^+, and Ca^{2+} on glutamine synthesis and distribution in rat brain cortex slices: a possible linkage of glutamine synthetase with cerebral transport processes and energetics in the astrocytes. J Neurochem 48:1157–1164

Bennay M, Langer J, Meier SD, Kafitz KW, Rose CR (2008) Sodium signals in cerebellar Purkinje neurons and Bergmann glial cells evoked by glutamatergic synaptic transmission. Glia 56:1138–1149

Berridge MJ, Lipp P, Bootman MD (2000) The versatility and universality of calcium signalling. Nat Rev Mol Cell Biol 1:11–21

Boison D (2008) Adenosine as a neuromodulator in neurological diseases. Curr Opin Pharmacol 8:2–7

Boison D, Chen JF, Fredholm BB (2010) Adenosine signaling and function in glial cells. Cell Death Differ 17:1071–1082

Bordey A, Sontheimer H (2000) Ion channel expression by astrocytes in situ: comparison of different CNS regions. Glia 30:27–38

Brasko C, Hawkins V, De La Rocha IC, Butt AM (2017) Expression of $K_{ir}4.1$ and $K_{ir}5.1$ inwardly rectifying potassium channels in oligodendrocytes, the myelinating cells of the CNS. Brain Struct Funct 222:41–59

Breslin K, Wade JJ, Wong-Lin K, Harkin J, Flanagan B, Van Zalinge H, Hall S, Walker M, Verkhratsky A, McDaid L (2018) Potassium and sodium microdomains in thin astroglial processes: a computational model study. PLoS Comput Biol 14:e1006151

Bridges RJ, Natale NR, Patel SA (2012) System xc(−) cystine/glutamate antiporter: an update on molecular pharmacology and roles within the CNS. Br J Pharmacol 165:20–34

Buffo A, Rossi F (2013) Origin, lineage and function of cerebellar glia. Prog Neurobiol 109:42–63

Burdakov D, Petersen OH, Verkhratsky A (2005) Intraluminal calcium as a primary regulator of endoplasmic reticulum function. Cell Calcium 38:303–310

Burnstock G, Verkhratsky A (2012) Purinergic signalling and the nervous system. Springer, Heidelberg

Bushong EA, Martone ME, Jones YZ, Ellisman MH (2002) Protoplasmic astrocytes in CA1 stratum radiatum occupy separate anatomical domains. J Neurosci 22:183–192

Butenko O, Dzamba D, Benesova J, Honsa P, Benfenati V, Rusnakova V, Ferroni S, Anderova M (2012) The increased activity of TRPV4 channel in the astrocytes of the adult rat hippocampus after cerebral hypoxia/ischemia. PLoS One 7:e39959

Butt AM, Kalsi A (2006) Inwardly rectifying potassium channels (K_{ir}) in central nervous system glia: a special role for $K_{ir}4.1$ in glial functions. J Cell Mol Med 10:33–44

Butt AM, Colquhoun K, Tutton M, Berry M (1994) Three-dimensional morphology of astrocytes and oligodendrocytes in the intact mouse optic nerve. J Neurocytol 23:469–485

Cahoy JD, Emery B, Kaushal A, Foo LC, Zamanian JL, Christopherson KS, Xing Y, Lubischer JL, Krieg PA, Krupenko SA, Thompson WJ, Barres BA (2008) A transcriptome database for astrocytes, neurons, and oligodendrocytes: a new resource for understanding brain development and function. J Neurosci 28:264–278

Carafoli E (2002) Calcium signaling: a tale for all seasons. Proc Natl Acad Sci U S A 99:1115–1122

Chan-Palay V, Palay SL (1972) The form of velate astrocytes in the cerebellar cortex of monkey and rat: high voltage electron microscopy of rapid Golgi preparations. Z Anat Entwicklungsgesch 138:1–19

Chen Y, Swanson RA (2003) The glutamate transporters EAAT2 and EAAT3 mediate cysteine uptake in cortical neuron cultures. J Neurochem 84:1332–1339

Chesler M (2003) Regulation and modulation of pH in the brain. Physiol Rev 83:1183–1221

Chvatal A, Verkhratsky A (2018) Early history of neuroglial research: personalities. Neuroglia 1:245–281

Colombo JA (2017) The interlaminar glia: from serendipity to hypothesis. Brain Struct Funct 222:1109–1129

Cornell-Bell AH, Finkbeiner SM, Cooper MS, Smith SJ (1990) Glutamate induces calcium waves in cultured astrocytes: long-range glial signaling. Science 247:470–473

Danbolt NC (2001) Glutamate uptake. Prog Neurobiol 65:1–105

Deitmer JW, Rose CR (1996) pH regulation and proton signalling by glial cells. Prog Neurobiol 48:73–103

Deitmer JW, Rose CR (2010) Ion changes and signalling in perisynaptic glia. Brain Res Rev 63:113–129

Ding F, O'Donnell J, Thrane AS, Zeppenfeld D, Kang H, Xie L, Wang F, Nedergaard M (2013) α_1-adrenergic receptors mediate coordinated Ca^{2+} signaling of cortical astrocytes in awake, behaving mice. Cell Calcium 54:387–394

Ding F, O'Donnell J, Xu Q, Kang N, Goldman N, Nedergaard M (2016) Changes in the composition of brain interstitial ions control the sleep-wake cycle. Science 352:550–555

Dong JH, Chen X, Cui M, Yu X, Pang Q, Sun JP (2012) β_2-adrenergic receptor and astrocyte glucose metabolism. J Mol Neurosci 48:456–463

Dringen R, Pfeiffer B, Hamprecht B (1999) Synthesis of the antioxidant glutathione in neurons: supply by astrocytes of CysGly as precursor for neuronal glutathione. J Neurosci 19:562–569

Dringen R, Gutterer JM, Hirrlinger J (2000) Glutathione metabolism in brain metabolic interaction between astrocytes and neurons in the defense against reactive oxygen species. Eur J Biochem 267:4912–4916

Eaton MJ, Skatchkov SN, Brune A, Biedermann B, Veh RW, Reichenbach A (2002) SURI and $K_{ir}6.1$ subunits of K_{ATP}-channels are co-localized in retinal glial (Muller) cells. Neuroreport 13:57–60

Edwards L, Nashmi R, Jones O, Backx P, Ackerley C, Becker L, Fehlings MG (2002) Upregulation of K_v 1.4 protein and gene expression after chronic spinal cord injury. J Comp Neurol 443:154–167

Egawa K, Yamada J, Furukawa T, Yanagawa Y, Fukuda A (2013) Cl^- homeodynamics in gap junction-coupled astrocytic networks on activation of GABAergic synapses. J Physiol 591:3901–3917

Enkvist MO, Holopainen I, Akerman KE (1989) Glutamate receptor-linked changes in membrane potential and intracellular Ca^{2+} in primary rat astrocytes. Glia 2:397–402

Eroglu C, Barres BA (2010) Regulation of synaptic connectivity by glia. Nature 468:223–231

Eroglu C, Allen NJ, Susman MW, O'Rourke NA, Park CY, Ozkan E, Chakraborty C, Mulinyawe SB, Annis DS, Huberman AD, Green EM, Lawler J, Dolmetsch R, Garcia KC, Smith SJ, Luo ZD, Rosenthal A, Mosher DF, Barres BA (2009) Gabapentin receptor $\alpha 2\delta$-1 is a neuronal thrombospondin receptor responsible for excitatory CNS synaptogenesis. Cell 139:380–392
Feig SL, Haberly LB (2011) Surface-associated astrocytes, not endfeet, form the glia limitans in posterior piriform cortex and have a spatially distributed, not a domain, organization. J Comp Neurol 519:1952–1969
Filosa JA, Bonev AD, Straub SV, Meredith AL, Wilkerson MK, Aldrich RW, Nelson MT (2006) Local potassium signaling couples neuronal activity to vasodilation in the brain. Nat Neurosci 9:1397–1403
Giaume C, Leybaert L, Naus CC, Saez JC (2013) Connexin and pannexin hemichannels in brain glial cells: properties, pharmacology, and roles. Front Pharmacol 4:88
Giaume C, Naus CC, Saez JC, Leybaert L (2021) Glial connexins and pannexins in the healthy and diseased brain. Physiol Rev 101:93–145
Gibbs ME, Shleper M, Mustafa T, Burnstock G, Bowser DN (2011) ATP derived from astrocytes modulates memory in the chick. Neuron Glia Biol 7:177–186
Gotz S, Bribian A, Lopez-Mascaraque L, Gotz M, Grothe B, Kunz L (2021) Heterogeneity of astrocytes: electrophysiological properties of juxtavascular astrocytes before and after brain injury. Glia 69:346–361
Grosche J, Matyash V, Moller T, Verkhratsky A, Reichenbach A, Kettenmann H (1999) Microdomains for neuron-glia interaction: parallel fiber signaling to Bergmann glial cells. Nat Neurosci 2:139–143
Halestrap AP (2012) The monocarboxylate transporter family – structure and functional characterization. IUBMB Life 64:1–9
Haydon PG (2017) Astrocytes and the modulation of sleep. Curr Opin Neurobiol 44:28–33
Hertz L (1965) Possible role of neuroglia: a potassium-mediated neuronal--neuroglial--neuronal impulse transmission system. Nature 206:1091–1094
Hertz L (2013) The glutamate-glutamine (GABA) cycle: importance of late postnatal development and potential reciprocal interactions between biosynthesis and degradation. Front Endocrinol (Lausanne) 4:59
Hertz L, Chen Y (2016) Importance of astrocytes for potassium ion (K^+) homeostasis in brain and glial effects of K^+ and its transporters on learning. Neurosci Biobehav Rev 71:484–505
Hertz L, Dringen R, Schousboe A, Robinson SR (1999) Astrocytes: glutamate producers for neurons. J Neurosci Res 57:417–428
Hertz L, Chen Y, Gibbs ME, Zang P, Peng L (2004) Astrocytic adrenoceptors: a major drug target in neurological and psychiatric disorders? Curr Drug Targets CNS Neurol Disord 3:239–267
Hertz L, Gerkau NJ, Xu J, Durry S, Song D, Rose CR, Peng L (2015) Roles of astrocytic Na^+, K^+-ATPase and glycogenolysis for K^+ homeostasis in mammalian brain. J Neurosci Res 93:1019–1030
Hibino H, Fujita A, Iwai K, Yamada M, Kurachi Y (2004) Differential assembly of inwardly rectifying K^+ channel subunits, $K_{ir}4.1$ and $K_{ir}5.1$, in brain astrocytes. J Biol Chem 279:44065–44073
Hiyama TY, Yoshida M, Matsumoto M, Suzuki R, Matsuda T, Watanabe E, Noda M (2013) Endothelin-3 expression in the subfornical organ enhances the sensitivity of Na_x, the brain sodium-level sensor, to suppress salt intake. Cell Metab 17:507–519
Hutchinson DS, Summers RJ, Gibbs ME (2007) β_2- and β_3-adrenoceptors activate glucose uptake in chick astrocytes by distinct mechanisms: a mechanism for memory enhancement? J Neurochem 103:997–1008
Iadecola C (2017) The neurovascular unit coming of age: a journey through neurovascular coupling in health and disease. Neuron 96:17–42
Iliff JJ, Wang M, Liao Y, Plogg BA, Peng W, Gundersen GA, Benveniste H, Vates GE, Deane R, Goldman SA, Nagelhus EA, Nedergaard M (2012) A paravascular pathway facilitates CSF

flow through the brain parenchyma and the clearance of interstitial solutes, including amyloid beta. Sci Transl Med 4:147ra111
Illes P, Verkhratsky A, Burnstock G, Franke H (2012) P2X receptors and their roles in astroglia in the central and peripheral nervous system. Neuroscientist 18:422–438
Kalsi AS, Greenwood K, Wilkin G, Butt AM (2004) $K_{ir}4.1$ expression by astrocytes and oligodendrocytes in CNS white matter: a developmental study in the rat optic nerve. J Anat 204:475–485
Kavanaugh MP, Arriza JL, North RA, Amara SG (1992) Electrogenic uptake of γ-aminobutyric acid by a cloned transporter expressed in Xenopus oocytes. J Biol Chem 267:22007–22009
Kettenmann H, Verkhratsky A (2008) Neuroglia: the 150 years after. Trends Neurosci 31:653–659
Kettenmann H, Backus KH, Schachner M (1987) γ-aminobutyric acid opens Cl^- channels in cultured astrocytes. Brain Res 404:1–9
Kettenmann H, Kirchhoff F, Verkhratsky A (2013) Microglia: new roles for the synaptic stripper. Neuron 77:10–18
Khakh BS, Sofroniew MV (2015) Diversity of astrocyte functions and phenotypes in neural circuits. Nat Neurosci 18:942–952
King AE, Ackley MA, Cass CE, Young JD, Baldwin SA (2006) Nucleoside transporters: from scavengers to novel therapeutic targets. Trends Pharmacol Sci 27:416–425
Kirischuk S, Tuschick S, Verkhratsky A, Kettenmann H (1996) Calcium signalling in mouse Bergmann glial cells mediated by α_1-adrenoreceptors and H_1 histamine receptors. Eur J Neurosci 8:1198–1208
Kirischuk S, Kettenmann H, Verkhratsky A (1997) Na^+/Ca^{2+} exchanger modulates kainate-triggered Ca^{2+} signaling in Bergmann glial cells *in situ*. FASEB J 11:566–572
Kirischuk S, Kirchhoff F, Matyash V, Kettenmann H, Verkhratsky A (1999) Glutamate-triggered calcium signalling in mouse bergmann glial cells in situ: role of inositol-1,4,5-trisphosphate-mediated intracellular calcium release. Neuroscience 92:1051–1059
Kirischuk S, Kettenmann H, Verkhratsky A (2007) Membrane currents and cytoplasmic sodium transients generated by glutamate transport in Bergmann glial cells. Pflugers Arch 454:245–252
Kirischuk S, Parpura V, Verkhratsky A (2012) Sodium dynamics: another key to astroglial excitability? Trends Neurosci 35:497–506
Kofuji P, Newman EA (2004) Potassium buffering in the central nervous system. Neuroscience 129:1045–1056
Kölliker AV (1896) Handbuch der Gewebelehre des Menschen. 6 Aufl. Engelmann, Leipzig
Kriauciunaite K, Kausyle A, Pajarskiene J, Tunaitis V, Lim D, Verkhratsky A, Pivoriunas A (2021) Immortalised hippocampal astrocytes from 3xTG-AD mice fail to support BBB integrity in vitro: role of extracellular vesicles in glial-endothelial communication. Cell Mol Neurobiol 41:551–562
Kucukdereli H, Allen NJ, Lee AT, Feng A, Ozlu MI, Conatser LM, Chakraborty C, Workman G, Weaver M, Sage EH, Barres BA, Eroglu C (2011) Control of excitatory CNS synaptogenesis by astrocyte-secreted proteins Hevin and SPARC. Proc Natl Acad Sci U S A 108:E440–E449
Kuffler SW, Nicholls JG (1966) The physiology of neuroglial cells. Ergeb Physiol 57:1–90
Kwon J, An H, Sa M, Won J, Shin JI, Lee CJ (2017) Orai1 and Orai3 in combination with Stim1 mediate the majority of store-operated calcium entry in astrocytes. Exp Neurobiol 26:42–54
Lalo U, Pankratov Y, Kirchhoff F, North RA, Verkhratsky A (2006) NMDA receptors mediate neuron-to-glia signaling in mouse cortical astrocytes. J Neurosci 26:2673–2683
Lalo U, Pankratov Y, Wichert SP, Rossner MJ, North RA, Kirchhoff F, Verkhratsky A (2008) $P2X_1$ and $P2X_5$ subunits form the functional P2X receptor in mouse cortical astrocytes. J Neurosci 28:5473–5480
Lalo U, Pankratov Y, Parpura V, Verkhratsky A (2011) Ionotropic receptors in neuronal-astroglial signalling: what is the role of "excitable" molecules in non-excitable cells. Biochim Biophys Acta 1813:992–1002
Langer J, Rose CR (2009) Synaptically induced sodium signals in hippocampal astrocytes in situ. J Physiol 587:5859–5877

Larsen BR, Assentoft M, Cotrina ML, Hua SZ, Nedergaard M, Kaila K, Voipio J, MacAulay N (2014) Contributions of the Na^+/K^+-ATPase, NKCC1, and $K_{ir}4.1$ to hippocampal K^+ clearance and volume responses. Glia 62:608–622

Larsen BR, Stoica A, MacAulay N (2016) Managing brain extracellular K^+ during neuronal activity: the physiological role of the Na^+/K^+-ATPase subunit isoforms. Front Physiol 7:141

Latour I, Hamid J, Beedle AM, Zamponi GW, Macvicar BA (2003) Expression of voltage-gated Ca^{2+} channel subtypes in cultured astrocytes. Glia 41:347–353

Lenhossék MV (1895) Der feinere Bau des Nervensystems im Lichte neuester Forschung, 2nd edn. Fischer's Medicinische Buchhandlung H. Kornfield, Berlin

Letellier M, Park YK, Chater TE, Chipman PH, Gautam SG, Oshima-Takago T, Goda Y (2016) Astrocytes regulate heterogeneity of presynaptic strengths in hippocampal networks. Proc Natl Acad Sci U S A 113:E2685–E2694

Li B, Gu L, Hertz L, Peng L (2013) Expression of nucleoside transporter in freshly isolated neurons and astrocytes from mouse brain. Neurochem Res 38:2351–2358

Liang S, Lu Y, Li Z, Li S, Chen B, Zhang M, Chen B, Ji M, Gong W, Xia M, Verkhratsky A, Wu X, Li B (2020) Iron aggravates the depressive phenotype of stressed mice by compromising the glymphatic system. Neurosci Bull 36:1542–1546

Lisjak M, Potokar M, Rituper B, Jorgacevski J, Zorec R (2017) AQP4e-based orthogonal arrays regulate rapid cell volume changes in astrocytes. J Neurosci 37:10748–10756

Longden TA, Dunn KM, Draheim HJ, Nelson MT, Weston AH, Edwards G (2011) Intermediate-conductance calcium-activated potassium channels participate in neurovascular coupling. Br J Pharmacol 164:922–933

Lu DC, Zhang H, Zador Z, Verkman AS (2008) Impaired olfaction in mice lacking aquaporin-4 water channels. FASEB J 22:3216–3223

Lundgaard I, Osorio MJ, Kress BT, Sanggaard S, Nedergaard M (2014) White matter astrocytes in health and disease. Neuroscience 276:161–173

Macaulay N, Zeuthen T (2012) Glial K^+ clearance and cell swelling: key roles for cotransporters and pumps. Neurochem Res 37:2299–2309

Magistretti PJ (2009) Neuroscience. Low-cost travel in neurons. Science 325:1349–1351

Magnotti LM, Goodenough DA, Paul DL (2011) Functional heterotypic interactions between astrocyte and oligodendrocyte connexins. Glia 59:26–34

Makar TK, Nedergaard M, Preuss A, Gelbard AS, Perumal AS, Cooper AJ (1994) Vitamin E, ascorbate, glutathione, glutathione disulfide, and enzymes of glutathione metabolism in cultures of chick astrocytes and neurons: evidence that astrocytes play an important role in antioxidative processes in the brain. J Neurochem 62:45–53

Malarkey EB, Ni Y, Parpura V (2008) Ca^{2+} entry through TRPC1 channels contributes to intracellular Ca^{2+} dynamics and consequent glutamate release from rat astrocytes. Glia 56:821–835

Marcaggi P, Attwell D (2004) Role of glial amino acid transporters in synaptic transmission and brain energetics. Glia 47:217–225

Marina N, Christie IN, Korsak A, Doronin M, Brazhe A, Hosford PS, Wells JA, Sheikhbahaei S, Humoud I, Paton JFR, Lythgoe MF, Semyanov A, Kasparov S, Gourine AV (2020) Astrocytes monitor cerebral perfusion and control systemic circulation to maintain brain blood flow. Nat Commun 11:131

Mathiisen TM, Lehre KP, Danbolt NC, Ottersen OP (2010) The perivascular astroglial sheath provides a complete covering of the brain microvessels: an electron microscopic 3D reconstruction. Glia 58:1094–1103

Mauch DH, Nagler K, Schumacher S, Goritz C, Muller EC, Otto A, Pfrieger FW (2001) CNS synaptogenesis promoted by glia-derived cholesterol. Science 294:1354–1357

McCarthy KD, Salm AK (1991) Pharmacologically-distinct subsets of astroglia can be identified by their calcium response to neuroligands. Neuroscience 41:325–333

Mestre H, Mori Y, Nedergaard M (2020) The Brain's glymphatic system: current controversies. Trends Neurosci 43:458–466

Miller RL, Loewy AD (2013) ENaC γ-expressing astrocytes in the circumventricular organs, white matter, and ventral medullary surface: sites for Na^+ regulation by glial cells. J Chem Neuroanat 53:72–80
Mosso A (1880) Sulla circolazione del sangue nel cervello dell'uomo. Mem Real Acc Lincei 5:237–358
Mulligan SJ, MacVicar BA (2004) Calcium transients in astrocyte endfeet cause cerebrovascular constrictions. Nature 431:195–199
Nagelhus EA, Ottersen OP (2013) Physiological roles of aquaporin-4 in brain. Physiol Rev 93:1543–1562
Nagy JI, Dudek FE, Rash JE (2004) Update on connexins and gap junctions in neurons and glia in the mammalian nervous system. Brain Res Brain Res Rev 47:191–215
Nagy JI, Lynn BD, Tress O, Willecke K, Rash JE (2011) Connexin26 expression in brain parenchymal cells demonstrated by targeted connexin ablation in transgenic mice. Eur J Neurosci 34:263–271
Nedergaard M (2013) Neuroscience. Garbage truck of the brain. Science 340:1529–1530
Nedergaard M, Goldman SA (2020) Glymphatic failure as a final common pathway to dementia. Science 370:50–56
Nedergaard M, Verkhratsky A (2012) Artifact versus reality – how astrocytes contribute to synaptic events. Glia 60:1013–1023
Newman EA, Frambach DA, Odette LL (1984) Control of extracellular potassium levels by retinal glial cell K^+ siphoning. Science 225:1174–1175
Nilsson M, Eriksson PS, Ronnback L, Hansson E (1993) GABA induces Ca^{2+} transients in astrocytes. Neuroscience 54:605–614
Noda M, Sakuta H (2013) Central regulation of body-fluid homeostasis. Trends Neurosci 36:661–673
Oberheim NA, Takano T, Han X, He W, Lin JH, Wang F, Xu Q, Wyatt JD, Pilcher W, Ojemann JG, Ransom BR, Goldman SA, Nedergaard M (2009) Uniquely hominid features of adult human astrocytes. J Neurosci 29:3276–3287
Olsen ML, Campbell SL, Sontheimer H (2007) Differential distribution of $K_{ir}4.1$ in spinal cord astrocytes suggests regional differences in K^+ homeostasis. J Neurophysiol 98:786–793
Orellana JA, Saez PJ, Shoji KF, Schalper KA, Palacios-Prado N, Velarde V, Giaume C, Bennett MV, Saez JC (2009) Modulation of brain hemichannels and gap junction channels by pro-inflammatory agents and their possible role in neurodegeneration. Antioxid Redox Signal 11:369–399
Orkand RK, Nicholls JG, Kuffler SW (1966) Effect of nerve impulses on the membrane potential of glial cells in the central nervous system of amphibia. J Neurophysiol 29:788–806
Orlowski J, Grinstein S (1997) Na^+/H^+ exchangers of mammalian cells. J Biol Chem 272:22373–22376
Orthmann-Murphy JL, Freidin M, Fischer E, Scherer SS, Abrams CK (2007) Two distinct heterotypic channels mediate gap junction coupling between astrocyte and oligodendrocyte connexins. J Neurosci 27:13949–13957
Palygin O, Lalo U, Verkhratsky A, Pankratov Y (2010) Ionotropic NMDA and $P2X_{1/5}$ receptors mediate synaptically induced Ca^{2+} signalling in cortical astrocytes. Cell Calcium 48:225–231
Pappalardo LW, Liu S, Black JA, Waxman SG (2014a) Dynamics of sodium channel $Na_v1.5$ expression in astrocytes in mouse models of multiple sclerosis. Neuroreport 25:1208–1215
Pappalardo LW, Samad OA, Black JA, Waxman SG (2014b) Voltage-gated sodium channel Na_v 1.5 contributes to astrogliosis in an in vitro model of glial injury via reverse Na^+ /Ca^{2+} exchange. Glia 62:1162–1175
Pappalardo LW, Black JA, Waxman SG (2016) Sodium channels in astroglia and microglia. Glia 64:1628–1645
Parri HR, Gould TM, Crunelli V (2001) Spontaneous astrocytic Ca^{2+} oscillations *in situ* drive NMDAR-mediated neuronal excitation. Nat Neurosci 4:803–812

Pastor A, Chvatal A, Sykova E, Kettenmann H (1995) Glycine- and GABA-activated currents in identified glial cells of the developing rat spinal cord slice. Eur J Neurosci 7:1188–1198
Perez-Alvarez A, Araque A (2013) Astrocyte-neuron interaction at tripartite synapses. Curr Drug Targets 14:1220–1224
Petersen OH, Verkhratsky A (2007) Endoplasmic reticulum calcium tunnels integrate signalling in polarised cells. Cell Calcium 42:373–378
Petersen OH, Michalak M, Verkhratsky A (2005) Calcium signalling: past, present and future. Cell Calcium 38:161–169
Pfrieger FW (2010) Role of glial cells in the formation and maintenance of synapses. Brain Res Rev 63:39–46
Pfrieger FW, Barres BA (1997) Synaptic efficacy enhanced by glial cells *in vitro*. Science 277:1684–1687
Pines G, Danbolt NC, Bjoras M, Zhang Y, Bendahan A, Eide L, Koepsell H, Storm-Mathisen J, Seeberg E, Kanner BI (1992) Cloning and expression of a rat brain L-glutamate transporter. Nature 360:464–467
Plattner H, Verkhratsky A (2015) The ancient roots of calcium signalling evolutionary tree. Cell Calcium 57:123–132
Popov A, Brazhe A, Denisov P, Sutyagina O, Lazareva N, Verkhratsky A, Semyanov A (2020) Astrocytes dystrophy in ageing brain parallels impaired synaptic plasticity. bioRxiv
Reichenbach A, Bringmann A (2017) Comparative anatomy of glial cells in mammals. In: Kaas J (ed) Evolution of nervous systems, 2nd edn. Academic Press/Elsevier, New York, pp 309–348
Reichenbach A, Derouiche A, Kirchhoff F (2010) Morphology and dynamics of perisynaptic glia. Brain Res Rev 63:11–25
Reyes RC, Verkhratsky A, Parpura V (2012) Plasmalemmal Na^+/Ca^{2+} exchanger modulates Ca^{2+}-dependent exocytotic release of glutamate from rat cortical astrocytes. ASN Neuro 4:00075
Reyes RC, Verkhratsky A, Parpura V (2013) TRPC1-mediated Ca^{2+} and Na^+ signalling in astroglia: differential filtering of extracellular cations. Cell Calcium 54:120–125
Rose CR, Ransom BR (1996) Intracellular sodium homeostasis in rat hippocampal astrocytes. J Physiol 491(Pt 2):291–305
Rose CR, Verkhratsky A (2016) Principles of sodium homeostasis and sodium signalling in astroglia. Glia 64:1611–1627
Rose CR, Ziemens D, Verkhratsky A (2020) On the special role of NCX in astrocytes: translating Na^+-transients into intracellular Ca^{2+} signals. Cell Calcium 86:102154
Roy CS, Sherrington CS (1890) On the regulation of the blood-supply of the brain. J Physiol (Lond) 11:85–108
Rusakov DA, Fine A (2003) Extracellular Ca^{2+} depletion contributes to fast activity-dependent modulation of synaptic transmission in the brain. Neuron 37:287–297
Santello M, Cali C, Bezzi P (2012) Gliotransmission and the tripartite synapse. Adv Exp Med Biol 970:307–331
Saura J, Kettler R, Da Prada M, Richards JG (1992) Quantitative enzyme radioautography with 3H-Ro 41-1049 and 3H-Ro 19-6327 in vitro: localization and abundance of MAO-A and MAO-B in rat CNS, peripheral organs, and human brain. J Neurosci 12:1977–1999
Scharfman HE, Binder DK (2013) Aquaporin-4 water channels and synaptic plasticity in the hippocampus. Neurochem Int 63:702–711
Seifert G, Huttmann K, Binder DK, Hartmann C, Wyczynski A, Neusch C, Steinhauser C (2009) Analysis of astroglial K^+ channel expression in the developing hippocampus reveals a predominant role of the $K_{ir}4.1$ subunit. J Neurosci 29:7474–7488
Sheikhbahaei S, Turovsky EA, Hosford PS, Hadjihambi A, Theparambil SM, Liu B, Marina N, Teschemacher AG, Kasparov S, Smith JC, Gourine AV (2018) Astrocytes modulate brainstem respiratory rhythm-generating circuits and determine exercise capacity. Nat Commun 9:370
Shibasaki K, Hosoi N, Kaneko R, Tominaga M, Yamada K (2017) Glycine release from astrocytes via functional reversal of GlyT1. J Neurochem 140:395–403

Shigetomi E, Tong X, Kwan KY, Corey DP, Khakh BS (2012) TRPA1 channels regulate astrocyte resting calcium and inhibitory synapse efficacy through GAT-3. Nat Neurosci 15:70–80

Shigetomi E, Jackson-Weaver O, Huckstepp RT, O'Dell TJ, Khakh BS (2013) TRPA1 channels are regulators of astrocyte basal calcium levels and long-term potentiation via constitutive D-serine release. J Neurosci 33:10143–10153

Skatchkov SN, Rojas L, Eaton MJ, Orkand RK, Biedermann B, Bringmann A, Pannicke T, Veh RW, Reichenbach A (2002) Functional expression of K_{ir} 6.1/SUR1-K_{ATP} channels in frog retinal Muller glial cells. Glia 38:256–267

Skucas VA, Mathews IB, Yang J, Cheng Q, Treister A, Duffy AM, Verkman AS, Hempstead BL, Wood MA, Binder DK, Scharfman HE (2011) Impairment of select forms of spatial memory and neurotrophin-dependent synaptic plasticity by deletion of glial aquaporin-4. J Neurosci 31:6392–6397

Sohl G, Odermatt B, Maxeiner S, Degen J, Willecke K (2004) New insights into the expression and function of neural connexins with transgenic mouse mutants. Brain Res Brain Res Rev 47:245–259

Solovyova N, Verkhratsky A (2002) Monitoring of free calcium in the neuronal endoplasmic reticulum: an overview of modern approaches. J Neurosci Methods 122:1–12

Solovyova N, Veselovsky N, Toescu EC, Verkhratsky A (2002) Ca^{2+} dynamics in the lumen of the endoplasmic reticulum in sensory neurons: direct visualization of Ca^{2+}-induced Ca^{2+} release triggered by physiological Ca^{2+} entry. EMBO J 21:622–630

Sontheimer H, Waxman SG (1992) Ion channels in spinal cord astrocytes in vitro. II. Biophysical and pharmacological analysis of two Na+ current types. J Neurophysiol 68:1001–1011

Sontheimer H, Minturn JE, Black JA, Ransom BR, Waxman SG (1991) Two types of Na^+-currents in cultured rat optic nerve astrocytes: changes with time in culture and with age of culture derivation. J Neurosci Res 30:275–287

Sontheimer H, Fernandez-Marques E, Ullrich N, Pappas CA, Waxman SG (1994) Astrocyte Na^+ channels are required for maintenance of Na^+/K^+-ATPase activity. J Neurosci 14:2464–2475

Sosunov AA, Wu X, Tsankova NM, Guilfoyle E, McKhann GM 2nd, Goldman JE (2014) Phenotypic heterogeneity and plasticity of isocortical and hippocampal astrocytes in the human brain. J Neurosci 34:2285–2298

Storck T, Schulte S, Hofmann K, Stoffel W (1992) Structure, expression, and functional analysis of a Na^+-dependent glutamate/aspartate transporter from rat brain. Proc Natl Acad Sci U S A 89:10955–10959

Sweeney MD, Zhao Z, Montagne A, Nelson AR, Zlokovic BV (2019) Blood-brain barrier: from physiology to disease and back. Physiol Rev 99:21–78

Takeda H, Inazu M, Matsumiya T (2002) Astroglial dopamine transport is mediated by norepinephrine transporter. Naunyn Schmiedeberg's Arch Pharmacol 366:620–623

Theparambil SM, Naoshin Z, Thyssen A, Deitmer JW (2015) Reversed electrogenic sodium bicarbonate cotransporter 1 is the major acid loader during recovery from cytosolic alkalosis in mouse cortical astrocytes. J Physiol 593:3533–3547

Todd AC, Marx MC, Hulme SR, Broer S, Billups B (2017) SNAT3-mediated glutamine transport in perisynaptic astrocytes in situ is regulated by intracellular sodium. Glia 65:900–916

Toth AB, Hori K, Novakovic MM, Bernstein NG, Lambot L, Prakriya M (2019) CRAC channels regulate astrocyte Ca^{2+} signaling and gliotransmitter release to modulate hippocampal GABAergic transmission. Sci Signal 12:5450

Tzingounis AV, Wadiche JI (2007) Glutamate transporters: confining runaway excitation by shaping synaptic transmission. Nat Rev Neurosci 8:935–947

Unichenko P, Myakhar O, Kirischuk S (2012) Intracellular Na^+ concentration influences short-term plasticity of glutamate transporter-mediated currents in neocortical astrocytes. Glia 60:605–614

Verkhratsky A (2010) Physiology of neuronal-glial networking. Neurochem Int 57:332–343

Verkhratsky A, Chvatal A (2020) NMDA receptors in astrocytes. Neurochem Res 45:122–133

Verkhratsky A, Kettenmann H (1996) Calcium signalling in glial cells. Trends Neurosci 19:346–352
Verkhratsky A, Nedergaard M (2014) Astroglial cradle in the life of the synapse. Philos Trans R Soc Lond Ser B Biol Sci 369:20130595
Verkhratsky A, Nedergaard M (2016) The homeostatic astroglia emerges from evolutionary specialization of neural cells. Philos Trans R Soc Lond Ser B Biol Sci 371:20150428
Verkhratsky A, Nedergaard M (2018) Physiology of astroglia. Physiol Rev 98:239–389
Verkhratsky A, Parpura V (2014) Store-operated calcium entry in neuroglia. Neurosci Bull 30:125–133
Verkhratsky A, Petersen OH (2002) The endoplasmic reticulum as an integrating signalling organelle: from neuronal signalling to neuronal death. Eur J Pharmacol 447:141–154
Verkhratsky A, Steinhauser C (2000) Ion channels in glial cells. Brain Res Brain Res Rev 32:380–412
Verkhratsky A, Orkand RK, Kettenmann H (1998) Glial calcium: homeostasis and signaling function. Physiol Rev 78:99–141
Verkhratsky A, Krishtal OA, Burnstock G (2009) Purinoceptors on neuroglia. Mol Neurobiol 39:190–208
Verkhratsky A, Rodriguez JJ, Parpura V (2012) Calcium signalling in astroglia. Mol Cell Endocrinol 353:45–56
Verkhratsky A, Matteoli M, Parpura V, Mothet JP, Zorec R (2016) Astrocytes as secretory cells of the central nervous system: idiosyncrasies of vesicular secretion. EMBO J 35:239–257
Verkhratsky A, Trebak M, Perocchi F, Khananshvili D, Sekler I (2018) Crosslink between calcium and sodium signalling. Exp Physiol 103:157–169
Verkhratsky A, Semyanov A, Zorec R (2020) Physiology of astroglial excitability. Function 1:zqaa016
Wilson JX, Peters CE, Sitar SM, Daoust P, Gelb AW (2000) Glutamate stimulates ascorbate transport by astrocytes. Brain Res 858:61–66
Witcher MR, Kirov SA, Harris KM (2007) Plasticity of perisynaptic astroglia during synaptogenesis in the mature rat hippocampus. Glia 55:13–23
Zanotti S, Charles A (1997) Extracellular calcium sensing by glial cells: low extracellular calcium induces intracellular calcium release and intercellular signaling. J Neurochem 69:594–602
Zerangue N, Kavanaugh MP (1996) Flux coupling in a neuronal glutamate transporter. Nature 383:634–637
Zhang Y, Chen K, Sloan SA, Bennett ML, Scholze AR, O'Keeffe S, Phatnani HP, Guarnieri P, Caneda C, Ruderisch N, Deng S, Liddelow SA, Zhang C, Daneman R, Maniatis T, Barres BA, Wu JQ (2014) An RNA-sequencing transcriptome and splicing database of glia, neurons, and vascular cells of the cerebral cortex. J Neurosci 34:11929–11947
Zhou Y, Danbolt NC (2013) GABA and glutamate transporters in brain. Front Endocrinol (Lausanne) 4:165
Zhou M, Xu G, Xie M, Zhang X, Schools GP, Ma L, Kimelberg HK, Chen H (2009) TWIK-1 and TREK-1 are potassium channels contributing significantly to astrocyte passive conductance in rat hippocampal slices. J Neurosci 29:8551–8564
Zhu H, Zhao Y, Wu H, Jiang N, Wang Z, Lin W, Jin J, Ji Y (2016) Remarkable alterations of Nav1.6 in reactive astrogliosis during epileptogenesis. Sci Rep 6:38108
Zonta M, Angulo MC, Gobbo S, Rosengarten B, Hossmann KA, Pozzan T, Carmignoto G (2003) Neuron-to-astrocyte signaling is central to the dynamic control of brain microcirculation. Nat Neurosci 6:43–50

Principles of Astrogliopathology

Alexei Verkhratsky, Baoman Li, Caterina Scuderi, and Vladimir Parpura

Pathological Potential of Neuroglia

The pathological potential of neuroglia was widely recognised and acknowledged by neurologists and neuroanatomists at end of the nineteenth and the beginning of the twentieth century. Contribution of neuroglia to the diseases was described, and numerous pathological morphological types of glial cells have been characterised in detail (Achucarro 1910; Alzheimer 1910; Frommann 1878; Nissl 1899). By 1920 the universal involvement of neuroglia in neuropathology was universally accepted; neurologists agreed that "the appearance of neuroglia serves as a delicate indicator of the action of noxious influences upon the central nervous system" (del Río-Hortega and Penfield 1927); the concept of reactive gliosis has been formulated and

A. Verkhratsky (✉)
Faculty of Biology, Medicine and Health, The University of Manchester, Manchester, UK

Achucarro Center for Neuroscience, IKERBASQUE, Basque Foundation for Science, Bilbao, Spain
e-mail: alexej.verkhratsky@manchester.ac.uk

B. Li
Practical Teaching Center, School of Forensic Medicine, China Medical University, Shenyang, China
e-mail: bmli@cmu.edu.cn

C. Scuderi
Department of Physiology and Pharmacology "Vittorio Erspamer", SAPIENZA University of Rome, Rome, Italy
e-mail: caterina.scuderi@uniroma1.it

V. Parpura
Department of Neurobiology, The University of Alabama at Birmingham, Birmingham, AL, USA
e-mail: vlad@uab.edu

B. Li et al. (eds.), *Astrocytes in Psychiatric Disorders*, Advances in Neurobiology 26, https://doi.org/10.1007/978-3-030-77375-5_3

generally recognised (del Río-Hortega and Penfield 1927; Penfield 1928b). The widespread role and importance of neuroglia in neurological and neuropsychiatric diseases were somewhat forgotten in the course of the twentieth century. However, the recently passed decade witnessed much revival in the interest in glia in neuropathology as the neuroglial cells are firmly considered as key players in pathophysiology of all disorders of the nervous system (both central and peripheral), and neuropharmacology regards neuroglia as a legitimate target for new therapeutic strategies (Burda and Sofroniew 2014; Ferrer 2018; Giaume et al. 2007; Parpura et al. 2012; Pekny et al. 2016; Sofroniew 2014b; Verkhratsky et al. 2012b, 2016a, 2017; Verkhratsky and Parpura 2016; Zeidan-Chulia et al. 2014).

Astrogliopathology: General Principles

Astrocytes are primary homeostatic cells of the central nervous system (CNS; see previous chapter); in addition, astrocytes contribute to brain defence. Astrocytic contribution to neuropathology can be primary (when cell-autonomous changes drive the pathologic progression) or secondary, when astrocytes respond to lesions or to various pathological changes in the nervous tissue. Current classification (Fig. 1) distinguishes the following forms of astrogliopathology: (i) reactive astrogliosis, (ii) astrocytic atrophy with loss of function, (iii) pathological remodelling of astrocytes and (iv) astrodegeneration (Fig. 1, (Verkhratsky et al. 2017). These pathological groups cover multiple pathological phenotypes which are yet to be fully characterised; furthermore pathological changes in astrocytes can occur together or in isolation; they are sometimes specific to disease stages and they are affected by age and systemic pathologies.

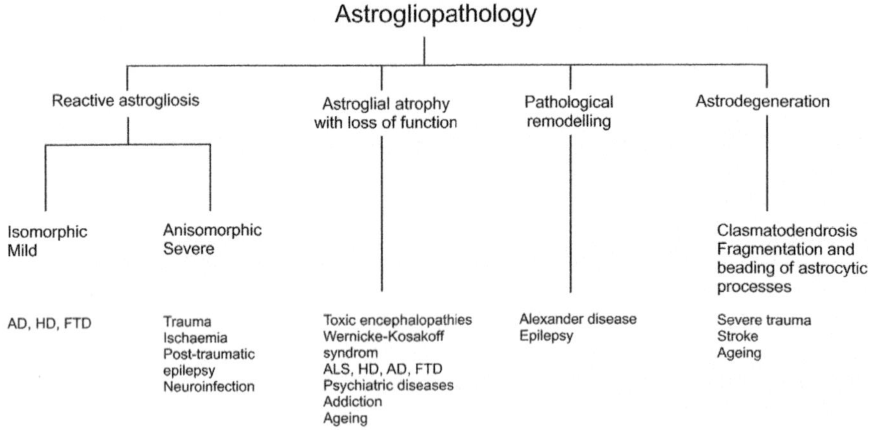

Fig. 1 Classification of astrocytic pathological changes. AD Alzheimer's disease, ALS amyotrophic lateral sclerosis, FTD fronto-temporal dementia, HD Huntington's disease

Reactive Astrogliosis

Reactive astrogliosis is a specific and evolutionary conserved (from arthropods to humans) response of astrocytes to polyaetiological brain lesions, from trauma and infection to neurodegeneration. Reactive astrogliosis is a process whereby, in response to pathology, astrocytes launch genetic programmes that result in biochemical, morphological, metabolic and physiological remodelling (Escartin et al. 2021). This remodelling leads to either gain or loss or modification of astrocytic functions, all aimed at neuroprotection and preservation of the nervous tissue integrity. Astrogliotic remodelling of astrocytes leads to an emergence of multiple context-specific reactive phenotypes, characteristic for particular, age, type of pathology and brain region. These multiple phenotypes differ in specific molecular profile, functions and distinct impact on diseases (Pekny et al. 2016; Sofroniew 2014a; Sofroniew 2020; Verkhratsky et al. 2017). Reactive astrogliosis is flexible to adapt functional and biochemical reprogramming of astrocytes to the nature and strength of the insult with an ultimate goal to mount maximal protection. Within the framework of the same pathology and even within the same affected areas, astrocytes remain heterogeneous in their expression of transcription factors, inflammatory agents and signalling molecules, arguably associated with distinct reactive phenotypes (Garcia et al. 2010; Herrmann et al. 2008).

Reactive astrogliosis contributes to many neurological diseases. In particular, prominent astrogliosis occurs in disorders associated with direct lesion to the nervous tissue by physical, biological or chemical agents. These conditions include neurotrauma (Burda et al. 2016; Faulkner et al. 2004), systemic inflammation and sepsis (Shulyatnikova and Verkhratsky 2019; Tremblay et al. 2020), microbial or viral neuroinfection (Soung and Klein 2018; Zorec et al. 2019), toxic encephalopathies (Li et al. 2021; O'Callaghan et al. 2014), autoimmune inflammation of the nervous tissue including multiple sclerosis (Voskuhl et al. 2009; Wheeler and Quintana 2019), cancerous growth (Henrik Heiland et al. 2019) and neurodegenerative diseases (Verkhratsky et al. 2010). Histopathologically reactive astrogliosis is characterised by morphological hypertrophy, changes in the thickness of processes, sometimes associated with retraction of distal leaflets (Plata et al. 2018); furthermore, reactivity is manifested by an up-regulation of two major cytoskeletal intermediate filaments, glial fibrillary acidic protein (GFAP) and vimentin (Hol and Pekny 2015; Pekny and Pekna 2014; Sofroniew 2014a).

There are several classifications of reactive astrogliosis (Fig. 2). According to morphological changes, astrogliosis is classified into isomorphic and anisomorphic astrogliosis. In isomorphic astrogliosis, astrocytes become hypertrophic; however, they do not move and do not proliferate and the reach of their individual territorial domains remains unchanged (Escartin et al. 2021; Wilhelmsson et al. 2006). Isomorphic astrogliosis is fully reversible, and after the resolution of pathology, astrocytes return to physiological morphology; the isomorphic astrogliosis is indispensable for post-lesion regeneration (Anderson et al. 2016). In anisomorphic astrogliosis astrocytes start to proliferate, and they migrate towards the site of lesion,

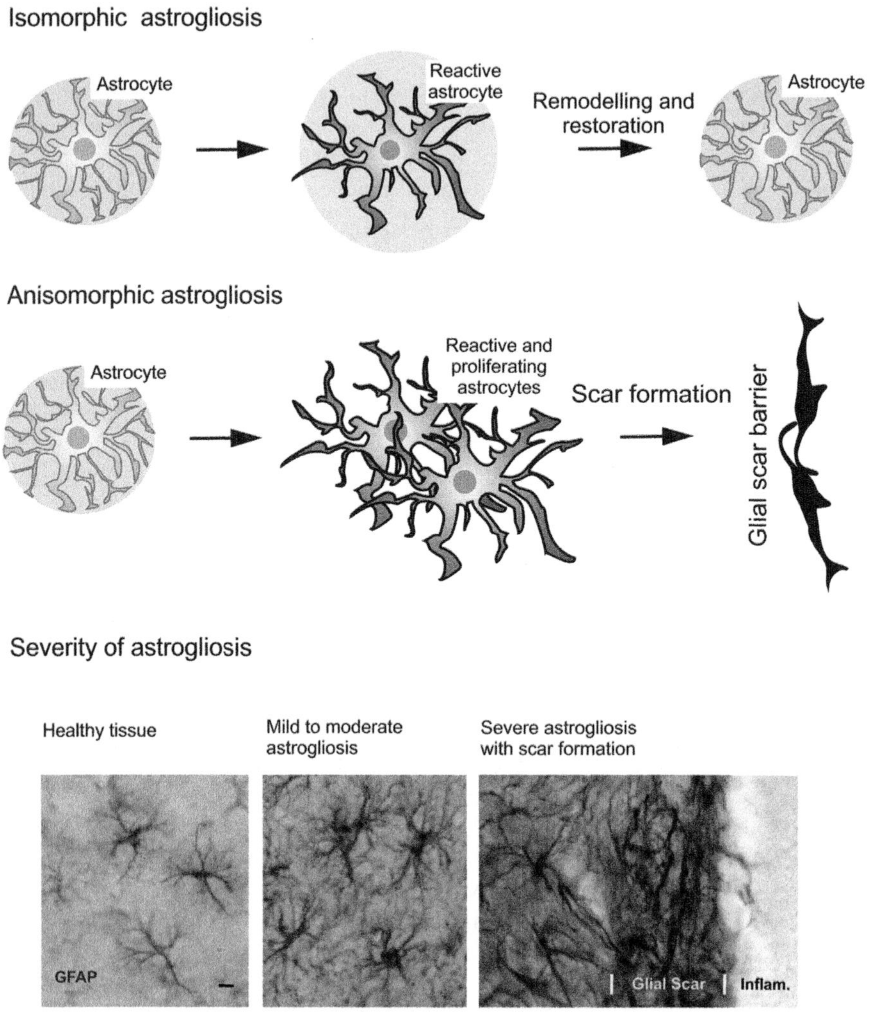

Fig. 2 Classification of reactive astrogliosis; see text for explanation. (Modified from Verkhratsky and Butt (2013) and Sofroniew (2009))

assemble into astroglial palisades and form the glial scar (Pekny et al. 2016; Sofroniew 2020). Another classification divides astrogliosis according to the severity of changes. According to this classification astrogliosis is classified into (i) mild to moderate astrogliosis, (ii) severe diffuse astrogliosis and (iii) severe astrogliosis with compact scar formation (Sofroniew 2009, 2014a). Although morphological presentation of reactive astrocytes can be similar in different pathological contexts generally following the above classification, their molecular signatures are quite distinct and disease-specific. Different astrocytic transcriptomes associate with different conditions and diseases including neurotrauma (Anderson et al. 2016), stroke

(Zamanian et al. 2012), animal models of multiple sclerosis (Itoh et al. 2018) or neurodegenerations; in the latter group astrocytes in Huntington's disease (Al-Dalahmah et al. 2020) are distinct from astrocytes in Alzheimer's disease (Kamphuis et al. 2015). Similarly, astrocytic reactive phenotypes can be different in different stages of the same disease (Wheeler et al. 2020; Zamanian et al. 2012).

Fundamentally, astrogliosis is a defensive response of astrocytes aimed at (i) neuroprotection and trophic support of neural cells tissue, (ii) isolation of the lesioned area, (iii) reconstruction of the compromised blood-brain barrier and (iv) facilitating the post-lesion regeneration of the nervous tissue (Sofroniew 2020). The ultimate result of severe astrogliosis, the scar formation, is essentially defensive reponse to isolate the damaged part of the nervous tissue and save the whole at the expense of its part (Pekny et al. 2016; Verkhratsky et al. 2017; Verkhratsky and Butt 2013). Inhibition of astroglial reactivity often exacerbates the damage to the nervous tissue and worsens neurological outcomes (Pekny et al. 2016). For example, suppression of astrogliotic response increases the size of the traumatic lesions and augments neurological deficit (Okada et al. 2006). Genetic deletion of GFAP and vimentin, both of which are critical for mounting reactive astrocyte remodelling, facilitates the evolution of brain ischaemia (Li et al. 2008) and potentiates posttraumatic synaptic loss (Pekny et al. 1999). Furthermore, inhibition of astroglial reactivity results in higher accumulation of β-amyloid and reduced microglial association with senile plaques in the animal model of Alzheimer's disease (AD); all these changes seem to exacerbate AD-type pathology (Kraft et al. 2013).

Reactive astrogliosis is instigated by multiple factors. Conceptually, astrocytes may sense and integrate numerous molecular cues that signal the damage and provide some information about the nature of this damage. Such molecular cues can have multiple origin and nature. They can be released by damaged cells, they can be associated with accumulation of pathological material (β-amyloid being a well-known example), they can be blood-borne (blood cells or proteins, such as albumin or thrombin) and they can be associated with invading pathogens (bacteria, viruses or prions) or systemic immune factors. Astrogliosis can also be stimulated by certain neurotransmitters and hormones (Fig. 3; (Pekny et al. 2016, Sofroniew 2020)). Conceptually, all factors associated with the instigation of astrogliosis can be classified into (Tang et al. 2012) damage-associated molecular patterns (DAMPs) or pathogen-associated molecular patterns (PAMPs). The DAMPs are molecules released from immune-responsive microglia or other stressed, damaged or dying cells or factors coming from the circulation through the compromised blood-brain barrier. These factors may include cytokines, chemokines, endothelins, blood-borne proteins, etc. Astrocytes express a wide pattern of receptors that can be activated by DAMPs (Verkhratsky and Nedergaard 2018). The archetypal DAMP is represented by ATP, which is massively released from damaged cells; in pathological contexts, ATP mainly acts on astrocytes through activation of $P2X_7$ purinoceptors, although other classes of purinoceptors may also contribute (Franke et al. 2012). The PAMPs are exogenous agents associated with pathogens such as bacteria, viruses or prions; these factors stimulate Toll-like receptors (TLRs) widely expressed in astrocytes (Jack et al. 2005; Kielian 2006). In addition, astrocytes express nucleotide-binding

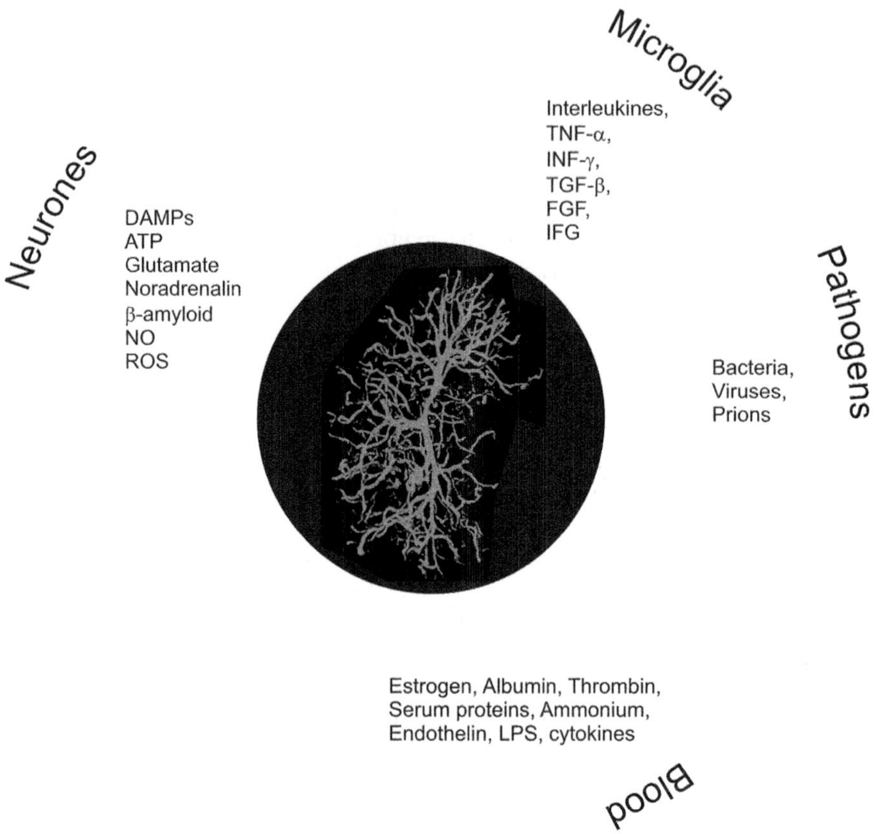

Fig. 3 Instigators of reactive astrogliosis. Numerous agents, including damage-associated molecular patterns (DAMPs) and pathogen-associated molecular patterns (PAMPS, pathogens), the former originating from various cells in the nervous tissue or from blood. All these agents can activate various astrocytic receptors which launch astrogliotic programmes. Abbreviations: TNF-α tumour necrosis factor α, INF-γ interferon γ, TGF-β transforming growth factor β, FGF fibroblast growth factor, IFG insulin growth factor, NO nitric oxide, ROS reactive oxygen species, LPS lipopolysaccharide

oligomerisation domain (NOD)-like receptors (NLRs), double-stranded RNA-dependent protein kinase, scavenger receptors, mannose receptor and receptors for complement components and mediators, such as CXCL10, CCL2, interleukin-6 and B-cell-activating factor of the TNF family, all of which are contributing to the regulation of reactive astrogliosis (Farina et al. 2007).

Intracellularly, initiation of reactive astrogliosis is associated with Ca^{2+} signalling. This signalling is an important part of astrocytic intracellular excitability, mediated by cytosolic ions and second messengers (Verkhratsky et al. 2020b, c). Exposure of astrocytes to various DAMPs and PAMPs is frequently associated with initiation of Ca^{2+} signals mainly originating from Ca^{2+} release from the intracellular endoplasmic reticulum (ER) Ca^{2+} store. This release is mediated by inositol-1,4,5,-trisphosphate ($InsP_3$) receptor type 2, which is predominant in astrocytes (Verkhratsky et al. 2012a). Similarly, pharmacological inhibition of Ca^{2+} release from the ER suppressed astrocytic reactivity in response to β-amyloid (Alberdi et al. 2013)

Despite being an intrinsically defensive response, reactive astrocytes may, in certain conditions, acquire maladaptive features which may exacerbate or even cause damage to the nervous tissue (Pekny et al. 2016; Sofroniew 2020). First, astrocytic reactivity may interfere and downregulate essential homeostatic functions such as K^+ buffering or glutamate homeostasis. In particular, failure of glutamate homeostasis seems to be a converging point in the pathophysiology of various neurological diseases, such as toxic encephalopathies (Li et al. 2021), hepatic encephalopathy (Montana et al. 2014; Obara-Michlewska et al. 2015), epilepsy (Bedner et al. 2015) or amyotrophic lateral sclerosis (Rossi et al. 2008; Valori et al. 2014). In addition, reactive astrocytes may be associated with the release of potentially damaging molecules through pathological gain of function, when existing homeostatic cascades start to overproduce particular agents. For example, in Alzheimer's disease astrocytes overexpress monoamine oxidase-B (MAO-B) to produce GABA from puterscin; this overproduction of GABA counteracts neuronal hyperexcitability closely associated with AD progression (Garaschuk and Verkhratsky 2019; Ghatak et al. 2019). Increase in MAO-B activity, however, results in overproduction of hydrogen peroxide that initiates neuronal damage and death (Chun et al. 2020). Similarly, astrocytic overproduction of complement C3 (which otherwise is a legitimate physiological ligand) leads to morphological and functional neuronal defects (Lian et al. 2015).

To summarise, reactive astrogliosis is an intrinsic physiological astrocyte programme aimed at neuroprotection, at maintenance of tissue homeostasis and at preservation of integrity of nervous tissue. In certain conditions, however, and in particular in conditions of chronic and severe stress, reactive astrocytes may acquire maladaptive properties contributing to the damage of the CNS. In both conditions, reactive astrocytes remain an important part of disease progression often defining the neurological outcome of neuropathological process.

Pathological Remodelling of Astrocytes

The second group of astrogliopathologies is represented by pathological remodelling of astrocytes. This class of pathological changes covers astrocytic abnormalities associated with an acquisition of aberrant molecular cascades or functional properties, which drive pathology (Ferrer 2018; Pekny et al. 2016). The best examples of pathological astrocytic remodelling are represented by primary genetic astrogliopathies linked to expression of mutated genes. Alexander disease, a genetic leukomalacia, stems from astrocytic expression of sporadically mutated GFAP gene, which affects, in a yet unknown way, astrocyte function which ultimately results in severe damage to the white matter (Messing et al. 2012). Another example of pathological remodelling of astrocytes occurs in Duchenne muscular dystrophy (DMD) associated with expression of mutated dystrophin gene. Although major clinical presentation of DMD is associated with muscular weakness and cardiomyopathy, most of the patients show psychosocial abnormalities and impaired cognitive abilities. In the CNS dystrophin is expressed mainly in astrocytes (Hendriksen et al. 2016), and its mutations are linked to aberrant CNS cyto-architecture, abnormalities in dendrites and loss of neurones. All these cytopathologies cause a general detrimental neurobehavioural profile, including attention-deficit/hyperactivity disorder (ADHD), autism spectrum disorders and obsessive-compulsive disorder (Anderson et al. 2012; Hendriksen et al. 2018; Ricotti et al. 2015). At the cellular level, expression of mutant dystrophin gene resulted in an aberrant cytoskeleton arrangement and deficient homeostatic capabilities of astrocytes derived from stem cells isolated from DMD patients. In particular, glutamate clearance was severely affected in these astrocytes (Patel et al. 2019). Astroglial pathological remodelling is also central for several other leukodystrophies including vanishing white matter disease, megalencephalic leukoencephalopathy with subcortical cysts and Aicardi-Goutières syndrome (Brignone et al. 2014; Dooves et al. 2016; Jorge and Bugiani 2019). Finally, pathological remodelling of astrocytes has been suggested to occur in mesial temporal lobe epilepsy, characterised by aberrant astrocytic morphology, reduced gap junctional coupling and downregulation of $K_{ir}4.1$ channel expression; all these changes converge into deficient K^+ homeostasis that facilitates generation of seizures (Bedner et al. 2015).

Astroglial Atrophy, Asthenia and Loss of Function

This class of pathological changes includes cell-autonomous astrocytic changes, which do not involve reactivity (i.e. they are not instigated by lesion to the CNS) while being associated with diminished astrocytic function. First, this astrocytic insufficiency is linked to cellular atrophy manifested by decrease in astrocytic morphological profile, with corresponding decrease in astrocytic territorial domain and diminished astrocytic synaptic coverage. This morphological atrophy associated with decrease astrocytic homeostatic support is observed in numerous neuropathological contexts. In particular, morphological atrophy of astrocytes has been detected in diseases of cognition such as neurodegenerative diseases (Heneka et al. 2010; Rodriguez et al. 2009; Verkhratsky et al. 2010) and psychiatric diseases (Dietz et al. 2020; Verkhratsky et al. 2014; Verkhratsky and Parpura 2016; Windrem et al. 2017). Major neuropsychiatric disorders, such as schizophrenia, major depressive disorder and addictive disorders are all associated with reduction of astrocytic density and decrease in astrocytic morphological profiles as revealed by multiple markers (Cotter et al. 2001; Czeh and Di Benedetto 2013; Czeh and Nagy 2018; Miguel-Hidalgo 2009; Rajkowska et al. 2002; Rajkowska and Stockmeier

2013; Scofield et al. 2016). Another pathological feature, the astrocytic asthenia, which is manifested by failures of astroglial homeostatic cascades, is also frequently present in diseases of the brain. In particular, severe decrease in glutamate clearance due to ~80% decrease in expression of astrocytic plasmalemmal glutamate transporters is a leading cause of Wernicke-Korsakoff encephalopathy, associated with massive excitotoxic neuronal death (Hazell 2009; Hazell et al. 2009). Deficits in astroglial glutamate clearance and failure in glutamate-glutamine/GABA shuttle are likely responsible for abnormal neurotransmission as well as for excitotoxic neuronal death, both resulting in psychotic symptoms (Sanacora and Banasr 2013). Decreased expression of plasmalemmal glutamate transporters and decreased glutamate clearance from the extracellular space/synaptic cleft are common features of many addictive disorders, with astrocytic plasmalemmal glutamate transporters representing a promising drug target (Roberts-Wolfe and Kalivas 2015). Neuronal death in amyotrophic lateral sclerosis similarly reflects astrocytic loss of function being a consequence of insufficient astroglial function in extracellular glutamate clearance (Rossi et al. 2008; Valori et al. 2014).

Atrophy of astrocytes linked to decreased synaptic connectivity and synaptic efficacy contributes to cognitive deficiency in both normal ageing and senile dementia. Ageing is the main risk factor for neurodegenerative diseases underlying senile dementia, including Alzheimer's disease. At the same time normal physiological brain ageing with mostly preserved cognitive capacity differs fundamentally from neurodegenerative pathology: in the former the number of neurones is largely preserved, whereas in the latter neurones undergo massive death, which underlies severe cognitive impairment (Pakkenberg and Gundersen 1997; Verkhratsky et al. 2004; von Bartheld et al. 2016; West 1993). Astrocytic numbers seem to be preserved in physiological ageing, whereas the data on astrocytic morphology are controversial and detailed analysis of astrocytic profiles is scarce (Olabarria et al. 2010; Pakkenberg and Gundersen 1997; Verkhratsky et al. 2020a). Most of our knowledge of the state of astrocytes in the ageing brain rests on the analysis of the expression of GFAP, the presumed universal marker of astrocytes (Hol and Pekny 2015). Expression of GFAP is generally increased in the aged brain, which was considered as a sign of astrogliosis and age-dependent inflammation (David et al. 1997; Goss et al. 1991; Hardy et al. 2018; Nichols et al. 1993). Morphometry of aged astrocytes, however, revealed rather contradictory results with both increase and decrease in size and complexity of GFAP-positive astrocytic profiles being observed (see Verkhratsky et al. (2020a) for details and references). All these results, however, need a critical revisit, because GFAP is not an ideal marker of astrocytes (see Verkhratsky and Nedergaard (2018) for detailed discussion). First, in a healthy brain, the majority of astrocytes do not express GFAP at the level of immunocytochemical detection. Second, the proportion of GFAP-positive cells depends on age and brain region. Third, increases in GFAP immunorecativity does not necessarily report reactive changes; in the suprachiasmatic nucleus and the intergeniculate leaflet, for example, GFAP expression undergoes substantial circadian changes (Moriya et al. 2000). Fourth, GFAP labels only cytoskeleton associated with primary astrocytic processes; the peripheral leaflets are always GFAP-negative, and therefore GFAP cannot accurately reveal astrocytic morphology. Finally, GFAP expression changes under various types of environmental stimulation: physical exercise or environmental enrichment increases GFAP-positive profiles, and this increase is beneficial for nervous tissue (Diniz et al. 2016; Rodriguez et al. 2013; Sampedro-Piquero et al. 2014). Thus, age-dependent changes in GFAP expression and GFAP-positive profiles do not reveal much about astrocytic ageing.

Labelling of astrocytes with other markers showed more complex age-dependent changes. Staining of astrocytes with Golgi black reaction did not identify age-dependent morphological changes (Castiglioni Jr. et al. 1991). Immunohistochemical analysis of astroglial profiles labelled with antibodies against GFAP, glutamine synthetase and protein s100β demonstrated complex region- and marker-dependent and age-dependent changes ranging from atrophy to hypertrophy (Rodriguez et al. 2014). The GFAP-labelled astrocytes showed hypertrophy in the CA1 region and in the dentate gyrus of old hippocampus but marked atrophy in the entorhinal cortex (EC). Astrocytes positive for glutamine synthetase were smaller in old hippocampus but larger in the old entorhinal cortex, while s100β-positive profiles from old animals demonstrated an increase in the entorhinal cortex and almost no change in the dentate gyrus and no changes in the CA1 region (Fig. 4).

Morphology of astrocytes probed with intracellular injection of the fluorescent dye Alexa Fluor® 594 revealed age-dependent changes in astrocytic morphology. Two-photon imaging with subsequent 3D reconstruction of astrocytes perfused with the dye showed a significant increase in the size and complexity of astrocytes in

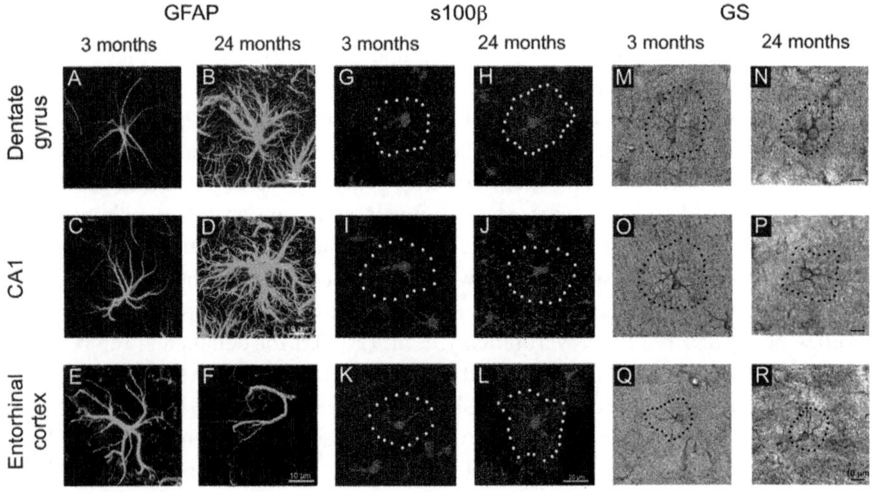

Fig. 4 Age-dependent remodelling of astroglial profiles in different brain areas. Confocal images showing glial fibrillary acidic protein (GFAP) (A to F), s100β (G to L) and glutamine synthetase (GS) (M to R) immunolabelled astrocytes in the dentate gyrus and CA1 hippocampal areas as well as in the entorhinal cortex of mice at 3 and 24 months. (Reproduced, with permission from Verkhratsky et al. (2020a))

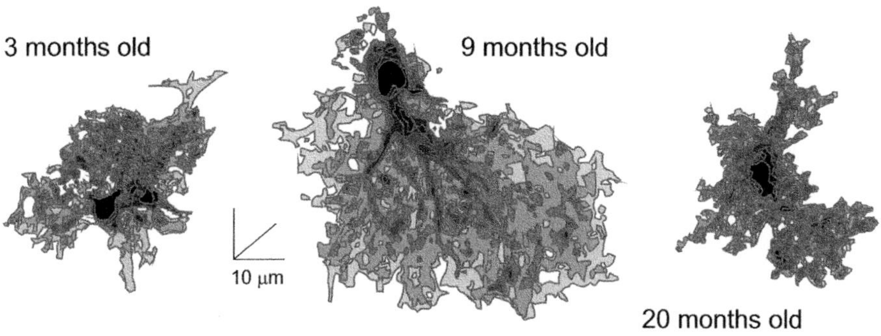

Fig. 5 Reconstructions of hippocampal protoplasmic astrocytes from young, adult and old mice. (Reproduced, with permission from Verkhratsky et al. (2020a))

development from youth to adulthood, whereas astrocytes in the old brains were smaller and less complex and significant decrease in size and complexity of astrocytes in old animals, with substantial reduction on the volume of peripheral processes (Fig. 5, (Popov et al. 2020)). These changes in peripheral processes affected synaptic coverage and synaptic homeostasis; in particular, astrocytic extracellular glutamate and K^+ clearance are both compromised in old animals, leading to depression of long-term potentiation reflecting on deficient memory (Popov et al. 2020).

Astrocytic atrophy is also present in neurodegenerative diseases. In AD, atrophic astrocytes appear in the brain together with reactive astrocytes. Subpopulations of atrophic astrocytes have been found in transgenic AD mouse models (Beauquis et al. 2013; Olabarria et al. 2010) and confirmed in stem cell-derived astrocytes from AD patients in vitro (Jones et al. 2017; Mohamet et al. 2018) and in vivo in derived astrocytes grafted in the mouse brain (Preman et al. 2020). In AD mouse models the total number of astrocytes does not change with age (Olabarria et al. 2010, 2011). At the same time at the early, pre-plaque, stages, astrocytes in entorhinal and prefrontal cortices and hippocampus demonstrate morphological atrophy (Beauquis et al. 2013; Kulijewicz-Nawrot et al. 2012; Olabarria et al. 2010; Yeh et al. 2012). There is a specific temporal pattern in the emergence of atrophic astrocytes in mouse AD models (Rodriguez et al. 2016; Verkhratsky et al. 2019). First, atrophic astrocytic profiles (as visualised by antibodies against GFAP, s100β or glutamine synthetase) appear in the entorhinal cortex (they are present already in 1-month-old mice); subsequently, atrophic astrocytes appear in the prefrontal cortex (3–4 months of age) and finally in the hippocampus (in 6- to 9-month-old animals). Appearance of atrophic astrocytes thus precedes formation of β-amyloid deposits.

At the later stages (12- to 18-month-old animals) of AD, the emergence of β-amyloid plaques in the hippocampus instigates astrogliotic remodelling; reactive astrocytes migrate towards and surround senile plaques and β-amyloid infested blood vessels; at the same time atrophic astrocytes are positioned distantly to β-amyloid depositions (Olabarria et al. 2010; Verkhratsky et al. 2016b). Conversely, in entorhinal and prefrontal cortices, extracellular β-amyloid depositions are not accompanied with astroglial reactivity (Verkhratsky et al. 2016b). Failed astrogliosis represents a loss of function, which defines vulnerability of different brain regions to AD pathology. Indeed, in humans AD starts in entorhinal and prefrontal cortices before this disease spreads to the hippocampus.

Astrocytic atrophy and loss of function can contribute to AD pathophysiology being responsible for early synaptic dysfunction and cognitive deficits. Atrophic astrocytes provide diminished synaptic coverage, which translates in decreased support of synapses by astroglial cradle. First and foremost, this affects K^+ buffering and glutamate homeostasis which both are critical for normal synaptic connectivity. In addition, astrocytes are fundamental for synaptogenesis not only in the developing but also in the adult brain, and astrocytic atrophy may impair the formation of new synapses associated with learning and neuronal plasticity. Early stages of AD are associated with synaptopathy (Coleman et al. 2004; Terry 2000), which might be directly linked to diminished astrocytic support. Astroglial asthenia and loss of function may also account for deficient support associated with the lactate shuttle. Finally, a failure of astrogliotic defence together with a loss of homeostatic capacity of astrocytes (the glial paralysis) can be directly linked to neuronal death and brain atrophy clinically manifested as senile dementia (Verkhratsky et al. 2015).

Astrodegeneration or Clasmatodendrosis

Insults to the brain as well as chronic brain pathologies stress astrocytes, which can undergo degenerative changes and necrotic or apoptotic death. Morphologically, astrodegeneration is manifested by clasmatodendrosis (from Greek "κλάσμα", fragment, "δένδρον", tree, "ωσις", process). This process has been initially characterised by Alois Alzheimer and also described and named by Santiago Ramón y Cajal (Penfield 1928a). Clasmatodendrosis appears as fragmentation of astroglial processes, vanishing of distal processes, and swelling and vacuolisation of the cell body. Clasmatodendrosis has been visualised in vitro and in tissue, and it was observed in several forms of neuropathology including ischaemia, infectious encephalopathies, stroke, dementia and psychiatric diseases (Hulse et al. 2001; Sahlas et al. 2002; Tachibana et al. 2019). Clasmatodendrotic astrocytes have been also identified in the brains of old mice (Mercatelli et al. 2016).

Envoi

We have outlined the presently ascribed roles of astroglia in nervous system pathology. As per human need to organize and stratify, we pigeonholed the roles into the present-day classification of astrogliopathology. While we have no doubt that this classification will develop further, likely by the time one reads these lines as the volume gets published, we deem it necessary and sufficient for further discussion of more detailed chapters that follow in this volume and delve into the role of astroglia in a variety of psychiatric conditions and diseases.

Acknowledgements BL's work is supported by the National Natural Science Foundation of China (grant number 8187185), LiaoNing Revitalization Talents Program (grant number XLYC1807137), Scientific Research Foundation for Returned Scholars of Education Ministry of China (grant number 20151098), LiaoNing Thousand Talents Program (grant number 202078) and "ChunHui" Program of Education Ministry of China (grant number 2020703). CS's work is supported by a grant from the Italian Ministry of Education, University and Research (2015KP7T2Y_002) and a grant from Sapienza University of Rome (RM11916B7A8D0225). VP's work is supported by a grant from the National Institute of General Medical Sciences of the National Institutes of Health (R01GM123971). VP is an Honorary Professor at University of Rijeka, Croatia.

References

Achucarro N (1910) Some pathological findings in the neuroglia and in the ganglion cells of the cortex in senile conditions. Bull Gov Hosp Insane 2:81–90

Al-Dalahmah O, Sosunov AA, Shaik A, Ofori K, Liu Y, Vonsattel JP, Adorjan I, Menon V, Goldman JE (2020) Single-nucleus RNA-seq identifies Huntington disease astrocyte states. Acta Neuropathol Commun 8:19

Alberdi E, Wyssenbach A, Alberdi M, Sanchez-Gomez MV, Cavaliere F, Rodriguez JJ, Verkhratsky A, Matute C (2013) Ca^{2+} -dependent endoplasmic reticulum stress correlates with astrogliosis in oligomeric amyloid beta-treated astrocytes and in a model of Alzheimer's disease. Aging Cell 12:292–302

Alzheimer A (1910) Beiträge zur Kenntnis der pathologischen Neuroglia und ihrer Beziehungen zu den Abbauvorgängen im Nervengewebe. In: Nissl F, Alzheimer A (eds) Histologische und histopathologische Arbeiten über die Grosshirnrinde mit besonderer Berücksichtigung der pathologischen Anatomie der Geisteskrankheiten, vol 3. Gustav Fischer, Jena, pp 401–562

Anderson JL, Head SI, Morley JW (2012) Duchenne muscular dystrophy and brain function. In: Muscular dystrophy. InTech

Anderson MA, Burda JE, Ren Y, Ao Y, O'Shea TM, Kawaguchi R, Coppola G, Khakh BS, Deming TJ, Sofroniew MV (2016) Astrocyte scar formation aids central nervous system axon regeneration. Nature 532:195–200

Beauquis J, Pavia P, Pomilio C, Vinuesa A, Podlutskaya N, Galvan V, Saravia F (2013) Environmental enrichment prevents astroglial pathological changes in the hippocampus of APP transgenic mice, model of Alzheimer's disease. Exp Neurol 239:28–37

Bedner P, Dupper A, Huttmann K, Muller J, Herde MK, Dublin P, Deshpande T, Schramm J, Haussler U, Haas CA, Henneberger C, Theis M, Steinhauser C (2015) Astrocyte uncoupling as a cause of human temporal lobe epilepsy. Brain 138:1208–1222

Brignone MS, Lanciotti A, Visentin S, De Nuccio C, Molinari P, Camerini S, Diociaiuti M, Petrini S, Minnone G, Crescenzi M, Laudiero LB, Bertini E, Petrucci TC, Ambrosini E (2014) Megalencephalic leukoencephalopathy with subcortical cysts protein-1 modulates endosomal pH and protein trafficking in astrocytes: relevance to MLC disease pathogenesis. Neurobiol Dis 66:1–18

Burda JE, Bernstein AM, Sofroniew MV (2016) Astrocyte roles in traumatic brain injury. Exp Neurol 275(Pt 3):305–315

Burda JE, Sofroniew MV (2014) Reactive gliosis and the multicellular response to CNS damage and disease. Neuron 81:229–248

Castiglioni AJ Jr, Legare ME et al (1991) Morphological changes in astrocytes of aging mice fed normal or caloric restricted diets. Age 14:102–106

Chun H, Im H, Kang YJ, Kim Y, Shin JH, Won W, Lim J, Ju Y, Park YM, Kim S, Lee SE, Lee J, Woo J, Hwang Y, Cho H, Jo S, Park JH, Kim D, Kim DY, Seo JS, Gwag BJ, Kim YS, Park KD, Kaang BK, Cho H, Ryu H, Lee CJ (2020) Severe reactive astrocytes precipitate pathological hallmarks of Alzheimer's disease via $H_2O_2^-$ production. Nat Neurosci 23:1555–1566

Coleman P, Federoff H, Kurlan R (2004) A focus on the synapse for neuroprotection in Alzheimer disease and other dementias. Neurology 63:1155–1162

Cotter D, Mackay D, Landau S, Kerwin R, Everall I (2001) Reduced glial cell density and neuronal size in the anterior cingulate cortex in major depressive disorder. Arch Gen Psychiatry 58:545–553

Czeh B, Di Benedetto B (2013) Antidepressants act directly on astrocytes: evidences and functional consequences. Eur Neuropsychopharmacol 23:171–185

Czeh B, Nagy SA (2018) Clinical findings documenting cellular and molecular abnormalities of glia in depressive disorders. Front Mol Neurosci 11:56

David JP, Ghozali F, Fallet-Bianco C, Wattez A, Delaine S, Boniface B, Di Menza C, Delacourte A (1997) Glial reaction in the hippocampal formation is highly correlated with aging in human brain. Neurosci Lett 235:53–56

del Río-Hortega P, Penfield WG (1927) Cerebral cicatrix: the reaction of neuroglia and microglia to brain wounds. Bull Johns Hopkins Hosp 41:278–303

Dietz AG, Goldman SA, Nedergaard M (2020) Glial cells in schizophrenia: a unified hypothesis. Lancet Psychiatry 7:272–281

Diniz DG, de Oliveira MA, de Lima CM, Foro CA, Sosthenes MC, Bento-Torres J, da Costa Vasconcelos PF, Anthony DC, Diniz CW (2016) Age, environment, object recognition and morphological diversity of GFAP-immunolabeled astrocytes. Behav Brain Funct 12:28

Dooves S, Bugiani M, Postma NL, Polder E, Land N, Horan ST, van Deijk AL, van de Kreeke A, Jacobs G, Vuong C, Klooster J, Kamermans M, Wortel J, Loos M, Wisse LE, Scheper GC, Abbink TE, Heine VM, van der Knaap MS (2016) Astrocytes are central in the pathomechanisms of vanishing white matter. J Clin Invest 126:1512–1524

Escartin C, Galea E, Lakatos A, O'Callaghan JP, Petzold GC, Serrano-Pozo A, Steinhauser C, Volterra A, Carmignoto G, Agarwal A, Allen NJ, Araque A, Luis Barbeito L, Barzilai A, Bergles DE, Bonvento G, Butt AM, Chen W-T, Cohen-Salmon M, Cunningham C, Deneen B, De Strooper B, Díaz-Castro B, Farina C, Freeman M, Gallo V, Goldman JE, Goldman SA, Götz M, Gutiérrez A, Haydon PG, Heiland DH, Hol EM, Holt MG, Iino M, Kastanenka KV, Kettenmann H, Khakh BS, Koizumi S, Lee CJ, Liddelow SA, MacVicar BA, Magistretti P, Messing A, Mishra A, Molofsky AV, Murai K, Norris CM, Okada S, SHR O, Oliveira JF, Panatier A, Parpura V, Pekna M, Pekny M, Pellerin L, Perea G, Pérez-Nievas BG, Pfrieger FW, Poskanzer KE, Quintana FJ, Ransohoff RM, Riquelme-Perez M, Robel S, Rose CR, Rothstein J, Rouach N, Rowitch DH, Semyanov A, Sirko S, Sontheimer H, Swanson RA, Vitorica J, Wanner I-B, Wood LB, Wu J, Zheng B, Zimmer ER, Zorec R, Sofroniew MV, Verkhratsky A (2021) Reactive astrocyte nomenclature, definitions and future directions. Nat Neurosci 24(3):312–325

Farina C, Aloisi F, Meinl E (2007) Astrocytes are active players in cerebral innate immunity. Trends Immunol 28:138–145

Faulkner JR, Herrmann JE, Woo MJ, Tansey KE, Doan NB, Sofroniew MV (2004) Reactive astrocytes protect tissue and preserve function after spinal cord injury. J Neurosci 24:2143–2155

Ferrer I (2018) Astrogliopathy in Tauopathies. Neuroglia 1:126–150

Franke H, Verkhratsky A, Burnstock G, Illes P (2012) Pathophysiology of astroglial purinergic signalling. Purinergic Signal 8:629–657

Frommann C (1878) Untersuchungen über die Gewebsveränderungen bei der Multiplen Sklerose des Gehirns und Rückenmarks. Verlag von Gustav Fischer, Jena

Garaschuk O, Verkhratsky A (2019) GABAergic astrocytes in Alzheimer's disease. Aging (Albany NY) 11:1602–1604

Garcia AD, Petrova R, Eng L, Joyner AL (2010) Sonic hedgehog regulates discrete populations of astrocytes in the adult mouse forebrain. J Neurosci 30:13597–13608

Ghatak S, Dolatabadi N, Trudler D, Zhang X, Wu Y, Mohata M, Ambasudhan R, Talantova M, Lipton SA (2019) Mechanisms of hyperexcitability in Alzheimer's disease hiPSC-derived neurons and cerebral organoids vs isogenic controls. Elife 8:e50333

Giaume C, Kirchhoff F, Matute C, Reichenbach A, Verkhratsky A (2007) Glia: the fulcrum of brain diseases. Cell Death Differ 14:1324–1335

Goss JR, Finch CE, Morgan DG (1991) Age-related changes in glial fibrillary acidic protein mRNA in the mouse brain. Neurobiol Aging 12:165–170

Hardy RN, Simsek ZD, Curry B, Core SL, Beltz T, Xue B, Johnson AK, Thunhorst RL, Curtis KS (2018) Aging affects isoproterenol-induced water drinking, astrocyte density, and central neuronal activation in female Brown Norway rats. Physiol Behav 192:90–97

Hazell AS (2009) Astrocytes are a major target in thiamine deficiency and Wernicke's encephalopathy. Neurochem Int 55:129–135

Hazell AS, Sheedy D, Oanea R, Aghourian M, Sun S, Jung JY, Wang D, Wang C (2009) Loss of astrocytic glutamate transporters in Wernicke encephalopathy. Glia 58:148–156

Hendriksen RGF, Schipper S, Hoogland G, Schijns OEMG, Dings JTA, Aalbers MW, Vles JSH (2016) Dystrophin distribution and expression in human and experimental temporal lobe epilepsy. Front Cell Neurosci 10:174

Hendriksen RGF, Vles JSH, Aalbers MW, Chin RFM, Hendriksen JGM (2018) Brain-related comorbidities in boys and men with Duchenne muscular dystrophy: a descriptive study. Eur J Paediatr Neurol 22:488–497

Heneka MT, Rodriguez JJ, Verkhratsky A (2010) Neuroglia in neurodegeneration. Brain Res Rev 63:189–211

Henrik Heiland D, Ravi VM, Behringer SP, Frenking JH, Wurm J, Joseph K, Garrelfs NWC, Strahle J, Heynckes S, Grauvogel J, Franco P, Mader I, Schneider M, Potthoff AL, Delev D, Hofmann UG, Fung C, Beck J, Sankowski R, Prinz M, Schnell O (2019) Tumor-associated reactive astrocytes aid the evolution of immunosuppressive environment in glioblastoma. Nat Commun 10:2541

Herrmann JE, Imura T, Song B, Qi J, Ao Y, Nguyen TK, Korsak RA, Takeda K, Akira S, Sofroniew MV (2008) STAT3 is a critical regulator of astrogliosis and scar formation after spinal cord injury. J Neurosci 28:7231–7243

Hol EM, Pekny M (2015) Glial fibrillary acidic protein (GFAP) and the astrocyte intermediate filament system in diseases of the central nervous system. Curr Opin Cell Biol 32:121–130

Hulse RE, Winterfield J, Kunkler PE, Kraig RP (2001) Astrocytic clasmatodendrosis in hippocampal organ culture. Glia 33:169–179

Itoh N, Itoh Y, Tassoni A, Ren E, Kaito M, Ohno A, Ao Y, Farkhondeh V, Johnsonbaugh H, Burda J, Sofroniew MV, Voskuhl RR (2018) Cell-specific and region-specific transcriptomics in the multiple sclerosis model: focus on astrocytes. Proc Natl Acad Sci U S A 115:E302–E309

Jack CS, Arbour N, Manusow J, Montgrain V, Blain M, McCrea E, Shapiro A, Antel JP (2005) TLR signaling tailors innate immune responses in human microglia and astrocytes. J Immunol 175:4320–4330

Jones VC, Atkinson-Dell R, Verkhratsky A, Mohamet L (2017) Aberrant iPSC-derived human astrocytes in Alzheimer's disease. Cell Death Dis 8:e2696

Jorge MS, Bugiani M (2019) Astroglia in leukodystrophies. Adv Exp Med Biol 1175:199–225

Kamphuis W, Kooijman L, Orre M, Stassen O, Pekny M, Hol EM (2015) GFAP and vimentin deficiency alters gene expression in astrocytes and microglia in wild-type mice and changes the transcriptional response of reactive glia in mouse model for Alzheimer's disease. Glia 63:1036–1056

Kielian T (2006) Toll-like receptors in central nervous system glial inflammation and homeostasis. J Neurosci Res 83:711–730

Kraft AW, Hu X, Yoon H, Yan P, Xiao Q, Wang Y, Gil SC, Brown J, Wilhelmsson U, Restivo JL, Cirrito JR, Holtzman DM, Kim J, Pekny M, Lee JM (2013) Attenuating astrocyte activation accelerates plaque pathogenesis in APP/PS1 mice. FASEB J 27:187–198

Kulijewicz-Nawrot M, Verkhratsky A, Chvatal A, Sykova E, Rodriguez JJ (2012) Astrocytic cytoskeletal atrophy in the medial prefrontal cortex of a triple transgenic mouse model of Alzheimer's disease. J Anat 221:252–262

Li B, Xia M, Zorec R, Parpura V, Verkhratsky A (2021) Astrocytes in heavy metal neurotoxicity and neurodegeneration. Brain Res 1752:147234

Li L, Lundkvist A, Andersson D, Wilhelmsson U, Nagai N, Pardo AC, Nodin C, Stahlberg A, Aprico K, Larsson K, Yabe T, Moons L, Fotheringham A, Davies I, Carmeliet P, Schwartz JP, Pekna M, Kubista M, Blomstrand F, Maragakis N, Nilsson M, Pekny M (2008) Protective role of reactive astrocytes in brain ischemia. J Cereb Blood Flow Metab 28:468–481

Lian H, Yang L, Cole A, Sun L, Chiang AC, Fowler SW, Shim DJ, Rodriguez-Rivera J, Taglialatela G, Jankowsky JL, Lu HC, Zheng H (2015) NFκB-activated astroglial release of complement C3 compromises neuronal morphology and function associated with Alzheimer's disease. Neuron 85:101–115

Mercatelli R, Lana D, Bucciantini M, Giovannini MG, Cerbai F, Quercioli F, Zecchi-Orlandini S, Delfino G, Wenk GL, Nosi D (2016) Clasmatodendrosis and beta-amyloidosis in aging hippocampus. FASEB J 30:1480–1491

Messing A, Brenner M, Feany MB, Nedergaard M, Goldman JE (2012) Alexander disease. J Neurosci 32:5017–5023

Miguel-Hidalgo JJ (2009) The role of glial cells in drug abuse. Curr Drug Abuse Rev 2:76–82

Mohamet L, Jones VC, Dayanithi G, Verkhratsky A (2018) Pathological human astroglia in Alzheimer's disease: opening new horizons with stem cell technology. Future Neurology 13:87–99

Montana V, Verkhratsky A, Parpura V (2014) Pathological role for exocytotic glutamate release from astrocytes in hepatic encephalopathy. Curr Neuropharmacol 12:324–333

Moriya T, Yoshinobu Y, Kouzu Y, Katoh A, Gomi H, Ikeda M, Yoshioka T, Itohara S, Shibata S (2000) Involvement of glial fibrillary acidic protein (GFAP) expressed in astroglial cells in circadian rhythm under constant lighting conditions in mice. J Neurosci Res 60:212–218

Nichols NR, Day JR, Laping NJ, Johnson SA, Finch CE (1993) GFAP mRNA increases with age in rat and human brain. Neurobiol Aging 14:421–429

Nissl F (1899) Über einige Beziehungen zwischen Nervenzellerkrankungen und gliösen Erscheinungen bei verschiedenen Psychosen. Arch Psychiatr 32:1–21

O'Callaghan JP, Kelly KA, VanGilder RL, Sofroniew MV, Miller DB (2014) Early activation of STAT3 regulates reactive astrogliosis induced by diverse forms of neurotoxicity. PLoS One 9:e102003

Obara-Michlewska M, Ruszkiewicz J, Zielinska M, Verkhratsky A, Albrecht J (2015) Astroglial NMDA receptors inhibit expression of $K_{ir}4.1$ channels in glutamate-overexposed astrocytes in vitro and in the brain of rats with acute liver failure. Neurochem Int 88:20–25

Okada S, Nakamura M, Katoh H, Miyao T, Shimazaki T, Ishii K, Yamane J, Yoshimura A, Iwamoto Y, Toyama Y, Okano H (2006) Conditional ablation of Stat3 or Socs3 discloses a dual role for reactive astrocytes after spinal cord injury. Nat Med 12:829–834

Olabarria M, Noristani HN, Verkhratsky A, Rodriguez JJ (2010) Concomitant astroglial atrophy and astrogliosis in a triple transgenic animal model of Alzheimer's disease. Glia 58:831–838

Olabarria M, Noristani HN, Verkhratsky A, Rodriguez JJ (2011) Age-dependent decrease in glutamine synthetase expression in the hippocampal astroglia of the triple transgenic Alzheimer's disease mouse model: mechanism for deficient glutamatergic transmission? Mol Neurodegener 6:55

Pakkenberg B, Gundersen HJ (1997) Neocortical neuron number in humans: effect of sex and age. J Comp Neurol 384:312–320

Parpura V, Heneka MT, Montana V, Oliet SH, Schousboe A, Haydon PG, Stout RF Jr, Spray DC, Reichenbach A, Pannicke T, Pekny M, Pekna M, Zorec R, Verkhratsky A (2012) Glial cells in (patho)physiology. J Neurochem 121:4–27

Patel AM, Wierda K, Thorrez L, van Putten M, De Smedt J, Ribeiro L, Tricot T, Gajjar M, Duelen R, Van Damme P, De Waele L, Goemans N, Tanganyika-de Winter C, Costamagna D, Aartsma-Rus A, van Duyvenvoorde H, Sampaolesi M, Buyse GM, Verfaillie CM (2019) Dystrophin deficiency leads to dysfunctional glutamate clearance in iPSC derived astrocytes. Transl Psychiatry 9:200

Pekny M, Johansson CB, Eliasson C, Stakeberg J, Wallen A, Perlmann T, Lendahl U, Betsholtz C, Berthold CH, Frisen J (1999) Abnormal reaction to central nervous system injury in mice lacking glial fibrillary acidic protein and vimentin. J Cell Biol 145:503–514

Pekny M, Pekna M (2014) Astrocyte reactivity and reactive astrogliosis: costs and benefits. Physiol Rev 94:1077–1098

Pekny M, Pekna M, Messing A, Steinhauser C, Lee JM, Parpura V, Hol EM, Sofroniew MV, Verkhratsky A (2016) Astrocytes: a central element in neurological diseases. Acta Neuropathol 131:323–345

Penfield W (1928a) Neuroglia and microglia-the interstitial tissue of the central nervous system. In: Cowdry EV (ed) Special cytology, the form and function of the cell in health and disease. Hoeber, New York, pp 1033–1068

Penfield WG (1928b) Neuroglia and microglia. the interstitial tissue of the central nervous system. In: Cowdry EV (ed) Special cytology, vol 2. Paul B. Hober Inc., New York, pp 1033–1068

Plata A, Lebedeva A, Denisov P, Nosova O, Postnikova TY, Pimashkin A, Brazhe A, Zaitsev AV, Rusakov DA, Semyanov A (2018) Astrocytic atrophy following status epilepticus parallels reduced Ca^{2+} activity and impaired synaptic plasticity in the rat hippocampus. Front Mol Neurosci 11:215

Popov A, Brazhe A, Denisov P, Sutyagina O, Lazareva N, Verkhratsky A, Semyanov A (2020) Astrocytes dystrophy in ageing brain parallels impaired synaptic plasticity. Aging Cell 20(3):e13334

Preman P, Julia TCW, Calafate S, Snellinx A, Alfonso-Triguero M, Corthout N, Munck S, Thal DT, Goate AM, De Strooper B, Arranz AM (2020) Human iPSC-derived astrocytes transplanted into the mouse brain display three morphological responses to amyloid-β plaques. bioRxiv

Rajkowska G, Miguel-Hidalgo JJ, Makkos Z, Meltzer H, Overholser J, Stockmeier C (2002) Layer-specific reductions in GFAP-reactive astroglia in the dorsolateral prefrontal cortex in schizophrenia. Schizophr Res 57:127–138

Rajkowska G, Stockmeier CA (2013) Astrocyte pathology in major depressive disorder: insights from human postmortem brain tissue. Curr Drug Targets 14:1225–1236

Ricotti V, Mandy WPL, Scoto M, Pane M, Deconinck N, Messina S, Mercuri E, Skuse DH, Muntoni F (2015) Neurodevelopmental, emotional, and behavioural problems in Duchenne muscular dystrophy in relation to underlying dystrophin gene mutations. Dev Med Child Neurol 58:77–84

Roberts-Wolfe DJ, Kalivas PW (2015) Glutamate transporter GLT-1 as a therapeutic target for substance use disorders. CNS Neurol Disord Drug Targets 14:745–756

Rodriguez JJ, Butt AM, Gardenal E, Parpura V, Verkhratsky A (2016) Complex and differential glial responses in Alzheimer's disease and ageing. Curr Alzheimer Res 13:343–358

Rodriguez JJ, Olabarria M, Chvatal A, Verkhratsky A (2009) Astroglia in dementia and Alzheimer's disease. Cell Death Differ 16:378–385

Rodriguez JJ, Terzieva S, Olabarria M, Lanza RG, Verkhratsky A (2013) Enriched environment and physical activity reverse astrogliodegeneration in the hippocampus of AD transgenic mice. Cell Death Dis 4:e678

Rodriguez JJ, Yeh CY, Terzieva S, Olabarria M, Kulijewicz-Nawrot M, Verkhratsky A (2014) Complex and region-specific changes in astroglial markers in the aging brain. Neurobiol Aging 35:15–23

Rossi D, Brambilla L, Valori CF, Roncoroni C, Crugnola A, Yokota T, Bredesen DE, Volterra A (2008) Focal degeneration of astrocytes in amyotrophic lateral sclerosis. Cell Death Differ 15:1691–1700

Sahlas DJ, Bilbao JM, Swartz RH, Black SE (2002) Clasmatodendrosis correlating with periventricular hyperintensity in mixed dementia. Ann Neurol 52:378–381

Sampedro-Piquero P, De Bartolo P, Petrosini L, Zancada-Menendez C, Arias JL, Begega A (2014) Astrocytic plasticity as a possible mediator of the cognitive improvements after environmental enrichment in aged rats. Neurobiol Learn Mem 114:16–25

Sanacora G, Banasr M (2013) From pathophysiology to novel antidepressant drugs: glial contributions to the pathology and treatment of mood disorders. Biol Psychiatry 73:1172–1179

Scofield MD, Li H, Siemsen BM, Healey KL, Tran PK, Woronoff N, Boger HA, Kalivas PW, Reissner KJ (2016) Cocaine self-administration and extinction leads to reduced glial fibrillary acidic protein expression and morphometric features of astrocytes in the nucleus accumbens core. Biol Psychiatry 80:207–215

Shulyatnikova T, Verkhratsky A (2019) Astroglia in sepsis associated encephalopathy. Neurochem Res 45(1):83–99

Sofroniew MV (2009) Molecular dissection of reactive astrogliosis and glial scar formation. Trends Neurosci 32:638–647

Sofroniew MV (2014a) Astrogliosis. Cold Spring Harb Perspect Biol 7:a020420

Sofroniew MV (2014b) Multiple roles for astrocytes as effectors of cytokines and inflammatory mediators. Neuroscientist 20:160–172

Sofroniew MV (2020) Astrocyte reactivity: subtypes, states, and functions in CNS innate immunity. Trends Immunol 41:758–770

Soung A, Klein RS (2018) Viral encephalitis and neurologic diseases: focus on astrocytes. Trends Mol Med 24:950–962

Tachibana M, Mohri I, Hirata I, Kuwada A, Kimura-Ohba S, Kagitani-Shimono K, Fushimi H, Inoue T, Shiomi M, Kakuta Y, Takeuchi M, Murayama S, Nakayama M, Ozono K, Taniike M (2019) Clasmatodendrosis is associated with dendritic spines and does not represent autophagic astrocyte death in influenza-associated encephalopathy. Brain Dev 41:85–95

Tang D, Kang R, Coyne CB, Zeh HJ, Lotze MT (2012) PAMPs and DAMPs: signal 0s that spur autophagy and immunity. Immunol Rev 249:158–175

Terry RD (2000) Cell death or synaptic loss in Alzheimer disease. J Neuropathol Exp Neurol 59:1118–1119

Tremblay ME, Madore C, Bordeleau M, Tian L, Verkhratsky A (2020) Neuropathobiology of COVID-19: the role for glia. Front Cell Neurosci 14:592214

Valori CF, Brambilla L, Martorana F, Rossi D (2014) The multifaceted role of glial cells in amyotrophic lateral sclerosis. Cell Mol Life Sci 71:287–297

Verkhratsky A, Augusto-Oliveira M, Pivoriunas A, Popov A, Brazhe A, Semyanov A (2020a) Astroglial asthenia and loss of function, rather than reactivity, contribute to the ageing of the brain. Pflugers Arch 12:1–22

Verkhratsky A, Butt AM (2013) Glial physiology and pathophysiology. Wiley-Blackwell, Chichester

Verkhratsky A, Marutle A, Rodriguez-Arellano JJ, Nordberg A (2015) Glial asthenia and functional paralysis: a new perspective on neurodegeneration and Alzheimer's disease. Neuroscientist 21:552–568

Verkhratsky A, Mattson MP, Toescu EC (2004) Aging in the mind. Trends Neurosci 27:577–578

Verkhratsky A, Nedergaard M (2018) Physiology of astroglia. Physiol Rev 98:239–389

Verkhratsky A, Olabarria M, Noristani HN, Yeh CY, Rodriguez JJ (2010) Astrocytes in Alzheimer's disease. Neurotherapeutics 7:399–412

Verkhratsky A, Parpura V (2016) Astrogliopathology in neurological, neurodevelopmental and psychiatric disorders. Neurobiol Dis 85:254–261

Verkhratsky A, Rodrigues JJ, Pivoriunas A, Zorec R, Semyanov A (2019) Astroglial atrophy in Alzheimer's disease. Pflugers Arch 471:1247–1261

Verkhratsky A, Rodriguez JJ, Parpura V (2012a) Calcium signalling in astroglia. Mol Cell Endocrinol 353:45–56
Verkhratsky A, Rodriguez JJ, Steardo L (2014) Astrogliopathology: a central element of neuropsychiatric diseases? Neuroscientist 20:576–588
Verkhratsky A, Semyanov A, Zorec R (2020b) Physiology of astroglial excitability. Function 1:zqaa016
Verkhratsky A, Sofroniew MV, Messing A, deLanerolle NC, Rempe D, Rodriguez JJ, Nedergaard M (2012b) Neurological diseases as primary gliopathies: a reassessment of neurocentrism. ASN Neuro 4:AN20120010
Verkhratsky A, Steardo L, Parpura V, Montana V (2016a) Translational potential of astrocytes in brain disorders. Prog Neurobiol 144:188–205
Verkhratsky A, Untiet V, Rose CR (2020c) Ionic signalling in astroglia beyond calcium. J Physiol 598:1655–1670
Verkhratsky A, Zorec R, Parpura V (2017) Stratification of astrocytes in healthy and diseased brain. Brain Pathol 27:629–644
Verkhratsky A, Zorec R, Rodriguez JJ, Parpura V (2016b) Astroglia dynamics in ageing and Alzheimer's disease. Curr Opin Pharmacol 26:74–79
von Bartheld CS, Bahney J, Herculano-Houzel S (2016) The search for true numbers of neurons and glial cells in the human brain: a review of 150 years of cell counting. J Comp Neurol 524:3865–3895
Voskuhl RR, Peterson RS, Song B, Ao Y, Morales LB, Tiwari-Woodruff S, Sofroniew MV (2009) Reactive astrocytes form scar-like perivascular barriers to leukocytes during adaptive immune inflammation of the CNS. J Neurosci 29:11511–11522
West MJ (1993) Regionally specific loss of neurons in the aging human hippocampus. Neurobiol Aging 14:287–293
Wheeler MA, Clark IC, Tjon EC, Li Z, Zandee SEJ, Couturier CP, Watson BR, Scalisi G, Alkwai S, Rothhammer V, Rotem A, Heyman JA, Thaploo S, Sanmarco LM, Ragoussis J, Weitz DA, Petrecca K, Moffitt JR, Becher B, Antel JP, Prat A, Quintana FJ (2020) MAFG-driven astrocytes promote CNS inflammation. Nature 578:593–599
Wheeler MA, Quintana FJ (2019) Regulation of astrocyte functions in multiple. Sclerosis Cold Spring Harb Perspect Med 9(1):a029009
Wilhelmsson U, Bushong EA, Price DL, Smarr BL, Phung V, Terada M, Ellisman MH, Pekny M (2006) Redefining the concept of reactive astrocytes as cells that remain within their unique domains upon reaction to injury. Proc Natl Acad Sci U S A 103:17513–17518
Windrem MS, Osipovitch M, Liu Z, Bates J, Chandler-Militello D, Zou L, Munir J, Schanz S, McCoy K, Miller RH, Wang S, Nedergaard M, Findling RL, Tesar PJ, Goldman SA (2017) Human iPSC glial mouse chimeras reveal Glial contributions to schizophrenia. Cell Stem Cell 21(195-208):e196
Yeh CY, Vadhwana B, Verkhratsky A, Rodriguez JJ (2012) Early astrocytic atrophy in the entorhinal cortex of a triple transgenic animal model of Alzheimer's disease. ASN Neuro 3:271–279
Zamanian JL, Xu L, Foo LC, Nouri N, Zhou L, Giffard RG, Barres BA (2012) Genomic analysis of reactive astrogliosis. J Neurosci 32:6391–6410
Zeidan-Chulia F, Salmina AB, Malinovskaya NA, Noda M, Verkhratsky A, Moreira JC (2014) The glial perspective of autism spectrum disorders. Neurosci Biobehav Rev 38:160–172
Zorec R, Zupanc TA, Verkhratsky A (2019) Astrogliopathology in the infectious insults of the brain. Neurosci Lett 689:56–62

Morphological Features of Astrocytes in Health and Neuropsychiatric Disorders

Celia Roman, Eugenia Vivi, and Barbara Di Benedetto

Introduction

In the 1850s, Rudolf Virchow first coined the concept of "neuroglia" (or more simply "glia," from the Greek γλία and γλοία which means "glue"), referring to the passive connective fibers and intercellular masses, which embed neurons and appear like glue (Virchow 1858).

In the late nineteenth century, additional observations by the Italian physician and cytologist Camillo Golgi and the Spanish neurohistologist Ramon y Cajal revealed that neuroglia and nerve cells represent distinct cellular populations. Following this theory, they further described a variety of glial shapes and forms, suggesting the existence of different subtypes of glia cells, which may provide a corresponding diversity of functions within the CNS (Golgi 1871).

Yet the term "astrocyte" (from the Greek ἄστρον, ástron, "star" + κύτος, kútos, "cavity," "cell") was only first introduced in 1893 by the Hungarian anatomist and histologist Lenhossek, clearly referring to the stellate morphology of these cells (Lenhossek 1893).

Celia Roman and Eugenia Vivi contributed equally with all other contributors.

C. Roman · E. Vivi
Department of Psychiatry and Psychotherapy, University of Regensburg, Regensburg, Germany

B. Di Benedetto (✉)
Department of Psychiatry and Psychotherapy, University of Regensburg, Regensburg, Germany

Regensburg Center of Neuroscience, University of Regensburg, Regensburg, Germany
e-mail: Barbara.Di-Benedetto@ukr.de

B. Li et al. (eds.), *Astrocytes in Psychiatric Disorders*, Advances in Neurobiology 26, https://doi.org/10.1007/978-3-030-77375-5_4

With this new name, he advanced the concept that astrocytes cover a range of functions as important as nerve cells, thus getting away from the term "brain glue."

Although astrocytes are in close contact with neuronal structures, it was not clear whether they would be as diverse as neurons. Indeed, already the early histological studies from Ramon y Cajal's work revealed a high morphologic diversity of astrocytes, suggesting several functional specializations (De Filipe 2010). However, only in the past decade, accumulating evidences indeed showed that astrocytes display a high regional heterogeneity in the developed CNS. This heterogeneity is essential for maintaining, among others, a proper brain architecture, functional homeostasis, and an appropriate delivery of nutritional elements, due to the intimate interaction of astrocytes with a variety of distinct cell types in multiple brain areas and during different developmental or adult stages (John Lin et al. 2017; Kettenmann and Verkhratsky 2008; Liddelow et al. 2017).

Moreover, other studies have demonstrated inter-species morphological and functional distinctions in astrocytes, highlighting the potential contribution of astrocytes to species-specific neuronal processing (Oberheim et al. 2006; Oberheim et al. 2009).

Indeed, it is today recognized that astrocytes display a body of morphological and functional aspects that reflect the brain network in which they are embedded.

Astrocyte Morphological Heterogeneity

Astrocytes account for approximately 20–40% of the total number of brain cells, representing the most abundant glial cells populating the CNS (Herculano-Houzel 2014).

The most remarkable characteristic of astrocyte morphology, which makes them heterogeneous, derives from their process ramification complexity and irregularity. Additionally, astrocyte heterogeneity can be classified on the basis of development, reactive state, regional identity, and age-related diversities.

Since the late nineteenth century, astrocytes were already recognized as a morphologically complex population tiling the CNS. Accordingly, they were already divided into two major classes, still currently in use: protoplasmic astrocytes of the gray matter and fibrous astrocytes of the white matter (Ramón y Cajal 1897) (Kohler et al. 2019) (Fig. 1a, b).

Protoplasmic astrocytes are the more numerous type of astrocytes and are characterized by few thick processes, which branch out into several fine and short processes, widely and uniformly distributed in the gray matter. As a matter of fact, protoplasmic astrocytes can be composed of six to seven main branches originating from the soma, with various secondary branches that can spread into thousands of smaller processes, which establish interactions with other cell types. These fine small processes, termed peripheral processes or perisynaptic astrocyte processes (PAPs), are specialized and polarized structures. In accordance, a protoplasmic astrocyte can be characterized by a diameter of 40–60μm, with approximately 95%

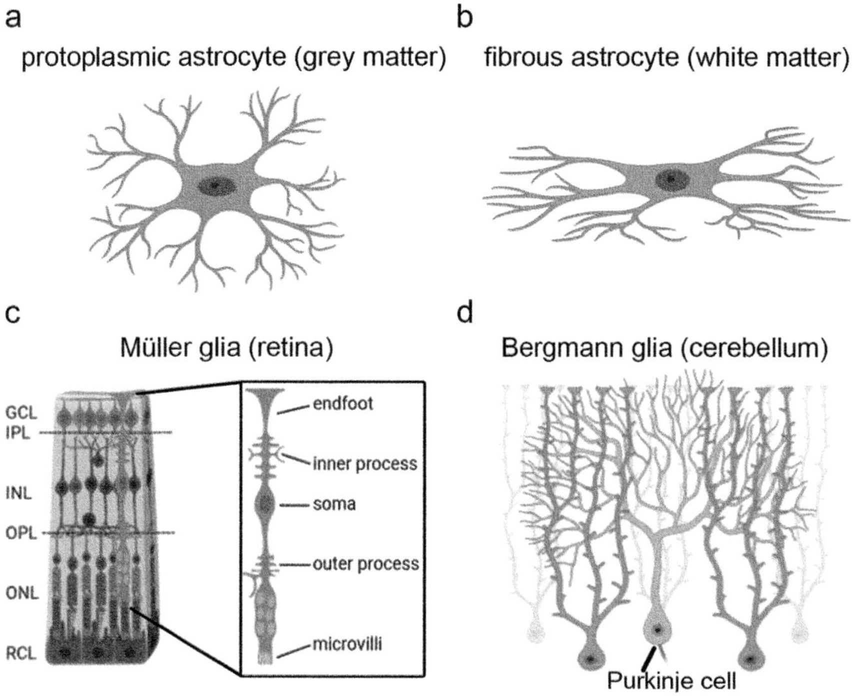

Fig. 1 Astroglia heterogeneity. Astroglia can be grouped in morphologically different subtypes: (**a**) protoplasmic astrocytes of the gray matter; (**b**) fibrous astrocytes of the white matter; (**c**) Müller glia, specific of the retina; and (**d**) Bergmann glia, typical of the cerebellum. GCL ganglion cell layer, IPL inner plexiform layer, INL inner nuclear layer, OPL outer plexiform layer, ONL outer nuclear layer, RCL retinal cell layer (pigment epithelium). (Created with BioRender.com)

of its surface composed of processes. Such a highly ramified body allows them to reach numerous synapses, especially in the hippocampus and cerebral cortex, where they are present in high numbers (Reichenbach et al. 2010).

Interestingly, compared to rodent astrocytes, human protoplasmic astrocytes are up to threefold larger in diameter and extend ten times more primary processes coming into contact with 100-fold more synapses than their rodent counterparts (Bushong et al. 2002; Oberheim et al. 2006).

Although human protoplasmic astrocytes are significantly larger and more complex than the rodent counterpart, they are still similarly organized in non-overlapping domains (Sofroniew and Vinters 2010).

The many fine processes of protoplasmic astrocytes also extend to the microvasculature, enwrapping the wall of blood vessels to form the blood-brain barrier (BBB) (Sofroniew and Vinters 2010). In this way, they develop what is called the "neurovascular unit" (NVU), a functionally complex structure juxtaposed between endothelial cells and neurons (Fig. 2). Moreover, the processes of two adjacent protoplasmic astrocytes occupy non-overlapping domains, with the domain

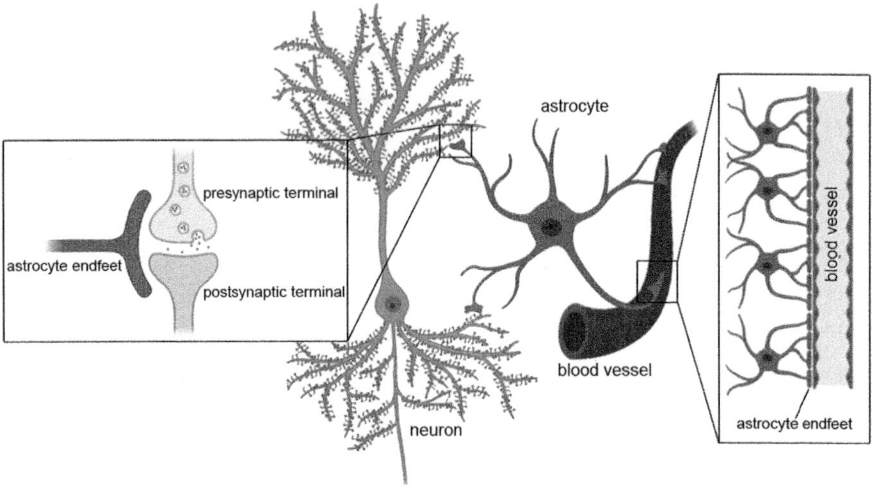

Fig. 2 The neurovascular unit. Astrocyte terminal processes (endfeet) interact on the one side with brain blood vessels to form the blood-brain barrier and on the other side enwrap synaptic terminals to modulate neurotransmission. (Created with BioRender.com)

of any single astrocyte coordinating specific clusters of synapses (Bushong et al. 2002; Halassa et al. 2007).

In contrast to the protoplasmic, the fibrous astrocytes populate the white matter, the optic nerve, and the retinal nerve fiber layer of mammals. They are morphologically characterized by elongated cell bodies with distinguishable processes, which give rise to long, thin, and less branched secondary/tertiary processes. In addition, they are longitudinally oriented along the plane of the fiber bundles (Miller and Raff 1984). The processes of fibrous astrocytes are typically longer then those of protoplasmic astrocytes. Additionally, the processes of protoplasmic astrocytes envelop synapses and blood vessels, whereas the processes of fibrous astrocytes are in contact with the nodes of Ranvier (Peters et al. 1991). This differentiation becomes relevant when thinking about the distinct functions that various cell types play in either the gray or the white matter. On the one hand, the gray matter's major task is to compute information at neuronal synapses, and for this reason protoplasmic astrocytes' fine processes interact with up to 200,000 synapses in mice and 2 million in humans to modulate synaptic neurotransmission (Oberheim et al. 2006; Sofroniew and Vinters 2010) (Fig. 2). On the other hand, information in the white matter is mainly transmitted along the myelinated axons, with the support of their functionally specialized units, the nodes of Ranvier; thus, fibrous astrocyte processes line up with the axons and connect to these units to modulate long-range axonal neurotransmission. Above all, many of these morphological differences are then further determined by the properties and functions of a given brain region.

In addition to these basic cell types, some specialized astrocytes in different areas of the brain, such as Müller cells and Bergmann glial cells, have been described (Fig. 1c, d).

Müller cells are the principal radial glia of the vertebrate retina (Bringmann et al. 2006). They are the only cells representing the macroglia family in the retina. The processes of Müller cells contact the extracellular clefts and virtually every neuronal and non-neuronal element, such as blood vessels. These specialized radial glial cells form an architectural supporting structure, which extends across the entire thickness of the neural retina, from the vitread surface to the subretinal space. Müller cell bodies are localized in the central nuclear layers of the retina. Their thick and thin processes project to both outer and inner limiting membrane. Interestingly, in the plexiform layers, Müller cell processes appear like those of protoplasmic astrocytes (Bringmann et al. 2006; Reichenbach and Bringmann 2013). Each of these cells ensheathes and supports a columnar "micro-unit" of retinal neurons. The size and shape of Müller cells can vary according to animal species and retinal topography.

Bergmann glia are radially oriented, unipolar astrocytes spanning through the entire thickness of the molecular layer of the cerebellum in all vertebrates (Reichenbach et al. 2010).Their cell bodies reside in between the Purkinje cell somata and extend their processes transversely across the molecular layer to terminate in the pia mater with their endfeet, without reaching the ventricular surface (Yamada and Watanabe 2002). In the pia mater, Bergmann processes are aligned parallel to the long axis of the folium. The processes of Bergmann glial cells display elaborated morphological and functional interactions with the dendritic synapses of the Purkinje cells. Indeed, longitudinal columns of Bergmann glial cells cover the dendritic arbors of each Purkinje cell. The soma of these glial cells are localized between those of Purkinje neurons and their process endfeet in the pia matter. The length of these processes is determined by the thickness of the molecular layer. Interestingly, the thickness of this layer diversifies greatly between different species. In small species the Bergmann glial cell processes are short, whereas in large species they are much longer but less dense (Grosche et al. 2002). It is estimated that on average there are about 8.1 Bergmann glial cells to every Purkinje neuron in the rodent cerebellum. Because of this intrinsic relationship, Bergmann glial cells are involved in the regulation and support of the Purkinje structure and function in the adult cerebellum (Palay and Chan-Palay 1974; Reichenbach et al. 1995).

Furthermore, there are more subtypes of cortical astrocytes exclusively found in human and nonhuman primates, such as the interlaminar astrocytes (ILAs) residing in the layer I and projecting long processes toward multiple laminae, terminating in layer II/IV, and varicose projection astrocytes that reside in layers V–VI (Colombo et al. 1997; Falcone et al. 2019; Oberheim et al. 2009). Interlaminar astrocytes are particularly packed in layer I and display oblong cell bodies with their processes organized in a columnar manner through the cortical layers.

Even though they are primate-specific cells, interlaminar human astrocytes are not only larger in number, but they also exhibit mild morphological differences with respect to their chimpanzee counterparts (Colombo et al. 1997; Oberheim et al. 2009).

Accordingly, human interlaminar astrocytes are characterized by spheroid cell bodies and several short processes that contribute to the pial glial limitans, creating a thick network of fibers. They extend one or two processes from layer I, resulting in numerous and tortuous millimeter-long processes. Electron microscopy analysis revealed that these endings possess a multilamellar structure and are enriched in mitochondria, suggesting a high metabolic activity (Colombo et al. 1997). The exact function of these cells remains obscure, but their morphological characteristics suggest that they might provide a network for long-distance calcium propagation (Oberheim et al. 2009).

Varicose projection astrocytes, observed in humans and chimpanzee only, are characterized by main processes shorter and straighter than those exhibited by typical protoplasmic astrocytes. Additionally, they frequently extend one to five long, less branched processes of up to 1 millimeter in length, which terminate in the neuropil or on blood vessels. Human varicose projection astrocytes are more complicated and larger than those described in chimpanzee. The processes of varicose projection astrocytes travel in all directions, frequently penetrating the domains of neighboring astrocytes (Colombo et al. 1995; Oberheim et al. 2009). Since varicose projection astrocytes are found only in human and higher-order primates, it is speculated that they might be involved in specific cognitive functions, even though their precise role is still unknown.

In conclusion, astrocytes exhibit a high regional and microenvironmental diversity. However, unlike neurons, it is still a lot more challenging to isolate diverse regional and/or astrocyte subclasses, as astrocytes homogeneously occupy the CNS microenvironment.

Astrocyte Molecular Heterogeneity

Different regions of the adult brain contain highly morphologically heterogeneous classes of astrocytes that define the CNS neuroanatomy (Emsley and Macklis 2006).

Their morphological characteristics and relationship with both neurons and blood vessels vastly influence brain functions in important, region-specific ways. Thus, astrocyte populations highly likely display additional unique molecular signatures, which may correlate with their morphological features. However, while studies on neurons and oligodendrocytes have been smoothed by the existence of subtype-specific markers, astrocyte corresponding molecular signature has not yet been defined. Nevertheless, the glial fibrillary acidic protein (GFAP), a member of the cytoskeletal protein family and thought to be important for modulation of astrocyte morphology and motility, is classically used to discriminate and detect astrocytes within the CNS (Eng 1985).

Even though GFAP has been widely used as a standard marker for astrocytes, several studies have reported the existence of astrocytes with undetectable levels of

GFAP. Additionally, not all cells in the CNS that express GFAP are astrocytes (Kimelberg 2004; Mishima and Hirase 2010). As an example, layer III and IV astrocytes are devoid of GFAP, although astrocytes appear homogeneously distributed in these regions (Gabbott and Stewart 1987). In accordance, fibrous astrocytes of the white matter express high levels of GFAP, whereas protoplasmic astrocytes of gray matter normally express little or no GFAP, except for few locations, such as the CA1 region of the hippocampus. While a number of proteins have been reported to be selectively expressed by astrocytes, not all of them entirely label all and only astrocytes, but are restricted to either their subcompartments or subgroups or are present in other cell types, too. For example, aquaporin 4 (AQP4), a water channel protein involved in maintaining water and ion homeostasis associated with neuronal activity and highly specific for astrocytes, is localized primarily to astrocytes endfeet; connexin 43 (Cx43), a transmembrane protein that forms gap junction channels and hemichannels, only labels some astrocytes and is also expressed in endothelial cells; finally, S100 calcium-binding protein ß (S100ß) involved in cell cycle regulation and cytoskeleton modification, classically thought to be principally expressed by astrocytes, is not found in all astrocytes. Indeed, S100ß is only expressed by a subgroup of mature astrocytes that ensheath blood vessels and by immature and mature oligodendrocytes (Hachem et al. 2005).

Other astrocyte markers commonly used to label astrocytes are: glutamate transporter GLT-1 (or human EAAT2), expressed in developing astrocyte precursors (Rothstein et al. 1994; Shibata et al. 1997); the glutamine synthase (GS), an enzyme that catalyzes the conversion of ammonia and glutamine to glutamate, which is considered highly specific, although it is also found in oligodendrocytes of the white and gray matter (D'Amelio et al. 1990); and the inwardly rectifying K^+ channels like Kir4.1, predominantly confined to the membrane of astrocyte processes surrounding synapses and blood vessels, thus suggesting that Kir channels can carry both K^+-uptake current in astrocytes at synaptic sites and K^+-extrusion current to the blood vessels (Higashi et al. 2001).

Moreover, the major understandings of astrocyte heterogeneity come from transcriptome wide sequencing experiments to reveal the cell molecular machinery. For instance, a seminal study from Doyle and colleagues examined the transcriptome heterogeneity of isolated astrocytes from different brain regions via TRAP (Translating Ribosome Affinity Purification) technique and showed that astrocytes form the cortex and the cerebellum display considerable different gene clustering sets (Doyle et al. 2008).

Similarly, in another study, Morel et al. profiled astroglial ribosome-associated mRNA in the cortex, hippocampus, nucleus accumbens, caudate putamen, thalamus, and hypothalamus. They showed that gene astroglial expression pattern was highly overlapping between cortex and hippocampus, while the nucleus accumbens, caudate putamen, thalamus, and hypothalamus exhibited a unique mRNA expression set not identified in astrocytes from other brain regions (Morel et al. 2017). As an example, Hevin, a pro-synaptogenic protein, and its mRNA were expressed at comparable levels across all regions. However, Sparc, which antagonizes the synaptogenic functions of Hevin, was enriched in thalamic and hypothalamic astrocytes. This might suggest that cortical or subcortical astrocytes differently promote synaptic development, following a gradient of astrocyte gene expression along the dorsoventral axis.

In addition, isolation of astrocytes from the cortical layers via intersectional labeling approaches has disclosed differences in gene expression between two groups of astrocytes, including MerTK, which mediates synapse elimination, and Sparc, highly expressed in upper-layer astrocytes and deeper-layer astrocytes, along with Chrdl1 expressed at a higher level in upper-layer astrocytes than deeper-layer astrocytes (Bayraktar et al. 2020; Lanjakornsiripan et al. 2018).

Therefore, the majority of evidences point towards a complex picture of astrocytes also on a gene and protein level, which reflect their morphological heterogeneity, even within a single brain region. Astrocytes may share common genes, but then again variation in transcriptomes produces distinctive astrocyte subtypes, specialized in performing different functions.

Current astrocyte markers are either not homogeneously expressed in all astrocytes or do not entirely label the astrocyte soma and processes, hence, the importance of identifying the most appropriate combination of astrocyte markers in different brain regions in order to better guide researchers in exploring astrocyte functions.

Morphological Changes in Neuropsychiatric Disorders

Major Depressive Disorder

As mentioned before, astrocytes have different functions during development and in the adult CNS, in order to facilitate the formation of neural networks and maintain brain homeostasis. They act by providing nutrients and neurotrophic factors, by regulating ion and neurotransmitter homeostasis and by removing waste metabolites (Parpura et al. 2012). They are closely associated with neuronal synapses, forming the so-called tripartite synapse (Araque et al. 1999). With the endfeet of their processes, which wrap around synapses, they regulate the formation and refinement of neural circuits through complex events such as synaptogenesis, synaptic maturation, and pruning (Roman et al., 2020) (Bosworth and Allen 2017).

Most of the morphological studies in astrocytes have used immunohistochemistry techniques to label them with GFAP, a well-established marker for astrocytes, which make their identification relatively easy, although with the limitations described before (Williams et al. 2014). Among other methods, the Golgi staining and cresyl violet have been extensively used in neuroanatomical studies to show glial morphology (Chana et al. 2003; Torres-Platas et al. 2011).

Changes in astrocyte morphology, density, protein content, and gene expression have been described in a broad spectrum of neuropsychiatric disorders (Di Benedetto and Rupprecht 2013; Zhou et al. 2019). Findings from postmortem human brain tissues have highlighted changes in glial cell morphology in mood disorders like major depressive disorder (MDD). Specifically, morphometric analysis revealed bigger glial cell nuclei, suggesting an enlargement of the glial cells which were located in the dorsolateral prefrontal cortex (dlPFC) of depressive patients (Rajkowska et al. 1999). In particular, 3-D reconstruction of gray and white matter astrocytes using Golgi staining revealed that fibrous astrocytes of the anterior cingulate cortex had bigger cell bodies, were more ramified, and presented longer lengths of their processes in depressed subjects who committed suicide, when compared to matched controls (Chana et al. 2003; Torres-Platas et al. 2011). Other evidences also suggested that changes occur in astrocyte volume in postmortem brain tissue of MDD patients in an age-dependent way, with older depressed patients (46–86 years old) showing larger astrocytes, when compared to their healthy age-matched controls, and younger depressed patients (30–45 years old) exhibiting smaller astrocyte areal fraction and packing densities when compared to their controls (Miguel-Hidalgo et al. 2000). However, no changes in glia clustering were found in the prefrontal area of postmortem brain tissue of MDD subjects, besides the reductions in glial cell densities (Cotter et al. 2002).

Astrocytes also play a role in establishing and modulating the integrity and functionality of the BBB, with astrocyte endfeet representing a major structural component of the BBB. In MDD, a decreased coverage of blood vessel walls by astrocytic endfeet positive for AQP4 ($AQP4^+$) was first described in 2013 (Rajkowska and Stockmeier 2013). The reduction reached a 50% level and was specifically restricted to the orbitofrontal gray matter of subjects with MDD, as compared to nonpsychiatric control subjects, but not in other brain areas. However, this effect was not accompanied by a similar difference when quantifying $GFAP^+$ astrocyte processes, suggesting a deficit related to AQP4 localization more than a simple structural impairment. Indeed, a functional impairment of the BBB, characterized by increased permeability, has been described in MDD, thereby supporting the idea that morphofunctional changes in astrocytes might strongly contribute to BBB dysfunctions (Rajkowska and Stockmeier 2013). In a similar way, Di Benedetto and colleagues confirmed a 60% reduction in coverage of blood vessels by AQP4 immunoreactive astrocyte endfeet in the adult prefrontal cortex (PFC) of high anxiety-like behavior rats (HAB), rats with an endogenous depressive-like phenotype, similar to the findings in MDD patients (Di Benedetto et al. 2016). Ex vivo and in vitro analysis additionally revealed that HAB astrocytes extend less processes than their control counterpart cells (Di Benedetto et al. 2016; Malik et al. 2020). Moreover, slightly different than in the human situation, GFAP staining additionally showed a more generalized lower morphological complexity of HAB astrocytes, when compared to control cells (nonselectively bred for anxiety behavior, NAB). Finally, antidepressant treatment with the commonly used antidepressant drug (AD) fluoxetine (FLX) increased the number of processes in healthy astrocytes, indicating this morphological change as a FLX pharmacological target. In addition, FLX restored the basal total number of processes per cell in HAB astrocytes, reversing the disease phenotype and suggesting such morphological changes as putative therapeutic targets of ADs (Fig. 3). However, the results showed that that

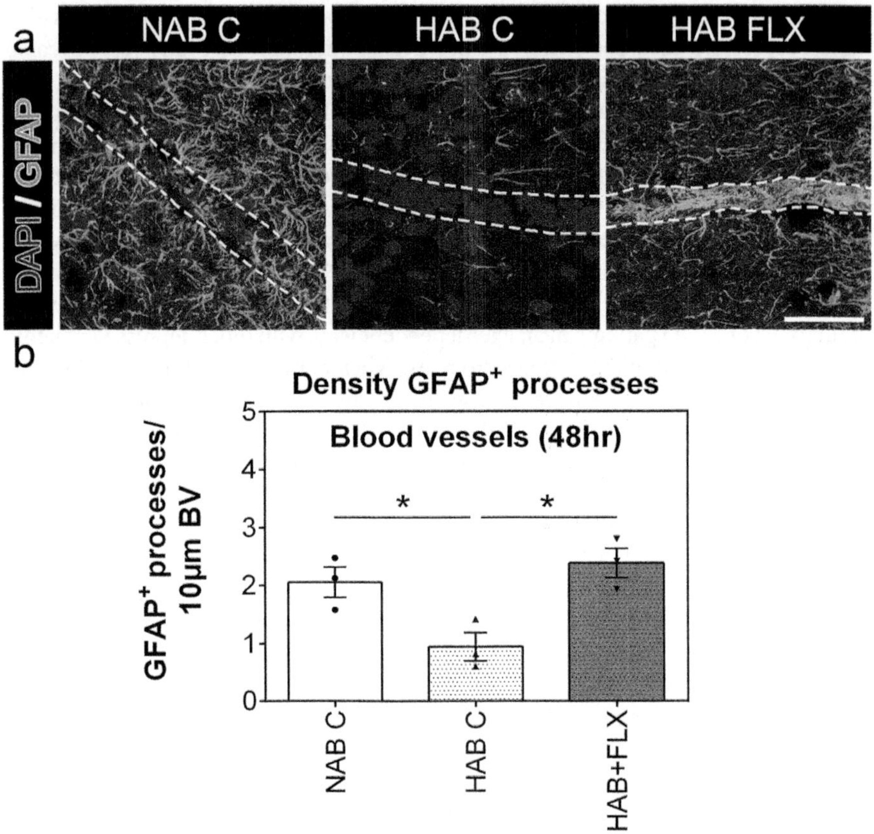

Fig. 3 Astroglia in MDD. (**a**) Representative photomicrographs of GFAP-stained brain slices from the PFC of NAB and HAB control (C) rats and HAB rats upon FLX treatment (HAB FLX). Stainings show a strong atrophy of $GFAP^+$ astrocytes in HAB brains, which could be rescued by short-term FLX treatment; (**b**) graph shows the quantification of $GFAP^+$ processes/10μm blood vessel (BV) and highlights the significant differences between NAB C/HAB C/HAB FLX. Data are presented as mean ± SEM, $^*p \leq 0.05$. Scale bar = 50μm

the short-term treatment was not sufficient to relocate AQP4 to the astrocytic endfeet and rescue the coverage of blood vessels by AQP4+ processes, thus suggesting that long-term treatments might be necessary to reach a functional rescue, whereas short times may be sufficient to improve the structural deficits (Di Benedetto et al. 2016).

Our results about the role of astrocytes in mediating AD effects (Di Benedetto et al. 2013; Di Benedetto et al. 2016; Di Benedetto and Rupprecht 2013; Tanasic et al. 2016) become even more relevant when thinking about the ability of stress, a known trigger of neuropsychiatric disorders, to induce structural remodeling of astrocytes (Menard et al. 2017). Several findings from animal models have consistently showed that astrocytes change their morphology and physiological characteristics after they are exposed to particular psychological stressors. 3-D reconstructions of astrocytes using high-resolution imaging methodologies highlighted profound changes in three main morphological parameters in GFAP+ astrocytes from the PFC after chronic stress: astrocytes presented, in average, a ~40% reduction of process length, a ~56% decrease in process volume, and a ~58% decline in the number of process branches (Tynan et al. 2013). Interestingly, repeated chronic immobilization stress for 2 h/day for 10 consecutive days decreased the neuropil volume occupied by the astrocytes in the basal amygdala, but not in the CA3 area of the hippocampus. Measurements of neuropil volumes are relevant to map the physical extent of an association between astrocytes and neurons. Therefore, the changes found in the amygdala were interpreted as a reflection of a declined volume occupied by fine astrocytic protrusions, which are functionally interacting with neuronal structures (Naskar and Chattarji 2019).

Furthermore, using a chronic psychosocial stress paradigm for 5 weeks, it was found that this particular stress produced a 25% significant decrease in both the number and somal volume of astroglia in the hippocampus. These changes in astroglia plasticity correlated with the stress-induced hippocampal volume reduction observed in vivo and might account for hippocampal volume changes observed in the brains of patients suffering from neuropsychiatric disorders. Interestingly, FLX treatment prevented the stress-induced numerical decrease of astrocytes, but had no counteracting effect on somal volume shrinkage (Czeh et al. 2006; Czeh and Di Benedetto 2012). A similar finding was also described in C57/Bl6 mice after chronic treatment with corticosterone. Administration of 20 mg/kg of corticosterone once per day produced a significant decrease in number, somal volume, and process length of astrocytes in the hippocampus. This treatment additionally triggered a depressive-like behavior in male mice, accompanied by a decreased hippocampal volume, which may result from the abovementioned morphological changes associated with an impaired astrocyte branching complexity (Zhang et al. 2015).

Chronic mild unpredictable stress (CMUS) paradigm was also used in mice to induce stress-related reactions by using auditory and olfactory stimuli. Astrocytes in the hippocampal subfields were differently affected: those in the molecular layer were the mostly impaired, with a surface area which was significantly reduced in the CMUS-treated group, whereas CA1 and CA3 astrocytes were much less affected (Virmani et al. 2020). Interestingly, the same CMUS paradigm using different stressors, such as restricted space, exposure to overnight illumination, or food and water deprivations, did not significantly change the morphology of astrocytes in any hippocampal area. In any case, chronic treatment with the ADs FLX and imipramine promoted a transient reduction of astrocytic complexity after the 6-week protocol exposure to stress. However, those ADs facilitated restoration of the physiological astrocytic complexity when animals were re-exposed to stress, thus suggesting a protective effect of ADs to further potentially damaging conditions (Machado-Santos et al. 2019).

The examination of animal models of depression, such as the Flinders Sensitive Line (FSL) rats, a rat line selected for an endogenous depressive phenotype, showed a reduction in size of GFAP+ astrocytes and in the branching patterns of astrocytic processes in the stratum radiatum of hippocampal area CA1. Moreover, astrocytes in the molecular layer of the dentate gyrus (DG) were significantly smaller than their corresponding ones in the control group. Administration of ketamine, a drug with fast-acting AD effects, showed rapid and conspicuous rescuing effects on both parameters, the soma size and astrocyte arborization, in CA1 and DG areas (Ardalan et al. 2017).

In the stress-hyperresponsive Fischer rats, an early life isolation stress in pups produced a moderate but consistent reduction in the density of GFAP-immunoreactive astrocytes in areas implicated in both stress-related and depressive-like behaviors, such us the PFC, the CA1 and dorsal DG of the hippocampus, the cingulate cortex, and the amygdala (Leventopoulos et al. 2007). Another early life stress, namely, maternal deprivation, produced consequences in the postnatal (PN) development of rat pups (day PN15), with a reduction in astrocytic density and a shortening of astrocytic processes in the hippocampus, without affecting cell soma (Roque et al. 2016; Saavedra et al. 2017). Interestingly, no changes in astrocyte morphology have been reported related to long-term effects of maternal deprivation (Burke et al. 2013; Gondre-Lewis et al. 2016).

In an animal model of post-traumatic stress disorder (PTSD), induced by inescapable footshock, rats showed notable changes in hippocampal astrocytes, whereas astrocytes in the medial amygdala were left unaffected, when compared to controls. The astrocytic branching and number of primary processes were reduced in the lateral quadrants of the hippocampus in the PTSD group, with lateral quadrants defined as the portions of tissue parallel to the stratum pyramidale where astrocytes extend their processes. The analysis of the orientation index of astrocyte processes, determined by the quotient of the number of lateral intersections of the astrocytic processes divided by the number of their central intersections, indicated that PTSD model alters the polarity of hippocampal astrocytes, inducing a more fusiform shape (Saur et al. 2016).

In Rhesus macaques showing a self-injurious behavior (SIB), a complex neuropsychiatric condition characterized by a spectrum of abnormal neuropsychological and locomotor behaviors, astrocytic cell bodies in the white matter presented an atrophy, with significantly smaller cell bodies. Moreover, the total volume of astrocytic processes was significantly decreased in both white and gray matter astrocytes in the SIB group. Specifically, the astrocyte complexity was decreased in frontal cortical areas of SIB animals, with reduced numbers of bifurcations and process end points within the first 30μm radii around the cell body in both white and gray matter astrocytes. Finally, the total length of the cell arbor observed in white matter astrocytes was significantly decreased compared to controls. Sholl analysis is the most commonly used method in the previous studies. This type of analysis has been used to examine astrocyte and neuron arborizations. This method is appropriate to analyze the cell complexity, by quantifying the number of process branches and intersections and giving as a result branching profiles of dye-filled cells (Naskar and Chattarji 2019).

Schizophrenia and Bipolar Disorder

Interestingly, bigger nuclear sizes of glial cell seem to be specific of depressive disorders, with no change in nuclear glial sizes reported in schizophrenia (Rajkowska and Miguel-Hidalgo 2007). In one study, schizophrenic brains only present a trend increase in mean glial size in occipital areas 9 and 17 (Rajkowska et al. 1998). Another study in humans used electron microscopy to reveal ultrastructural abnormalities of astrocytes in the CA3 region of the hippocampus in young (<50 years old) and old (>50 years old) schizophrenic cases, with lower values of total area covered by the single astrocyte (Kolomeets and Uranova 2010).

A subgroup of patients with schizophrenia, which was included in the "high inflammatory" subgroup, was characterized by a prevalent occurrence of hypertrophic astrocytes with an associated higher expression of inflammatory markers, similar to results observed in rodent model of neuroinflammation (Catts et al. 2014).

In the white matter of brains from people suffering from schizophrenia and bipolar disorder, GFAP-labeled astrocytes exhibited significantly altered spatial cell distribution. Astrocytes were more clustered in the psychiatric groups than in the control group, suggesting that a disruption of a structural or metabolic support of axons might be among the triggers of the disease. This altered spatial distribution might depend on changes in astrocyte morphology, providing further evidence for the morphological abnormalities found in the white matter of schizophrenic patients (Hercher et al. 2014).

Data from the analysis of astrocytes in the subgenual cingulate cortex of schizophrenia patients have identified two types of astrocytes based on their morphology: fibrillary astrocytes, which were distinguished by multiple visibly stained processes, and gemistocytic astrocytes which presented larger cell bodies and either one or no stained processes. The study revealed an overall decrease in fibrillary astrocyte density in the cingulate white matter of patients, while gemistocytic astrocytes remained unaltered (Williams et al. 2014).

A study using glial progenitor cells derived from schizophrenic patients revealed alterations in the astrocytic phenotype, with specific deficits in both phenotypic maturation and differentiation processes. In this study glial progenitor cells (GPCs) produced from induced pluripotent stem cells (iPSCs) derived from patients with childhood-onset schizophrenia were further implanted into mice. The schizophrenia glial chimeras also showed delayed astrocytic differentiation and abnormal astrocytic morphologies. Sholl analysis of those chimeras further revealed that astrocytes derived from people suffering from schizophrenia had fewer primary processes, less proximal branching, longer distal fibers, and less coherent domain structure compared to the control-derived counterparts (Windrem et al. 2017). RNAseq also revealed a disrupted expression of genes associated with glial differentiation and synaptic development. For instance, the levels of neuroligins (NL)1, NL2, and NL3 have been found to be reduced in glial progenitor cells from schizophrenic patients when compared to controls (Windrem et al. 2017). Astrocytic

NL2 has been shown to control the balance of excitatory and inhibitory synaptic connections, as well as astrocytic NL1-3 has been seen to guide the establishment of astrocyte morphology. Silencing studies in transfecting astrocytes with short hairpin RNAs (shRNAs) against NL1-3 revealed that the neuron-induced astrocyte complexity in vitro was completely blocked. Those astrocytes were not able to develop the main branches during development at P7 and showed a delayed elaboration of their thinner processes into the neuropil by P21 with respect to control transfected astrocytes. Specifically, the knockdown of NL1 markedly reduced astrocytic neuropil infiltration volume at P7, which was reversed only later in development by P21. NL2 knockdown was much more severe, with a restricted volume infiltration at both time points. Conversely, silencing NL3 severely affected astrocyte growth only at P21. Moreover, silencing each individual astrocytic NL impaired astrocyte morphology by partially diminishing astrocyte arborization (Stogsdill et al. 2017).

Less is known, instead, about astrocytes in bipolar disorder. Histological studies of brains derived from people suffering from bipolar disorder have revealed alterations in the shape of the glial nucleus to a less rounded conformation in the dorsolateral prefrontal area 9 as well as a glial enlargement in layers I and III (Rajkowska et al. 2001).

Acknowledgments This work was supported by intramural funding from the University of Regensburg and by the German Research Council (DFG-GRK2174 "Neurobiology of Emotion Dysfunction" (SP1)) to BDB. The sponsors did not have any role in writing the report and in the decision to submit the article for publication.

CR and EV contributed to literature screening and drafting the manuscript. BDB contributed to editing and revising the final version of the manuscript. All authors read and approved the final version of the manuscript.

The authors declare that the research was conducted in the absence of any commercial or financial relationships that could be construed as a potential competing interest.

References

Araque A, Parpura V, Sanzgiri RP, Haydon PG (1999) Tripartite synapses: glia, the unacknowledged partner. Trends Neurosci 22(5):208–215

Ardalan M, Rafati AH, Nyengaard JR, Wegener G (2017) Rapid antidepressant effect of ketamine correlates with astroglial plasticity in the hippocampus. Br J Pharmacol 174(6):483–492. https://doi.org/10.1111/bph.13714

Bayraktar OA, Bartels T, Holmqvist S, Kleshchevnikov V, Martirosyan A, Polioudakis D et al (2020) Astrocyte layers in the mammalian cerebral cortex revealed by a single-cell in situ transcriptomic map. Nat Neurosci 23(4):500–509. https://doi.org/10.1038/s41593-020-0602-1

Bosworth AP, Allen NJ (2017) The diverse actions of astrocytes during synaptic development. Curr Opin Neurobiol 47:38–43. https://doi.org/10.1016/j.conb.2017.08.017

Bringmann A, Pannicke T, Grosche J, Francke M, Wiedemann P, Skatchkov SN et al (2006) Muller cells in the healthy and diseased retina. Prog Retin Eye Res 25(4):397–424. https://doi.org/10.1016/j.preteyeres.2006.05.003

Burke NN, Llorente R, Marco EM, Tong K, Finn DP, Viveros MP et al (2013) Maternal deprivation is associated with sex-dependent alterations in nociceptive behavior and neuroinflammatory mediators in the rat following peripheral nerve injury. J Pain 14(10):1173–1184. https://doi.org/10.1016/j.jpain.2013.05.003

Bushong EA, Martone ME, Jones YZ, Ellisman MH (2002) Protoplasmic astrocytes in CA1 stratum radiatum occupy separate anatomical domains. J Neurosci 22(1):183–192

Catts VS, Wong J, Fillman SG, Fung SJ, Shannon WC (2014) Increased expression of astrocyte markers in schizophrenia: association with neuroinflammation. Aust N Z J Psychiatry 48(8):722–734. https://doi.org/10.1177/0004867414531078

Chana G, Landau S, Beasley C, Everall IP, Cotter D (2003) Two-dimensional assessment of cytoarchitecture in the anterior cingulate cortex in major depressive disorder, bipolar disorder, and schizophrenia: evidence for decreased neuronal somal size and increased neuronal density. Biol Psychiatry 53(12):1086–1098. https://doi.org/10.1016/s0006-3223(03)00114-8

Colombo JA, Yanez A, Puissant V, Lipina S (1995) Long, interlaminar astroglial cell processes in the cortex of adult monkeys. J Neurosci Res 40(4):551–556. https://doi.org/10.1002/jnr.490400414

Colombo JA, Gayol S, Yanez A, Marco P (1997) Immunocytochemical and electron microscope observations on astroglial interlaminar processes in the primate neocortex. J Neurosci Res 48(4):352–357

Cotter D, Mackay D, Chana G, Beasley C, Landau S, Everall IP (2002) Reduced neuronal size and glial cell density in area 9 of the dorsolateral prefrontal cortex in subjects with major depressive disorder. Cereb Cortex 12(4):386–394. https://doi.org/10.1093/cercor/12.4.386

Czeh B, Di Benedetto B (2012) Antidepressants act directly on astrocytes: evidences and functional consequences. Eur Neuropsychopharmacol. https://doi.org/10.1016/j.euroneuro.2012.04.017

Czeh B, Simon M, Schmelting B, Hiemke C, Fuchs E (2006) Astroglial plasticity in the hippocampus is affected by chronic psychosocial stress and concomitant fluoxetine treatment. Neuropsychopharmacology 31(8):1616–1626. https://doi.org/10.1038/sj.npp.1300982

D'Amelio F, Eng LF, Gibbs MA (1990) Glutamine synthetase immunoreactivity is present in oligodendroglia of various regions of the central nervous system. Glia 3(5):335–341. https://doi.org/10.1002/glia.440030504

De Filipe J (2010) Cajal's butterflies of the soul: science and art

Di Benedetto B, Rupprecht R (2013) Targeting glia cells: novel perspectives for the treatment of neuropsychiatric diseases. Curr Neuropharmacol 11(2):171–185. https://doi.org/10.2174/1570159X11311020004

Di Benedetto B, Radecke J, Schmidt MV, Rupprecht R (2013) Acute antidepressant treatment differently modulates ERK/MAPK activation in neurons and astrocytes of the adult mouse prefrontal cortex. Neuroscience 232:161–168. https://doi.org/10.1016/j.neuroscience.2012.11.061

Di Benedetto B, Malik VA, Begum S, Jablonowski L, Gómez-González GB, Neumann ID et al (2016) Fluoxetine requires the endfeet protein aquaporin-4 to enhance plasticity of astrocyte processes. Front Cell Neurosci 10(8):62–73. https://doi.org/10.3389/fncel.2016.00008

Doyle JP, Dougherty JD, Heiman M, Schmidt EF, Stevens TR, Ma G et al (2008) Application of a translational profiling approach for the comparative analysis of CNS cell types. Cell 135(4):749–762. https://doi.org/10.1016/j.cell.2008.10.029

Emsley JG, Macklis JD (2006) Astroglial heterogeneity closely reflects the neuronal-defined anatomy of the adult murine CNS. Neuron Glia Biol 2(3):175–186. https://doi.org/10.1017/S1740925X06000202

Eng LF (1985) Glial fibrillary acidic protein (GFAP): the major protein of glial intermediate filaments in differentiated astrocytes. J Neuroimmunol 8(4-6):203–214. https://doi.org/10.1016/s0165-5728(85)80063-1

Falcone C, Wolf-Ochoa M, Amina S, Hong T, Vakilzadeh G, Hopkins WD et al (2019) Cortical interlaminar astrocytes across the therian mammal radiation. J Comp Neurol 527(10):1654–1674. https://doi.org/10.1002/cne.24605

Gabbott PL, Stewart MG (1987) Distribution of neurons and glia in the visual cortex (area 17) of the adult albino rat: a quantitative description. Neuroscience 21(3):833–845. https://doi.org/10.1016/0306-4522(87)90040-6

Golgi C (1871) Contribuzione alla fina Anatomia degli organi centrali del sistema nervoso. Tipi Fava e Garagnani

Gondre-Lewis MC, Darius PJ, Wang H, Allard JS (2016) Stereological analyses of reward system nuclei in maternally deprived/separated alcohol drinking rats. J Chem Neuroanat 76(Pt B):122–132. https://doi.org/10.1016/j.jchemneu.2016.02.004

Grosche J, Kettenmann H, Reichenbach A (2002) Bergmann glial cells form distinct morphological structures to interact with cerebellar neurons. J Neurosci Res 68(2):138–149. https://doi.org/10.1002/jnr.10197

Hachem S, Aguirre A, Vives V, Marks A, Gallo V, Legraverend C (2005) Spatial and temporal expression of S100B in cells of oligodendrocyte lineage. Glia 51(2):81–97. https://doi.org/10.1002/glia.20184

Halassa MM, Fellin T, Takano H, Dong JH, Haydon PG (2007) Synaptic islands defined by the territory of a single astrocyte. J Neurosci 27(24):6473–6477. https://doi.org/10.1523/JNEUROSCI.1419-07.2007

Hercher C, Chopra V, Beasley CL (2014) Evidence for morphological alterations in prefrontal white matter glia in schizophrenia and bipolar disorder. J Psychiatry Neurosci 39(6):376–385. https://doi.org/10.1503/jpn.130277

Herculano-Houzel S (2014) The glia/neuron ratio: how it varies uniformly across brain structures and species and what that means for brain physiology and evolution. Glia 62(9):1377–1391. https://doi.org/10.1002/glia.22683

Higashi K, Fujita A, Inanobe A, Tanemoto M, Doi K, Kubo T et al (2001) An inwardly rectifying K(+) channel, Kir4.1, expressed in astrocytes surrounds synapses and blood vessels in brain. Am J Physiol Cell Physiol 281(3):C922–C931. https://doi.org/10.1152/ajpcell.2001.281.3.C922

John Lin CC, Yu K, Hatcher A, Huang TW, Lee HK, Carlson J et al (2017) Identification of diverse astrocyte populations and their malignant analogs. Nat Neurosci 20(3):396–405. https://doi.org/10.1038/nn.4493

Kettenmann H, Verkhratsky A (2008) Neuroglia: the 150 years after. Trends Neurosci 31(12):653–659. https://doi.org/10.1016/j.tins.2008.09.003

Kimelberg HK (2004) The problem of astrocyte identity. Neurochem Int 45(2-3):191–202. https://doi.org/10.1016/j.neuint.2003.08.015

Kohler S, Winkler U, Hirrlinger J (2019) Heterogeneity of astrocytes in grey and white matter. Neurochem Res 46(1):3–14. https://doi.org/10.1007/s11064-019-02926-x

Kolomeets NS, Uranova N (2010) Ultrastructural abnormalities of astrocytes in the hippocampus in schizophrenia and duration of illness: a postortem morphometric study. World J Biol Psychiatry 11(2 Pt 2):282–292. https://doi.org/10.1080/15622970902806124

Lanjakornsiripan D, Pior BJ, Kawaguchi D, Furutachi S, Tahara T, Katsuyama Y et al (2018) Layer-specific morphological and molecular differences in neocortical astrocytes and their dependence on neuronal layers. Nat Commun 9(1):1623. https://doi.org/10.1038/s41467-018-03940-3

Lenhossek M (1893) Der feinere Bau des Nervensystems im Lichte neuester Forschung

Leventopoulos M, Ruedi-Bettschen D, Knuesel I, Feldon J, Pryce CR, Opacka-Juffry J (2007) Long-term effects of early life deprivation on brain glia in Fischer rats. Brain Res 1142:119–126. https://doi.org/10.1016/j.brainres.2007.01.039

Liddelow SA, Guttenplan KA, Clarke LE, Bennett FC, Bohlen CJ, Schirmer L et al (2017) Neurotoxic reactive astrocytes are induced by activated microglia. Nature 541(7638):481–487. https://doi.org/10.1038/nature21029

Machado-Santos AR, Alves ND, Araujo B, Correia JS, Patricio P, Mateus-Pinheiro A et al (2019) Astrocytic plasticity at the dorsal dentate gyrus on an animal model of recurrent depression. Neuroscience. https://doi.org/10.1016/j.neuroscience.2019.10.032

Malik VA, Zajicek F, Mittmann LA, Klaus J, Unterseer S, Rajkumar S et al (2020) GDF15 promotes simultaneous astrocyte remodeling and tight junction strengthening at the blood-brain barrier. J Neurosci Res 98(7):1433–1456. https://doi.org/10.1002/jnr.24611
Menard C, Pfau ML, Hodes GE, Kana V, Wang VX, Bouchard S et al (2017) Social stress induces neurovascular pathology promoting depression. Nat Neurosci 20(12):1752–1760. https://doi.org/10.1038/s41593-017-0010-3
Miguel-Hidalgo JJ, Baucom C, Dilley G, Overholser JC, Meltzer HY, Stockmeier CA et al (2000) Glial fibrillary acidic protein immunoreactivity in the prefrontal cortex distinguishes younger from older adults in major depressive disorder. Biol Psychiatry 48(8):861–873. https://doi.org/10.1016/s0006-3223(00)00999-9
Miller RH, Raff MC (1984) Fibrous and protoplasmic astrocytes are biochemically and developmentally distinct. J Neurosci 4(2):585–592
Mishima T, Hirase H (2010) In vivo intracellular recording suggests that gray matter astrocytes in mature cerebral cortex and hippocampus are electrophysiologically homogeneous. J Neurosci 30(8):3093–3100. https://doi.org/10.1523/JNEUROSCI.5065-09.2010
Morel L, Chiang MSR, Higashimori H, Shoneye T, Iyer LK, Yelick J et al (2017) Molecular and functional properties of regional astrocytes in the adult brain. J Neurosci 37(36):8706–8717. https://doi.org/10.1523/JNEUROSCI.3956-16.2017
Naskar S, Chattarji S (2019) Stress elicits contrasting effects on the structure and number of astrocytes in the amygdala versus hippocampus. eNeuro 6(1). https://doi.org/10.1523/ENEURO.0338-18.2019.
Oberheim NA, Wang X, Goldman S, Nedergaard M (2006) Astrocytic complexity distinguishes the human brain. Trends Neurosci 29(10):547–553. https://doi.org/10.1016/j.tins.2006.08.004
Oberheim NA, Takano T, Han X, He W, Lin JH, Wang F et al (2009) Uniquely hominid features of adult human astrocytes. J Neurosci 29(10):3276–3287. https://doi.org/10.1523/JNEUROSCI.4707-08.2009
Palay SL, Chan-Palay V (1974) Cerebellar cortex. Cytology and organization
Parpura V, Heneka MT, Montana V, Oliet SH, Schousboe A, Haydon PG et al (2012) Glial cells in (patho) physiology. J Neurochem 121(1):4–27. https://doi.org/10.1111/j.1471-4159.2012.07664.x
Peters A, Palay SL, Webster HD (1991) The fine structure of the nervous system
Rajkowska G, Miguel-Hidalgo JJ (2007) Gliogenesis and glial pathology in depression. CNS Neurol Disord Drug Targets 6(3):219–233
Rajkowska G, Stockmeier CA (2013) Astrocyte pathology in major depressive disorder: insights from human postmortem brain tissue. Curr Drug Targets 14(11):1225–1236
Rajkowska G, Selemon LD, Goldman-Rakic PS (1998) Neuronal and glial somal size in the prefrontal cortex: a postmortem morphometric study of schizophrenia and Huntington disease. Arch Gen Psychiatry 55(3):215–224. https://doi.org/10.1001/archpsyc.55.3.215
Rajkowska G, Miguel-Hidalgo JJ, Wei J, Dilley G, Pittman SD, Meltzer HY et al (1999) Morphometric evidence for neuronal and glial prefrontal cell pathology in major depression. Biol Psychiatry 45(9):1085–1098
Rajkowska G, Halaris A, Selemon LD (2001) Reductions in neuronal and glial density characterize the dorsolateral prefrontal cortex in bipolar disorder. Biol Psychiatry 49(9):741–752. https://doi.org/10.1016/s0006-3223(01)01080-0
Reichenbach A, Bringmann A (2013) New functions of Muller cells. Glia 61(5):651–678. https://doi.org/10.1002/glia.22477
Reichenbach A, Siegel A, Rickmann M, Wolff JR, Noone D, Robinson SR (1995) Distribution of Bergmann glial somata and processes: implications for function. J Hirnforschung 36(4):509–517
Reichenbach A, Derouiche A, Kirchhoff F (2010) Morphology and dynamics of perisynaptic glia. Brain Res Rev 63(1-2):11–25. https://doi.org/10.1016/j.brainresrev.2010.02.003
Roque A, Ochoa-Zarzosa A, Torner L (2016) Maternal separation activates microglial cells and induces an inflammatory response in the hippocampus of male rat pups, independently

of hypothalamic and peripheral cytokine levels. Brain Behav Immun 55:39–48. https://doi.org/10.1016/j.bbi.2015.09.017
Rothstein JD, Martin L, Levey AI, Dykes-Hoberg M, Jin L, Wu D et al (1994) Localization of neuronal and glial glutamate transporters. Neuron 13(3):713–725. https://doi.org/10.1016/0896-6273(94)90038-8
Saavedra LM, Fenton Navarro B, Torner L (2017) Early life stress activates glial cells in the hippocampus but attenuates cytokine secretion in response to an immune challenge in rat pups. Neuroimmunomodulation 24(4-5):242–255. https://doi.org/10.1159/000485383
Saur L, Baptista PP, Bagatini PB, Neves LT, de Oliveira RM, Vaz SP et al (2016) Experimental post-traumatic stress disorder decreases astrocyte density and changes astrocytic polarity in the CA1 hippocampus of male rats. Neurochem Res 41(4):892–904. https://doi.org/10.1007/s11064-015-1770-3
Shibata T, Yamada K, Watanabe M, Ikenaka K, Wada K, Tanaka K et al (1997) Glutamate transporter GLAST is expressed in the radial glia-astrocyte lineage of developing mouse spinal cord. J Neurosci 17(23):9212–9219
Sofroniew MV, Vinters HV (2010) Astrocytes: biology and pathology. Acta Neuropathol 119(1):7–35. https://doi.org/10.1007/s00401-009-0619-8
Stogsdill JA, Ramirez J, Liu D, Kim YH, Baldwin KT, Enustun E et al (2017) Astrocytic neuroligins control astrocyte morphogenesis and synaptogenesis. Nature 551(7679):192–197. https://doi.org/10.1038/nature24638
Tanasic S, Mattusch C, Wagner EM, Eder M, Rupprecht R, Rammes G et al (2016) Desipramine targets astrocytes to attenuate synaptic plasticity via modulation of the ephrinA3/EphA4 signalling. Neuropharmacology 105:154–163. https://doi.org/10.1016/j.neuropharm.2016.01.021
Torres-Platas SG, Hercher C, Davoli MA, Maussion G, Labonte B, Turecki G et al (2011) Astrocytic hypertrophy in anterior cingulate white matter of depressed suicides. Neuropsychopharmacology 36(13):2650–2658. https://doi.org/10.1038/npp.2011.154
Tynan RJ, Beynon SB, Hinwood M, Johnson SJ, Nilsson M, Woods JJ et al (2013) Chronic stress-induced disruption of the astrocyte network is driven by structural atrophy and not loss of astrocytes. Acta Neuropathol 126(1):75–91. https://doi.org/10.1007/s00401-013-1102-0
Virchow RLK (1858) Die cellularpathologie in ihrer begründung auf physiologische und pathologische gewebelehre. Good Press
Virmani G, D'almeida P, Nandi A, Marathe S (2020) Subfield-specific effects of chronic mild unpredictable stress on hippocampal astrocytes. bioRxiv. https://doi.org/10.1101/2020.02.07.938472
Williams M, Pearce RK, Hirsch SR, Ansorge O, Thom M, Maier M (2014) Fibrillary astrocytes are decreased in the subgenual cingulate in schizophrenia. Eur Arch Psychiatry Clin Neurosci 264(4):357–362. https://doi.org/10.1007/s00406-013-0482-4
Windrem MS, Osipovitch M, Liu Z, Bates J, Chandler-Militello D, Zou L et al (2017) Human iPSC glial mouse chimeras reveal glial contributions to schizophrenia. Cell Stem Cell 21(2):195–208. e6. https://doi.org/10.1016/j.stem.2017.06.012
Yamada K, Watanabe M (2002) Cytodifferentiation of Bergmann glia and its relationship with Purkinje cells. Anat Sci Int 77(2):94–108. https://doi.org/10.1046/j.0022-7722.2002.00021.x
Zhang H, Zhao Y, Wang Z (2015) Chronic corticosterone exposure reduces hippocampal astrocyte structural plasticity and induces hippocampal atrophy in mice. Neurosci Lett 592:76–81. https://doi.org/10.1016/j.neulet.2015.03.006
Zhou X, Xiao Q, Xie L, Yang F, Wang L, Tu J (2019) Astrocyte, a promising target for mood disorder interventions. Front Mol Neurosci 12:136. https://doi.org/10.3389/fnmol.2019.00136

Part II
Astroglia in Mood Disorders

Astrocytes in Bipolar Disorder

Arthur M. Butt and Andrea D. Rivera

Introduction

Bipolar disorder (BD) is a complex group of neuropsychiatric disorders, typically comprising both manic and depressive episodes, termed bipolar I disorder, whilst manic attacks without depressive episodes are classified as bipolar II disorder (Carvalho et al. 2020). BD affects 1 in 100 people and approximately 45 million people worldwide, with a typical onset around the age of 20 years (McIntyre et al. 2020). The heritability of BD is high (≥70%) and genome-wide association studies (GWAS) have identified potential risk genes, although they have very small effect sizes (Gordovez and McMahon 2020). Genes consistently associated with BD include *Ank3*, an important scaffolding protein that anchors sodium channels essential for action potential generation at nodes of Ranvier and the axonal initial segment (Hughes et al. 2018; Ferreira et al. 2008), together with *Cacna1c* (Sklar et al. 2008; Ferreira et al. 2008), which encodes voltage-gated calcium channel Cav1.2 and supports evidence of altered calcium signalling being central to changes in synaptic function and plasticity in BD (Berridge 2014; Moon et al. 2018). In addition, BD is associated with major disruption of metabolism (Kato 2019) and circadian rhythms (Geoffroy et al. 2016; McClung 2007); multiple circadian gene associations have been identified, including *Clock, Gsk3b* and *Per2*, that are regulated by lithium treatment. Furthermore, neuroimaging and post-mortem studies consistently show altered glutamatergic signalling in BD, and the NMDAR antagonist ketamine is an effective anti-depressive and anti-suicidal treatment in BD (Li et al. 2018; De Sousa et al. 2017; Wilkowska et al. 2020). The underlying neuropathology

A. M. Butt (✉)
School of Pharmacy and Biomedical Science, University of Portsmouth, Portsmouth, UK
e-mail: arthur.butt@port.ac.uk

A. D. Rivera
Department of Neuroscience, Institute of Human Anatomy, University of Padua, Padua, Italy

B. Li et al. (eds.), *Astrocytes in Psychiatric Disorders*, Advances in Neurobiology 26, https://doi.org/10.1007/978-3-030-77375-5_5

of BD is not well established, but imaging and post-mortem studies have highlighted strong associations of BD with thinning of cortical grey matter (GM) and white matter (WM), with progressive atrophy being associated with more severe illness episodes, which is alleviated by lithium treatment (Hibar et al. 2018; Favre et al. 2019; Serafini et al. 2021; Passos et al. 2016). Cortical thinning is associated with the regional loss of neurones and astrocytes (Harrison et al. 2020; Peng et al. 2016). Astrocyte atrophy has major effects on synaptic signalling and plasticity, together with regulation of glutamate, calcium signalling, metabolism and sleep, all of which are disrupted in BD (Fig. 1) (Steardo Jr. et al. 2019; Peng et al. 2016). The gold standard treatment for BD remains the mood stabilizer lithium, together with valproic acid (VPA) and carbamazepine (CBZ) (Alda 2015; McIntyre et al. 2020; Carvalho et al. 2020). The therapeutic actions of lithium are due at least in part to its potent inhibition of glycogen synthase kinase 3β (GSK3β), both directly and indirectly by its activation of Akt, which inhibits GSK3β activity (Fig. 2) (Alda 2015). Inhibition of GSK3β and activation of Akt by lithium have been demonstrated in BD-relevant brain regions of the frontal cortex, striatum and hippocampus activity (Freland and Beaulieu 2012). At a cellular level, lithium is neuroprotective and, as the primary neuroprotective cells of the CNS, there is increasing evidence that astrocytes are prominently involved in BD and are therapeutic targets of lithium and other frontline and adjunct therapies in BD (Keshavarz 2017; Jia et al. 2018; Rivera and Butt 2019; Peng et al. 2016; Stenovec et al. 2020). In addition, genetic and

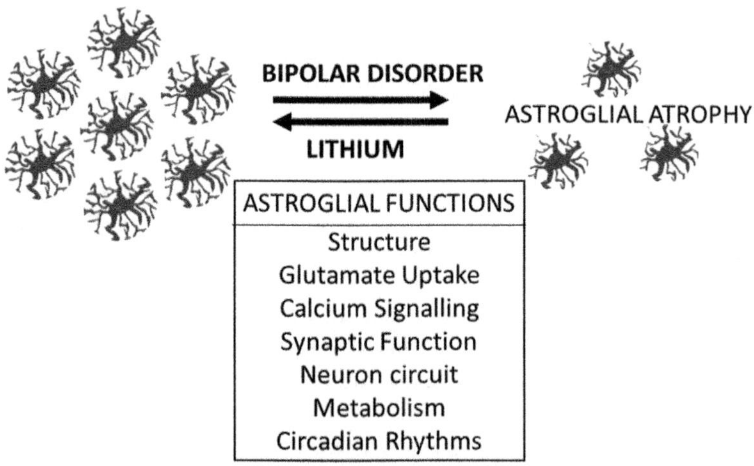

Fig. 1 Astrocyte atrophy and disruption of homeostatic functions in BD. Brain imaging and post-mortem studies consistently demonstrate decreased volume of grey matter (GM) and white matter (WM) in BD, due to astroglial atrophy and loss of specific neuronal populations. Astrocyte atrophy results in dysregulation of diverse astroglial functions, including structural changes, glutamate regulation, synaptic and neural circuit plasticity, together with metabolism and circadian rhythms, which are severely disrupted in BD. Notably, lithium has been shown to have positive effects on astroglial numbers, morphological plasticity and homeostatic functions, which underpin the protective effect of lithium on GM and WM thinning and its beneficial therapeutic effects in BD

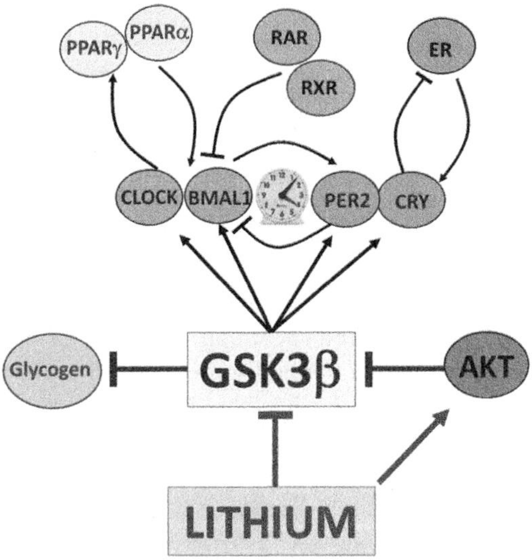

Fig. 2 GSK3β is a primary astroglial target of lithium in BD. GSK3β is a negative regulator of diverse cellular signalling pathways. Lithium is a potent inhibitor of GSK3β, both directly and indirectly by its activation of Akt, and this is a major factor in its beneficial therapeutic actions in BD. Astrocytes, like most cells, express an internal clock regulated by BMAL1/CLOCK and PER2/CRY and involving nuclear receptors (peroxisome proliferator-activated receptor gamma, PPAR; retinoic acid receptors, RAR/RXR; and estrogen receptors, ER), which are key BD-associated genes. Astrocytes are the main glycogen store in the brain and provide metabolic support to neurons via glycogenolysis and the astrocyte-neuron lactate shuttle; these astroglial functions are regulated by lithium via its actions on GSK3β, which inhibits the eponymous glycogen synthase. In addition to circadian rhythms and cellular metabolism, the cellular clock controls diverse astroglial functions that are severely disrupted in BD, including synaptogenesis

lithium-mediated inhibition of astroglial GSK3β has been shown to increase astroglial numbers in situ (Jung et al. 2016; Rivera and Butt 2019), consistent with the beneficial effects of lithium on attenuating thinning of GM and WM in BD. Notably, GSK3β is a master regulator of the circadian clock and cellular metabolism (Iitaka et al. 2005), which are disturbed in BD and are regulated lithium (Kato and Kato 2000; Geoffroy et al. 2014), and astrocytes play critical roles in maintaining circadian rhythms and brain energy balance (Niu et al. 2016; Brown and Ransom 2007; Brancaccio et al. 2019; Barca-Mayo et al. 2020). These studies demonstrate the importance of astrocytes in BD and highlight astroglial regulation of circadian rhythms and metabolism as major factors in the beneficial therapeutic effects of lithium in BD.

Neuropathology of BD

Although the results of BD neuropathology studies are variable, a consistent finding is the volume reduction of specific GM brain regions, such as the prefrontal cortex, and progressive changes have been associated with more severe illness episodes, which have been termed 'neuroprogression' (Serafini et al. 2021). The hypothesis is that a vicious cycle of repeated episodes results in the cumulative loss of neurones and glia, which in turn leads to further episodes and progressive brain atrophy and dysfunction. The importance of neuroprogression in BD remains controversial, and there is variability in the literature concerning regional GM atrophy in BD, most likely reflecting the heterogeneity of the disorder, as well as the impact of medication, whereby lithium may increase GM volume (Passos et al. 2016; Serafini et al. 2021). Nonetheless, brain imaging studies consistently demonstrate decreased volume of frontal, temporal and parietal GM in BD (Wang et al. 2020; Cotovio et al. 2020), and MRI analysis of 6503 individuals from the ENIGMA-BD Working Group found cortical GM thinning was associated with longer duration of illness and was alleviated by lithium treatment (Hibar et al. 2018). In addition, widespread white matter (WM) abnormalities are one of the most consistent findings in BD, for example as resolved by the ENIGMA-BD Working Group large cohort analysis of diffusion tensor imaging (DTI) data (Favre et al. 2019). Altered WM connectivity in frontocortical-striatal-thalamic circuits is most strongly associated with mood dysregulation in BD, and WM disruption is associated with a more severe course of illness (Lee et al. 2020; Mahon et al. 2010). A consistent feature of BD is shrinkage of the corpus callosum, the largest WM tract in the brain that connects the two cerebellar hemispheres, and these changes are associated with illness duration and are ameliorated by lithium treatment (Walterfang et al. 2009; Benedetti et al. 2013). Overall, the cognitive, behavioural, circadian and metabolic disturbances that characterize BD are associated with altered interconnectivity between brain regions. The balance of evidence indicates disease progression is associated with the loss of specific populations of neurones in several BD-relevant brain regions (Rajkowska et al. 2001; Bouras et al. 2001; Harrison et al. 2020; Mühleisen et al. 2014). However, BD is not characterized by neurodegeneration or neuroinflammation, and astrocyte reactivity or microglial activation are not characteristic features of BD (Giridharan et al. 2020; Harrison et al. 2020; Gigase et al. 2019). The most consistent finding in BD is astrocyte atrophy in specific cortical regions (Öngür et al. 1998; Rajkowska 2000; Toro et al. 2006; Hercher et al. 2014; Harrison et al. 2020), and associated dysregulation of astroglial homeostatic functions would have significant adverse effects on multiple aspects of BD neuropathology (Fig. 1).

Astrocyte Changes in BD

Recent systematic reviews highlight the variability of reports on astroglial changes in BD, depending on the brain regions analysed, and affected by differences in sample size and techniques used, as well as disease progression and medication (Giridharan et al. 2020; Gigase et al. 2019; Harrison et al. 2020). Nonetheless, it is evident that neurodegeneration, neuroinflammation and astrocyte 'reactivity' are not significant features of BD, the latter defined by increased density of cells (gliosis), cellular hypertrophy and increased expression of the astrocyte marker glial fibrillary acidic protein (GFAP) (Escartin et al. 2021). On the contrary, there is a consensus that cortical astrocytes are diminished in BD, both in numerical density and possibly morphological atrophy (Öngür et al. 1998; Rajkowska 2000; Hercher et al. 2014; Webster et al. 2005). These astroglial changes appear to be region specific, with decreased astrocyte density being a consistent feature of the prefrontal and cingulate cortex, as determined by Nissl staining (Todtenkopf et al. 2005; Öngür et al. 1998; Rajkowska et al. 2001), and by GFAP expression measured by immunostaining, immunoautoradiography and in situ hybridization (Hercher et al. 2014; Toro et al. 2006; Webster et al. 2005). The GFAP immunostaining area fraction was significantly decreased in BD (Hercher et al. 2014), indicating cellular atrophy, which has been reported in the ageing brain and a mouse model of Alzheimer's disease (AD) (Rodríguez et al. 2014). In contrast, no differences in astrocytes were evident in the amygdala, using Nissl staining and GFAP or S100β expression (Hamidi et al. 2004; Bezchlibnyk et al. 2007; Altshuler et al. 2010; Pantazopoulos et al. 2010). In the hippocampus, two studies using Nissl staining and GFAP immunolabelling found no changes in astrocytes (Malchow et al. 2015; Webster et al. 2001), whereas another study found S100β-immunopositive astrocytes were decreased in the hippocampus (Gos et al. 2013). Whilst recognizing that GFAP labels only a subset of astrocytes, and that there are multiple subtypes of GFAP-immunoreactive astrocytes in the human cortex (Verkhratsky et al. 2018; Escartin et al. 2021), the balance of evidence indicates that BD is associated with astrocyte atrophy in the prefrontal cortex, in particular the anterior cingulate cortex, which displays significant thinning in BD (Bouras et al. 2001). These studies are supported by transcriptomic analyses that revealed significant changes in astroglial genes in BD cortex (Ramaker et al. 2017; Toker et al. 2018), and changes in astrocytes may be present early in the illness course (Kaur et al. 2005). Overall, the observed astroglial atrophy is consistent with dysregulation of their homeostatic functions playing key roles in the clinical manifestations of BD (Fig. 1) (Peng et al. 2016).

Astroglial Glutamate Regulation in BD

Regulation of glutamate at synapses is a key function of astrocytes that is often altered in pathology (Escartin et al. 2021), and abnormalities in glutamatergic signalling are implicated in BD (Scotti-Muzzi et al. 2021; Li et al. 2018; De Sousa et al. 2017). GWAS studies associate glutamate signalling genes with BD (Nurnberger Jr. et al. 2014), and post-mortem studies have revealed reduced expression of NMDA, AMPA and KA glutamate receptors in BD (Beneyto and Meador-Woodruff 2008; Beneyto et al. 2007). Significantly, astrocyte atrophy in the prefrontal and anterior cingulate cortex correlates with abnormal glutamatergic activity in these regions in BD (Ongür et al. 2008; Eastwood and Harrison 2010). Astroglial reuptake of glutamate at synapses is via excitatory amino acid transporters (EAAT) 1 and 2 (encoded by the *SLC1A1/2* genes) and genetic variants of SLC1A2 are associated with BD (Fiorentino et al. 2015; Veldic et al. 2019). Furthermore, dysregulation of glutamate homeostasis in WM results in the loss of oligodendrocytes and myelin, resulting in WM thinning and disruption of connectivity that characterizes BD (Butt et al. 2014). The life-long generation of oligodendrocytes and myelin is the function of adult oligodendrocyte progenitor cells (OPCs), often called NG2-glia (Butt et al. 2002), and is essential for myelin repair (Ortiz et al. 2019), which is promoted by lithium treatment (Azim and Butt 2011). Glutamatergic signalling and neuronal activity drive OPC differentiation and myelination that is essential for learning and behaviour (Wake et al. 2011; Chen et al. 2018; Gibson et al. 2014; McKenzie et al. 2014; Xiao et al. 2016). Notably, disruption of OPCs has been shown to result in altered glutamatergic synaptic activity and to cause depressive-like behaviour in mice (Sakry et al. 2014; Birey et al. 2015), indicating the importance of bidirectional OPC-neuron interactions in cognitive function and behaviour (Rivera et al. 2016). Glutamatergic signalling in WM also stimulates oligodendrocyte-axon metabolic support, which is critical for maintaining WM connectivity (Rivera et al. 2021a), and pathological changes in oligodendrocytes and/or myelin ultimately lead to loss of axonal integrity and neuronal survival (Saab and Nave 2017), as observed in neuroprogression in BD. Hence, dysregulation of astroglial glutamate regulation has diverse negative effects on GM and WM function that would play major roles in the pathophysiology of BD.

Direct evidence of changes in astroglial glutamate regulation in BD is currently lacking. Expression levels of *SLC1A2*/EAAT2 mRNA in BD are largely unaltered in most studies (Medina et al. 2013; McCullumsmith and Meador-Woodruff 2002; Bernard et al. 2011; Zhang et al. 2021), and in one study, they were increased (Shao and Vawter 2008). Also, in BD prefrontal cortex where GFAP was reduced, Toro and colleagues observed no changes in glutamine synthetase (GS), the astroglial enzyme that recycles synaptic glutamate (Toro et al. 2006). In addition, a review of proton magnetic resonance spectroscopy (H-MRS) studies in BD failed to detect a significant correlation between lithium treatment and brain levels of glutamate, or of N-acetylaspartate (NAA), a marker for neuronal loss (Szulc et al. 2018). In contrast, VPA has been shown to increase mRNA and protein levels of EAAT1/2 and to stimulate glutamate uptake in astrocytes (Johnson Jr. et al. 2018). Furthermore, drugs that target glutamate activity are promising adjunct therapies in BD, including the NMDA blocker ketamine and riluzole, which increase glutamate-glutamine cycling in BD (Diazgranados et al. 2010; Park et al. 2017; Sakurai et al. 2019; Brennan et al. 2010). Significantly, riluzole has been shown to increase astroglial *SLC1A2*/EAAT2 activity (Carbone et al. 2012), supporting its positive effects on glutamate cycling and depressive symptoms in BD. In addition, ketamine regulates astroglial cAMP signalling to attenuate their release of neurotransmitters (Stenovec et al. 2020). This is significant because cAMP signalling is dysregulated in BD (Dwivedi and Pandey 2008), and *ADCY2* (Adenylate Cyclase2, AC2), a key enzyme in cAMP signalling, has been identified as a risk factor in BD (Mühleisen et al. 2014), indicating the diverse effects of ketamine on astrocytes are therapeutically relevant to BD. Interestingly, treatment of astrocytes with VPA altered the excitatory-inhibitory balance in vitro by reducing synapse formation in inhibitory neurons (Takeda et al. 2021), which may be important in its mood-stabilizing effects in BD, although systematic analyses of H-MRS measurements of GABA in BD patients have been inconclusive (Schür et al. 2016). Thus, there is growing indirect evidence that altered astroglial glutamate regulation may play an important role in BD, and further studies are required to determine their importance in the pathophysiology and treatment responsiveness of this disorder.

Astrocytes and Disruption of Circadian Rhythms in BD

BD is characterized by marked disturbances in circadian rhythms, which are regulated by a master circadian pacemaker in the suprachiasmatic nuclei (SCN) of the ventral hypothalamus (Jagannath et al. 2013). The circadian clock consists of a transcription-translation negative feedback loop in which the transcription factors CLOCK and BMAL1 drive the expression of clock genes Period *(Per)* and Cryptochrome *(Cry)* that in turn feedback to regulate their own expression (Reppert and Weaver 2002). It is significant, therefore, that multiple circadian genes are associated with BD, including *Clock, Gsk3b* and *Per2* (McClung 2007; Akula et al. 2014). Moreover, lithium has been shown to affect behavioural and molecular circadian rhythms in humans and animal models, and the lithium response is associated with multiple circadian gene polymorphisms (Geoffroy et al. 2014). In addition to driving bodily circadian rhythms, most cells express an internal clock regulated by BMAL1/CLOCK which controls diverse cellular functions, including cellular metabolism, and changes in glucose are a particularly potent cue for cellular clocks, involving nuclear receptors that act as cellular nutrient sensors, including peroxisome proliferator-activated receptor gamma (PPAR-γ), retinoic acid receptors (RAR) and estrogen receptors (ER) (Li et al. 2013; Picard and Auwerx 2002). Genetic associations have been identified between lithium response and *PPARGC1A* and *RORA*, which are involved in both circadian rhythms and cellular metabolism (Geoffroy et al. 2016). Moreover, GSK3β is a master regulator of the circadian clock and cellular metabolism (Iitaka et al. 2005), and synaptic plasticity of hippocampal neurones is dependent on GSK3β-mediated regulation of circadian rhythmicity (Besing et al. 2017), consistent with evidence that the cellular clock plays a key role in altered neuronal synaptic plasticity in BD (Vadnie and McClung 2017). Astrocytes possess functional circadian clocks that regulate their response to daily oscillations in brain activity and metabolism (McKee et al. 2020). Moreover, GSK3β is a direct target of lithium in astrocytes and acts at least in part to regulate PPAR-γ (Rivera and Butt 2019). Furthermore, astrocytes have been shown to autonomously sustain circadian patterns of neuronal activity in the SCN (Brancaccio et al. 2019), and astrocyte-specific deletion of *BMAL1* is sufficient to disrupt brain glucose homeostasis (Barca-Mayo et al. 2020). Overall, disruption of astroglial regulation of circadian rhythms and metabolism is likely to be a major factor in BD and in the beneficial therapeutic effects of lithium (Fig. 2).

Astrocytes and Altered Metabolism in BD

Abnormal metabolism is a consistent finding in BD (Kato and Kato 2000), and this strongly implicates astrocytes, which are key players in brain energy regulation. Astrocytes provide metabolic support to neurons via the astrocyte-neuron lactate shuttle, which is tightly regulated by neuronal activity (Hertz et al. 2006; Magistretti and Allaman 2018; Barros 2013). Astrocytes take up glucose via glucose transporters and store it as glycogen and pyruvate, which can be converted by lactate dehydrogenase 5 (LDH5) to lactate, that can be transferred via monocarboxylate transporters (MCTs) to neurons, where lactate is converted back to pyruvate by LDH1 and used for energy production. Notably, decreased glucose and elevated lactate in BD suggests a shift towards glycolysis (Dager et al. 2004; Hosokawa et al. 2009). Furthermore, lithium is a major inhibitor of GSK3β, which regulates cellular metabolism through glycogen synthase and pyruvate dehydrogenase (Jope 2011), and regulates glutamate metabolism and glycogen synthesis in the brain (Gould et al. 2004; Kohno et al. 2007) and in astrocytes (Fan et al. 2010; De Almeida Souza et al. 2010). Furthermore, treatment with lithium, CBZ or VPA has been shown to regulate GSK3β signalling and glycogen production in astrocytes (Jia et al. 2018). Notably, astrocytes also provide metabolic support for WM axons in the form of lactate (Niu et al. 2016; Brown and Ransom 2007) and provide glucose to oligodendrocytes, which in turn deliver lactate to axons via MCTs (Lee et al. 2012; Funfschilling et al. 2012). Disruption of astroglial or oligodendroglial axonal metabolic support results in altered axonal activity (Trevisiol et al. 2017), and ultimately a loss of axonal integrity (Alexandra et al. 2018). These studies emphasize that the role of astrocytes in altered metabolism and circadian rhythms in BD could underpin changes in GM and WM integrity and is an area of research that merits considerable attention.

Lithium-Responsive Astroglial Gene Networks

Lithium remains the main therapy for BD, and genomic analysis of the effects of lithium is considered crucially important for the identification of novel lithium-responsive genes in BD, as well as identifying pathophysiological mechanisms and therapeutic targets (Papiol et al. 2018). As noted above, lithium regulates multiple astroglial functions that are implicated in BD, but the mechanisms of actions are unresolved. To help fill this gap in our knowledge, we examined the effects of lithium in WM astrocytes, using a combined neurobiological and systems biological approach (Rivera and Butt 2019). A key finding was that the ECM-modifying enzyme Lysyl oxidase (LOX) was identified as a novel target of lithium that promotes astrocyte proliferation and profoundly affects their morphology, which is consistent with the positive effects of lithium in alleviating GM and WM thinning in BD. The structural role of astrocytes and their regulation of the ECM is essential for the maintenance and remodelling of synaptic and neural circuit function (Fig. 3), which are impaired when astrocytes undergo atrophy, as seen in ageing and pathology (Pirttimaki and Parri 2013; Sampedro-Piquero et al. 2014; Rodríguez et al. 2014). Hence, the identification of LOX as a BD-associated lithium-responsive astroglial gene that regulates astrocyte morphological remodelling requires further investigation to examine its role in astrocyte atrophy and its impact on neuronal circuit activity in BD, as well as developing new therapies that regulate LOX activity (Rivera and Butt 2019). Analysis of lithium-responsive astroglial genes revealed

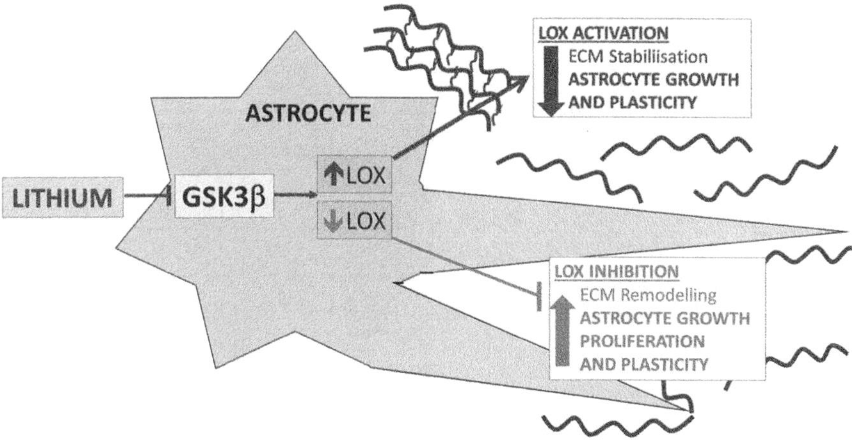

Fig. 3 Lysyl oxidase (LOX) is a novel target of lithium in astrocytes. Combined neurobiological and systems biology approaches identified the extracellular matrix (ECM)-modifying enzyme LOX as a major target of lithium in astrocytes (Rivera and Butt 2019). LOX activation promotes stabilization of the ECM to impede cellular growth and plasticity. Lithium inhibits LOX activity and increases astrocyte numbers and promotes morphological plasticity, which is essential for the maintenance and remodelling of synapses and neural circuity. These actions of astroglial LOX are consistent with the positive effects of lithium in rescuing astrocyte atrophy and GM/WM thinning in BD

up-regulation of neuron-supporting pathways that may be central to the positive therapeutic outcomes of lithium on rescuing GM and WM atrophy (Rivera and Butt 2019). Meta-analysis of BD-associated astroglial genes identified key genes with recognized importance in BD, including *Ank3, Cacna1, cAMP* and *Per2*, as well as multiple GluR and GABAR (Fig. 4). These analyses highlight diverse astroglial gene networks that are implicated in BD, including P2X7R that plays a role in depressive behaviour and has been suggested to have a role in BD in relation

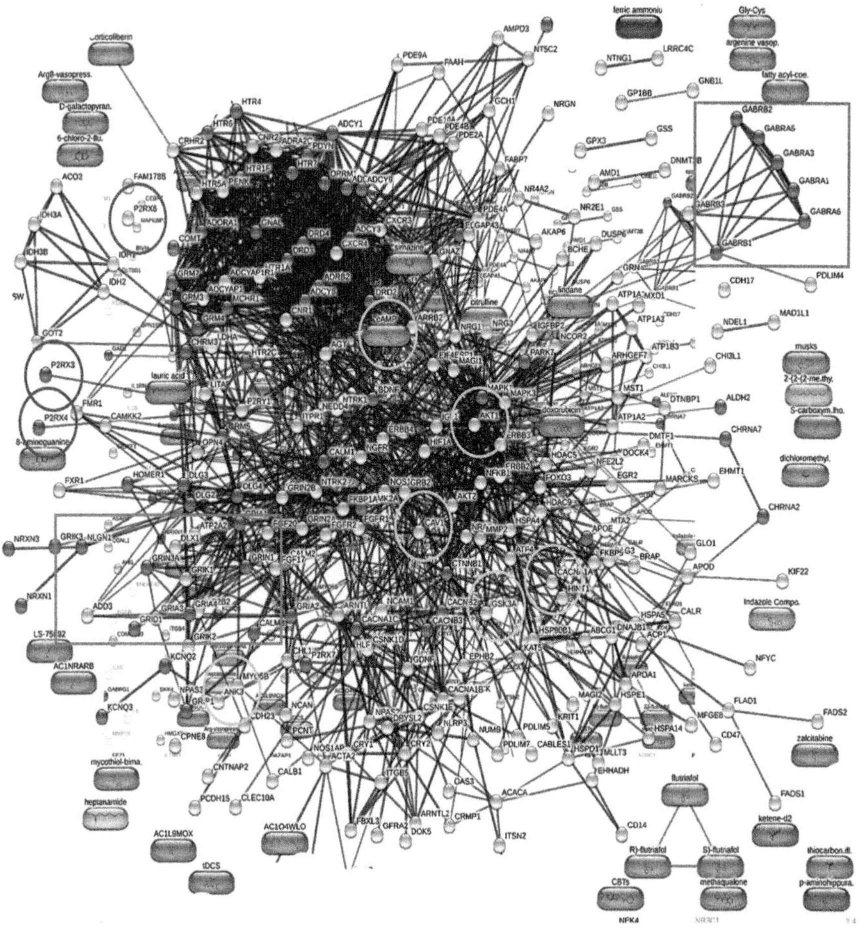

Fig. 4 BD-associated astroglial gene networks. Functional protein-protein network analysis of BD-associated astrocyte genes (Rivera and Butt 2019; Rivera et al. 2021b) resolved astroglial gene networks that have recognized importance in BD (highlighted in yellow circles, e.g. *Ank3, Cacna1, cAMP, Gsk3*), as well as multiple GluR (highlighted in green rectangle, e.g. *Grin1, Gria2, Grik3*), GABAR (highlighted in blue rectangle, e.g. *Gabra1, Gabrab1*) and P2R (highlighted in red circles, e.g. *P2ry1, P2rx3 P2rx7*). Many of these gene networks are involved in neuron-astrocyte signalling and regulation of cellular metabolism, which have important implications for BD

inflammation (Illes et al. 2020). In addition, P2X7R has a well-established role in astroglial calcium signalling and release of neurotransmitters (Butt 2011; Hamilton et al. 2008) and regulation of cellular metabolism (Rivera et al. 2021a), which have important implications for BD (De Sousa et al. 2017). Pharmacogenomic analysis identified pioglitazone and flavanols as small molecules that target BD-associated lithium-responsive astroglial genes (Rivera and Butt 2019), which is of interest because piogliatazone acts on PPAR-γ, whilst flavanols activate ER (De Almeida et al. 2020; Rivera and Butt 2019), which are integral to regulation of circadian rhythms and metabolism and are dysregulated in BD. Notably, pioglitazone has been shown to have promising results for its anti-depressant activity in clinical studies (Nierenberg et al. 2018; Brusotti et al. 2017), and flavanols have been shown to have positive effects in anxiety, mood disorder, and cognitive decline in animal models (Stringer et al. 2015; Wang et al. 2012). These combined neurobiological and pharmacogenomic strategies are beginning to identify novel astroglial targets that are promising therapies in BD.

Conclusions

To date, the neuropathology of BD has not been fully defined, but a consistent replicated finding is that cortical thickness is significantly decreased in BD and is associated with loss of astrocytes and subpopulations of neurons. An important negative finding is that astrocyte reactive gliosis and neuroinflammation are not major features of BD. The effects of lithium and other treatments have provided further evidence that disruption of astroglial homeostatic functions is of primary importance in BD neuropathology. These studies highlight the importance of GSK3β in astroglial regulation of circadian rhythms and metabolism in BD. Moreover, BD is highly specific in its response to lithium, and lithium-responsive astroglial gene networks have a distinct signature that provides an insight into the beneficial therapeutic effects of lithium in BD. Astrocytes are important therapeutic targets in BD, highlighting the critical need for further research into specific astrocyte-neuron interactions that are dysregulated in BD.

Acknowledgements We would like to thank the support of the BBSRC and MRC (AB), and the MSCA Seal of Excellence @ UNIPD (AR).

References

Akula N, Barb J, Jiang X, Wendland JR, Choi KH, Sen SK, Hou L, Chen DTW, Laje G, Johnson K, Lipska BK, Kleinman JE, Corrada-Bravo H, Detera-Wadleigh S, Munson PJ, McMahon FJ (2014) RNA-sequencing of the brain transcriptome implicates dysregulation of neuroplasticity, circadian rhythms and GTPase binding in bipolar disorder. Mol Psychiatry 19:1179–1185

Alda M (2015) Lithium in the treatment of bipolar disorder: pharmacology and pharmacogenetics. Mol Psychiatry 20:661–670

Alexandra IM, Constanze D, Klaus-Armin N (2018) An emerging role of dysfunctional axon-oligodendrocyte coupling in neurodegenerative diseases. Dialogues Clin Neurosci 20:283–292

Altshuler LL, Abulseoud OA, Foland-Ross L, Bartzokis G, Chang S, Mintz J, Hellemann G, Vinters HV (2010) Amygdala astrocyte reduction in subjects with major depressive disorder but not bipolar disorder. Bipolar Disord 12:541–549

Azim K, Butt AM (2011) GSK3beta negatively regulates oligodendrocyte differentiation and myelination in vivo. Glia 59:540–553

Barca-Mayo O, Boender AJ, Armirotti A, De Pietri Tonelli D (2020) Deletion of astrocytic BMAL1 results in metabolic imbalance and shorter lifespan in mice. Glia 68:1131–1147

Barros LF (2013) Metabolic signaling by lactate in the brain. Trends Neurosci 36:396–404

Benedetti F, Bollettini I, Barberi I, Radaelli D, Poletti S, Locatelli C, Pirovano A, Lorenzi C, Falini A, Colombo C, Smeraldi E (2013) Lithium and GSK3-beta promoter gene variants influence white matter microstructure in bipolar disorder. Neuropsychopharmacology 38:313–327

Beneyto M, Meador-Woodruff JH (2008) Lamina-specific abnormalities of NMDA receptor-associated postsynaptic protein transcripts in the prefrontal cortex in Schizophrenia and bipolar disorder. Neuropsychopharmacology 33:2175–2186

Beneyto M, Kristiansen LV, Oni-Orisan A, McCullumsmith RE, Meador-Woodruff JH (2007) Abnormal glutamate receptor expression in the medial temporal lobe in Schizophrenia and mood disorders. Neuropsychopharmacology 32:1888–1902

Bernard R, Kerman IA, Thompson RC, Jones EG, Bunney WE, Barchas JD, Schatzberg AF, Myers RM, Akil H, Watson SJ (2011) Altered expression of glutamate signaling, growth factor, and glia genes in the locus coeruleus of patients with major depression. Mol Psychiatry 16:634–646

Berridge MJ (2014) Calcium signalling and psychiatric disease: bipolar disorder and schizophrenia. Cell Tissue Res 357:477–492

Besing RC, Rogers CO, Paul JR, Hablitz LM, Johnson RL, McMahon LL, Gamble KL (2017) GSK3 activity regulates rhythms in hippocampal clock gene expression and synaptic plasticity. Hippocampus 27:890–898

Bezchlibnyk YB, Sun X, Wang J-F, MacQueen GM, McEwen BS, Young LT (2007) Neuron somal size is decreased in the lateral amygdalar nucleus of subjects with bipolar disorder. J Psychiatr Neurosci 32:203–210

Birey F, Kloc M, Chavali M, Hussein I, Wilson M, Christoffel DJ, Chen T, Frohman MA, Robinson JK, Russo SJ, Maffei A, Aguirre A (2015) Genetic and stress-induced loss of NG2 Glia triggers emergence of depressive-like behaviors through reduced secretion of FGF2. Neuron 88:941–956

Bouras C, Kövari E, Hof PR, Riederer BM, Giannakopoulos P (2001) Anterior cingulate cortex pathology in schizophrenia and bipolar disorder. Acta Neuropathol 102:373–379

Brancaccio M, Edwards MD, Patton AP, Smyllie NJ, Chesham JE, Maywood ES, Hastings MH (2019) Cell-autonomous clock of astrocytes drives circadian behavior in mammals. Science 363:187

Brennan BP, Hudson JI, Jensen JE, McCarthy J, Roberts JL, Prescot AP, Cohen BM, Pope HG Jr, Renshaw PF, Ongür D (2010) Rapid enhancement of glutamatergic neurotransmission in bipolar depression following treatment with riluzole. Neuropsychopharmacology 35:834–846

Brown AM, Ransom BR (2007) Astrocyte glycogen and brain energy metabolism. Glia 55:1263–1271

Brusotti G, Montanari R, Capelli D, Cattaneo G, Laghezza A, Tortorella P, Loiodice F, Peiretti F, Bonardo B, Paiardini A, Calleri E, Pochetti G (2017) Betulinic acid is a PPARγ antagonist that improves glucose uptake, promotes osteogenesis and inhibits adipogenesis. Sci Rep 7:5777

Butt AM (2011) ATP: a ubiquitous gliotransmitter integrating neuron-glial networks. Semin Cell Dev Biol 22:205–213

Butt AM, Kiff J, Hubbard P, Berry M (2002) Synantocytes: new functions for novel NG2 expressing glia. J Neurocytol 31:551–565

Butt AM, Fern RF, Matute C (2014) Neurotransmitter signaling in white matter. Glia 62:1762–1779
Carbone M, Duty S, Rattray M (2012) Riluzole elevates GLT-1 activity and levels in striatal astrocytes. Neurochem Int 60:31–38
Carvalho AF, Firth J, Vieta E (2020) Bipolar Disorder. N Engl J Med 383:58–66
Chen TJ, Kula B, Nagy B, Barzan R, Gall A, Ehrlich I, Kukley M (2018) In vivo regulation of oligodendrocyte precursor cell proliferation and differentiation by the AMPA-receptor subunit GluA2. Cell Rep 25:852-861.e7
Cotovio G, Talmasov D, Barahona-Corrêa JB, Hsu J, Senova S, Ribeiro R, Soussand L, Velosa A, Silva VCE, Rost N, Wu O, Cohen AL, Oliveira-Maia AJ, Fox MD (2020) Mapping mania symptoms based on focal brain damage. J Clin Invest 130:5209–5222
Dager SR, Friedman SD, Parow A, Demopulos C, Stoll AL, Lyoo IK, Dunner DL, Renshaw PF (2004) Brain metabolic alterations in medication-free patients with bipolar disordeR. Arch Gen Psychiatry 61:450–458
De Almeida Souza A, Da Silva GSS, Velez BS, Santoro ABM, Montero-Lomelí M (2010) Glycogen synthesis in brain and astrocytes is inhibited by chronic lithium treatment. Neurosci Lett 482:128–132
De Almeida MMA, Souza CDS, Dourado NS, Da Silva AB, Ferreira RS, David JM, David JP, Costa MFD, Da Silva VDA, Butt AM, Costa SL (2020) Phytoestrogen Agathisflavone Ameliorates Neuroinflammation-induced by LPS and IL-1β and protects neurons in Cocultures of Glia/neurons. Biomolecules 10:562
De Sousa RT, Loch AA, Carvalho AF, Brunoni AR, Haddad MR, Henter ID, Zarate CA, Machado-Vieira R (2017) Genetic studies on the tripartite glutamate synapse in the pathophysiology and therapeutics of mood disorders. Neuropsychopharmacology 42:787–800
Diazgranados N, Ibrahim L, Brutsche NE, Newberg A, Kronstein P, Khalife S, Kammerer WA, Quezado Z, Luckenbaugh DA, Salvadore G, Machado-Vieira R, Manji HK, Zarate CA Jr (2010) A randomized add-on trial of an N-methyl-D-aspartate antagonist in treatment-resistant bipolar depression. Arch Gen Psychiatry 67:793–802
Dwivedi Y, Pandey GN (2008) Adenylyl cyclase-cyclicAMP signaling in mood disorders: role of the crucial phosphorylating enzyme protein kinase A. Neuropsychiatr Dis Treat 4:161–176
Eastwood SL, Harrison PJ (2010) Markers of glutamate synaptic transmission and plasticity are increased in the anterior cingulate cortex in bipolar disorder. Biol Psychiatry 67:1010–1016
Escartin C, Galea E, Lakatos A, O'Callaghan JP, Petzold GC, Serrano-Pozo A, Steinhäuser C, Volterra A, Carmignoto G, Agarwal A, Allen NJ, Araque A, Barbeito L, Barzilai A, Bergles DE, Bonvento G, Butt AM, Chen W-T, Cohen-Salmon M, Cunningham C, Deneen B, De Strooper B, Díaz-Castro B, Farina C, Freeman M, Gallo V, Goldman JE, Goldman SA, Götz M, Gutiérrez A, Haydon PG, Heiland DH, Hol EM, Holt MG, Iino M, Kastanenka KV, Kettenmann H, Khakh BS, Koizumi S, Lee CJ, Liddelow SA, MacVicar BA, Magistretti P, Messing A, Mishra A, Molofsky AV, Murai KK, Norris CM, Okada S, Oliet SHR, Oliveira JF, Panatier A, Parpura V, Pekna M, Pekny M, Pellerin L, Perea G, Pérez-Nievas BG, Pfrieger FW, Poskanzer KE, Quintana FJ, Ransohoff RM, Riquelme-Perez M, Robel S, Rose CR, Rothstein JD, Rouach N, Rowitch DH, Semyanov A, Sirko S, Sontheimer H, Swanson RA, Vitorica J, Wanner I-B, Wood LB, Wu J, Zheng B, Zimmer ER, Zorec R, Sofroniew MV, Verkhratsky A (2021) Reactive astrocyte nomenclature, definitions, and future directions. Nat Neurosci 24:312–325
Fan TWM, Yuan P, Lane AN, Higashi RM, Wang Y, Hamidi AB, Zhou R, Guitart X, Chen G, Manji HK, Kaddurah-Daouk R (2010) Stable isotope-resolved metabolomic analysis of lithium effects on glial-neuronal metabolism and interactions. Metabolomics 6:165–179
Favre P, Pauling M, Stout J, Hozer F, Sarrazin S, Abé C, Alda M, Alloza C, Alonso-Lana S, Andreassen OA, Baune BT, Benedetti F, Busatto GF, Canales-Rodríguez EJ, Caseras X, Chaim-Avancini TM, Ching CRK, Dannlowski U, Deppe M, Eyler LT, Fatjo-Vilas M, Foley SF, Grotegerd D, Hajek T, Haukvik UK, Howells FM, Jahanshad N, Kugel H, Lagerberg TV, Lawrie SM, Linke JO, McIntosh A, Melloni EMT, Mitchell PB, Polosan M, Pomarol-Clotet E, Repple J, Roberts G, Roos A, Rosa PGP, Salvador R, Sarró S, Schofield PR, Serpa MH, Sim K, Stein DJ, Sussmann JE, Temmingh HS, Thompson PM, Verdolini N, Vieta E, Wessa M,

Whalley HC, Zanetti MV, Leboyer M, Mangin J-F, Henry C, Duchesnay E, Houenou J, for the, E. B. D. W. G (2019) Widespread white matter microstructural abnormalities in bipolar disorder: evidence from mega- and meta-analyses across 3033 individuals. Neuropsychopharmacology 44:2285–2293

Ferreira MA, O'Donovan MC, Meng YA, Jones IR, Ruderfer DM, Jones L, Fan J, Kirov G, Perlis RH, Green EK, Smoller JW, Grozeva D, Stone J, Nikolov I, Chambert K, Hamshere ML, Nimgaonkar VL, Moskvina V, Thase ME, Caesar S, Sachs GS, Franklin J, Gordon-Smith K, Ardlie KG, Gabriel SB, Fraser C, Blumenstiel B, Defelice M, Breen G, Gill M, Morris DW, Elkin A, Muir WJ, McGhee KA, Williamson R, MacIntyre DJ, MacLean AW, St CD, Robinson M, Van Beck M, Pereira AC, Kandaswamy R, McQuillin A, Collier DA, Bass NJ, Young AH, Lawrence J, Ferrier IN, Anjorin A, Farmer A, Curtis D, Scolnick EM, McGuffin P, Daly MJ, Corvin AP, Holmans PA, Blackwood DH, Gurling HM, Owen MJ, Purcell SM, Sklar P, Craddock N (2008) Collaborative genome-wide association analysis supports a role for ANK3 and CACNA1C in bipolar disorder. Nat Genet 40:1056–1058

Fiorentino A, Sharp SI, McQuillin A (2015) Association of rare variation in the glutamate receptor gene SLC1A2 with susceptibility to bipolar disorder and schizophrenia. Eur J Hum Genet 23:1200–1206

Freland L, Beaulieu J-M (2012) Inhibition of GSK3 by lithium, from single molecules to signaling networks. Front Mol Neurosci 5:14–14

Funfschilling U, Supplie LM, Mahad D, Boretius S, Saab AS, Edgar J, Brinkmann BG, Kassmann CM, Tzvetanova ID, Mobius W, Diaz F, Meijer D, Suter U, Hamprecht B, Sereda MW, Moraes CT, Frahm J, Goebbels S, Nave KA (2012) Glycolytic oligodendrocytes maintain myelin and long-term axonal integrity. Nature 485:517–521

Geoffroy PA, Etain B, Sportiche S, Bellivier F (2014) Circadian biomarkers in patients with bipolar disorder: promising putative predictors of lithium response. Int J Bipolar Disorders 2:5

Geoffroy PA, Etain B, Lajnef M, Zerdazi EH, Brichant-Petitjean C, Heilbronner U, Hou L, Degenhardt F, Rietschel M, McMahon FJ, Schulze TG, Jamain S, Marie-Claire C, Bellivier F (2016) Circadian genes and lithium response in bipolar disorders: associations with PPARGC1A (PGC-1α) and RORA. Genes Brain Behav 15:660–668

Gibson EM, Purger D, Mount CW, Goldstein AK, Lin GL, Wood LS, Inema I, Miller SE, Bieri G, Zuchero JB, Barres BA, Woo PJ, Vogel H, Monje M (2014) Neuronal activity promotes oligodendrogenesis and adaptive myelination in the mammalian brain. Science 344:1252304

Gigase FAJ, Snijders G, Boks MP, De Witte LD (2019) Neurons and glial cells in bipolar disorder: a systematic review of postmortem brain studies of cell number and size. Neurosci Biobehav Rev 103:150–162

Giridharan VV, Sayana P, Pinjari OF, Ahmad N, Da Rosa MI, Quevedo J, Barichello T (2020) Postmortem evidence of brain inflammatory markers in bipolar disorder: a systematic review. Mol Psychiatry 25:94–113

Gordovez FJA, McMahon FJ (2020) The genetics of bipolar disorder. Mol Psychiatry 25:544–559

Gos T, Schroeter ML, Lessel W, Bernstein HG, Dobrowolny H, Schiltz K, Bogerts B, Steiner J (2013) S100B-immunopositive astrocytes and oligodendrocytes in the hippocampus are differentially afflicted in unipolar and bipolar depression: a postmortem study. J Psychiatr Res 47:1694–1699

Gould TD, Chen G, Manji HK (2004) In vivo evidence in the brain for Lithium inhibition of glycogen synthase Kinase-3. Neuropsychopharmacology 29:32–38

Hamidi M, Drevets WC, Price JL (2004) Glial reduction in amygdala in major depressive disorder is due to oligodendrocytes. Biol Psychiatry 55:563–569

Hamilton N, Vayro S, Kirchhoff F, Verkhratsky A, Robbins J, Gorecki DC, Butt AM (2008) Mechanisms of ATP- and glutamate-mediated calcium signaling in white matter astrocytes. Glia 56:734–749

Harrison PJ, Colbourne L, Harrison CH (2020) The neuropathology of bipolar disorder: systematic review and meta-analysis. Mol Psychiatry 25:1787–1808

Hercher C, Chopra V, Beasley CL (2014) Evidence for morphological alterations in prefrontal white matter glia in schizophrenia and bipolar disorder. J Psychiatry Neurosci 39:376–385
Hertz L, Peng L, Dienel GA (2006) Energy metabolism in astrocytes: high rate of oxidative metabolism and spatiotemporal dependence on glycolysis/Glycogenolysis. J Cereb Blood Flow Metab 27:219–249
Hibar DP, Westlye LT, Doan NT, Jahanshad N, Cheung JW, Ching CRK, Versace A, Bilderbeck AC, Uhlmann A, Mwangi B, Krämer B, Overs B, Hartberg CB, Abé C, Dima D, Grotegerd D, Sprooten E, Bøen E, Jimenez E, Howells FM, Delvecchio G, Temmingh H, Starke J, Almeida JRC, Goikolea JM, Houenou J, Beard LM, Rauer L, Abramovic L, Bonnin M, Ponteduro MF, Keil M, Rive MM, Yao N, Yalin N, Najt P, Rosa PG, Redlich R, Trost S, Hagenaars S, Fears SC, Alonso-Lana S, Van Erp TGM, Nickson T, Chaim-Avancini TM, Meier TB, Elvsåshagen T, Haukvik UK, Lee WH, Schene AH, Lloyd AJ, Young AH, Nugent A, Dale AM, Pfennig A, McIntosh AM, Lafer B, Baune BT, Ekman CJ, Zarate CA, Bearden CE, Henry C, Simhandl C, McDonald C, Bourne C, Stein DJ, Wolf DH, Cannon DM, Glahn DC, Veltman DJ, Pomarol-Clotet E, Vieta E, Canales-Rodriguez EJ, Nery FG, Duran FLS, Busatto GF, Roberts G, Pearlson GD, Goodwin GM, Kugel H, Whalley HC, Ruhe HG, Soares JC, Fullerton JM, Rybakowski JK, Savitz J, Chaim KT, Fatjó-Vilas M, Soeiro-De-Souza MG, Boks MP, Zanetti MV, Otaduy MCG, Schaufelberger MS, Alda M, Ingvar M, Phillips ML, Kempton MJ, Bauer M, Landén M, Lawrence NS et al (2018) Cortical abnormalities in bipolar disorder: an MRI analysis of 6503 individuals from the ENIGMA Bipolar Disorder Working Group. Mol Psychiatry 23:932–942
Hosokawa T, Momose T, Kasai K (2009) Brain glucose metabolism difference between bipolar and unipolar mood disorders in depressed and euthymic states. Prog Neuro-Psychopharmacol Biol Psychiatry 33:243–250
Hughes T, Sønderby IE, Polushina T, Hansson L, Holmgren A, Athanasiu L, Melbø-Jørgensen C, Hassani S, Hoeffding LK, Herms S, Bergen SE, Karlsson R, Song J, Rietschel M, Nöthen MM, Forstner AJ, Hoffmann P, Hultman CM, Landén M, Cichon S, Werge T, Andreassen OA, Le Hellard S, Djurovic S (2018) Elevated expression of a minor isoform of ANK3 is a risk factor for bipolar disorder. Transl Psychiatry 8:210
Iitaka C, Miyazaki K, Akaike T, Ishida N (2005) A role for glycogen synthase kinase-3beta in the mammalian circadian clock. J Biol Chem 280:29397–29402
Illes P, Verkhratsky A, Tang Y (2020) Pathological ATPergic signaling in major depression and bipolar disorder. Front Mol Neurosci 12:331
Jagannath A, Peirson SN, Foster RG (2013) Sleep and circadian rhythm disruption in neuropsychiatric illness. Curr Opin Neurobiol 23:888–894
Jia S, Li B, Huang J, Verkhratsky A, Peng L (2018) Regulation of glycogen content in astrocytes via Cav-1/PTEN/AKT/GSK-3β pathway by three anti-bipolar drugs. Neurochem Res 43:1692–1701
Johnson J Jr, Pajarillo E, Karki P, Kim J, Son DS, Aschner M, Lee E (2018) Valproic acid attenuates manganese-induced reduction in expression of GLT-1 and GLAST with concomitant changes in murine dopaminergic neurotoxicity. Neurotoxicology 67:112–120
Jope R (2011) Glycogen synthase Kinase-3 in the etiology and treatment of mood disorders. Front Mol Neurosci 4:16
Jung E-M, Ka M, Kim W-Y (2016) Loss of GSK-3 causes abnormal astrogenesis and behavior in mice. Mol Neurobiol 53:3954–3966
Kato T (2019) Current understanding of bipolar disorder: toward integration of biological basis and treatment strategies. Psychiatry Clin Neurosci 73:526–540
Kato T, Kato N (2000) Mitochondrial dysfunction in bipolar disorder. Bipolar Disord 2:180–190
Kaur S, Sassi RB, Axelson D, Nicoletti M, Brambilla P, Monkul ES, Hatch JP, Keshavan MS, Ryan N, Birmaher B, Soares JC (2005) Cingulate cortex anatomical abnormalities in children and adolescents with bipolar disorder. Am J Psychiatr 162:1637–1643
Keshavarz M (2017) Glial cells as key elements in the pathophysiology and treatment of bipolar disorder. Acta Neuropsychiatr 29:140–152

Kohno T, Shiga T, Toyomaki A, Kusumi I, Matsuyama T, Inoue T, Katoh C, Koyama T, Tamaki N (2007) Effects of lithium on brain glucose metabolism in healthy men. J Clin Psychopharmacol 27:698

Lee Y, Morrison BM, Li Y, Lengacher S, Farah MH, Hoffman PN, Liu Y, Tsingalia A, Jin L, Zhang PW, Pellerin L, Magistretti PJ, Rothstein JD (2012) Oligodendroglia metabolically support axons and contribute to neurodegeneration. Nature 487:443–448

Lee D-K, Lee H, Park K, Joh E, Kim C-E, Ryu S (2020) Common gray and white matter abnormalities in schizophrenia and bipolar disorder. PLoS One 15:e0232826

Li M-D, Ruan H-B, Hughes ME, Lee J-S, Singh JP, Jones SP, Nitabach MN, Yang X (2013) O-GlcNAc signaling entrains the circadian Clock by inhibiting BMAL1/CLOCK ubiquitination. Cell Metab 17:303–310

Li CT, Yang KC, Lin WC (2018) Glutamatergic dysfunction and glutamatergic compounds for major psychiatric disorders: evidence from clinical neuroimaging studies. Front Psych 9:767

Magistretti PJ, Allaman I (2018) Lactate in the brain: from metabolic end-product to signalling molecule. Nat Rev Neurosci 19:235–249

Mahon K, Burdick KE, Szeszko PR (2010) A role for white matter abnormalities in the pathophysiology of bipolar disorder. Neurosci Biobehav Rev 34:533–554

Malchow B, Strocka S, Frank F, Bernstein H-G, Steiner J, Schneider-Axmann T, Hasan A, Reich-Erkelenz D, Schmitz C, Bogerts B, Falkai P, Schmitt A (2015) Stereological investigation of the posterior hippocampus in affective disorders. J Neural Transm 122:1019–1033

McClung CA (2007) Clock genes and bipolar disorder: implications for therapy. Pharmacogenomics 8:1097–1100

McCullumsmith RE, Meador-Woodruff JH (2002) Striatal excitatory amino acid transporter transcript expression in schizophrenia, bipolar disorder, and major depressive disorder. Neuropsychopharmacology 26:368–375

McIntyre RS, Berk M, Brietzke E, Goldstein BI, López-Jaramillo C, Kessing LV, Malhi GS, Nierenberg AA, Rosenblat JD, Majeed A, Vieta E, Vinberg M, Young AH, Mansur RB (2020) Bipolar disorders. Lancet 396:1841–1856

McKee CA, Lananna BV, Musiek ES (2020) Circadian regulation of astrocyte function: implications for Alzheimer's disease. Cell Mol Life Sci 77:1049–1058

McKenzie IA, Ohayon D, Li H, De Faria JP, Emery B, Tohyama K, Richardson WD (2014) Motor skill learning requires active central myelination. Science 346:318–322

Medina A, Burke S, Thompson RC, Bunney W Jr, Myers RM, Schatzberg A, Akil H, Watson SJ (2013) Glutamate transporters: a key piece in the glutamate puzzle of major depressive disorder. J Psychiatr Res 47:1150–1156

Moon AL, Haan N, Wilkinson LS, Thomas KL, Hall J (2018) CACNA1C: association with psychiatric disorders, behavior, and neurogenesis. Schizophr Bull 44:958–965

Mühleisen TW, Leber M, Schulze TG, Strohmaier J, Degenhardt F, Treutlein J, Mattheisen M, Forstner AJ, Schumacher J, Breuer R, Meier S, Herms S, Hoffmann P, Lacour A, Witt SH, Reif A, Müller-Myhsok B, Lucae S, Maier W, Schwarz M, Vedder H, Kammerer-Ciernioch J, Pfennig A, Bauer M, Hautzinger M, Moebus S, Priebe L, Czerski PM, Hauser J, Lissowska J, Szeszenia-Dabrowska N, Brennan P, McKay JD, Wright A, Mitchell PB, Fullerton JM, Schofield PR, Montgomery GW, Medland SE, Gordon SD, Martin NG, Krasnow V, Chuchalin A, Babadjanova G, Pantelejeva G, Abramova LI, Tiganov AS, Polonikov A, Khusnutdinova E, Alda M, Grof P, Rouleau GA, Turecki G, Laprise C, Rivas F, Mayoral F, Kogevinas M, Grigoroiu-Serbanescu M, Propping P, Becker T, Rietschel M, Nöthen MM, Cichon S (2014) Genome-wide association study reveals two new risk loci for bipolar disorder. Nat Commun 5:3339

Nierenberg AA, Ghaznavi SA, Sande Mathias I, Ellard KK, Janos JA, Sylvia LG (2018) Peroxisome proliferator-activated receptor gamma Coactivator-1 alpha as a novel target for bipolar disorder and other neuropsychiatric disorders. Biol Psychiatry 83:761

Niu J, Li T, Yi C, Huang N, Koulakoff A, Weng C, Li C, Zhao CJ, Giaume C, Xiao L (2016) Connexin-based channels contribute to metabolic pathways in the oligodendroglial lineage. J Cell Sci 129:1902–1914

Nurnberger JI Jr, Koller DL, Jung J, Edenberg HJ, Foroud T, Guella I, Vawter MP, Kelsoe JR, For The Psychiatric Genomics Consortium Bipolar, G (2014) Identification of pathways for bipolar disorder: a meta-analysis. JAMA Psychiat 71:657–664

Öngür D, Drevets WC, Price JL (1998) Glial reduction in the subgenual prefrontal cortex in mood disorders. Proc Natl Acad Sci 95:13290

Ongür D, Jensen JE, Prescot AP, Stork C, Lundy M, Cohen BM, Renshaw PF (2008) Abnormal glutamatergic neurotransmission and neuronal-glial interactions in acute mania. Biol Psychiatry 64:718–726

Ortiz FC, Habermacher C, Graciarena M, Houry PY, Nishiyama A, Oumesmar BN, Angulo MC (2019) Neuronal activity in vivo enhances functional myelin repair. JCI Insight 5:e123434

Pantazopoulos H, Woo T-UW, Lim MP, Lange N, Berretta S (2010) Extracellular matrix-glial abnormalities in the amygdala and entorhinal cortex of subjects diagnosed with schizophrenia. Arch Gen Psychiatry 67:155–166

Papiol S, Schulze TG, Alda M (2018) Genetics of lithium response in bipolar disorder. Pharmacopsychiatry 51:206–211

Park LT, Lener MS, Hopkins M, Iadorola N, Machado-Vieira R, Ballard E, Nugent A, Zarate CA Jr (2017) A double-blind, placebo-controlled, pilot study of Riluzole monotherapy for acute bipolar depression. J Clin Psychopharmacol 37:355–358

Passos IC, Mwangi B, Vieta E, Berk M, Kapczinski F (2016) Areas of controversy in neuroprogression in bipolar disorder. Acta Psychiatr Scand 134:91–103

Peng L, Li B, Verkhratsky A (2016) Targeting astrocytes in bipolar disorder. Expert Rev Neurother 16:649–657

Picard F, Auwerx J (2002) PPAR(gamma) and glucose homeostasis. Annu Rev Nutr 22:167–197

Pirttimaki TM, Parri HR (2013) Astrocyte plasticity: implications for synaptic and neuronal activity. Neuroscientist 19:604–615

Rajkowska G (2000) Postmortem studies in mood disorders indicate altered numbers of neurons and glial cells. Biol Psychiatry 48:766–777

Rajkowska G, Halaris A, Selemon LD (2001) Reductions in neuronal and glial density characterize the dorsolateral prefrontal cortex in bipolar disorder. Biol Psychiatry 49:741–752

Ramaker RC, Bowling KM, Lasseigne BN, Hagenauer MH, Hardigan AA, Davis NS, Gertz J, Cartagena PM, Walsh DM, Vawter MP, Jones EG, Schatzberg AF, Barchas JD, Watson SJ, Bunney BG, Akil H, Bunney WE, Li JZ, Cooper SJ, Myers RM (2017) Post-mortem molecular profiling of three psychiatric disorders. Genome Med 9:72–72

Reppert SM, Weaver DR (2002) Coordination of circadian timing in mammals. Nature 418:935–941

Rivera AD, Butt AM (2019) Astrocytes are direct cellular targets of lithium treatment: novel roles for lysyl oxidase and peroxisome-proliferator activated receptor-γ as astroglial targets of lithium. Transl Psychiatry 9:211

Rivera A, Vanzuli I, Arellano JJ, Butt A (2016) Decreased regenerative capacity of oligodendrocyte progenitor cells (NG2-Glia) in the ageing brain: a vicious cycle of synaptic dysfunction, myelin loss and neuronal disruption? Curr Alzheimer Res 13:413–418

Rivera AD, Chacon-De-La-Rocha I, Pieropan F, Papanikolau M, Azim K, Butt AM (2021a) Keeping the ageing brain wired: a role for purine signalling in regulating cellular metabolism in oligodendrocyte progenitors. Pflugers Arch - Eur J Physiol 473:775

Rivera AD, Pieropan F, Chacon-De-La-Rocha I, Lecca D, Abbracchio MP, Azim K, Butt AM (2021b) Functional genomic analyses highlight a shift in Gpr17-regulated cellular processes in oligodendrocyte progenitor cells and underlying myelin dysregulation in the aged mouse cerebrum. Aging Cell n/a:e13335

Rodríguez JJ, Yeh C-Y, Terzieva S, Olabarria M, Kulijewicz-Nawrot M, Verkhratsky A (2014) Complex and region-specific changes in astroglial markers in the aging brain. Neurobiol Aging 35:15–23

Saab AS, Nave KA (2017) Myelin dynamics: protecting and shaping neuronal functions. Curr Opin Neurobiol 47:104–112

Sakry D, Neitz A, Singh J, Frischknecht R, Marongiu D, Biname F, Perera SS, Endres K, Lutz B, Radyushkin K, Trotter J, Mittmann T (2014) Oligodendrocyte precursor cells modulate the neuronal network by activity-dependent ectodomain cleavage of glial NG2. PLoS Biol 12:e1001993

Sakurai H, Dording C, Yeung A, Foster S, Jain F, Chang T, Trinh N-H, Bernard R, Boyden S, Iqbal SZ, Wilkinson ST, Mathew SJ, Mischoulon D, Fava M, Cusin C (2019) Longer-term open-label study of adjunctive riluzole in treatment-resistant depression. J Affect Disord 258:102–108

Sampedro-Piquero P, De Bartolo P, Petrosini L, Zancada-Menendez C, Arias JL, Begega A (2014) Astrocytic plasticity as a possible mediator of the cognitive improvements after environmental enrichment in aged rats. Neurobiol Learn Mem 114:16–25

Schür RR, Draisma LW, Wijnen JP, Boks MP, Koevoets MG, Joëls M, Klomp DW, Kahn RS, Vinkers CH (2016) Brain GABA levels across psychiatric disorders: a systematic literature review and meta-analysis of (1) H-MRS studies. Hum Brain Mapp 37:3337–3352

Scotti-Muzzi E, Umla-Runge K, Soeiro-De-Souza MG (2021) Anterior cingulate cortex neurometabolites in bipolar disorder are influenced by mood state and medication: a meta-analysis of (1)H-MRS studies. Eur Neuropsychopharmacol 47:62

Serafini G, Pardini M, Monacelli F, Orso B, Girtler N, Brugnolo A, Amore M, Nobili F, Team on Dementia of the IRCCS Ospedale Policlinico San Martino, D. M (2021) Neuroprogression as an illness trajectory in bipolar disorder: a selective review of the current literature. Brain Sci 11:276

Shao L, Vawter MP (2008) Shared gene expression alterations in schizophrenia and bipolar disorder. Biol Psychiatry 64:89–97

Sklar P, Smoller JW, Fan J, Ferreira MA, Perlis RH, Chambert K, Nimgaonkar VL, McQueen MB, Faraone SV, Kirby A, De Bakker PI, Ogdie MN, Thase ME, Sachs GS, Todd-Brown K, Gabriel SB, Sougnez C, Gates C, Blumenstiel B, Defelice M, Ardlie KG, Franklin J, Muir WJ, McGhee KA, MacIntyre DJ, McLean A, Vanbeck M, McQuillin A, Bass NJ, Robinson M, Lawrence J, Anjorin A, Curtis D, Scolnick EM, Daly MJ, Blackwood DH, Gurling HM, Purcell SM (2008) Whole-genome association study of bipolar disorder. Mol Psychiatry 13:558–569

Steardo L Jr, De Filippis R, Carbone EA, Segura-Garcia C, Verkhratsky A, De Fazio P (2019) Sleep disturbance in bipolar disorder: neuroglia and circadian rhythms. Front Psych 10:501

Stenovec M, Li B, Verkhratsky A, Zorec R (2020) Astrocytes in rapid ketamine antidepressant action. Neuropharmacology 173:108158

Stringer TP, Guerrieri D, Vivar C, Van Praag H (2015) Plant-derived flavanol (−)epicatechin mitigates anxiety in association with elevated hippocampal monoamine and BDNF levels, but does not influence pattern separation in mice. Transl Psychiatry 5:e493

Szulc A, Wiedlocha M, Waszkiewicz N, Galińska-Skok B, Marcinowicz P, Gierus J, Mosiolek A (2018) Proton magnetic resonance spectroscopy changes after lithium treatment. Systematic review. Psychiatry Res Neuroimaging 273:1–8

Takeda K, Watanabe T, Oyabu K, Tsukamoto S, Oba Y, Nakano T, Kubota K, Katsurabayashi S, Iwasaki K (2021) Valproic acid-exposed astrocytes impair inhibitory synapse formation and function. Sci Rep 11:23

Todtenkopf MS, Vincent SL, Benes FM (2005) A cross-study meta-analysis and three-dimensional comparison of cell counting in the anterior cingulate cortex of schizophrenic and bipolar brain. Schizophr Res 73:79–89

Toker L, Mancarci BO, Tripathy S, Pavlidis P (2018) Transcriptomic evidence for alterations in astrocytes and Parvalbumin interneurons in subjects with bipolar disorder and Schizophrenia. Biol Psychiatry 84:787–796

Toro CT, Hallak JE, Dunham JS, Deakin JF (2006) Glial fibrillary acidic protein and glutamine synthetase in subregions of prefrontal cortex in schizophrenia and mood disorder. Neurosci Lett 404:276–281

Trevisiol A, Saab AS, Winkler U, Marx G, Imamura H, Mobius W, Kusch K, Nave KA, Hirrlinger J (2017) Monitoring ATP dynamics in electrically active white matter tracts. elife 6:e24241

Vadnie CA, McClung CA (2017) Circadian rhythm disturbances in mood disorders: insights into the role of the suprachiasmatic nucleus. Neural Plast 2017:1504507

Veldic M, Millischer V, Port JD, Ho AM-C, Jia Y-F, Geske JR, Biernacka JM, Backlund L, McElroy SL, Bond DJ, Villaescusa JC, Skime M, Choi D-S, Lavebratt C, Schalling M, Frye MA (2019) Genetic variant in SLC1A2 is associated with elevated anterior cingulate cortex glutamate and lifetime history of rapid cycling. Transl Psychiatry 9:149

Verkhratsky A, Bush NA, Nedergaard M, Butt A (2018) The special case of human astrocytes. Neuroglia 1:21

Wake H, Lee PR, Fields RD (2011) Control of local protein synthesis and initial events in myelination by action potentials. Science 333:1647–1651

Walterfang M, Wood AG, Barton S, Velakoulis D, Chen J, Reutens DC, Kempton MJ, Haldane M, Pantelis C, Frangou S (2009) Corpus callosum size and shape alterations in individuals with bipolar disorder and their first-degree relatives. Prog Neuro-Psychopharmacol Biol Psychiatry 33:1050–1057

Wang J, Ferruzzi MG, Ho L, Blount J, Janle EM, Gong B, Pan Y, Gowda GA, Raftery D, Arrieta-Cruz I, Sharma V, Cooper B, Lobo J, Simon JE, Zhang C, Cheng A, Qian X, Ono K, Teplow DB, Pavlides C, Dixon RA, Pasinetti GM (2012) Brain-targeted proanthocyanidin metabolites for Alzheimer's disease treatment. J Neurosci 32:5144–5150

Wang J, Liu P, Zhang A, Yang C, Liu S, Wang J, Xu Y, Sun N (2020) Specific gray matter volume changes of the brain in unipolar and bipolar depression. Front Hum Neurosci 14:592419

Webster MJ, Knable MB, Johnston-Wilson N, Nagata K, Inagaki M, Yolken RH (2001) Immunohistochemical localization of phosphorylated glial fibrillary acidic protein in the prefrontal cortex and hippocampus from patients with schizophrenia, bipolar disorder, and depression. Brain Behav Immun 15:388–400

Webster MJ, O'Grady J, Kleinman JE, Weickert CS (2005) Glial fibrillary acidic protein mRNA levels in the cingulate cortex of individuals with depression, bipolar disorder and schizophrenia. Neuroscience 133:453–461

Wilkowska A, Szałach Ł, Cubała WJ (2020) Ketamine in Bipolar Disorder: A Review. Neuropsychiatr Dis Treat 16:2707–2717

Xiao L, Ohayon D, McKenzie IA, Sinclair-Wilson A, Wright JL, Fudge AD, Emery B, Li H, Richardson WD (2016) Rapid production of new oligodendrocytes is required in the earliest stages of motor-skill learning. Nat Neurosci 19:1210–1217

Zhang X, Alnafisah RS, Hamoud A-RA, Shukla R, Wen Z, McCullumsmith RE, O'Donovan SM (2021) Role of astrocytes in major neuropsychiatric disorders. Neurochem Res doi: 10.1007/s11064-020-03212-x. Online ahead of print

Astroglia Abnormalities in Post-stroke Mood Disorders

Tracey Singer, Sarah Ding, and Shinghua Ding

Overview of Stroke

Stroke is the leading cause of serious, long-term disability in the USA. Stroke can be categorized into two different types—ischemic stroke and hemorrhagic stroke. According to a recent report from American Heart Association (AHA), of all the strokes in the USA, 87% are considered ischemic stroke, 10% are intracerebral hemorrhage (ICH), and 3% are subarachnoid hemorrhage (SAH) (Virani et al. 2020).

Ischemic stroke occurs when the brain's blood vessels become narrowed or blocked (Moskowitz et al. 2010). This is problematic for the brain because it depends on arteries to carry fresh blood to it in order to bring in oxygen and glucose (as well as other nutrients) and remove carbon dioxide and cellular waste. If the arteries are blocked for too long, neurons, the most important cell type in central nervous system (CNS), cannot produce enough energy and will stop functioning and die. Ischemic stroke is most often caused by atherosclerosis, which is the buildup of fats, cholesterol, calcium, and other substances called plaque on artery walls (Campbell et al. 2019; Moskowitz et al. 2010; Lo et al. 2003). Atherosclerosis is a progressive disease and begins with damage to the inner layer of an artery. Over time, the plaque will harden, and arteries will become narrower, limiting blood flow and eventually forming blood clots, further narrowing arteries and limiting the flow of oxygen-rich blood. A blood clot may completely block blood flow or break apart and trigger a stroke. Although the cause of atherosclerosis is not known, factors

T. Singer · S. Ding
Dalton Cardiovascular Research Center, Columbia, MO, USA

S. Ding (✉)
Dalton Cardiovascular Research Center, Columbia, MO, USA

Department of Biomedical, Biological and Chemical Engineering, University of Missouri, Columbia, MO, USA
e-mail: dings@missouri.edu

B. Li et al. (eds.), *Astrocytes in Psychiatric Disorders*, Advances in Neurobiology 26, https://doi.org/10.1007/978-3-030-77375-5_6

such as high blood pressure, high cholesterol, obesity, diabetes, and tobacco may increase risk for the disease.

Ischemic stroke can be further categorized into thrombotic stroke and embolic stroke (Barber and Demchuk 2003; Dirnagl et al. 1999). Thrombotic stroke, or cerebral thrombosis, is caused by a blood clot (thrombus) that blocks blood flow in a damaged cerebral artery leading to or within the brain. Large-vessel thrombosis occurs when arteries such as the carotid or middle cerebral are blocked, and small-vessel thrombosis occurs when the brain's smaller, yet deeper, penetrating arteries are blocked. Embolic stroke, or cerebral embolism, is caused by a wandering blood clot (embolus) in an artery somewhere other than the brain such as the neck or heart. The bloodstream carries the clot, which blocks a blood vessel in or leading to the brain (embolism). Again, this restricts the flow of blood to the brain and results in near-immediate physical and neurological problems. The main cause of embolism is atrial fibrillation, an irregular heartbeat which has the potential to develop blood clots within the heart. These clots can then circulate to the brain, causing ischemic stroke. The risk of stroke from atrial fibrillation depends on many factors including age, high blood pressure, and diabetes.

Hemorrhagic stroke, on the other hand, is caused by the leaking or bursting of blood vessels (Moskowitz et al. 2010). The leaked blood from vessels in the brain puts pressure on brain cells and damages them. Hemorrhagic strokes can be a result of conditions affecting blood vessels, including high blood pressure, overtreatment of blood thinners, aneurysms, trauma, cerebral amyloid angiopathy (protein deposits in blood vessel walls that weaken vessel walls), and arteriovenous malformation (the rupture of an abnormal tangle of thin-walled blood vessels). Intracerebral hemorrhage is the most common type of hemorrhagic stroke, occurring when brain arteries burst and flood surrounding tissue with blood. Subarachnoid hemorrhage refers to bleeding in the subarachnoid space, which is the area between the brain and tissues that surround the brain.

There are a number of detectable clinical symptoms and pre-stroke warning signs (Saengsuwan et al. 2017; Mochari-Greenberger et al. 2014) including sudden unilateral weakness (paralysis of face, arm, or leg on one side of the body), sudden trouble with speaking and understanding speech, sudden trouble with walking, loss of balance, sudden blackened or blurred vision in one or both eyes, double vision, and sudden and severe headaches perhaps accompanied by vomiting, dizziness, or altered consciousness. These symptoms are dependent on which side and region of the brain are affected and how severely the brain is damaged. It is often the case that only one side of the body is affected.

After a stroke, some patients will recover fully, others will be disabled long term or for the rest of their lives, and some will die if the damage to their brain is too severe. Approximately 3% of males and 2% of females reported that they were disabled because of stroke (Virani et al. 2020). Some problems that patients may experience after stroke can be managed by medical professionals through different types of rehabilitation therapy (Campbell et al. 2019). Regaining the activities of daily living become the focus of rehabilitation after stroke. For stroke survivors, the first 3 months post stroke are the most important period for recovery and when patients show the most improvement (Grefkes and Fink 2020; Ronning and Guldvog 1998; Poulin et al. 2016). The goal of rehabilitation is to restore function as close as possible to "normal" or find strategies to work around new limitations. For example, speech therapy can help those who have trouble understanding or producing speech. Physical therapy helps with relearning movement and coordination skills lost after stroke which may include paralysis and/or weakness on one or both sides of the body. Occupational therapy improves daily activities such as eating, drinking, dressing, bathing, reading, and writing. Stroke patients may have trouble with chewing and swallowing or bladder and bowel control, so occupational therapists may also help them regain these skills. Even though rehabilitation psychologists and neuropsychologists can help, many patients still suffer long-term complications (effects). These complications are not only physical symptoms such as weakness, paralysis, difficulty swallowing, gait instability, and falls and fractures but also pain syndromes, depression, anxiety, pseudobulbar affect, epilepsy, cognitive impairment, and dementia (Virani et al. 2020).

Stroke has a large impact on public health and significant financial costs. A comprehensive study was conducted on 97,374 hospitalizations with a primary or secondary diagnosis of stroke from 2006 to 2008. It found that ischemic, hemorrhagic, and other types of strokes had average hospitalization costs of $18,963, $32,035, and $19,248, respectively. Overall, hemorrhagic stroke cost $14,499 more than ischemic stroke. In addition, the direct medical cost of stroke in 2008 in the USA was estimated to be $18.8 billion, and for the same year, the estimated per-person expenditure for stroke care was $7657 (Wang et al. 2014a). These costs largely increased based on the 2020 update of Heart Disease and Stroke Statistics from America Heart Association (AHA). According to the update, the average annual direct medical cost of stroke in the USA was $28.0 billion and direct care was estimated at $7902 in 2014 and 2015 (Virani et al. 2020). The costs of hospitalizations involving stroke are high and vary greatly by type of stroke, diagnosis status, and comorbidities.

In this chapter, before briefly describing the pathological processes of ischemic stroke, we will focus on the role of astrocytes in post-stroke mood disorders.

Pathological Processes of Ischemic Stroke

Acute Responses and Cell Death After Ischemic Stroke

Healthy brain tissue has a high level of consumption of oxygen and glucose. Neuronal function relies on continuous ATP production which requires oxygen and glucose from the blood. When a focal ischemia stroke takes place, cerebral blood flow is less than 20% of the normal standard at the ischemic core (IC) (Dirnagl et al. 1999; Lo et al. 2003). Impairment of cerebral blood flow restricts delivery of glucose and oxygen, impairing the energetics required for ion gradients and leading to a loss of membrane potential in glia and neurons.

In focal ischemic stroke (FIS), the regions determine the pattern of cell death. Necrosis occurs in the IC in the acute phase, causing irreversible tissue loss while apoptosis occurs in the penumbra or peri-infarct region (PIR), which is a region between the lethally damaged IC and the normal brain. The PIR has partially preserved energy metabolism (moderately hypo-perfused) and retains structural integrity but has lost or impaired function. While cells in the IC can undergo permanent anoxic depolarization, cells in the PIR can repolarize with more energy consumption. This region can progress to infarction due to persistent excitotoxicity, spreading depolarization, inflammation, and apoptosis (Dirnagl et al. 1999). Therefore, the goal of neuroprotection is to salvage the ischemic penumbra. Alternative blood flow pathways, called collaterals, can sustain viability in penumbral regions, although the extent of collateral flow varies between individuals (Dirnagl et al. 1999). Patients with good collateral blood flow have infarcts that progress slower, while those with poor collateral blood flow display a more rapid progression of infarction.

Anoxic depolarization develops minutes after ischemia in neurons. As a result of the loss of membrane potential, voltage-dependent Ca^{2+} channels are activated, and excitatory amino acids, including glutamate, are released into the extracellular space (Campbell et al. 2019; Lo et al. 2003). Since energy-dependent reuptake is impeded, glutamate accumulation in the extracellular space is exacerbated. *N*-methyl-D-aspartate receptors (NMDARs) are blocked by extracellular Mg^{2+} under normal conditions; however, upon depolarization, the Mg^{2+} is removed, which leads to substantially higher conduction (Nowak et al. 1984) and subsequent Ca^{2+} influx and release from intracellular stores (Soriano et al. 2008; Wu and Tymianski 2018; Mayer and Miller 1990). Glutamate accumulation and subsequent Ca^{2+} overloading have diverse consequences. Ca^{2+} overloading in neurons following bioenergetic failure activates numerous Ca^{2+}-dependent enzymatic reactions and subsequently induces lipolysis, proteolysis, breakdown of ion homeostasis, and disaggregation of microtubules, causing cell death and tissue damage in the IC during acute phase after ischemic stroke (Folbergrova et al. 1995). The Ca^{2+} increase also activates neuronal nitric oxide synthase (NOS) and results in free radical production and cell death processes such as apoptosis, necrosis, necroptosis, and autophagy. NO reacts with a superoxide anion to form peroxynitrite that promotes tissue damage due to its highly reactive nature (Iadecola et al. 1997; Barber and Demchuk 2003). Additionally, the activation of phospholipase A_2 and cyclooxygenase generates free-radical species leading to lipid peroxidation and membrane damage. An increase in free radicals leads to a leaky mitochondrial membrane and subsequent cytochrome C release, which triggers apoptosis. Studies have shown that mice whose iNOS gene expression was induced with mRNA had larger infarcts and motor deficits produced by occlusion compared to iNOS knockout mice (Iadecola et al. 1997).

Ca^{2+} increase was also observed in astrocytes in the acute phase following FIS. It was reported that astrocytes exhibited enhanced Ca^{2+} signaling in IC and PIR after photothrombosis (Ding et al. 2009), which may in turn induce more glutamate release from astrocytes. Deletion of IP_3R2 in astrocytes could reduce brain infarction and limit motor function deficits (Li et al. 2015; Dong et al. 2013). These results suggest that, in addition to neuronal Ca^{2+} signaling, astrocytic Ca^{2+} increase also contributes to brain damage through a non-cell autonomous effect.

There are two major modes of ischemic cell death: necrotic and apoptotic. Necrosis only occurs after exogenous insults. Following acute, permanent vascular occlusion, necrosis is the predominant form of cell death taking place in the IC, where the cells shrink and become very electron-dense (Lipton 1999). A hallmark of necrosis is the double-stranded breakdown of DNA into nucleosomal segments, which manifests as DNA laddering with fragments that are multiples of around 200 bp (Bonfoco 1995; Orrenius 1995). Overactivation of poly(ADP-ribose) polymerase (PARP), an NAD^+-consuming enzyme, has been proposed to lead to necrosis by excessive energy loss (Eliasson et al. 1997; Yuan 2009). Deletion of PARP-1 or use of PARP-1 inhibitors is neuroprotective in vitro and in vivo following ischemia (Eliasson et al. 1997; Zhang et al. 1994). Calpain is a cytosolic Ca^{2+}-activated protease and is known to be highly upregulated after ischemia (Vosler et al. 2011; Bano et al. 2005). Its activation from Ca^{2+} overloading under ischemic conditions contributes to neuronal cell death by cleaving multiple substrates such as cytoskeletal and associated proteins, kinases, phosphatases, membrane receptors, and transporters. Group I mGluR inhibitors 2-methyl-6-(phenylethynyl)pyridine (MPEP) and LY36738 could inhibit chaplain activation and thus reduce brain infarction (Li et al. 2013a).

Different from the necrotic death in the IC, neurons in the PIR may undergo apoptosis several hours or days after the onset of ischemic stroke. The protein families of Bcl-2 and caspase play critical roles in the activation, signal transduction, and execution of apoptosis (Broughton et al. 2009; Yuan 2009). Among the identified caspases, caspases 1, 8, and 9 appear to have key roles in ischemia-related apoptosis (Broughton et al. 2009; Yuan 2009). Caspases are activated when cytochrome C is released from the mitochondria and activates an apoptosome complex in the presence of dATP (Broughton et al. 2009). Activated caspases are aspartate-specific cysteine proteases that cleave enzymes and modify proteins which affect homeostasis and repair. Caspase activity can be blocked by the administration of small peptides which bind covalently to the catalytic pocket and alkylate the cysteine at position 70. Caspase inhibitors have been shown to decrease the volume of dead tissue in ischemia as well as neurological deficits (Hara et al. 1997). When ischemia is mild, inhibitors are particularly effective since they can be coupled with MK801, an NMDA-receptor antagonist, or growth hormones like fibroblast growth factor. While NMDA-receptor antagonists are typically administered before or immediately after ischemia, caspase inhibitors may still reduce injury even when injected many hours later. TNF1α, a cytokine, is also involved in apoptosis. TNF1α can exacerbate injury by causing a rapid decrease in mitochondrial membrane potential and thus impair mitochondrial function and increase caspase 8 activity, resulting in the release of cytochrome c from the mitochondria (Doll et al. 2015). Apoptosis-inducing factor (AIF) translocation from mitochondria to nuclei is involved in caspase-independent apoptosis (Wang et al. 2016; Broughton et al. 2009). NAD^+

repletion or neuronal overexpression of NAMPT, the rate-limiting enzyme in the NAD^+ salvage pathway, can reduce glutamate- and OGD-induced apoptosis through suppressing AIF translocation (Wang et al. 2014b, 2016). It is worth mentioning that one insult may lead to more than one mode of death in the same cell population, and cells may also manifest signs of multiple forms of death (Wang et al. 2016).

Astrocytes and Reactive Astrogliosis After Ischemic Stroke

Astrocytes are starlike cells and the most abundant glial cell type in the CNS. Classically, based on morphology and specific protein markers, there are two major types of astrocytes in the adult brain: fibrous astrocytes, present in white matter tracts such as the corpus callosum, and protoplasmic astrocytes, present in gray matter such as the cortex. Mature astrocytes have the following eight criteria (Kimelberg 2010): (a) electrically non-excitable; (b) a very negative membrane potential determined by the transmembrane K^+ gradient; (c) expression of functional transporters for glutamate and GABA uptake; (d) a large number of intermediate filament bundles, which are the sites of the astrocyte-specific protein glial fibrillary acidic protein (GFAP); (e) having glycogen granules; (f) processes from each cell surrounding blood vessels; (g) many more processes from each cell surrounding synapses; and (h) linkage to other astrocytes by gap junctions consisting of connexins (CX) 43 and 30.

Astrocytes are traditionally considered as supporting cells for maintaining ionic homeostasis and providing growth factors for neurons and structural support in the CNS. Protoplasmic astrocytes exhibit the following functions under normal conditions (Kimelberg and Nedergaard 2010; Kimelberg 2010): (a) extracellular K^+ buffering, (b) control of extracellular H^+ and brain pH, (c) uptake of glutamate and GABA with their transporters, (d) mobilizing intracellular Ca^{2+} stores by activation of G-protein-coupled receptors (GPCRs) such as mGLuR5 and P2Y, (e) regulation of cerebral blood flow, (f) control of water transport by aquaporin (AQP) water channel, (g) astrocyte-neuron lactate shuttle, and (h) modulation and control of synaptic activity.

Although many astrocytes conform to the aforementioned criteria and functional roles, they are heterogeneous in morphology, molecular expression, and physiological function (Zhang and Barres 2010; Matyash and Kettenmann 2010). Morphologically, a protoplasmic astrocyte is highly branched and has several primary processes each with elaborated sub-branched fine process arborizations to form a bush-like territory with little overlap between neighboring astrocytes (Wilhelmsson et al. 2006; Bushong et al. 2002). A fibrous astrocyte has thicker and less branched processes with a high degree of overlap with neighboring astrocytes. Intermediate filament protein GFAP is primarily expressed in the thick main processes in astrocytes and has been considered as a "pan-astrocyte" marker. Transcriptomic study has revealed that the Aldh1L1 gene is the most widely and homogeneously expressed in astrocytes and that the Aldh1L1 protein is highly expressed in the cell body and extensive processes of an astrocyte (Cahoy et al. 2008). Therefore, Aldh1L1 is now considered as a new "pan-astrocyte" marker. Growing evidence indicates that astrocytes also play an active role through the "tripartite synapse," which includes the pre- and post-synaptic neuron as well as the surrounding astrocyte. Astrocytes can modulate synaptic function, wakefulness, cognition, and memory (Araque et al. 1999; Halassa et al. 2007, 2009; Haydon 2001; Volterra and Meldolesi 2005; Khakh and Sofroniew 2015; Santello et al. 2019; Cui et al. 2018; Suzuki et al. 2011). Therefore, astrocytes are global controllers in the CNS which coordinate both local responses and those over larger distances.

Astrocytes are also involved in neurological disorders, including ischemic stroke (Seifert et al. 2006; Nedergaard and Dirnagl 2005; Maragakis and Rothstein 2006; Ding et al. 2009; Choudhury and Ding 2016; Ding 2014; Li et al. 2015). Ischemia induces a variety of molecular and cellular changes in astrocytes at the PIR, including cellular proliferation, morphology, and gene expression in a temporally and spatially dependent manner. Reactive astrogliosis and subsequent formation of a glial scar are the hallmarks of focal ischemic stroke (Lo 2008; Choudhury and Ding 2016; Ding 2014). The activated astrocytes are therefore called reactive astrocytes (RAs) that can be detected by enhanced expression of GFAP and many other proteins, as well as dramatic morphological changes (Choudhury and Ding 2016; Li et al. 2014; Sofroniew and Vinters 2010).

RAs in the PIR exhibit spatiotemporally dependent changes in morphology (hypotrophy), proliferation capacity, function, and gene expression during the subacute phase (Barreto et al. 2011; Li et al. 2014; Choudhury and Ding 2016). Morphologically, RAs become hypertrophic with large processes that can be revealed by GFAP staining. After a prolonged time following focal ischemic stroke, the morphology of RAs remains stable, the proliferation of RAs ceases, and a glial scar is formed (Barreto et al. 2011; Li et al. 2014; Ding 2014; Choudhury and Ding 2016; Burda and Sofroniew 2014; Voskuhl et al. 2009). Reactive astrogliosis and glial scar formation eventually cause substantial tissue remodeling and permanent structural changes in the penumbra. Analysis of immunostaining for GFAP and individually dyed cells in astrogliosis induced in the cortex or hippocampus by electrically induced lesion revealed that astrocyte processes become thicker and bushier (Wilhelmsson et al. 2006; Li et al. 2014; Zhang et al. 2020b).

The developing scar contains extracellular matrix components, and this extracellular proteoglycan deposition is part of the initial response that likely limits the spread of damage. The scar narrows as it matures, but persisting features remain, such as the upregulation of GFAP in astrocytes and their tightly intertwined processes. Selective impairment of scar formation leads to a greater spread of tissue damage and worse neurological outcomes. The impaired restoration of the blood-brain barrier leads to greater infiltration of leukocytes, inflammatory cell spread, and increased neuronal loss, which were observed when features of glial scar maturation were disrupted by transcription 3 (STAT3) in astrocytes (Wanner et al. 2013). Treatment with DAPT, which inhibits the Notch-activating enzyme γ -secretase, reduces damage to the ischemic brain, possibly by causing a reduction in the proliferation of reactive astrocytes. From this treatment or conditional knockout of the receptor, mice with impaired Notch signaling exhibited more microglia and macrophage invasion (Shimada et al. 2011; LeComte et al. 2015). It has been suggested that proliferation and differentiation of reactive astrocytes is promoted by Notch signaling via the translocation of the transcription factor Olig2 from the cytosol to the nucleus (LeComte et al. 2015).

On the other hand, the proliferation rate reaches a peak 3–4 days after FIS and recedes thereafter (Li et al. 2014; Barreto et al. 2011). Thus, metabolically, RAs are highly active on day 3–4 post-stroke to meet high energetic and biosynthetic demands for proliferation. Proliferation of RAs occurs mostly within 200μm from the IC and allows for the development of a glial scar after ischemia (Choudhury and Ding 2016; Zhang et al. 2020b; Li et al. 2014). Some astrocytes in the tissue adjacent to the infarct are actually derived from neural stem cells which migrate from the subventricular zone, and some survive as a part of the mature glial scar weeks after stroke (Faiz et al. 2015). Other cells, including microglia and/or macrophages, peri-vascular pericytes, or stromal cells, also proliferate (Barreto et al. 2011; Li et al. 2014; Fernandez-Klett et al. 2012). Many signaling pathways, including mitogen-activated protein kinases (MAPK), Notch signaling pathway, STAT3, and transforming growth factor beta (TGF-β) signaling, may participate reactive astrogliosis after ischemic stroke (Choudhury and Ding 2016).

Most changes in gene expression in RAs are due to altered cellular responses in the PIR, based on gene array analysis of RAs acutely isolated from brain tissue following MCA occlusion (Zamanian et al. 2012). For many genes, increased expression peaked 1 to 3 days after the onset of stroke, while a small proportion of genes including GFAP and some chemokines had increasing expression up to 1 week after stroke. Evidence also suggests that astroglia are influenced by specific conditions associated with the insult rather than an all-or-none response (Zamanian et al. 2012).

RAs have been demonstrated to have protective effects after FIS and injury (Linnerbauer and Rothhammer 2020; Myer et al. 2006; Faulkner et al. 2004; Anderson et al. 2016; Choudhury and Ding 2016). Proliferating RAs may impact tissue preservation, repair/remodeling, and functional outcome. Recent studies indicate RAs are also involved in depression, anxiety, cognition, and post-stroke mood shifting.

Post-stroke Symptoms and Post-stroke Mood Disorder (PSMD)

Mood disturbances are also frequent symptoms in stroke survivors (Kim 2016). After a stroke, many patients not only have some physical disability including difficulties in moving, speaking, and seeing, but it is common that patients may also exhibit changes in mood or emotion even after rehabilitation therapy (Schottke and Giabbiconi 2015). Anger, frustration, lack of motivation, and crying or laughing for the wrong reasons are also common for stroke patients. Post-stroke depression (PSD) (Robinson and Jorge 2015), post-stroke anxiety (PSA) (Maaijwee et al. 2016), and pseudobulbar affect (PBA) (Gillespie et al. 2016) are common post-stroke mood disorders (PSMD).

Symptoms of PSD include depressed mood, anhedonia, loss of energy, decreased concentration, and psychic retardation. It is characterized by feelings of overarching sadness, lack of pleasure, or changes in eating and sleeping patterns. Multiple studies have found that PSD affects between one-third to two-thirds of stroke survivors (Schottke and Giabbiconi 2015; Fang et al. 2017; Ayerbe et al. 2013; Kim 2016; Robinson and Jorge 2015; Virani et al. 2020), although its prevalence decreases over time. A systematic review of patients found that depression is associated with increased disability and mortality (Ayerbe et al. 2013).

PSA affects about 20% of survivors and occurs when they focus on worries and feel anxious for no particular reason (Schottke and Giabbiconi 2015; Wright et al. 2017). The core symptoms of PSA are excessive anxiousness or worry and difficulty in controlling worries, restlessness, decreased energy, poor concentration, irritation, nervous tension, and insomnia (Maaijwee et al. 2016). Meta-analyses have suggested that there is a significant correlation between PSD and PSA (Schottke and Giabbiconi 2015; Wright et al. 2017).

PBA is characterized by a mismatch between feelings and expression. Individuals may cry or laugh in an uncontrolled manner. Crying is a more common PBA presentation following stroke than laughing (Gillespie et al. 2016). PBA affects approximately one in five stroke survivors at the acute and post-acute phases and one in eighth survivors 6 months or more post-stroke. Research indicates that PBA is more common in survivors of brainstem stroke, but it can also occur with other types of strokes (Balakrishnan and Rosen 2008).

Tools for cognitive and mood assessment have been suggested (Quinn et al. 2018). A framework for assessment emphasizes the need for differing approaches to testing at differing points in the stroke pathway rather than a comprehensive critique of all cognitive and mood assessment tools. Cognitive and mood problems are both associated with poor outcomes of stroke. Clinically, a number of drug and psychosocial treatments have been assessed, but the results have been disappointing. These drug trials are generally of poor quality and do not provide sufficient information to judge their true costs and benefits (Hackett et al. 2008). Psychosocial interventions are popular with patients, but there is conflicting evidence for their effectiveness in either treating or preventing anxiety and depression (Gao et al. 2016; Wu et al. 2012). The problem-based and behavioral therapies seem to be promising for patients with stroke (Hill et al. 2019). Improved coping skills should result in reduced psychological distress and rates of depression. A licensed mental health practitioner (therapist or nurse) and the individual actively worked together toward recovery based on a psychological assessment. Standardized measures of mood (28-item General Health Questionnaire/GHQ-28), cognitive state (mini-mental state examination), and function (Barthel ADL Index, Frenchay Activities Index) were taken at different times. Hill et al. reported that 6 months later, all psychological and activity measures favored problem-solving therapy. At 12 months, patients in the problem-solving therapy group had significantly lower GHQ-28 scores and lower median Present State Examination symptom scores (Hill et al. 2019). In addition, the problem-solving therapy group was more satisfied with some aspects of care.

Astrocytes in Major Depressive Disorder (MDD) and PSMD

MDD is the most prevalent form of depression, often manifested with a long-lasting and recurrent psychiatric condition. It is reported that MDD affects 20% of the population throughout their lifetimes (Kessler and Bromet 2013), and it is now considered the leading cause of disability worldwide. The symptoms of MDD included depressed (low) mood; anhedonia; feelings of hopelessness (despair), worthlessness, or guilt; changes in appetite, weight, and sleep; an inability to feel pleasure (anhedonia); fatigue; and suicidal ideation. Studies suggest that the neural activities of specific brain circuits are altered in response to external stimuli, such as stress, as a result of maladaptive molecular and cellular changes in MDD. Astrocytes interact intimately with neurons to support and regulate essential functions and mediate information processing through tripartite synapses in the brain, where astrocyte processes wrap tightly around pre-synaptic and post-synaptic sites (Araque et al. 1999; Halassa et al. 2007; Haydon 2001; Volterra and Meldolesi 2005; Khakh and Sofroniew 2015). Therefore, astrocytes have been receiving increasing attention in mood disorders since significant abnormalities were observed in the postmortem brain of MDD patients. Indeed, studies using postmortem brain specimen and animal models of depression provide a large body of evidence that astrocytes play an important role in MDD and PSMD.

Astrocyte in MDD

Studies from Human Postmortem Brain Specimens

Brain imaging studies have revealed marked volume reductions in the hippocampus and medial prefrontal cortex (mPFC) in MDD (Drevets 2000; Sheline 2003). Studies from human specimens showed that abnormalities in glial cells may alter normal brain function and likely contribute to mood disorder development. Profound alterations of astrocytes connecting to mood disorders were observed from postmortem brain specimen. These include changes in cell number and cell morphology in patients who had mood disorders. Prominent reductions in glial cell number and packing density have been reported in independent laboratories based on studies using postmortem brains from subjects with mood disorders in different regions of the PFC, the anterior cingulate cortex (Gittins and Harrison 2011; Cotter et al. 2001a), and the amygdala (Rajkowska 2000; Altshuler et al. 2010; Ongur et al. 1998; Bowley et al. 2002). Such a striking cellular deficit in glial number suggests that glia may be unique targets for novel strategies in the treatment of mood disorders.

Altered expression of astrocyte-specific biomarkers was also observed in postmortem brain specimens of individuals that had suffered from mood disorders, for example, low levels of GFAP have been found in the hippocampus, PFC, anterior cingulate, and amygdala (Webster et al. 2001; Gittins and Harrison 2011; Altshuler et al. 2010). In postmortem brains of suicide completers, reduced GFAP mRNA and protein in the mediodorsal thalamus and caudate nucleus were observed in depression-related suicides compared with controls, suicides not linked to depression (Torres-Platas et al. 2016). Furthermore, a regional comparison revealed that GFAP expression in both subcortical regions was, on average, between 11- and 15-fold greater than in the cerebellum and neocortex. Examining astrocyte morphology by immunohistochemistry showed that astrocytes in both thalamus and caudate displayed larger cell bodies and extended more ramified processes across larger domains than the previously described cortical astrocytes. This study reveals that astrocytic abnormalities are not brain wide and suggests that they are restricted to cortical and subcortical networks known to be affected in mood disorders.

S100B proteins are a family of acidic proteins that influence cellular responses along calcium signal transduction pathways primarily produced by astrocytes. It was found that S100B was elevated in the blood serum and cerebrospinal fluid of MDD patients, and antidepressant treatment could lower S100B levels in parallel with reduced depressive symptoms (Matthias et al. 2013; Schroeter et al. 2002). In addition, a systematic and quantitative meta-analysis demonstrated that both young and older subjects suffering from mood disorders showed elevated S100B values in blood serum compared to control subjects. Also, higher levels of S100B were found in older compared with younger adult subjects, indicating that glial pathology is modified by age in mood disorders (Schroeter et al. 2011). Thus, S100B can be considered as a new marker for MDD.

Changes to gene transcription and protein expressions relating to normal astrocyte function were also found in patients diagnosed with mood disorders. For example, gene and protein expression of some astrocytic function-related proteins, including glutamine synthetase, glutamate transporters, and even gap junction proteins, was downregulated in patients with depression (Sequeira et al. 2009; Bernard et al. 2011). Altered cortical glutamatergic and GABAergic signal transmission in depression is associated with downregulation of high-affinity glutamate transporters GLT1 and glutamine synthetase (Choudary et al. 2005). Using immunostaining, a reduction of blood vessel coverage by AQP4 astrocytic endfeet was reported in MDD subjects (Rajkowska et al. 2013).

Postmortem histopathologic studies of depressed patients strongly suggest that the abnormalities in astrocytic functions, such as regulation of water homeostasis, blood flow, glucose transport and metabolism, the blood-brain barrier, glutamate and GABA turnover, and synaptic plasticity, may contribute to the pathophysiology of MDD.

Studies from Animal Models

While studies using postmortem brain specimens provide insights into pathological changes of astrocytes in MDD, studies using animal models may provide the underlying molecular mechanisms by which astrocytes participate in disease etiology. Stressful life events including acute and chronic stress increase the risk for MDD in humans (Arsenault-Lapierre et al. 2004). Thus, in animal experimental studies on MDD, mice and rats are exposed to acute stress (e.g., tail suspension test, forced swim test, unavoidable foot shock test, social defeat test, or restraint stress) or chronic unpredictable stress (CUS) to induce depressive-like states (Henn and Vollmayr 2005).

Due to the nature of tripartite synapses, astrocyte-derived substances such as glutamate, D-serine, lactate, and ATP can actively modulate synaptic function. Meanwhile, the changes in the expression of astrocyte-specific glutamate transporters, such as GLT1 and GLAST1; glutamate receptors, such as mGluR5; enzymes, such as glutamate synthesis and lactate dehydrogenase; and ion channels and gap junctions may affect/regulate ion homeostasis, synaptic transmission, pH, redox state, etc. These multiscale and multimodal interactions of astrocytes with synaptic circuits are likely to contribute to behavior. Because neuronal synaptic changes are a hallmark of depression and astrocytes are integral to tripartite synapses, astrocytes are very likely to participate in the pathology of depression and anxiety disorders. Indeed, a large number of studies from rodent model have provided evidence of astrocytic involvement in MDD. For extensive reviews of astrocytes in this area, readers are advised to consult a few detailed reviews (Wang et al. 2017; Bender et al. 2016; Park and Lee 2020). Here, I will briefly summarize the recent progress in this area.

An attractive hypothesis for MDD is that impaired ATP release from astrocytes causes MDD (Illes et al. 2020; Cao et al. 2013). Cao et al. identified ATP as a key factor involved in astrocytic modulation of depressive-like behavior in the hippocampus and prefrontal cortex of adult mice with chronic social defeat stress (Cao et al. 2013). Low ATP abundance was observed in the brains of mice that were susceptible to chronic social defeat. Furthermore, they found that the administration of ATP induced a rapid antidepressant-like effect in these mice. Both a lack of inositol 1,4,5-trisphosphate receptor type 2 (IP_3R2) and transgenic blockage of vesicular gliotransmission-induced deficiencies in astrocytic ATP release caused depressive-like behaviors that could be rescued via the administration of ATP. Using transgenic mice that express a Gq G-protein-coupled receptor (GPCR) only in astrocytes to enable selective activation of astrocytic Ca^{2+} signaling, they also found that stimulating endogenous ATP release from astrocytes induced antidepressant-like effects in mouse models of depression. Moreover, P2X2 receptors in the medial prefrontal cortex mediated the antidepressant-like effects of ATP. Their results highlight astrocytic ATP release as a biological mechanism of MDD. However, the data above using IP_3R2-deficient mice is inconsistent with a study by Petravicz et al. (Petravicz et al. 2014), who found that the mice did not show any abnormalities in the open field test or the tail suspension test. Therefore, it remains difficult to clarify the relationship between astrocytic Ca^{2+} signaling and depression, suggesting a complex nature of depression and the lack of decisive behavior tests to evaluate it.

The lateral habenula (LHb) is a nucleus that relays information from the limbic forebrain to multiple monoamine centers and has recently emerged as a key brain region in regulating negatively motivated behavior and the pathophysiology of major depression (Cui et al. 2018; Li et al. 2013b; Hu et al. 2020; Baker et al. 2016; Yang et al. 2018). In a rat depression model of congenitally learned helplessness, Cui et al. recently discovered that Kir4.1 in astrocytic membrane processes was upregulated in the lateral habenula (LHb) (Cui et al. 2018). Their study using electrophysiology and modeling data shows that the level of Kir4.1 in astrocytes tightly regulates the degree of membrane hyperpolarization and the amount of bursting activity of LHb neurons. Astrocyte-specific gain and loss of Kir4.1 in the LHb bidirectionally regulates neuronal bursting and depression-like symptoms. Together, their results show that a glia-neuron interaction at the perisomatic space of LHb is involved in setting the neuronal firing mode in models of a major psychiatric disease. Therefore, targeting Kir4.1 in the LHb might be a potential strategy for treating clinical depression.

Neuroinflammation contributes to the cognitive impairments accompanying many neurological disorders. Chronic stress and neuroinflammation are considered to be fundamental in the etiology of MDD. Astrocyte dysfunction and inflammation have been proven to be associated with the pathogenesis of MDD. Using mouse depression models of 6 weeks of chronic unpredictable mild stress (CUMS) or 10 days of lipopolysaccharide (LPS) intraperitoneal injection, Leng et al. explored intermediary components that are modulated by stress and neuroinflammation (Leng et al. 2018). They found that multiple endocrine neoplasia type 1 (Men1; protein: menin) expression is attenuated in the brain of mice exposed to CUMS or LPS. Astrocyte-specific deletion of Men1 (GcKO) led to depressive-like behaviors in mice and enhanced IL-1β production through NF-kB activation. Further, they observed that depressive-like behaviors in GcKO mice could be restored by an NF-kB inhibitor or an IL-1β receptor antagonist. Importantly, they identified a SNP in human MEN1, where G503D substitution is associated with a higher risk of MDD onset. G503D substitution abolished menin-p65 interactions, thereby enhancing NF-kB activation and IL-1β production. The study revealed a distinct role of tumor suppressor menin in astrocytes in regulating astrocytic inflammation in depression, and menin may be an attractive therapeutic target in MDD.

IL-10 is a key cytokine that is mainly produced by astrocytes and microglia and that represses excessive inflammatory responses. In the CNS, IL-10 is upregulated after various insults. Astrocytes isolated from transgenic mice deficient with IL-10 are prone to characteristic A1 reactive astrocytes and these transgenic mice exhibited increased immobility time in the forced swim test and defective learning and memory behavior in the Morris water maze test (Zhang et al. 2020a), suggesting IL-10 contributes to the depression-like behavior and memory deficits.

It was found that local application of TFNα at pathological levels activates astrocyte TNF receptor type 1 (TNFR1), which in turn triggers an astrocyte-neuron signaling cascade that results in persistent functional modification of hippocampal excitatory synapses. Astrocytic TNFR1 signaling induced hippocampal synaptic alteration and contextual learning-memory impairment. This process may contribute to the pathogenesis of cognitive disturbances (Habbas et al. 2015). High extracellular levels of TNFα can trigger release of glutamate from astrocytes (Santello et al. 2011). Thus, the mechanism by which the cytokine TNFα affects cognitive disturbances upon CNS inflammation is via a local increase of TNFα in the hippocampal dentate gyrus that activates TNFR1, which then triggers an astrocyte-neuron signaling cascade resulting in a persisting modification of hippocampal excitatory synapses (Habbas et al. 2015). Astrocytes are not only key to normal cognitive functions, but also the modification of synapses following neuroinflammation.

Astrocytes provide metabolic support to neurons. One mechanism is through the astrocyte-neuron lactate shuttle (Pellerin and Magistretti 1994; Magistretti and Pellerin 1996; Belanger et al. 2011; Magistretti and Allaman 2015). It is reported that peripheral administration of lactate exerts antidepressant-like behavioral effects (Carrard et al. 2018). These behavioral effects of lactate may be brought about by its ability to increase synaptic excitability.

The glutamatergic predominance in the excitatory-inhibitory balance is postulated to be involved in the pathogenesis of depression. Such an imbalance may be induced by astrocyte ablation which reduces glutamate uptake and increases glutamate levels in the synaptic cleft, causing depression-like behaviors. A causal relation between astrocytic dysfunction and depression is also provided by animal studies. The selective destruction of frontocortical astrocytes with the gliotoxin L-α-aminoadipic acid (L-AA) is sufficient to trigger a depressive-like phenotype (Banasr and Duman 2008; Domin et al. 2014). Both the behavioral and GFAP level changes could be prevented by injection of 3-((2-methyl-4-thiazolyl)ethynyl)pyridine (MTEP), a mGluR5 antagonist, through the inhibition of glutamatergic transmission (Domin et al. 2014). The results demonstrate that glial ablation in the PFC is sufficient to induce depressive-like behaviors similar to chronic stress, supporting the hypothesis that loss of glia contributes to the core symptoms of depression. Astrocyte dysfunction therefore may lead to excitatory-inhibitory imbalances, resulting in mood disorders such as depression and anxiety. Astroglial degeneration in the prefrontal cortex is a useful rat depression model.

Gap junctional communication is a main determinant of astrocytic function. Rats exposed to CUS, a rodent model of depression, showed behavioral deficits in sucrose preference test (SPT) and novelty suppressed feeding test (NSFT) and exhibited significant decreases in diffusion of gap junction channel-permeable dye and expression of Cx43, a major component of astrocyte gap junctions, and abnormal gap junction ultrastructure in the PFC (Sun et al. 2012). The results showed that infusions of gap junction blocker carbenoxolone (CBX) induced anhedonia in SPT and anxiety in NSFT, a core symptom of depression. Gap junction dysfunction contributes to the pathophysiology of depression.

Water channel AQP4 is expressed in astrocytes in the brain and is also involved in the pathogenesis of depression. Using a mouse model of depression induced by repeated corticosterone injections (Zhao et al. 2008), AQP4 knockout mice exhibit exacerbated depression-like behaviors based on forced swimming test (FST) and tail suspension test (TST) (Kong et al. 2014). These knockout mice also show a significant loss of astrocytes; aggravated downregulation of excitatory amino acid transporter 2 (EAAT2), synapsin-1, and glial cell line-derived neurotrophic factor (GDNF); and less hippocampal neurogenesis. Thus, astrocytic AQP4 can modulate astrocytic function and adult neurogenesis during the pathogenesis of depression and might be a potential target for the treatment for depression.

Blockade of astrocytic glutamate transporter GLT1 induces depressive-like behaviors in rats (Bechtholt-Gompf et al. 2010; John et al. 2012). Dysfunction or imbalance of the monoamine- or L-glutamate (L-Glu)-mediated synaptic transmission is known to be a pathogenic cause of many disorders. Astrocytes take up synaptic L-Glu during synaptic transmission through excitatory amino acid transporters GLT1 and GLAST. L-Glu is subsequently metabolized to L-glutamine (L-Gln) by glutamine synthetase (GS). Then, release of astrocytic L-Gln is used as a neuronal L-Glu precursor, making up the glutamine cycle. Expression levels of astrocyte transporters and metabolizing enzymes are dynamically regulated by synaptic activity, which empowers astrocytes to support synaptic transmission. Pharmacological blocking of astrocyte glutamate buffering by blocking GLT1 triggered depressive behaviors in the form of latency to drink sucrose solution, increased intracranial self-stimulation, or decreased social interaction (John et al. 2012). Decreased GLT1 and GLAST expression may cause impaired L-Glu turnover, which contributes to depression. Evidence that riluzole, which activates L-Glu transporters, reverses the decreased GFAP expression in rat prefrontal cortex and improves depressive-like behavior supports the involvement of impaired L-Glu turnover in depression pathogenesis (Banasr and Duman 2008). The role of astrocytes in the excitatory-inhibitory imbalance hypothesis of depression and anxiety is related to changes in glutamate receptor function.

Depression and anxiety are associated with NMDAR, and its antagonist ketamine was shown to produce a rapid antidepressive effect (Yang et al. 2018; Li et al. 2010). NMDARs are activated after binding of the agonist glutamate to the NR2 subunit along with a co-agonist, either L-glycine or D-serine, to the NR1 subunit at a glycine modulatory site to function as a gatekeeper. L-serine is synthesized by astrocytes, which is then transported to neurons for conversion to D-serine by serine racemase (SR). Substantial evidence suggests that D-serine is the most relevant co-agonist in forebrain regions, supporting its role in fear conditioning and anxiety disorders (Wolosker and Balu 2020). Transgenic mice overexpressing SR which increase the availability of astrocyte D-serine led to a reduced depressive phenotype based on the forced swim, novelty suppression of feeding and olfactory bulbectomy paradigms; chronic dietary D-serine supplement mimics the depression-related behavioral phenotype observed in SR transgenic mice (Otte et al. 2013). The acquisition and extinction of fear memory engages the SR/D-serine system in the mouse amygdala, and D-serine administration facilitates fear extinction (Balu et al. 2018).

Overall, the aforementioned studies indicate that astrocytes play a role in depression-like behaviors using different mechanisms, and impaired astrocytic function is necessary and sufficient to induce depression-like behaviors in animal experimental studies.

Astrocytes in PSMD

Although there were few functional studies on the role and mechanism depression in humans, studies from human specimens and animal models of MDD may provide insights into the mechanism and therapeutic strategies for PSMD. Generally, there are two causes of mood swings after stroke: biological changes and lifestyle changes. Biological changes are caused by damage in the emotion centers of the brain. Few studies have been done on the relationship between biological changes and PSMSD at cellular and molecular levels. Data from experimental studies using rodent models might provide insight on PSMDs, depression, and anxiety in general. Since MRI studies have found significant alterations in different brain regions in MDD patients, such as in the frontal lobe, hippocampus, temporal lobe, thalamus, striatum, and amygdala (Pandya et al. 2012; Zhang et al. 2018), whether stroke patients develop PSMD might be dependent on the region that is injured. On the other hand, ischemic stroke causes changes in morphology, metabolism, and gene expression in reactive astrocytes. Any changes related to glutamate synthesis, transportation, and degradation after stroke may affect mood swings.

The monoaminergic hypothesis of depression postulates symptoms to be a result of an imbalance in the central monoaminergic system including serotonergic, dopaminergic, and/or noradrenergic neurotransmission (Marathe et al. 2018). Astrocytes express transporters for norepinephrine and serotonin (Inazu et al. 2003; Hirst et al. 1998), which are the targets of several classical antidepressant drugs. This raises a possibility that antidepressants can have direct effects on astrocytes. These studies suggest that astrocytes may exert control of serotonergic and noradrenergic transmission and are thus cellular targets for antidepressant drugs that block the reuptake of monoamines by astrocytes.

Wang et al. developed rat model for post-stroke depression (PSD) using middle cerebral artery occlusion (MCAO), followed by an 18-day chronic mild stress (CMS), and assessed depression-like behavior and the effects of the antidepressant citalopram (Wang et al. 2008). Using the open-field test (OFT) and the sucrose consumption test, they found that citalopram could ameliorate the behavioral abnormalities, suggesting that the ischemic rat CMS model is an appropriate model for PSD.

Inflammation and alterations in glutamate neurotransmission are two novel pathways to pathophysiology in mood disorders (Haroon et al. 2017). Stroke causes dysfunction of glutamate transporters and reduced glutamate uptake by astrocyte in acute phase. Increasing data indicate that inflammation causes impaired astrocytic glutamate uptake, and patients with depression and anxiety have increased inflammation (Haroon et al. 2017). Since inflammation also occurs in ischemic stroke, inflammation-induced impairment of glutamate uptake may contribute to the development of or enhanced predisposition for PSMD. Elevated glucocorticoids due to illness-related stress could also downregulate glial activity and exacerbate or predispose patients to psychiatric illness via enhanced excitotoxicity (Cotter et al. 2001b). Since excitotoxicity is a common acute consequence of ischemic stroke, this may contribute to the subsequent development of mood disorders such as post-stroke depression and anxiety.

Chen et al. showed that post-stroke enrichment environment (EE) increased high-mobility group box-1 (HMGB1) and interleukin-6 (IL-6) expression in astrocytes, led to decreased depression and anxiety-like behavior, and promoted angiogenesis and functional recovery compared to standard environment (Chen et al. 2017). EE mice treated with glycyrrhizin decreased, whereas EE mice treated with recombinant HMGB1 (rHMGB1) increased in the levels of IL-6 and p-AKT. Their study highlighted the role of the astrocytic HMGB1-IL6 pathway in PSMD in animal models.

Yu et al. found that GLT1 was downregulated in astrocytes in the post-stroke rat model (Yu et al. 2019b). Reduced expression of GLT1 in PSD astrocytes inhibited the formation of functional synapses by influencing glutamate metabolism. Another study from the same group investigated the effect of ceftriaxone, which increases GLT1 expression, on depression-like behaviors of rats after MCAO (Yu et al. 2019a). They found that treatment of ceftriaxone gradually increased GLT1 both in transcription and translation levels and inhibited depression-like behaviors by increasing locomotor and rearing activity and improving anhedonia of the rats. Moreover, ceftriaxone can promote glutamate circulation and synaptic plasticity by increasing astrocytic GLT1 levels. Their study indicates that reduction of GLT1 in astrocytes is one potential mechanism of pathogenesis of post-stroke depression.

Rats injected with the astrocyte-specific toxin L-AA in the mPFC, which is anatomically and functionally linked with cognitive and emotional processing, resulted in a pronounced loss of astrocytes in the region and adversely affected set-shifting, working memory, and reversal learning functions (Lima et al. 2014). The lesion sites also showed progressive neuronal loss and dendritic atrophy in surviving neurons, suggesting that L-AA-induced astrocytic loss in this brain region leads to neuronal damage that results in cognitive impairment. Therefore, it is clear that astrocytes play a key role in cognitive impairment, which also occurs after stroke.

The neurotrophic hypothesis in depression and anxiety disorders was proposed because lower serum levels of neurotrophic factors like brain-derived neurotrophic factor (BDNF) are often observed in patients with depression, while increased expression of BDNF and GDNF is observed in patients as a response to antidepressant treatment (Pisoni et al. 2018). BDNF overexpression in hippocampal astrocytes leads to antidepressant-like activity and increased neurogenesis in mice (Quesseveur et al. 2013). Recently, a study demonstrated that RAs expressed increased GDNF (Zhang et al. 2020b), and this suggests that RAs might play a role in improving PSMD through GDNF release.

D-Serine has also been found to play a role in post-traumatic stress disorder (PTSD) (Wolosker and Balu 2020). It was found that the single nucleotide polymorphism, rs4523957, in human serine racemase gene, is associated with PTSD based on postmortem human brain study (Balu et al. 2018). D-Serine might play a role in PSMD since stroke and traumatic injury have similar mechanisms of brain injury.

Overall, these data suggest that astrocytes/reactive astrocytes are an integral part of depression and anxiety through different mechanisms and may be a potential therapeutic target for PSMD.

Concluding Remarks

Ischemic stroke is a leading cause of human disability, but recovery of behavioral function in animal stroke models and young patients can be remarkable, much of which is due to neuroplasticity that involves the strengthening of existing synapses, synaptogenesis, or local sprouting. However, these effects are often accompanied by neurological problems because of failure to recruit larger and more diffuse networks to function normally. Beyond the disability in normal motor function, it is common that stroke survivors have depression and anxiety-like mood disorders. Although a large body of evidence indicates that astrocytes play a role in MDD from studies of both human postmortem specimen and animal models, there are fewer studies regarding the participation of astrocytes or reactive astrocytes in PSMD. Considering that astrocytes can influence neuronal function through a variety of mechanisms, future studies on the mechanisms by which astrocytes/reactive astrocytes play a role in PSMD should focus on the context of astrocyte-neuron interactions in stroke models.

Acknowledgments This work was supported by the National Institutes of Health [National Institute of Neurological Disorders and Stroke (NINDS) grants R01NS069726 and R01NS094539 to SD] and the America Heart Association [Midwest Affiliate Grant-in-Aid (16GRNT31280014) and NCRG-IRG 16IRG27780023 to SD].

Author Contributions Tracey Singer, original draft preparation and editing; Sarah Ding, original draft preparation; Shinghua Ding, original draft preparation, editing, and supervision.

Conflict of Interest The authors declare no conflict of interest.

References

Altshuler LL, Abulseoud OA, Foland-Ross L et al (2010) Amygdala astrocyte reduction in subjects with major depressive disorder but not bipolar disorder. Bipolar Disord 12:541–549

Anderson MA, Burda JE, Ren Y et al (2016) Astrocyte scar formation aids central nervous system axon regeneration. Nature 532:195–200

Araque A, Parpura V, Sanzgiri RP, Haydon PG (1999) Tripartite synapses: glia, the unacknowledged partner. [Review] [61 refs]. Trends Neurosci 22:208–215

Arsenault-Lapierre GV, Kim C, Turecki G (2004) Psychiatric diagnoses in 3275 suicides: a meta-analysis. BMC Psychiatry 4:37

Ayerbe L, Ayis S, Wolfe CDA, Rudd AG (2013) Natural history, predictors and outcomes of depression after stroke: systematic review and meta-analysis. Br J Psychiatry 202:14–21

Baker PM, Jhou T, Li B et al (2016) The lateral Habenula circuitry: reward processing and cognitive control. J Neurosci 36:11482
Balakrishnan P, Rosen H (2008) The causes and treatment of pseudobulbar affect in ischemic stroke. Curr Treat Options Cardiovasc Med 10:216
Balu DT, Presti KT, Huang CCY et al (2018) Serine racemase and D-serine in the amygdala are dynamically involved in fear learning. Biol Psychiatry 83:273–283
Banasr M, Duman RS (2008) Glial loss in the prefrontal cortex is sufficient to induce depressive-like behaviors. Biol Psychiatry 64:863–870
Bano D, Young KW, Guerin CJ, LeFeuvre R, Rothwell NJ, Naldini L, Rizzuto R, Carafoli E, Nicotera P (2005) Cleavage of the plasma membrane Na^+/Ca^{2+} exchanger in excitotoxicity. Cell 120:275–285
Barber PA, Demchuk AMH (2003) Biochemistry ischemic stroke. Adv Neurol 92:151–164
Barreto GE, Sun X, Xu L, Giffard RG (2011) Astrocyte proliferation following stroke in the mouse depends on distance from the infarct. PLoS One 6:e27881
Bechtholt-Gompf AJ, Walther HV, Adams MA et al (2010) Blockade of astrocytic glutamate uptake in rats induces signs of anhedonia and impaired spatial memory. Neuropsychopharmacology 35:2049–2059
Belanger M, Allaman I, Magistretti P (2011) Brain energy metabolism: focus on astrocyte-neuron metabolic cooperation. Cell Metab 14:724–738
Bender CL, Calfa GD, Molina VA (2016) Astrocyte plasticity induced by emotional stress: a new partner in psychiatric physiopathology? Prog Neuro-Psychopharmacol Biol Psychiatry 65:68–77
Bernard R, Kerman IA, Thompson RC et al (2011) Altered expression of glutamate signaling, growth factor, and glia genes in the locus coeruleus of patients with major depression. Mol Psychiatry 16:634–646
Bonfoco E (1995) Apoptosis and necrosis, two distinct events induced, respectively, by mild and intense insults with N-methyl-D-aspartate or nitric oxide/superoxide in cortical cell culture. Proc Natl Acad Sci U S A 92:7162–7166
Bowley MP, Drevets WC, Ongur D, Price JL (2002) Low glial numbers in the amygdala in major depressive disorder. Biol Psychiatry 52:404–412
Broughton BRS, Reutens DC, Sobey CG (2009) Apoptotic mechanisms after cerebral ischemia. Stroke 40:e331–e339
Burda J, Sofroniew M (2014) Reactive gliosis and the multicellular response to CNS damage and disease. Neuron 81:229–248
Bushong EA, Martone ME, Jones YZ, Ellisman MH (2002) Protoplasmic astrocytes in CA1 stratum radiatum occupy separate anatomical domains. J Neurosci 22:183–192
Cahoy JD, Emery B, Kaushal A et al (2008) A transcriptome database for astrocytes, neurons, and oligodendrocytes: a new resource for understanding brain development and function. J Neurosci 28:264–278
Campbell BCV, De Silva DA, Macleod MR et al (2019) Ischemic stroke. Nat Rev Dis Primers 5:70
Cao X, Li LP, Wang Q et al (2013) Astrocyte-derived ATP modulates depressive-like behaviors. Nat Med 19:773–777
Carrard A, Elsayed M, Margineanu M et al (2018) Peripheral administration of lactate produces antidepressant-like effects. Mol Psychiatry 23:392–399
Chen JY, Yu Y, Yuan Y et al (2017) Enriched housing promotes post-stroke functional recovery through astrocytic HMGB1-IL-6-mediated angiogenesis. Cell Death Discov 3:17054
Choudary PV, Molnar M, Evans SJ et al (2005) Altered cortical glutamatergic and GABAergic signal transmission with glial involvement in depression. PNAS 102:15653
Choudhury GR, Ding S (2016) Reactive astrocytes and therapeutic potential in focal ischemic stroke. Neurobiol Dis 85:234–244
Cotter D, Mackay D, Landau S et al (2001a) Reduced glial cell density and neuronal size in the anterior cingulate cortex in major depressive disorder. Arch Gen Psychiatry 58:545–553

Cotter DR, Pariante CM, Everall IP (2001b) Glial cell abnormalities in major psychiatric disorders: the evidence and implications. Brain Res Bull 55:585–595

Cui Y, Yang Y, Ni Z et al (2018) Astroglial Kir4.1 in the lateral habenula drives neuronal bursts in depression. Nature 554:323–327

Ding S (2014) Dynamic reactive astrocytes after focal ischemia. Neural Regen Res 9:2048–2052

Ding S, Wang T, Cui W, Haydon PG (2009) Photothrombosis ischemia stimulates a sustained astrocytic Ca^{2+} signaling in vivo. Glia 57:767–776

Dirnagl U, Iadecola C, Moskowitz MA (1999) Pathobiology of ischaemic stroke: an integrated view. Trends Neurosci 22:391–397

Doll DN, Rellick SL, Barr TL et al (2015) Rapid mitochondrial dysfunction mediates TNF-alpha-induced neurotoxicity. J Neurochem 132:443–451

Domin H, Szewczyk B, Wozniak M et al (2014) Antidepressant-like effect of the mGluR5 antagonist MTEP in an astroglial degeneration model of depression. Behav Brain Res 273:23–33

Dong Q, He J, Chai Z (2013) Astrocytic Ca^{2+} waves mediate activation of extrasynaptic NMDA receptors in hippocampal neurons to aggravate brain damage during ischemia. Neurobiol Dis 58:68–75

Drevets WC (2000) Functional anatomical abnormalities in limbic and prefrontal cortical structures in major depression. Prog Brain Res 126:413–431

Eliasson MJ et al (1997) Poly(ADP-ribose) polymerase gene disruption renders mice resistant to cerebral ischemia. Nat Med 3:1089–1095

Faiz M, Sachewsky N, Gascon S et al (2015) Adult neural stem cells from the subventricular zone give rise to reactive astrocytes in the cortex after stroke. Cell Stem Cell 17:624–634

Fang Y, Mpofu E, Athanasou J (2017) Reducing depressive or anxiety symptoms in post-stroke patients: pilot trial of a constructive integrative psychosocial intervention. Int J Health Sci 11:53–58

Faulkner JR, Herrmann JE, Woo MJ et al (2004) Reactive astrocytes protect tissue and preserve function after spinal cord injury. J Neurosci 24:2143–2155

Fernandez-Klett F, Potas JR, Hilpert D et al (2012) Early loss of pericytes and perivascular stromal cell-induced scar formation after stroke. J Cereb Blood Flow Metab 33:428–439

Folbergrova J, Zhao Q, Katsura K, Siesjo BK (1995) N-tert-butyl-α-phenylnitrone improves recovery of brain energy state in rats following transient focal ischemia. PNAS 92:5057–5061

Gao J, Lin M, Zhao J et al (2016) Different interventions for post-ischaemic stroke depression in different time periods: a single-blind randomized controlled trial with stratification by time after stroke. Clin Rehabil 31:71–81

Gillespie DC, Cadden AP, Lees R et al (2016) Prevalence of pseudobulbar affect following stroke: a systematic review and meta-analysis. J Stroke Cerebrovasc Dis 25:688–694

Gittins RA, Harrison PJ (2011) A morphometric study of glia and neurons in the anterior cingulate cortex in mood disorder. J Affect Disord 133:328–332

Grefkes C, Fink GR (2020) Recovery from stroke: current concepts and future perspectives. Neurol Res Pract 2:17

Habbas S, Santello M, Becker D et al (2015) Neuroinflammatory TNFα impairs memory via astrocyte signaling. Cell 163:1730–1741

Hackett ML, Anderson CS, House A, Halteh C (2008) Interventions for preventing depression after stroke. Cochrane Database Syst Rev. https://doi.org/10.1002/14651858.CD003689.pub3

Halassa MM, Fellin T, Haydon PG (2007) The tripartite synapse: roles for gliotransmission in health and disease. Trends Mol Med 13:54–63

Halassa MM, Florian C, Fellin T et al (2009) Astrocytic modulation of sleep homeostasis and cognitive consequences of sleep loss. Neuron 61:213–219

Hara H, Friedlander RM, Gagliardini V, Ayata C, Fink K, Huang Z, Shimizu-Sasamata M, Yuan J, Moskowitz MA (1997) Inhibition of interleukin 1β converting enzyme family proteases reduces ischemic and excitotoxic neuronal damage. PNAS 94:2007–2012

Haroon E, Miller AH, Sanacora G (2017) Inflammation, glutamate, and glia: a trio of trouble in mood disorders. Neuropsychopharmacology 42:193–215

Haydon PG (2001) GLIA: listening and talking to the synapse. Nat Rev Neurosci 2:185–193
Henn FA, Vollmayr B (2005) Stress models of depression: forming genetically vulnerable strains. Neurosci Biobehav Rev 29:799–804
Hill K, House A, Knapp P et al (2019) Prevention of mood disorder after stroke: a randomised controlled trial of problem solving therapy versus volunteer support. BMC Neurol 19:128
Hirst WD, Price GW, Rattray M, Wilkin GP (1998) Serotonin transporters in adult rat brain astrocytes revealed by [3H]5-HT uptake into glial plasmalemmal vesicles. Neurochem Int 33:11–22
Hu H, Cui Y, Yang Y (2020) Circuits and functions of the lateral habenula in health and in disease. Nat Rev Neurosci 21:277–295
Iadecola C, Zhang F, Casey R et al (1997) Delayed reduction of ischemic brain injury and neurological deficits in mice lacking the inducible nitric oxide synthase gene. J Neurosci 17:9157–9164
Illes P, Rubini P, Yin H, Tang Y (2020) Impaired ATP release from brain astrocytes may be a cause of major depression. Neurosci Bull 36:1281–1284
Inazu M, Takeda H, Matsumiya T (2003) Functional expression of the norepinephrine transporter in cultured rat astrocytes. J Neurochem 84:136–144
John CS, Smith KL, Van'T Veer A et al (2012) Blockade of astrocytic glutamate uptake in the prefrontal cortex induces anhedonia. Neuropsychopharmacology 37:2467–2475
Kessler RC, Bromet EJ (2013) The epidemiology of depression across cultures. Annu Rev Public Health 34:119–138
Khakh BS, Sofroniew MV (2015) Diversity of astrocyte functions and phenotypes in neural circuits. Nat Neurosci 18:942–952
Kim JS (2016) Post-stroke mood and emotional disturbances: pharmacological therapy based on mechanisms. J Stroke 18:244–255
Kimelberg HK (2010) Functions of mature mammalian astrocytes: a current view. Neuroscientist 16:79–106
Kimelberg H, Nedergaard M (2010) Functions of astrocytes and their potential as therapeutic targets. Neurotherapeutics 7:338–353
Kong H, Zeng XN, Fan Y et al (2014) Aquaporin-4 knockout exacerbates corticosterone-induced depression by inhibiting astrocyte function and hippocampal neurogenesis. CNS Neurosci Ther 20:391–402
LeComte MD, Shimada IS, Sherwin C, Spees JL (2015) Notch1-TAT3-ETBR signaling axis controls reactive astrocyte proliferation after brain injury. PNAS 112:8726–8731
Leng L et al (2018) Menin deficiency leads to depressive-like behaviors in mice by modulating astrocyte-mediated neuroinflammation. Neuron 100:551–563
Li N, Lee B, Liu RJ et al (2010) mTOR-dependent synapse formation underlies the rapid antidepressant effects of NMDA antagonists. Science 329:959–964
Li H, Zhang N, Sun G, Ding S (2013a) Inhibition of the group I mGluRs reduces acute brain damage and improves long-term histological outcomes after photothrombosis-induced ischaemia. ASN Neuro 5:195–207
Li K, Zhou T, Liao L et al (2013b) βCaMKII in lateral Habenula mediates core symptoms of depression. Science 341:1016–1020
Li H, Zhang N, Lin H et al (2014) Histological, cellular and behavioral assessments of stroke outcomes after photothrombosis-induced ischemia in adult mice. BMC Neurosci 15:58
Li H, Xie Y, Zhang N et al (2015) Disruption of IP_3R2-mediated Ca^{2+} signaling pathway in astrocytes ameliorates neuronal death and brain damage while reducing behavioral deficits after focal ischemic stroke. Cell Calcium 58:565–576
Lima A, Sardinha VM, Oliveira AF et al (2014) Astrocyte pathology in the prefrontal cortex impairs the cognitive function of rats. Mol Psychiatry 19:834–841
Linnerbauer M, Rothhammer V (2020) Protective functions of reactive astrocytes following central nervous system insult. Front Immunol 11:573256
Lipton P (1999) Ischemic cell death in brain neurons. Physiol Rev 79:1431–1568
Lo EH (2008) A new penumbra: transitioning from injury into repair after stroke. Nat Med 14:497–500

Lo EH, Dalkara T, Moskowitz MA (2003) Mechanisms, challenges and opportunities in stroke. Nat Rev Neurosci 4:399–415

Maaijwee NAMM, Tendolkar I, Rutten-Jacobs LCA et al (2016) Long-term depressive symptoms and anxiety after transient ischaemic attack or ischaemic stroke in young adults. Eur J Neurol 23:1262–1268

Magistretti P, Allaman I (2015) A cellular perspective on brain energy metabolism and functional imaging. Neuron 86:883–901

Magistretti PJ, Pellerin L (1996) Cellular bases of brain energy metabolism and their relevance to functional brain imaging: evidence for a prominent role of astrocytes. Cereb Cortex 6:50–61

Maragakis NJ, Rothstein JD (2006) Mechanisms of disease: astrocytes in neurodegenerative disease. Nat Clin Pract Neurol 2:679–689

Marathe SV, D'almeida PL, Virmani G et al (2018) Effects of monoamines and antidepressants on astrocyte physiology: implications for monoamine hypothesis of depression. J Exp Neurosci 12:1–7

Matthias LS, Julia S, Johann S (2013) Serum S100B represents a new biomarker for mood disorders. Curr Drug Targets 14:1237–1248

Matyash V, Kettenmann H (2010) Heterogeneity in astrocyte morphology and physiology. Brain Res Rev 63:2–10

Mayer ML, Miller RJ (1990) Excitatory amino acid receptors, second messengers and regulation of intracellular Ca^{2+} in mammalian neurons. Trends Pharmacol Sci 11:254–260

Mochari-Greenberger H, Towfighi A, Mosca L (2014) National women's knowledge of stroke warning signs, overall and by race/ethnic group. Stroke 45:1180–1182

Moskowitz MA, Lo EH, Iadecola C (2010) The science of stroke: mechanisms in search of treatments. Neuron 67:181–198

Myer DJ, Gurkoff GG, Lee SM et al (2006) Essential protective roles of reactive astrocytes in traumatic brain injury. Brain 129:2761–2772

Nedergaard M, Dirnagl U (2005) Role of glial cells in cerebral ischemia. Glia 50:281–286

Nowak L, Bregestovski P, Ascher P et al (1984) Magnesium gates glutamate-activated channels in mouse central neurones. Nature 307:462–465

Ongur D, Drevets WC, Price JL (1998) Glial reduction in the subgenual prefrontal cortex in mood disorders. PNAS 95:13290

Orrenius S (1995) Apoptosis: molecular mechanisms and implications for human disease. J Intern Med 237:529–536

Otte DM, Barcena de Arellano ML et al (2013) Effects of chronic D-serine elevation on animal models of depression and anxiety-related behavior. PLoS One 8:e67131

Pandya M, Altinay M, Malone DA, Anand A (2012) Where in the brain is depression? Curr Psychiatry Rep 14:634–642

Park K, Lee SJ (2020) Deciphering the star codings: astrocyte manipulation alters mouse behavior. Exp Mol Med 52:1028–1038

Pellerin L, Magistretti PJ (1994) Glutamate uptake into astrocytes stimulates aerobic glycolysis: a mechanism coupling neuronal activity to glucose utilization. Proc Natl Acad Sci U S A 91:10625–10629

Petravicz J, Boyt KM, McCarthy KD (2014) Astrocyte IP_3R2-dependent Ca^{2+} signaling is not a major modulator of neuronal pathways governing behavior. Front Behav Neurosci 8:384

Pisoni A, Strawbridge R, Hodsoll J et al (2018) Growth factor proteins and treatment-resistant depression: a place on the path to precision. Front Psych 9:386

Poulin VR, Korner-Bitensky N, Bherer L et al (2016) Comparison of two cognitive interventions for adults experiencing executive dysfunction post-stroke: a pilot study. Disabil Rehabil 39:1–13

Quesseveur G, David DJ, Gaillard MC et al (2013) BDNF overexpression in mouse hippocampal astrocytes promotes local neurogenesis and elicits anxiolytic-like activities. Transl Psychiatry 3:e253

Quinn TJ, Elliott E, Langhorne P (2018) Cognitive and mood assessment tools for use in stroke. Stroke 49:483–490

Rajkowska G (2000) Postmortem studies in mood disorders indicate altered numbers of neurons and glial cells. Biol Psychiatry 48:766–777
Rajkowska G, Hughes J, Stockmeier CA et al (2013) Coverage of blood vessels by astrocytic endfeet is reduced in major depressive disorder. Biol Psychiatry 73:613–621
Robinson RG, Jorge RE (2015) Post-stroke depression: a review. Am J Psychiatry 173:221–231
Ronning OM, Guldvog B (1998) Outcome of subacute stroke rehabilitation: a randomized controlled trial. Stroke 29:779–784
Saengsuwan J, Suangpho P, Tiamkao S (2017) Knowledge of stroke risk factors and warning signs in patients with recurrent stroke or recurrent transient ischaemic attack in Thailand. Neurol Res Int 2017:8215726
Santello M, Bezzi P, Volterra A (2011) TNFα controls glutamatergic gliotransmission in the hippocampal dentate gyrus. Neuron 69:988–1001
Santello M, Toni N, Volterra A (2019) Astrocyte function from information processing to cognition and cognitive impairment. Nat Neurosci 22:154–166
Schottke H, Giabbiconi CM (2015) Post-stroke depression and post-stroke anxiety: prevalence and predictors. Int Psychoger 27:1805–1812
Schroeter ML, Abdul-Khaliq H, Diefenbacher A, Blasig IE (2002) S100B is increased in mood disorders and may be reduced by antidepressive treatment. Neuroreport 13:1675–1678
Schroeter ML, Steiner J, Mueller K (2011) Glial pathology is modified by age in mood disorders-a systematic meta-analysis of serum S100B in vivo studies. J Affect Disord 134:32–38
Seifert G, Schilling K, Steinhauser C (2006) Astrocyte dysfunction in neurological disorders: a molecular perspective. Nat Rev Neurosci 7:194–206
Sequeira A, Mamdani F, Ernst C et al (2009) Global brain gene expression analysis links glutamatergic and GABAergic alterations to suicide and major depression. PLoS One 4:e6585
Sheline YI (2003) Neuroimaging studies of mood disorder effects on the brain. Biol Psychiatry 54:338–352
Shimada IS, Borders A, Aronshtam A, Spees JL (2011) Proliferating reactive astrocytes are regulated by notch-1 in the peri-infarct area after stroke. Stroke 42:3231–3237
Sofroniew M, Vinters H (2010) Astrocytes: biology and pathology. Acta Neuropathol 119:7–35
Soriano FX, Martel MA, Papadia S et al (2008) Specific targeting of pro-death NMDA receptor signals with differing reliance on the NR2B PDZ ligand. J Neurosci 28:10696–10710
Sun JD, Liu Y, Yuan YH et al (2012) Gap junction dysfunction in the prefrontal cortex induces depressive-like behaviors in rats. Neuropsychopharmacology 37:1305–1320
Suzuki A, Stern S, Bozdagi O et al (2011) Astrocyte-neuron lactate transport is required for long-term memory formation. Cell 144:810–823
Torres-Platas SG, Nagy C, Wakid M et al (2016) Glial fibrillary acidic protein is differentially expressed across cortical and subcortical regions in healthy brains and downregulated in the thalamus and caudate nucleus of depressed suicides. Mol Psychiatry 21:509–515
Virani SS et al (2020) Heart disease and stroke statistics-2020 update: a report from the American Heart Association. Circulation 141:e139–e596
Volterra A, Meldolesi J (2005) Astrocytes, from brain glue to communication elements: the revolution continues. Nat Rev Neurosci 6:626–640
Voskuhl RR, Peterson RS, Song B et al (2009) Reactive astrocytes form scar-like perivascular barriers to leukocytes during adaptive immune inflammation of the CNS. J Neurosci 29:11511–11522
Vosler PS, Gao Y, Brennan CS et al (2011) Ischemia-induced calpain activation causes eukaryotic (translation) initiation factor 4G1 (eIF4GI) degradation, protein synthesis inhibition, and neuronal death. PNAS 108:18102–18107
Wang SH, Zhang ZJ, Guo YJ et al (2008) Anhedonia and activity deficits in rats: impact of post-stroke depression. J Psychopharmacol 23:295–304
Wang G, Zhang Z, Ayala C et al (2014a) Costs of hospitalization for stroke patients aged 18-64 years in the United States. J Stroke Cerebrovasc Dis 23:861–868

Wang X, Li H, Ding S (2014b) The effects of NAD^+ on apoptotic neuronal death and mitochondrial biogenesis and function after glutamate excitotoxicity. Int J Mol Sci 15:1012–1022
Wang X, Li H, Ding S (2016) Pre-B-cell colony-enhancing factor protects against apoptotic neuronal death and mitochondrial damage in ischemia. Sci Rep 6:32416
Wang Q, Jie W, Liu JH et al (2017) An astroglial basis of major depressive disorder? An overview. Glia 65:1227–1250
Wanner IB, Anderson MA, Song B et al (2013) Glial scar borders are formed by newly proliferated, elongated astrocytes that interact to corral inflammatory and fibrotic cells via STAT3-dependent mechanisms after spinal cord injury. J Neurosci 33:12870–12886
Webster MJ, Knable MB, Johnston-Wilson N et al (2001) Immunohistochemical localization of phosphorylated glial fibrillary acidic protein in the prefrontal cortex and hippocampus from patients with schizophrenia, bipolar disorder, and depression. Brain Behav Immun 15:388–400
Wilhelmsson U, Bushong EA, Price DL et al (2006) Redefining the concept of reactive astrocytes as cells that remain within their unique domains upon reaction to injury. Proc Natl Acad Sci U S A 103:17513–17518
Wolosker H, Balu DT (2020) D-serine as the gatekeeper of NMDA receptor activity: implications for the pharmacologic management of anxiety disorders. Transl Psychiatry 10:184
Wright F, Wu S, Chun HYY, Mead G (2017) Factors associated with poststroke anxiety: a systematic review and meta-analysis. Stroke Res Treat 2017:2124743
Wu QJ, Tymianski M (2018) Targeting NMDA receptors in stroke: new hope in neuroprotection. Mol Brain 11:15
Wu DY, Guo M, Gao YS et al (2012) Clinical effects of comprehensive therapy of early psychological intervention and rehabilitation training on neurological rehabilitation of patients with acute stroke. Asian Pac J Trop Med 5:914–916
Yang Y, Cui Y, Sang K et al (2018) Ketamine blocks bursting in the lateral habenula to rapidly relieve depression. Nature 554:317–322
Yu D, Cheng Z, Ali AI et al (2019a) Chronic unexpected mild stress destroys synaptic plasticity of neurons through a glutamate transporter, GLT-1, of astrocytes in the ischemic stroke rat. Neural Plast 2019:1615925
Yu D, Cheng Z, Ali AI et al (2019b) Down-expressed GLT-1 in PSD astrocytes inhibits synaptic formation of NSC-derived neurons in vitro. Cell Cycle 18:105–114
Yuan J (2009) Neuroprotective strategies targeting apoptotic and necrotic cell death for stroke. Apoptosis 14:469–477
Zamanian JL, Xu L, Foo LC et al (2012) Genomic analysis of reactive astrogliosis. J Neurosci 32:6391–6410
Zhang Y, Barres BA (2010) Astrocyte heterogeneity: an underappreciated topic in neurobiology. Curr Opin Neurobiol 20:588–594
Zhang J, Dawson VL, Dawson TM, Snyder SH (1994) Nitric oxide activation of poly(ADP-ribose) synthetase in neurotoxicity. Science 263:687
Zhang FF, Peng W, Sweeney JA et al (2018) Brain structure alterations in depression: psychoradiological evidence. CNS Neurosci Ther 24:994–1003
Zhang HY, Wang Y, He Y et al (2020a) A1 astrocytes contribute to murine depression-like behavior and cognitive dysfunction, which can be alleviated by IL-10 or fluorocitrate treatment. J Neuroinflammation 17:200
Zhang N, Zhang Z, He R et al (2020b) GLAST-CreERT2 mediated deletion of GDNF increases brain damage and exacerbates long-term stroke outcomes after focal ischemic stroke in mouse model. Glia 68:2395–2414
Zhao Y, Ma R, Shen J et al (2008) A mouse model of depression induced by repeated corticosterone injections. Eur J Pharmacol 581:113–120

Astroglia and Obsessive Compulsive Disorder

Kohichi Tanaka

Introduction

Obsessive compulsive disorder (OCD) has a prevalence rate of 1–3% in the general population and has been ranked as one of the top ten leading causes of illness-related disability (American Psychiatric Association 2013; Kessler et al. 2005). OCD is characterized by persistent intrusive thoughts (obsessions) and repetitive behaviors (compulsions) (Leckman et al. 1997). There are various OCD-related disorders, including Tourette syndrome (TS), grooming disorders (e.g., skin-picking, trichotillomania), and autism spectrum disorders (ASD) that share considerable overlapping features with OCD (Browne et al. 2014). Although the neurobiological basis of OCD still remains obscure, neuroimaging studies in patients with OCD and OCD-related disorders have consistently identified hyperactivity in orbitofrontal cortex and striatum (Cerliani et al. 2015; Hou et al. 2014; Jung et al. 2017; Neuner et al. 2014). However, the cellular and synaptic abnormalities underlying this hyperactivity are unclear. The most prominent theory regarding the underlying mechanisms of OCD and OCD-related disorders is an increased excitation to inhibition (E/I) ratio due to increased glutamatergic excitation or reduced GABAergic inhibition (Albin and Mink 2006; Rubenstein and Merzenich 2003; Wu et al. 2012). A proper E/I ratio is achieved by factors expressed in neuron and glia. In astrocytes, both the glutamate transporter GLT1 and GABA transporter GAT-3 are critical for regulating the E/I balance (Aida et al. 2015; Aizawa et al. 2020; Boddum et al. 2016; Cui et al. 2014; Kersanté et al. 2013; Kiryk et al. 2008; Matos et al. 2018; Scimemi 2014; Sugimoto et al. 2018; Sugiyama et al. 2017; Tanaka et al. 1997; Zhao et al. 2018). Although astrocyte dysfunction has not been directly explored in OCD

K. Tanaka (✉)
Tokyo Medical and Dental University, Department of Molecular Neuroscience, Medical Research Institute, Tokyo, Japan
e-mail: tanaka.aud@mri.tmd.ac.jp

B. Li et al. (eds.), *Astrocytes in Psychiatric Disorders*, Advances in Neurobiology 26, https://doi.org/10.1007/978-3-030-77375-5_7

patients, several animal studies have found that astrocytes are involved in the pathophysiology of OCD. In this chapter, I highlight recent studies in which astrocyte dysfunction contributed to E/I imbalance, leading to pathological repetitive behaviors shared between patients with OCD, TS, and ASD.

Astrocytic Glutamate Transporter GLT1-Deficient Mice Exhibit Pathological Repetitive Behaviors

Glutamate is the major excitatory neurotransmitter in the mammalian central nervous system (CNS) and plays an important role in most aspects of normal brain function including cognition, learning, and memory (Nakanishi et al. 1998) Despite its importance as a neurotransmitter, excess glutamate is toxic to neurons, a phenomenon known as excitotoxicity (Choi 1994). Excitotoxicity is implicated in the pathophysiology of various neuropsychiatric diseases such as OCD, TS, and ASD (Albin and Mink 2006; Rubenstein and Merzenich 2003; Wu et al. 2012). Clearance of extracellular glutamate is critical for the maintenance of low extracellular glutamate concentrations and is achieved by the uptake of released glutamate by Na^+-dependent transporters (Tanaka 2000). There are five subtypes of Na^+-dependent glutamate transporters, GLAST (EAAT1), GLT1 (EAAT2), EAAC1 (EAAT3), EAAT4, and EAAT5. GLAST and GLT1 are predominantly expressed in glial cells, whereas EAAC1, EAAT4, and EAAT5 are expressed in neurons. GLT1 plays a critical role in the maintenance of extracellular glutamate homeostasis in the forebrain (Aida et al. 2015; Aizawa et al. 2020; Cui et al. 2014; Kiryk et al. 2008; Sugimoto et al. 2018; Tanaka et al. 1997). GLT1 null mice exhibited severe seizures and premature death (Tanaka et al. 1997). To overcome the premature lethality of GLT1 null mice, we generated astrocyte-specific GLT1-inducible knockdown (GLT1-KD) mice by crossing floxed-GLT1 mice ($GLT1^{flox/flox}$) (Cui et al. 2014), with mice expressing an inducible form of Cre under the astrocyte-specific, endogenous GLAST promoter ($GLAST^{CreERT2/+}$) (Mori et al. 2006). Tamoxifen was injected into $GLAST^{CreERT2/+}$; $GLT1^{flox/flox}$ mice from P19 for 5 days. In GLT1-KD mice, GLT1 protein levels were decreased by 60–80% in the cortex and striatum (Aida et al. 2015).

GLT1-KD mice show pathological repetitive behaviors including excessive self-grooming and tic-like head shakes. We investigated functional abnormalities of the whole brain network in GLT1-KD mice with functional magnetic resonance imaging (Abe et al. 2020). Blood oxygenation-level dependent-functional magnetic resonance imaging (BOLD-fMRI) is a tool for investigating whole brain functional connectivity and activity, based on the neurovascular coupling hypothesis (Drew 2019). A previous study showed that astrocytes could generate BOLD-fMRI responses without neuronal modulation (Takata et al. 2018). Indeed, GLT1 deletion affected the neurovascular coupling (Aizawa et al. 2020; Voutsinos-Porche et al. 2003). Thus, BOLD-fMRI may fail in GLT1-KD mice. In contrast, diffusion fMRI

(DfMRI) is an alternative functional imaging method to detect neural activity without interference from hemodynamics (Abe et al. 2017; Le Bihan et al. 2006; Tsurugizawa et al. 2013). We examined abnormalities of dynamic brain function in GLT1-KD mice by taking advantage of this DfMRI. Our DfMRI approaches analyzed three brain dynamic processes: resting state brain activity, functional connectivity (FC), and propagation of neural information. This approach revealed the hyperactive resting state activity and biased FC of the cortico-striatal-thalamic (CST) circuitry of GLT1-KD mice. In addition, we performed ignition-driven mean integration (IDMI) analysis, which examines how intrinsic local activation is propagated into whole brain regions (Deco and Kringelbach 2017; Deco et al. 2017). This analysis revealed that the phase-synchronized connectivity initiated from orbital area, primary motor area, and primary somatosensory area was increased in GLT1-KD mice, suggesting that GLT1-KD mice showed an unbalanced distribution of neural propagation in the CST circuitry. These global brain network functional abnormalities detected by DfMRI were associated with OCD-related behaviors in GLT1-KD mice.

Electrophysiological studies of basal corticostriatal synaptic transmission in GLT1-KD mice did not show any difference in the kinetics of either the AMPA or NMDA receptor-mediated excitatory postsynaptic currents (EPSCs), the AMPA/NMDA ratio, or paired-pulse ratio of evoked AMPA receptor-mediated EPSCs. However, a difference was observed in synaptic responses to a prolonged repetitive stimulation that triggers a massive glutamate release. In the control and GLT1-KD mice, the normalized EPSC amplitude decreased gradually during prolonged repetitive stimulation. Notably, this attenuation of EPSCs was significantly reduced in GLT1-KD mice (Aida et al. 2015). These data indicate that prolonged repetitive stimulation, but not a single episode of stimulation, leads to corticostriatal glutamatergic hyperactivity in GLT1-KD mice. The attenuation of EPSCs during repetitive stimulation is thought to reflect a presynaptic cycling process in which depleted docked vesicles are replenished by reserve pool vesicles (Hoshina et al. 2013). The presynaptic functions at corticostriatal synapses are intact in GLT1-KD mice because there is no difference in paired-pulse ratio between GLT1-KD and control mice. Therefore, the reduced attenuation of EPSCs in the striatum of GLT1-KD mice is likely induced by an increase in the extracellular glutamate concentration due to knockdown of the astrocytic glutamate transporter GLT1. Systemic administration of memantine, an NMDA receptor inhibitor, alleviated the pathological repetitive behaviors in GLT1-KD mice, whereas administration of fluoxetine, a selective serotonin reuptake inhibitor (SSRI), had no effect on repetitive behaviors in GLT1-KD mice. These results suggest that astrocytic GLT1 has a critical role in controlling activity and FC across the CST circuitry via the modulation of the synaptic efficacy, and its dysfunction is involved in the pathophysiology of pathological repetitive behaviors (Fig. 1).

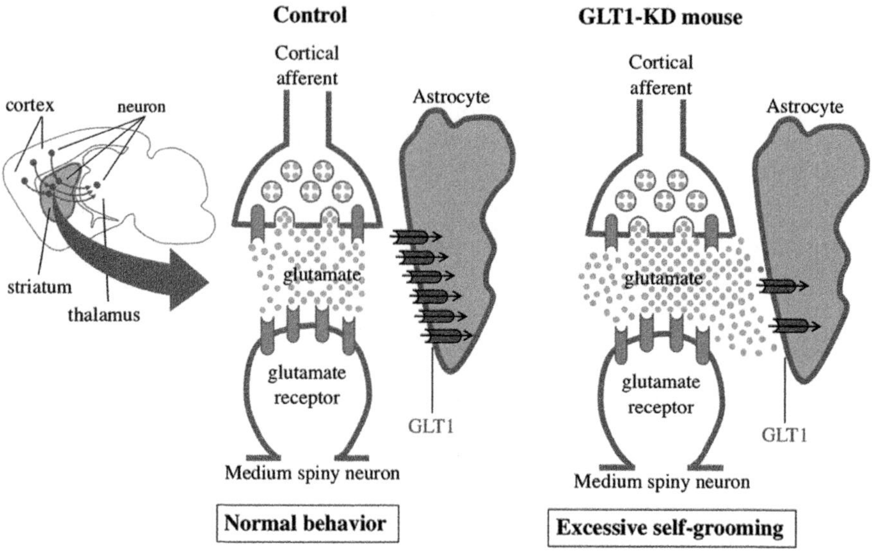

Fig. 1 GLT1-KD mice exhibit OCD-like behaviors via corticostriatal glutamatergic hyperactivity

GLT-1 Loss in *C. elegans* Drives Repetitive Behavior Through Presynaptic Activation of mGluR5

C. elegans is an excellent model system for studying how specific glia-neuron interactions control neuronal functions and animal behaviors because its nervous system is simple and extensively studied (Singhvi and Shaham 2019). Cephalic sheath (CEPsh) glia in *C. elegans* resemble vertebrate astrocytes in developmental, molecular, morphological, and functional properties. CEPsh glia express *ifa-4*, homologous to the glial fibrillary acidic protein (GFAP) and *glt-1*, homologous to the astrocytic glutamate transporter GLT1. GLT-1 loss, like postembryonic ablation of CEPsh glia, drives repetitive locomotor reversal behavior (Katz et al. 2019). These repetitive bouts of reversal locomotion are rescued by the loss of the vesicular glutamate transporter *eat-4*, suggesting that repetitive reversal behavior is dependent on presynaptic glutamate release. AVA is a major *C. elegans* interneuron regulating reversal behavior and RIM is a glutamatergic interneuron presynaptic to AVA. Thus, RIM-AVA synapses in *C. elegans* play a critical role in reversal behavior. CEPsh glia ensheath RIM-AVA synapses. GLT-1 mutants exhibit oscillatory glutamate release near postsynaptic sites of AVA and consequent oscillations in AVA activity. MGL-2, homologous to the metabotropic glutamate receptor 5 (mGluR5), is mainly expressed in RIM, and loss of MGL-2 suppresses the oscillation of AVA activity and repetitive reversal behavior in GLT-1 mutants. These data suggest that astrocytic GLT-1 loss allows glutamate to bind MGL-2 on presynaptic RIM, resulting in the induction of oscillatory glutamate release that leads to firing of the postsynaptic AVA and consequent repetitive reversal behavior (Fig. 2). In the mouse, GLT-1 loss

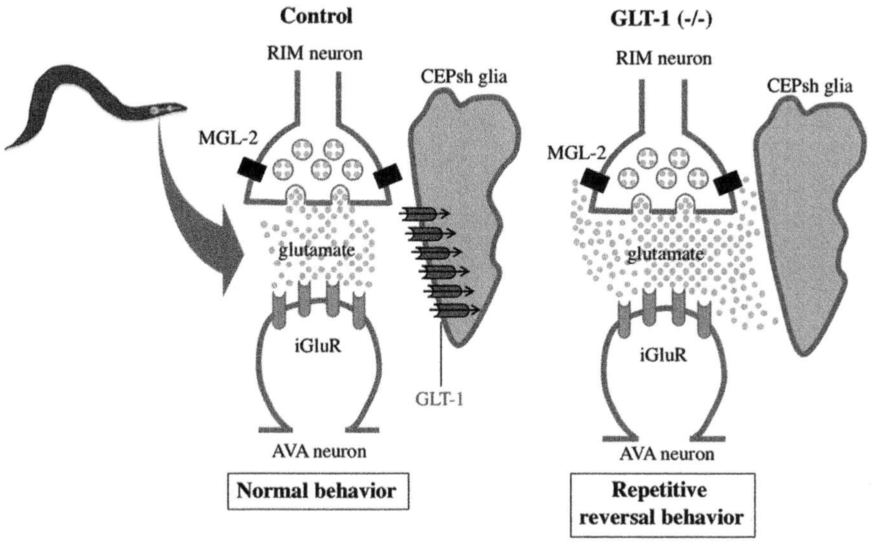

Fig. 2 GLT-1 knockout *C. elegans* exhibit repetitive behavior via presynaptic activation of MGL-2 CEPsh glia, cephalic sheath glia; iGluR, ionotropic glutamate receptor

induces repetitive behaviors (Aida et al. 2015). Thus, uncontrolled extracellular glutamate level due to astrocyte dysfunction is a driver of repetitive behaviors. In addition, excessive mGluR5 signaling is implicated in repetitive self-grooming behavior in mouse models for ASD and OCD (Ade et al. 2016; Silverman et al. 2010, 2012). That the same proteins act together in repetitive behaviors in *C. elegans* and mice suggests a common mechanism controlling repetitive behaviors.

Reduced Astrocyte Calcium Signaling Produces Abnormal Repetitive Behavior via Functional Upregulation of Astrocytic GABA Transporter GAT-3

Astrocytes display a form of excitability based on intracellular Ca^{2+} dynamics and have capabilities to regulate neuronal function (Bazargani and Attwell 2016; Shigetomi et al. 2016). In addition, recent studies suggest that astrocyte Ca^{2+} signaling is dysregulated in mouse models of neurological disorders (Khakh et al. 2017; Mustaly-Kalimi et al. 2018; Verkhratsky et al. 2017). The Khakh laboratory evaluated the consequences of reducing astrocyte calcium signaling in the adult mice (Yu et al. 2018). To reduce astrocyte Ca^{2+} signaling, they used a modified isoform of the human plasma membrane Ca^{2+} pump (hPMCA2w/b), which constitutively extrudes cytosolic Ca^{2+}. They delivered hPMCA2w/b to striatal astrocytes by injecting adeno-associated viruses with an astrocyte-specific GFAP promoter bilaterally into the dorsolateral striatum. hPMCA2w/b reduced the amplitude and shortened the duration of astrocyte Ca^{2+} signals and reduced astrocyte basal Ca^{2+} levels. hPMCA2w/b-expressing mice exhibited excessive self-grooming, whereas they did not show any anxiety or depression-like phenotypes. Although reduction of striatal astrocyte Ca^{2+} signaling did not affect fast inhibitory or excitatory synaptic transmission onto medium spiny neurons (MSNs), MSN tonic inhibition was reduced in hPMCA2w/b-expressing mice. Astrocyte GABA transporter 3 (GAT-3) plays a critical role in tonic inhibition by regulating ambient GABA levels (Kersanté et al. 2013; Scimemi 2014). In hPMCA2w/b-expressing mice, GAT-3 protein levels were upregulated, and SNAP5114, a selective GAT-3 blocker, partially rescued self-grooming behavior. In vivo imaging in freely behaving mice revealed alterations of MSN activity, including increased correlated activity between MSNs during non-grooming episodes, an increase in the frequency of MSN activity during non-grooming episodes, and a reduction in the frequency of MSN activity during self-grooming, in hPMCA2w/b-expressing mice. However, it is not known how these alterations of MSN activity contribute to self-grooming, because there is no microcircuit level model of self-grooming. Furthermore, in a mouse model of Huntington's disease, astrocyte Ca^{2+} signaling was reduced and was associated with MSN tonic inhibition and excessive self-grooming, which was relieved by blocking GAT-3. These results indicate that upregulation of GAT-3 in striatal astrocytes contributes to excessive self-grooming via tonic GABA-mediated neuromodulation (Fig. 3).

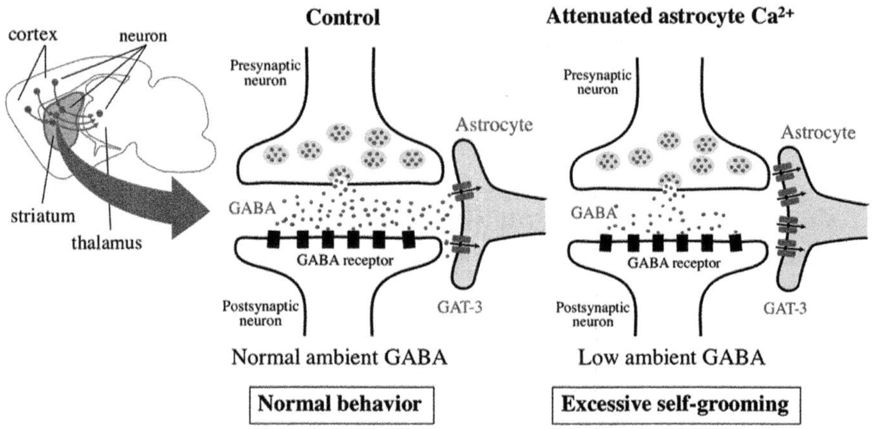

Fig. 3 Reduced astrocyte calcium signaling produces excessive self-grooming via functional upregulation of GAT-3

Conclusion

Neuroimaging studies in patients with OCD and OCD-related disorders have consistently identified hyperactivity in the corticostriatal network (Cerliani et al. 2015; Hou et al. 2014; Jung et al. 2017; Neuner et al. 2014). Studies on animal models of OCD and OCD-related disorders indicate that an increase in excitatory drive or a decrease in inhibitory drive in striatal microcircuitry may lead to enhanced MSN activity (Ahmari et al. 2013; Burguière et al. 2015). Because astrocytes play an important role in the modulation of neuronal excitability via the regulation of glutamate and GABA homeostasis, astrocytic dysfunction may play a critical role in the development of OCD. Although the roles of glutamatergic and GABAergic signaling in OCD have been investigated in patients, astrocytic alterations in patients have not received sufficient attention. Recent genetically engineered animals that exhibit excessive repetitive behaviors suggest that dysfunction of astrocytic GLT1 and GAT-3 plays a critical role in the development of pathological repetitive behaviors (Aida et al. 2015; Katz et al. 2019; Yu et al. 2018). Loss of astrocytic glutamate transporter GLT1 (GLT-1) results in high glutamate levels in the synaptic cleft causing aberrant excitatory transmission that contributes to excessive repetitive behaviors in mice and *C. elegans* (Aida et al. 2015; Katz et al. 2019). In addition, upregulation of astrocytic GABA transporter GAT-3 in the striatum results in loss of tonic inhibition via low ambient GABA levels in the synaptic cleft causing altered MSN activity that contributes to excessive self-grooming in mice (Yu et al. 2018). Although genome-wide association studies (GWAS) in patients with OCD reveal that significant associations were observed with neuronal genes encoding proteins involved in glutamatergic synaptic transmission, including DLGAP1 (disc large-association protein 1), PTPRD (protein tyrosine phosphatase delta), ISM1, and SLC1A1 (Arnold et al. 2006; Dickel et al. 2006; Mattheisen et al. 2015; Samuels et al. 2011; Stewart et al. 2013), no significant association was found in genes mainly expressed in astrocytes such as GLT1 and GAT-3. However, significant associations were observed with SLC1A2 (GLT1) in ASD, which accompanies repetitive behaviors (Autism Genome Project Consortium et al. 2007; Xu et al. 2008). Taken together, these studies suggest that astrocyte dysfunction may be involved in the pathophysiology of pathological repetitive behaviors in OCD and OCD-related disorders. Targeting glutamate or GABA uptake through astrocytes could lead to novel treatment options for patients with OCD and OCD-related disorders.

References

Abe Y, Tsurugizawa T, Le Bihan D (2017) Water diffusion closely reveals neural activity status in rat brain loci affected by anesthesia. PLoS Biol 15:e2001494

Abe Y, Takata N, Sakai Y, Hamada HT, Hiraoka Y, Aida T, Tanaka K, Bihan DL, Doya K, Tanaka KF (2020) Diffusion functional MRI reveals global brain network functional abnormalities driven by targeted local activity in a neuropsychiatric disease mouse model. NeuroImage 223:117318

Ade KK, Wan Y, Hamann HC, O'Hare JK, Guo W, Quian A, Kumar S, Bhagat S, Rodriguiz RM, Wetsel WC, Conn PJ, Dzirasa K, Huber KM, Calakos N (2016) Increased metabotropic glutamate receptor 5 signaling underlies obsessive-compulsive disorder-like behavioral and striatal circuit abnormalities in mice. Biol Psychiatry 80:522–533

Ahmari SE, Spellman T, Douglass NL, Kheirbek MA, Simpson HB, Deisseroth K, Gordon JA, Hen R (2013) Repeated cortico-striatal stimulation generates persistent OCD-like behavior. Science 340:1234–1239

Aida T, Yoshida J, Nomura M, Tanimura A, Iino Y, Soma M, Bai N, Ito Y, Cui W, Aizawa H, Yanagisawa M, Nagai T, Takata N, Tanaka KF, Takayanagi R, Kano M, Götz M, Hirase H, Tanaka K (2015) Astroglial glutamate transporter deficiency increases synaptic excitability and leads to pathological repetitive behaviors in mice. Neuropsychopharmacology 40:1569–7159

Aizawa H, Sun W, Sugiyama K, Itou Y, Aida T, Cui W, Toyoda S, Terai H, Yanagisawa M, Tanaka K (2020) Glial glutamate transporter GLT-1 determines susceptibility to spreading depression in the mouse cerebral cortex. Glia 68:2631–2642

Albin RL, Mink JW (2006) Recent advances in Tourette syndrome research. Trends Neurosci 29:175–182

American Psychiatric Association (2013) Diagnostic and statistical manual of mental disorders (DSM-V). American Psychiatric Publishing, Arington

Arnold PD, Sicard T, Burroughs E, Richter MA, Kennedy JL (2006) Glutamate transporter gene SLC1A1 associated with obsessive-compulsive disorder. Arch Gen Psychiatry 63:769–776

Autism Genome Project Consortium, Szatmari P, Paterson AD, Zwaigenbaum L, Roberts W, Brian J, Liu XQ, Vincent JB, Skaug JL, Thompson AP, Senman L, Feuk L, Qian C, Bryson SE, Jones MB, Marshall CR, Scherer SW, Vieland VJ, Bartlett C, Mangin LV, Goedken R, Segre A, Pericak-Vance MA, Cuccaro ML, Gilbert JR, Wright HH, Abramson RK, Betancur C, Bourgeron T, Gillberg C, Leboyer M, Buxbaum JD, Davis KL, Hollander E, Silverman JM, Hallmayer J, Lotspeich L, Sutcliffe JS, Haines JL, Folstein SE, Piven J, Wassink TH, Sheffield V, Geschwind DH, Bucan M, Brown WT, Cantor RM, Constantino JN, Gilliam TC, Herbert M, Lajonchere C, Ledbetter DH, Lese-Martin C, Miller J, Nelson S, Samango-Sprouse CA, Spence S, State M, Tanzi RE, Coon H, Dawson G, Devlin B, Estes A, Flodman P, Klei L, McMahon WM, Minshew N, Munson J, Korvatska E, Rodier PM, Schellenberg GD, Smith M, Spence MA, Stodgell C, Tepper PG, Wijsman EM, Yu CE, Rogé B, Mantoulan C, Wittemeyer K, Poustka A, Felder B, Klauck SM, Schuster C, Poustka F, Bölte S, Feineis-Matthews S, Herbrecht E, Schmötzer G, Tsiantis J, Papanikolaou K, Maestrini E, Bacchelli E, Blasi F, Carone S, Toma C, Van Engeland H, de Jonge M, Kemner C, Koop F, Langemeijer M, Hijmans C, Staal WG, Baird G, Bolton PF, Rutter ML, Weisblatt E, Green J, Aldred C, Wilkinson JA, Pickles A, Le Couteur A, Berney T, McConachie H, Bailey AJ, Francis K, Honeyman G, Hutchinson A, Parr JR, Wallace S, Monaco AP, Barnby G, Kobayashi K, Lamb JA, Sousa I, Sykes N, Cook EH, Guter SJ, Leventhal BL, Salt J, Lord C, Corsello C, Hus V, Weeks DE, Volkmar F, Tauber M, Fombonne E, Shih A, Meyer KJ (2007) Mapping autism risk loci using genetic linkage and chromosomal rearrangements. Nat Genet 39:319–328

Bazargani N, Attwell D (2016) Astrocyte calcium signaling: the third wave. Nat Neurosci 19:182–189

Boddum K, Jensen TP, Magloire V, Kristiansen U, Rusakov DA, Pavlov I, Walker MC (2016) Astrocytic GABA transporter activity modulates excitatory neurotransmission. Nat Commun 7:13572

Browne HA, Gair SL, Scharf JM, Grice DE (2014) Genetics of obsessive-compulsive disorder and related disorders. Psychiatr Clin North Am 37:319–335

Burguière E, Monteiro P, Mallet L, Feng G, Graybiel AM (2015) Striatal circuits, habits, and implications for obsessive-compulsive disorder. Curr Opin Neurobiol 30:59–65

Cerliani L, Mennes M, Thomas RM, Di Martino A, Thioux M, Keysers C (2015) Increased functional connectivity between subcortical and cortical resting-state networks in Autism Spectrum Disorder. JAMA Psychiat 72:767–777
Choi DW (1994) Glutamate receptors and the induction of excitotoxic neuronal death. Prog Brain Res 100:47–51
Cui W, Mizukami H, Yanagisawa M, Aida T, Nomura M, Isomura Y, Takayanagi R, Ozawa K, Tanaka K, Aizawa H (2014) Glial dysfunction in the mouse habenula causes depressive-like behaviors and sleep disturbance. J Neurosci 34:16273–11685
Deco G, Kringelbach ML (2017) Hierarchy of information processing in the brain: a novel 'Intrinsic Ignition' framework. Neuron 94:961–968
Deco G, Tagliazucchi E, Laufs H, Sanjuán A, Kringelbach ML (2017) Novel intrinsic ignition method measuring local-global integration characterizes wakefulness and deep sleep. eNeuro 4:ENEURO.0106–17.2017
Dickel DE, Veenstra-VanderWeele J, Cox NJ, Wu X, Fischer DJ, Van Etten-Lee M, Himle JA, Leventhal BL, Cook EH Jr, Hanna GL (2006) Association testing of the positional and functional candidate gene SLC1A1/EAAC1 in early-onset obsessive-compulsive disorder. Arch Gen Psychiatry 63:778–785
Drew PJ (2019) Vascular and neural basis of the BOLD signal. Curr Opin Neurobiol 58:61–69
Hoshina N, Tanimura A, Yamasaki M, Inoue T, Fukabori R, Kuroda T, Yokoyama K, Tezuka T, Sagara H, Hirano S, Kiyonari H, Takada M, Kobayashi K, Watanabe M, Kano M, Nakazawa T, Yamamoto T (2013) Protocadherin 17 regulates presynaptic assembly in topographic cortico-basal Ganglia circuits. Neuron 78:839–854
Hou JM, Zhao M, Zhang W, Song LH, Wu WJ, Wang J, Zhou DQ, Xie B, He M, Guo JW, Qu W, Li HT (2014) Resting-state functional connectivity abnormalities in patients with obsessive-compulsive disorder and their healthy first-degree relatives. J Psychiatry Neurosci 39:304–311
Jung WH, Yücel M, Yun JY, Yoon YB, Cho KI, Parkes L, Kim SN, Kwon JS (2017) Altered functional network architecture in orbitofronto-striato-thalamic circuit of unmedicated patients with obsessive-compulsive disorder. Hum Brain Mapp 38:109–119
Katz M, Corson F, Keil W, Singhal A, Bae A, Lu Y, Liang Y, Shaham S (2019) Glutamate spillover in C. elegans triggers repetitive behavior through presynaptic activation of MGL-2/mGluR5. Nat Commun 10:1882
Kersanté F, Rowley SC, Pavlov I, Gutièrrez-Mecinas M, Semyanov A, Reul JM, Walker MC, Linthorst AC (2013) A functional role for both -aminobutyric acid (GABA) transporter-1 and GABA transporter-3 in the modulation of extracellular GABA and GABAergic tonic conductances in the rat hippocampus. J Physiol 591:2429–2441
Kessler RC, Berglund P, Demler O, Jin R, Merikangas KR, Walters EE (2005) Lifetime prevalence and age-of-onset distributions of DSM-IV disorders in the National Comorbidity Survey Replication. Arch Gen Psychiatry 62:593–602
Khakh BS, Beaumont V, Cachope R, Munoz-Sanjuan I, Goldman SA, Grantyn R (2017) Unravelling and exploiting astrocyte dysfunction in Huntington's disease. Trends Neurosci 40:422–437
Kiryk A, Aida T, Tanaka K, Banerjee P, Wilczynski GM, Meyza K, Knapska E, Filipkowski RK, Kaczmarek L, Danysz W (2008) Behavioral characterization of GLT1 (+/−) mice as a model of mild glutamatergic hyperfunction. Neurotox Res 13:19–30
Le Bihan D, Urayama S, Aso T, Hanakawa T, Fukuyama H (2006) Direct and fast detection of neuronal activation in the human brain with diffusion MRI. Proc Natl Acad Sci U S A 103:8263–8268
Leckman JF, Grice DE, Boardman J, Zhang H, Vitale A, Bondi C, Alsobrook J, Peterson BS, Cohen DJ, Rasmussen SA, Goodman WK, McDougle CJ, Pauls DL (1997) Symptoms of obsessive-compulsive disorder. Am J Psychiatry 154:911–917
Matos M, Bosson A, Riebe I, Reynell C, Vallée J, Laplante I, Panatier A, Robitaille R, Lacaille JC (2018) Astrocytes detect and upregulate transmission at inhibitory synapses of somatostatin interneurons onto pyramidal cells. Nat Commun 9:4254

Mattheisen M, Samuels JF, Wang Y, Greenberg BD, Fyer AJ, McCracken JT, Geller DA, Murphy DL, Knowles JA, Grados MA, Riddle MA, Rasmussen SA, McLaughlin NC, Nurmi EL, Askland KD, Qin HD, Cullen BA, Piacentini J, Pauls DL, Bienvenu OJ, Stewart SE, Liang KY, Goes FS, Maher B, Pulver AE, Shugart YY, Valle D, Lange C, Nestadt G (2015) Genome-wide association study in obsessive-compulsive disorder: results from the OCGAS. Mol Psychiatry 20:337–344

Mori T, Tanaka K, Buffo A, Wurst W, Kühn R, Götz M (2006) Inducible gene deletion in astroglia and radial glia-a valuable tool for functional and lineage analysis. Glia 54:21–34

Mustaly-Kalimi S, Littlefield AM, Stutzmann GE (2018) Calcium signaling deficits in glia and Autophagic pathways contributing to neurodegenerative disease. Antioxid Redox Signal 29:1158–1175

Nakanishi S, Nakajima Y, Masu M, Ueda Y, Nakahara K, Watanabe D, Yamaguchi S, Kawabata S, Okada M (1998) Glutamate receptors: brain function and signal transduction. Brain Res Brain Res Rev 26:230–235

Neuner I, Werner CJ, Arrubla J, Stöcker T, Ehlen C, Wegener HP, Schneider F, Shah NJ (2014) Imaging the where and when of tic generation and resting state networks in adult Tourette patients. Front Hum Neurosci 8:362

Rubenstein JL, Merzenich MM (2003) Model of autism: increased ratio of excitation/inhibition in key neural systems. Genes Brain Behav 2:255–267

Samuels J, Wang Y, Riddle MA, Greenberg BD, Fyer AJ, McCracken JT, Rauch SL, Murphy DL, Grados MA, Knowles JA, Piacentini J, Cullen B, Bienvenu OJ 3rd, Rasmussen SA, Geller D, Pauls DL, Liang KY, Shugart YY, Nestadt G (2011) Comprehensive family-based association study of the glutamate transporter gene SLC1A1 in obsessive-compulsive disorder. Am J Med Genet B Neuropsychiatr Genet 156B:472–477

Scimemi A (2014) Structure, function, and plasticity of GABA transporters. Front Cell Neurosci 8:161

Shigetomi E, Patel S, Khakh BS (2016) Probing the complexities of astrocyte calcium signaling. Trends Cell Biol 26:300–312

Silverman JL, Tolu SS, Barkan CL, Crawley JN (2010) Repetitive self-grooming behavior in the BTBR mouse model of autism is blocked by the mGluR5 antagonist MPEP. Neuropsychopharmacology 35:976–989

Silverman JL, Smith DG, Rizzo SJ, Karras MN, Turner SM, Tolu SS, Bryce DK, Smith DL, Fonseca K, Ring RH, Crawley JN (2012) Negative allosteric modulation of the mGluR5 receptor reduces repetitive behaviors and rescues social deficits in mouse models of autism. Sci Transl Med 4:131ra51

Singhvi A, Shaham S (2019) Glia-neuron interactions in *Caenorhabditis elegans*. Annu Rev Neurosci 42:149–168

Stewart SE, Yu D, Scharf JM, Neale BM, Fagerness JA, Mathews CA, Arnold PD, Evans PD, Gamazon ER, Davis LK, Osiecki L, McGrath L, Haddad S, Crane J, Hezel D, Illman C, Mayerfeld C, Konkashbaev A, Liu C, Pluzhnikov A, Tikhomirov A, Edlund CK, Rauch SL, Moessner R, Falkai P, Maier W, Ruhrmann S, Grabe HJ, Lennertz L, Wagner M, Bellodi L, Cavallini MC, Richter MA, Cook EH Jr, Kennedy JL, Rosenberg D, Stein DJ, Hemmings SM, Lochner C, Azzam A, Chavira DA, Fournier E, Garrido H, Sheppard B, Umaña P, Murphy DL, Wendland JR, Veenstra-VanderWeele J, Denys D, Blom R, Deforce D, Van Nieuwerburgh F, Westenberg HG, Walitza S, Egberts K, Renner T, Miguel EC, Cappi C, Hounie AG, Conceição do Rosário M, Sampaio AS, Vallada H, Nicolini H, Lanzagorta N, Camarena B, Delorme R, Leboyer M, Pato CN, Pato MT, Voyiaziakis E, Heutink P, Cath DC, Posthuma D, Smit JH, Samuels J, Bienvenu OJ, Cullen B, Fyer AJ, Grados MA, Greenberg BD, McCracken JT, Riddle MA, Wang Y, Coric V, Leckman JF, Bloch M, Pittenger C, Eapen V, Black DW, Ophoff RA, Strengman E, Cusi D, Turiel M, Frau F, Macciardi F, Gibbs JR, Cookson MR, Singleton A, North American Brain Expression Consortium, Hardy J, UK Brain Expression Database, Crenshaw AT, Parkin MA, Mirel DB, Conti DV, Purcell S, Nestadt G, Hanna GL, Jenike MA,

Knowles JA, Cox N, Pauls DL (2013) Genome-wide association study of obsessive-compulsive disorder. Mol Psychiatry 18:788–798

Sugimoto J, Tanaka M, Sugiyama K, Ito Y, Aizawa H, Soma M, Shimizu T, Mitani A, Tanaka K (2018) Region-specific deletions of the glutamate transporter GLT1 differentially affect seizure activity and neurodegeneration in mice. Glia 66:777–788

Sugiyama K, Aida T, Nomura M, Takayanagi R, Zeilhofer HU, Tanaka K (2017) Calpain-dependent degradation of nucleoporins contributes to motor neuron death in a mouse model of chronic excitotoxicity. J Neurosci 37:8830–8844

Takata N, Sugiura Y, Yoshida K, Koizumi M, Hiroshi N, Honda K, Yano R, Komaki Y, Matsui K, Suematsu M, Mimura M, Okano H, Tanaka KF (2018) Optogenetic astrocyte activation evokes BOLD fMRI response with oxygen consumption without neuronal activity modulation. Glia 66:2013–2023

Tanaka K (2000) Functions of glutamate transporters in the brain. Neurosci Res 37:15–19

Tanaka K, Watase K, Manabe T, Yamada K, Watanabe M, Takahashi K, Iwama H, Nishikawa T, Ichihara N, Kikuchi T, Okuyama S, Kawashima N, Hori S, Takimoto M, Wada K (1997) Epilepsy and exacerbation of brain injury in mice lacking the glutamate transporter GLT-1. Science 276:1699–1702

Tsurugizawa T, Ciobanu L, Le Bihan D (2013) Water diffusion in brain cortex closely tracks underlying neuronal activity. Proc Natl Acad Sci U S A 110:11636–11641

Verkhratsky A, Rodríguez-Arellano JJ, Parpura V, Zorec R (2017) Astroglial calcium signalling in Alzheimer's disease. Biochem Biophys Res Commun 483:1005–1012

Voutsinos-Porche B, Bonvento G, Tanaka K, Steiner P, Welker E, Chatton JY, Magistretti PJ, Pellerin L (2003) Glial glutamate transporters mediate a functional metabolic crosstalk between neurons and astrocytes in the mouse developing cortex. Neuron 37:275–286

Wu K, Hanna GL, Rosenberg DR, Arnold PD (2012) The role of glutamate signaling in the pathogenesis and treatment of obsessive-compulsive disorder. Phrmacol Biochem Behav 100:726–735

Xu S, Han JC, Morales A, Menzie CM, Williams K, Fan YS (2008) Characterization of 11p14-p12 deletion in WAGR syndrome by array CGH for identifying genes contributing to mental retardation and autism. Cytogenet Genome Res 122:181–187

Yu X, Taylor AMW, Nagai J, Golshani P, Evans CJ, Coppola G, Khakh BS (2018) Reducing astrocyte calcium signaling in vivo alters striatal microcircuits and causes repetitive behavior. Neuron 99:1170–1187

Zhao Z, Hiraoka Y, Ogawa H, Tanaka K (2018) Region-specific deletions of the glutamate transporter GLT1 differentially affect nerve injury-induced neuropathic pain in mice. Glia 66:1988–1998

Part III
Astroglia in Schizophrenia

Astrocytes in Neuropsychiatric Disorders: A Review of Postmortem Evidence

Xiaolu Zhang, Rawan S. Alnafisah, Abdul-Rizaq A. Hamoud, Rammohan Shukla, Robert E. McCullumsmith, and Sinead M. O'Donovan

Introduction

Glial cell types in the central nervous system (CNS) include microglia, oligodendrocytes, and, the most diverse type, astrocytes (Kettenmann and Ransom 2004). Astrocytes were named by Michael von Lenhossek at the end of the nineteenth century for the distinctive starlike shape of their complex arborization (Oberheim et al. 2012; Parpura and Verkhratsky 2012), and the term was later popularized by Santiago Ramon y Cajal (1913 and von Lenhossék (1895). Since the first astrocytes were observed, many different subtypes of astrocyte, displaying extensive morphological heterogeneity and different physiological properties, have been identified (Andriezen 1893). However, the primary function of this cell type is maintaining brain homeostasis (Andriezen 1893; Verkhratsky and Nedergaard 2018). An overview of different astrocyte subtypes is described in Table 1. Clinical and experimental evidence suggests critical roles for astrocytes in the pathogenesis of CNS disease. Studies examining astrocyte in postmortem brain tissues from patients diagnosed with neuropsychiatric disorders offer invaluable insight into the role of astrocytes in illnesses like schizophrenia (SCZ), major depressive disorder (MDD), and bipolar disorder (BPD) (Tarasov et al. 2019; Rajkowska and Stockmeier 2013; Peng et al. 2016).

X. Zhang · R. S. Alnafisah · A.-R. A. Hamoud · R. Shukla · S. M. O'Donovan (✉)
Department of Neurosciences, University of Toledo College of Medicine, Toledo, OH, USA
e-mail: Sinead.ODonovan@utoledo.edu

R. E. McCullumsmith
Department of Neurosciences, University of Toledo College of Medicine, Toledo, OH, USA

Neurosciences Institute, ProMedica, Toledo, OH, USA

B. Li et al. (eds.), *Astrocytes in Psychiatric Disorders*, Advances in Neurobiology 26, https://doi.org/10.1007/978-3-030-77375-5_8

Table 1 Summary of astrocyte subtypes. FC, frontal cortex. "Mammals" indicates astrocyte subtype is found in humans, primates, and rodents

Class	Location	Species	Functions
Protoplasmic astrocytes	FC layers III and IV hippocampus	Mammals	Guide neuronal development (Hatten and Mason 1990)
			Regulate extracellular ions, metabolites, and neurotransmitters (Vernadakis 1996)
			Support neuronal and synaptic function (Araque et al. 1999)
Fibrous astrocytes	White matter	Mammals	Vascularization (Marin-Padilla 1995)
Interlaminar astrocytes	FC layers I and II	Primate, human	Long-distance coordination of intracortical communication (Oberheim et al. 2009)
			Coordinate blood flow (Oberheim et al. 2009)
Varicose astrocytes	FC layers V and VI	Higher-order primates, human	Long-distance coordination of intracortical communication (Sosunov et al. 2014)
			Vascularization (Sosunov et al. 2014)
Polarized astrocytes	FC layers V and VI	Higher-order primates, human	Long-distance coordination of intracortical communication (Verkhratsky et al. 2017)
Radial glia	Only in embryogenesis	Mammals	Embryonic neurogenesis Primary neural stem cells (Falk and Gotz 2017)
Bergmann glia	Cerebellum	Mammals	Regulate synapsis of Purkinje cells (Edwards et al. 1990)
Tanycytes	Hypothalamus	Mammals	Regulate neuroendocrine function (Prevot et al. 2018)
Müller glia	Retina	Mammals	Regulate synaptic activity in the inner retina Protect photoreceptors and neurons (Verkhratsky et al. 2019)
Radial glia-like neural stem cells	Neurogenic niches	Mammals	Local homeostasis Neurogenesis (Bonzano et al. 2018)
Velate astrocytes	Cerebellum	Mammals	Support enwrapped granule neurons (Verkhratsky et al. 2019) Isolate synaptic structures (Verkhratsky and Nedergaard 2018)
Perivascular and marginal astrocytes	Close to pia mater	Mammals	Form the pial and glia limitans barrier (Verkhratsky and Nedergaard 2018)
Pituicytes	Pituitary	Mammals	Release hormone Supply for synaptic neurotransmission in neurohypophysis (Garcia-Segura and McCarthy 2004)
Gomori-positive astrocytes	Hypothalamus and hippocampus	Mammals	Positive for Gomori's chrome staining Supply heme to neurons for the synthesis of heme enzymes Provide specific metabolic needs of neurons through high glucose metabolism (Verkhratsky and Nedergaard 2018)
Surface-associated astrocytes	Cortical surface	Mammals	Form the glia limitans (Feig and Haberly 2011)
Ependymocytes, choroid plexus cells, and retinal pigment epithelial cells	Ventricles and subretinal	Mammals	Produce cerebrospinal fluid (Verkhratsky and Nedergaard 2018)

Postmortem Changes in Astrocytes in Schizophrenia (SCZ)

A recently posited gliocentric hypothesis of schizophrenia suggests that suppression of glial progenitor cell differentiation during fetal development impacts astrocyte maturation, leading to reduced synapse coverage and disruption of glutamatergic homeostasis (Dietz et al. 2020). Reactive changes in astrocytes in SCZ are proposed to occur later in disease as secondary responses to other factors like neuroinflammation and drug treatment and may contribute to inconsistencies found in postmortem analyses of astrocytes in SCZ (Dietz et al. 2020). This idea is derived from neuroimaging, experimental data, and postmortem neuropathological studies. These studies show that astrocyte density, number, and morphology vary in SCZ in a manner that is highly brain region dependent (Williams et al. 2013; Benes et al. 1986, Bruton et al. 1990; Rajkowska et al. 2002a). In the middle frontal gyrus, hypertrophic astrocyte morphology is found in subjects with SCZ (Catts et al. 2014a), but in the subgenual cingulate white matter, a decrease in astrocyte density is reported in SCZ subjects compared to controls (Williams et al. 2014a). Astrocyte number was reportedly downregulated (Pakkenberg 1990) or upregulated (Barley et al. 2009) in the mediodorsal nucleus of the thalamus in SCZ in different studies. Astrocyte density varies across different subregions of the dorsolateral prefrontal cortex (DLPFC) (Rajkowska et al. 2002b), but there was no significant change in astrocyte number and density in hippocampal subfields of SCZ subjects compared to controls following stereological counting of morphologically identified (Nissl stain) cells (Schmitt et al. 2009). Yet, in the subiculum and CA1 in the hippocampus, there was a positive association between astrocyte number and disease duration in SCZ patients (Schmitt et al. 2009). A similar association between subiculum astrocyte number and age was also found in control subjects, suggesting age, rather than diagnosis, has a greater effect on astrocyte expression in this brain region. Additionally, immunoreactive counts of vimentin, a structural protein used as an astrocyte marker, were unaffected in the ventromedial temporal, frontal, and calcarine cortices in SCZ (Arnold et al. 1996a). Overall, reports of astrocyte morphology and density in SCZ studies vary between brain regions and methodological approaches applied.

In addition to astrocyte morphology and cell counts, protein and mRNA expression of astrocyte markers are also used to assess astrocytes in postmortem tissue. The immunoreactivity and total protein and mRNA expression of the intermediate filament protein glial fibrillary acidic protein (GFAP) were increased in the olfactory tubercle at the mesolimbic system and ventromedial temporal, frontal, and calcarine cortices (Toro et al. 2006a; Markova et al. 2000), but decreased in the anterior cingulate cortex (ACC) and white matter of the subgenual cingulate cortex (Williams et al. 2013; Steffek et al. 2008; Williams et al. 2014b) in SCZ subjects. Other studies showed no change in GFAP expression in the DLPFC, primary visual cortex, and hippocampus (Catts et al. 2014a; Steffek et al. 2008; Trépanier et al. 2016). SCZ patients presenting with dementia had higher numbers of GFAP-positive astrocytes, increased GFAP mRNA and protein expression, enlargement of astrocyte soma, and thickening of astrocytic processes compared to SCZ patients without dementia

(Arnold et al. 1996b). Therefore, it appears that age and dementia are two essential factors associated with higher astrocyte number. Overall, GFAP expression is altered in SCZ subjects compared with controls depending on the brain area studied. A summary of GFAP expression changes in SCZ is shown in Table 2.

Other commonly used astrocyte markers include vimentin, excitatory amino acid transporter (EAAT1/2), aldolase C, and S100 calcium-binding protein B (S100B) (Hwang et al. 2013; Feresten et al. 2013a; Katsel et al. 2011a, b). The protein expression of aldolase C was upregulated (Johnston-Wilson et al. 2000a; Martins-de-Souza 2010) or downregulated (Prabakaran et al. 2004; English et al. 2009) in SCZ depending on the brain region studied. Conversely, vimentin protein expression was unchanged in studies of the prefrontal cortex (Feresten et al. 2013a) and ventromedial temporal, frontal, and calcarine cortices (Arnold et al. 1996a). Interestingly, S100B mRNA expression was unchanged in the superficial layers or the underlying white matter of the cingulate cortex in SCZ subjects; however, it was significantly reduced in the deep layers of the anterior cingulate gyrus (Katsel et al. 2011b), suggesting lamina-specific changes in astrocyte expression (Arnold et al. 1996a; Feresten et al. 2013a). S100B immunopositive cells were increased in several brain regions such as ACC, DLPFC, orbitofrontal cortex, mediodorsal nucleus of the thalamus, and pyramidal layer of the hippocampus in a subset of SCZ subjects (Steiner et al. 2008). White matter levels of S100B were significantly increased in paranoid SCZ relative to residual SCZ subjects; however, this effect was driven by oligodendrocyte rather than astrocyte labeled S100B immunopositive cells (Steiner et al. 2008). In the absence of co-labeling with additional astrocyte markers, S100B labeled cell counts should be interpreted cautiously (Steiner et al. 2007).

EAAT2 mRNA and protein, commonly used markers of astrocytes, were increased in the prefrontal cortex Brodmann areas (BA) 9 and 10 in subjects with SCZ (Matute et al. 2005). However, significant reductions in mRNA (Ohnuma et al. 1998), significant increases in mRNA (Shao and Vawter 2008), and no change in mRNA or protein in the DLPFC and ACC (Bauer et al. 2008) have also been reported in SCZ. EAAT1 mRNA was increased in the ACC but unchanged in the DLPFC (Bauer et al. 2008), while EAAT1 protein was decreased in the DLPFC but unchanged in the ACC in SCZ (Bauer et al. 2008; Shan et al. 2013). Posttranslational modification of EAAT2 by glycosylation, necessary for EAAT2 localization to the membrane, is reduced in the DLPFC but unchanged in the ACC in SCZ (Bauer et al. 2010). In other brain regions, such as the hippocampus and superior temporal gyrus, an overall downregulation in EAAT2 protein expression was found in SCZ (Shan et al. 2013). EAAT1 protein was also downregulated in the superior temporal gyrus but unchanged in the hippocampus (Bauer et al. 2008; Shan et al. 2013). Immunoblotting showed a significant reduction in EAAT2 protein expression in the mediodorsal and ventral tier nuclei in the thalamus in SCZ. Protein expression of EAAT1 was reduced in the mediodorsal nucleus in SCZ subjects, confirming findings of reduced EAAT1 mRNA expression in an enriched population of astrocytes cut from that brain region in the same study (McCullumsmith et al. 2016). Overall, EAAT1/2 expression is broadly unchanged or reduced in SCZ although expression varies based on the brain region and whether mRNA or protein is being measured

Table 2 GFAP expression in postmortem brain tissues of subjects diagnosed with schizophrenia (SCZ)

Study	Brain region	SCZ	Method
Pantazopoulos et al. (2010)	Amygdala, entorhinal cortex	NS ($n = 11$)	IHC, ICC
Damadzic et al. (2001)	Entorhinal cortex	NS ($n = 21$)	IHC
Dean et al. (2006)	BA9, 10, 46, 40	NS ($n = 20$)	WB, qPCR
Webster et al. (2001)	DLPFC and hippocampus	Astrocytes containing pGFAP in DLPFC ↓; NS in hippocampus (n = 15 per group)	IHC
Altshuler et al. (2010)	Amygdala	NS ($n = 9$)	IHC
Feresten et al. (2013b)	DLPFC (BA9)	GFAP protein↑ ($n = 35$)	WB
Toro et al. (2006b)	DLPFC (BA9, 32, 46) and OFC (BA11,12,47,45)	GFAP protein DLPFC ↑; GFAP protein OFC ↓ ($n = 15$)	IHC
Fatemi et al. (2004)	Lateral cerebellum	GFAP protein↓ ($n = 15$ per group)	WB
Hercher et al. (2014)	White matter adjacent to DLPFC (BA9)	GFAP area fraction↓ ($n = 20$)	IHC
Johnston-Wilson et al. (2000b)	BA 10	GFAP311 protein↓ ($n = 24$)	2-DE
Markova et al. (2000)	Olfactory tubercle at the mesolimbic system	GFAP ↑ immunoreactive astrocytes ($n = 12$)	3-D Golgi, IHC
Arnold et al. (1996a)	Ventromedial temporal, frontal, and calcarine cortices	GFAP immunoreactivity↑ in SZ patients with dementia ($n = 21$)	ICC
Steffek et al. (2008)	DLPFC, primary visual cortex, superior Temporal gyrus, ACC, and hippocampus	GFAP protein ↓ ACC ($n = 23$)	WB
Williams et al. (2014a)	Subgenual cingulate cortex (white matter)	GFAP immunoreactivity ↓ ($n = 11$)	ICC
Williams et al. (2013)	ACC	GFAP mRNA↓ ($n = 11$)	IHC
Catts et al. (2014a)	Middle frontal gyrus anterior to the premotor cortex (BA46)	NS ($n = 37$)	WB, qPCR
Steffek et al. (2008)	DLPFC, ACC, primary visual cortex, superior Temporal gyrus, and hippocampus	NS ($n = 23$)	WB

BA Brodmann area, *DLPFC* dorsolateral prefrontal cortex, *OFC* orbitofrontal cortex, *ACC* anterior cingulate cortex, *WB* western blotting, *ICC* immunocytochemistry, *IHC* immunohistochemistry, *NS* no significant change, *2-DE* two-dimensional gel electrophoresis.

(Hu et al. 2015). However, EAAT2 mRNA is also expressed in neurons, albeit at low levels. Neuronal EAAT2 gene expression is increased in SCZ and other neuropsychiatric and neurological disorders (O'Donovan et al. 2017), which should be taken into account when interpreting reports of EAAT2 mRNA expression changes in SCZ.

The use of different methods such as immunohistochemistry, enzyme-linked immunosorbent assay (ELISA), and western immunoblot analysis to detect astrocyte marker protein expression may contribute to inconsistencies in reported changes in astrocyte marker expression in SCZ (Catts et al. 2014a; Davalieva et al. 2016). Differences in genetic risk, subject age, and cause of death of individuals with SCZ may also be factors (Barley et al. 2009; Gosselin et al. 2009). Increased immune activation, as indicated by increased expression of immune-related genes in postmortem brain transcriptomic studies, may underlie the hypertrophic astrocyte morphology and increase in GFAP expression in a subset of SCZ subjects (Catts et al. 2014a; Fillman et al. 2013, 2014; Catts et al. 2014b).

Postmortem Changes in Astrocytes in Major Depressive Disorder (MDD)

Postmortem studies have revealed significantly reduced glial cell counts and reduced cell density in subjects with MDD. There was a marked reduction in astrocyte counts in the amygdala, layers II, III, and V of the DLPFC and layer VI of the ACC (Bowley et al. 2002; Ongur et al. 1998; Rajkowska et al. 1999; Cotter et al. 2001a, 2002). In a small study ($n = 6$ MDD subjects), there was a 29% reduction in astrocyte density in the amygdala and a 36% reduction in the glia to neuron ratio. The changes in the amygdala were primarily in the left hemisphere (~53%). However, after the effects of age and sex were taken into account, the differences in glia to neuron ratio and astrocyte density were not significant in MDD subjects (Bowley et al. 2002). Studies examining astrocytes in the prefrontal cortex of MDD subjects found reduced cell counts in the prefrontal subgenual region (~24%) and reduced cell density (20–30%) in layers II/III of the DLPFC (Ongur et al. 1998; Rajkowska et al. 1999; Cotter et al. 2002). Overall, decreased astrocyte cell count and density are found in diverse brain regions in postmortem MDD studies.

In MDD patients who died by suicide, protoplasmic (gray matter) and fibrous (white matter) astrocytes were reconstructed for morphometric analysis. The Golgi staining method revealed that astrocyte cell bodies in layer VI of the ACC were spherical with varicose thorny processes projecting in every direction. Fibrous astrocytes of the white matter in the same subjects presented oblong cell bodies with long, unramified varicose processes extending in opposite directions. These differences between white and gray cortical matter astrocyte activation may indicate local inflammation of the white matter leading to changes in fibrous astrocytes (Torres-Platas et al. 2011). The hypertrophy observed in the white matter astrocytes is in line with the neuroinflammation hypothesis of MDD (Maes et al. 2009). Reduced astrocyte count and density was identified in both the gray and white matter in different brain regions in non-suicide MDD subjects, while hypertrophy of astrocytes was observed only in the white matter of the ACC (Torres-Platas et al. 2011; Qi et al. 2019). Based on the relatively small number of postmortem studies conducted to date, we can posit that cortical and subcortical brain regions in MDD subjects consistently present reduced astrocyte count and density. While the hypertrophy of astrocytes reported in the ACC of MDD subjects may be attributed to local inflammation due to proinflammatory cytokines, differences in cell count and density are region-specific, and further studies are necessary to understand the role of these astrocyte changes. In summary, based on observations of altered astrocyte numbers from postmortem MDD histopathologic studies, it has been posited that abnormal astrocyte function may contribute to the pathophysiology of mood disorders (Cotter et al. 2001b; Rajkowska and Miguel-Hidalgo 2007; Hercher et al. 2009).

Protein expression of marker GFAP was reported to have decreased tenfold in the prefrontal cortex (BA 9, layers I–VI) in MDD subjects under 60 years of age. There was a positive correlation between age and GFAP expression in these subjects (Si et al. 2004). Immunoreactive astrocytes had decreased GFAP expression in the amygdala, cerebellum, thalamus, and caudate nucleus of subjects with MDD illustrating widespread astrocyte pathology in MDD (Zhao et al. 2016; Choudary et al. 2005; Altshuler et al. 2010; Fatemi et al. 2004; Qi et al. 2019; Torres-Platas et al. 2016). A summary of GFAP expression changes in MDD is shown in Table 3. EAAT1 protein expression and EAAT1 and EAAT2 immunoreactivity were decreased in the orbitofrontal cortex of MDD and MDD + alcohol-dependent subjects compared to controls (Miguel-Hidalgo et al. 2010). Other postmortem MDD studies have found decreased EAAT1 and EAAT2 mRNA and protein in the hippocampus, DLPFC, orbitofrontal cortex, locus coeruleus, and cortex (Medina et al. 2016; Zhao et al. 2016; Choudary et al. 2005; Miguel-Hidalgo et al. 2010; Chandley et al. 2014). No changes in the astrocyte-specific enzyme glutamine synthetase, which converts glutamate to glutamine, were found in any groups (Miguel-Hidalgo et al. 2010).

Connexin 43 and connexin 30 are gap junction membrane proteins that are essential for astrocyte communication. Protein expression of the predominant gap junction protein, connexin 43, was decreased in the orbitofrontal cortex in MDD subjects, in MDD + alcohol-dependent subjects, and in subjects diagnosed with alcohol dependence only. The average packing density of connexin 43 in the orbitofrontal cortex was decreased in alcohol-dependent subjects, while the area fraction of connexin 43 immunoreactivity and connexin 43 puncta packing density in the gray matter was significantly decreased in MDD and MDD + alcohol-dependent groups. Subject age positively correlated with connexin 43 puncta in MDD and control groups suggesting that connexin 43 expression and localization to gap junctions may be differentially effected by age (Miguel-Hidalgo et al. 2014). Connexin 30 puncta density was unchanged in the ACC of MDD subjects, including in MDD subjects with and without a history of child abuse who died by suicide. However, connexin 30 coupling to oligodendrocyte- and myelin-specific connexins

Table 3 GFAP expression in postmortem brain tissues of subjects diagnosed with major depressive disorder (MDD)

Study	Brain region	MDD	Method
Damadzic et al. (2001)	Entorhinal cortex	NS ($n = 13$)	IHC
Webster et al. (2001)	DLPFC and hippocampus	Astrocytes containing pGFAP in DLPFC ↓; NS in hippocampus ($n = 15$ per group)	IHC
Altshuler et al. (2010)	Amygdala	GFAP immunoreactive astrocyte density↓ ($n = 11$)	IHC
Toro et al. (2006b)	DLPFC (BA9, 32, 46) and OFC (BA11,12,47,45)	NS ($n = 15$)	IHC
Fatemi et al. (2004)	Lateral cerebellum	GFAP protein↓ ($n = 15$ per group)	WB
Muller et al. (2001)	Hippocampus	GFAP immunoreactivity↓ ($n = 15$)	ICC
Johnston-Wilson et al. (2000b)	BA 10	GFAP 76,119,311 protein↓ ($n = 19$)	2-DE
Qi et al. (2019)	DLPFC, ACC	NS ($n = 5$)	IHC, qPCR
Si et al. (2004)	Prefrontal cortex (gray matter, layers I-VI)	GFAP protein ↓ ($n = 15$)	WB

BA Brodmann area, *DLPFC* dorsolateral prefrontal cortex, *OFC* orbitofrontal cortex, *ACC* anterior cingulate cortex, *WB* western blotting, *ICC* immunocytochemistry, *IHC* immunohistochemistry, *NS* no significant change, *2-DE* two-dimensional gel electrophoresis.

(astrocyte-oligodendrocyte gap junctions) may be decreased in the ACC (Tanti et al. 2019). No correlation was found between connexin 30 coupling and age, postmortem interval, substance abuse, or antidepressant medications, suggesting a disease-related effect.

Postmortem Changes in Astrocytes in Bipolar Disorder (BPD)

BPD is characterized by elevated mood (mania or hypomania) with intervening periods of normal mood (euthymia) and recurrent episodes of depression (Strakowski et al. 2012). There is growing evidence that abnormalities in astrocytes, including altered cell density, morphology, and function, may contribute to the onset of BPD. As with other neuropsychiatric disorders, there is a paucity of effective treatments for BPD. Lithium is a first-line treatment and most effective in relapsed BPD, but it has many side effects and potential toxicity including renal failure (Close et al. 2014) and puerperal relapse (Wesseloo et al. 2016). Quetiapine, olanzapine, risperidone, ziprasidone, valproic acid, carbamazepine, and lamotrigine are considered second-line treatments (Miura et al. 2014); however, they show varying efficacy and tolerability (Vieta et al. 2010; Sidor and Macqueen 2011). In comparison to MDD, antidepressants have little or no efficacy for depressive episodes associated with BPD and are used only as an adjunct to mood stabilizers for bipolar depression (Hirschfeld 2014). Lithium and mood stabilizers also have limited efficacy in treating bipolar depression, the leading cause of morbidity in patients with BPD (Baldessarini et al. 2010). Little is known about the role of astrocytes in bipolar depression. Further investigations of astrocytes may offer a novel therapeutic substrate for the treatment of BPD.

Histology studies in postmortem human prefrontal cortex, DLPFC, orbitofrontal cortex, frontal cortex, amygdala, ACC, entorhinal cortex, and hippocampus tissues have identified altered astrocyte cell numbers, morphology, and density in brain regions implicated in BPD (Strakowski et al. 2012). A study using computer-assisted three-dimensional morphometric cell counting found reduced density of glial cells in the DLPFC of subjects diagnosed with BPD, coupled with enlargement and changes in shape of glial nuclei across multiple lamina, although the type of glial cells was not specified (Rajkowska et al. 2001).

Other studies have measured the expression of astrocyte marker GFAP to elucidate changes in astrocyte number and morphology in the brains of individuals with BPD. Increased GFAP protein levels were found in the DLPFC of BPD subjects compared to controls (Feresten et al. 2013b). In contrast, no change in the density of GFAP immunoreactive astrocytes was reported in the amygdala of a similar BPD cohort (Altshuler et al. 2010). Others have reported no change (Pantazopoulos et al. 2010; Damadzic et al. 2001; Dean et al. 2006; Webster et al. 2001), reduced expression (Toro et al. 2006b; Fatemi et al. 2004; Hercher et al. 2014; Johnston-Wilson et al. 2000b; Muller et al. 2001), or increased expression (Feresten et al. 2013b; Qi et al. 2019; Rao et al. 2010) of GFAP protein levels in the amygdala, entorhinal cortex, ACC, DLPFC, hippocampus, cerebellum, frontal cortex, and orbitofrontal cortex in BPD subjects using immunochemistry or immunoblotting approaches. Even within the same brain region, for example, the DLPFC, GFAP protein levels were increased or unchanged in different studies (Dean et al. 2006; Feresten et al. 2013b). In addition to GFAP, phosphorylated GFAP is also used as an astrocyte marker. In a large-scale proteomic study of frontal cortex tissue, four GFAP isoforms, related to four potential sites for phosphorylation of GFAP, were differentially expressed in subjects diagnosed with a psychiatric disorder compared to unaffected controls. The GFAP500 phosphorylated form of GFAP was decreased in BPD subjects (Johnston-Wilson et al. 2000b). Overall, investigations of astrocyte density changes using GFAP as an astrocyte marker in BPD have yielded inconsistent results. In addition to discrepancies in reports of GFAP expression, constitutively expressed GFAP levels vary dramatically between cells with different activation states, that is, increased GFAP levels may imply that astrocytes are activated in a specific brain region rather than an increase in cell number (Feresten et al. 2013b; Qi et al. 2019). A summary of GFAP expression changes in BPD is shown in Table 4.

EAAT2 mRNA was significantly upregulated in the DLPFC of subjects with BPD compared to controls (Shao and Vawter 2008). In contrast, others reported no significant change in EAAT2 or EAAT1 mRNA in the DLPFC (Zhao et al. 2016; Choudary et al. 2005; Medina et al. 2013), ACC (Choudary et al. 2005), striatum

Table 4 GFAP expression in postmortem brain tissues of subjects diagnosed with bipolar disorder (BPD)

Study	Brain region	BPD	Method
Pantazopoulos et al. (2010)	Amygdala, entorhinal cortex	NS (*n* = 11)	IHC, ICC
Damadzic et al. (2001)	Entorhinal cortex	NS (*n* = 14)	IHC
Dean et al. (2006)	BA9, 10, 46, 40	NS (*n* = 8)	WB, qPCR
Webster et al. (2001)	DLPFC and hippocampus	NS (*n* = 15)	IHC
Altshuler et al. (2010)	Amygdala	NS (*n* = 10)	IHC
Rao et al. (2010)	Frontal cortex	GFAP protein and mRNA↑ (*n* = 10)	WB, qPCR
Feresten et al. (2013b)	DLPFC (BA9)	GFAP protein↑ (*n* = 34)	WB
Toro et al. (2006b)	DLPFC (BA9, 32, 46) and OFC (BA11,12,47,45)	GFAP protein OFC ↓ (*n* = 15)	IHC
Fatemi et al. (2004)	Lateral cerebellum	GFAP protein↓ (*n* = 15 per group)	WB
Muller et al. (2001)	Hippocampus	GFAP immunoreactivity↓ (*n* = 15)	ICC
Hercher et al. (2014)	White matter adjacent to DLPFC (BA9)	GFAP area fraction↓ (*n* = 20)	IHC
Johnston-Wilson et al. (2000b)	BA 10	GFAP500 protein↓ (*n* = 23)	2-DE
Qi et al. (2019)	DLPFC, ACC	GFAP mRNA↑; GFAP immunoreactive astrocytes↓ (DLPFC, *n* = 9; ACC, *n* = 7)	IHC, qPCR

BA Brodmann area, *DLPFC* dorsolateral prefrontal cortex, *OFC* orbitofrontal cortex, *ACC* anterior cingulate cortex, *WB* western blotting, *ICC* immunocytochemistry, *IHC* immunohistochemistry, *NS* no significant change, *2-DE* two-dimensional gel electrophoresis.

(McCullumsmith and Meador-Woodruff 2002), hippocampus (Medina et al. 2013), or locus coeruleus (Bernard et al. 2011). In general, EAAT1 and EAAT2 mRNA expression was not altered in BPD. However, little is known about the protein expression of these astrocyte markers. Postmortem studies of BPD typically have small numbers of subjects or are conducted by pooling BPD and MDD subjects, limiting interpretation of some findings (Zhao et al. 2016).

Subjects with BPD have altered levels of other proteins related to astrocyte function. Neuropathological analysis found a bilateral decrease in the density of S100B immunopositive astrocytes in hippocampal CA1 pyramidal layers of subjects with BPD (Gos et al. 2013). Protein and mRNA levels of S100B were decreased in BA 9 of the DLPFC but increased in BA 40 of the parietal cortex in subjects with BPD (Dean et al. 2006). Interestingly, a meta-analysis (*n* = 104 patients) showed higher levels of serum S100B in patients with BPD (da Rosa et al. 2016).

Glial cell-derived neurotrophic factor (GDNF) is a neurotrophic factor that is typically produced by reactive astrocytes upon brain injury and provides important neuroprotection to dopaminergic neurons (Cunningham and Su 2002). Some studies have reported lower serum GDNF in patients with manic and/or depressive stages of BPD (Takebayashi et al. 2006; Zhang et al. 2010). An increase in serum GDNF was observed after 8 weeks of treatment with mood-stabilizing drugs (Zhang et al. 2010). Plasma GDNF levels were also higher in the euthymic stage of disease compared to the manic stage in BPD patients or in healthy controls, suggesting GDNF is associated with different mood states in BPD (Barbosa et al. 2011). However, no significant change in serum GDNF levels was also reported by several studies (Otsuki et al. 2008; Tunca et al. 2014; Rybakowski et al. 2013). Again, findings related to peripheral GDNF are inconsistent. Many of these studies did not account for confounding factors like patient renal function, which significantly impacts serum protein levels (Onodera et al. 1999). In addition, the source of circulating neurotrophic factors in serum is not limited to astrocytes. It is also unknown whether peripheral GDNF levels correlate with GDNF levels in the brain. Studies have shown that GDNF penetrates the blood-brain barrier very poorly (Kastin et al. 2003).

Overall, the findings of astrocyte-related changes in BPD are heterogeneous. This may reflect differences in the diverse brain regions studied, the effects of chronic mood-stabilizing medication treatment, as well as the clinical heterogeneity of this disorder.

Factors That Affect Astrocytes in Postmortem Studies of Neuropsychiatric Disorders

GFAP is the most commonly used marker of astrocytes in human CNS (Nichols et al. 1993). Many of the postmortem studies that assess astrocyte changes in neuropsychiatric disorders utilize GFAP as a marker; however, it is worth noting that not all astrocytes express GFAP (Liddelow and Barres 2017), and overreliance on a single or small number of markers may limit detection of changes in the heterogeneous astrocyte subtypes in the human brain in disease. In Fig. 1, we show the relative gene expression ($\log_{10}$ FPKM) of GFAP and other astrocyte markers in different cell types collected from human brain and reported in the Brain-RNAseq database (Zhang et al. 2016). Data from Brain-RNAseq suggests that some commonly used markers are not exclusively, or even predominantly, expressed in astrocytes at the mRNA level. This is in agreement with work suggesting that classical astrocyte markers like excitatory amino acid transporter (EAAT2) are also expressed in neurons and that expression is increased in SCZ (O'Donovan et al. 2015; McCullumsmith et al. 2015) and decreased in MDD (Medina et al. 2016; Zhao et al. 2016; Choudary et al. 2005; McCullumsmith and Meador-Woodruff 2002), supporting the importance of using validated markers to assay astrocytes in the brain.

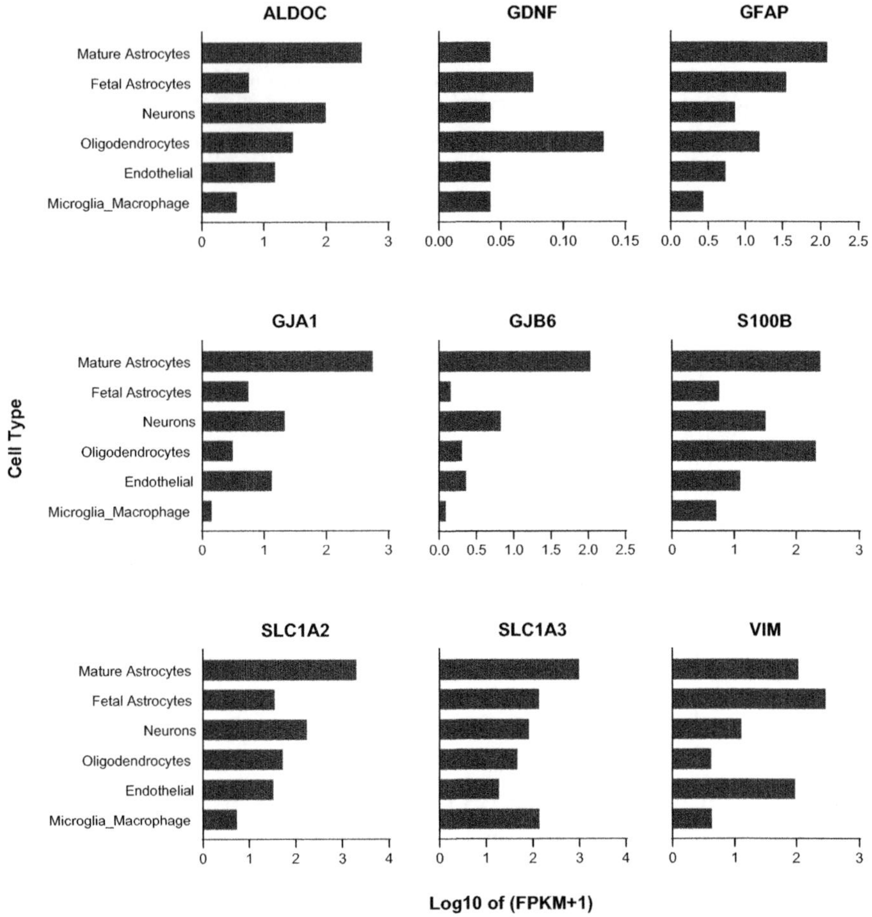

Fig. 1 Relative abundance and use of different astrocyte markers in human brain (Brain RNAseq) retrieved from Kaleidoscope. mRNA expression (Log10 of (FPKM+1)) for common astrocyte markers in six different cell types, fetal astrocytes, mature astrocytes, neurons, endothelial cells, microglia, and oligodendrocytes, was presented. These plots show that several commonly used astrocyte markers are also detected in other cell types at the mRNA level. ALDOC, aldolase C; GDNF, glial-derived neurotrophic factor; GFAP, glial fibrillary acidic protein; GJA1, connexin 43; GJB6, connexin 30; S100B, S100 calcium-binding protein B SLC1A3, excitatory amino acid transporter 1 (EAAT1); SLC1A2, excitatory amino acid transporter 2 (EAAT2); VIM, vimentin

Postmortem studies offer invaluable insight into astrocyte expression in complex neuropsychiatric disorders like SCZ, MDD, and BPD. However, several factors must be considered when interpreting data obtained from postmortem studies, and they may contribute to inconsistent reports of astrocyte expression changes in the brain in severe mental illness. The length of time until tissue is stored (postmortem interval), the different methodologies applied to assay protein and gene expression, the heterogeneity of astrocyte cell types in different brain regions and lamina assayed, subject age and duration of illness, and differences in disease stage may contribute to the lack of agreement in astrocyte expression changes reported in different neuropsychiatric disorders. In addition, postmortem studies often have limited sample sizes and cannot always account for additional variables including the effects of suicide or history of substance abuse on astrocyte expression in the brains of patients diagnosed with these disorders. Another important factor to consider is the use of medication; antipsychotics, antidepressants, and mood-stabilizing medications can significantly affect astrocyte expression (Peng et al. 2016; Trépanier et al. 2016; Fatemi et al. 2004; Tarasov et al. 2020).

Effects of Psychotropic Medications on Astrocytes in Postmortem Tissue

Few studies have directly examined the effects of antipsychotics and antidepressants on astrocyte expression in human brain. Chronic antipsychotic use is associated with reduced gray and white matter volumes in SCZ patients (Ho et al. 2011), including reductions in gray matter volume in patients administered the antipsychotic haloperidol following first-episode psychosis (Lieberman et al. 2005). Gray and white matter volume changes in SCZ patients may reflect reduced glial cell density or counts induced by antipsychotic medication. There are several reports suggesting no association between antidepressant use and changes in glial density, cell count, or astrocyte marker expression in MDD studies (Ongur et al. 1998; Rajkowska et al. 1999; Cotter et al. 2001a; Torres-Platas et al. 2016; Miguel-Hidalgo et al. 2010), although antidepressant use was correlated with reduced GFAP protein expression in the cerebellum of subjects with MDD (Fatemi et al. 2004). Importantly, many studies were not sufficiently powered to detect a medication effect, and further research will be required to determine the effects of antidepressants on astrocyte expression in MDD. Little is known about the effects of lithium, a first-line treatment for BPD, on astrocyte expression in human brain; however, transcriptomic analysis of mouse astrocytes found lithium-responsive genes in primary cultured mouse astrocytes, suggesting that astrocytes are a direct cellular and molecular target of lithium (Rivera and Butt 2019). Lithium administration also resulted in an increase in astrocyte density in the hippocampus of adult mice (Rajkowska et al. 2016) and induced a highly polarized morphology in mouse ex vivo cultured astrocytes which is typical of reactive astrogliosis (Rivera and Butt 2019). Taken together, these studies raise the possibility that astrocyte expression changes seen in the brain in SCZ, MDD, and BPD may reflect the effects of chronic medication use.

In summary, a growing number of postmortem studies highlight a change in astrocyte morphology and the expression of functionally important astrocyte markers and how they are altered in the brain in neuropsychiatric disorders. Further studies are necessary to provide a greater understanding of the role of astrocytes in the pathophysiology of these disorders and may lead to the development of novel therapeutic targets for these often devastating mental illnesses.

References

Altshuler LL, Abulseoud OA, Foland-Ross L, Bartzokis G, Chang S, Mintz J, Hellemann G, Vinters HV (2010) Amygdala astrocyte reduction in subjects with major depressive disorder but not bipolar disorder. Bipolar Disord 12:541–549

Andriezen WL (1893) The neuroglia elements in the human brain. Br Med J 2:227–230

Araque A, Parpura V, Sanzgiri RP, Haydon PG (1999) Tripartite synapses: glia, the unacknowledged partner. Trends Neurosci 22:208–215

Arnold SE, Franz BR, Trojanowski JQ, Moberg PJ, Gur RE (1996a) Glial fibrillary acidic protein-immunoreactive astrocytosis in elderly patients with schizophrenia and dementia. Acta Neuropathol 91:269–277

Arnold SE, Franz BR, Trojanowski JQ, Moberg PJ, Gur RE (1996b) Glial fibrillary acidic protein-immunoreactive astrocytosis in elderly patients with schizophrenia and dementia. Acta Neuropathol 91:269–277

Baldessarini RJ, Vieta E, Calabrese JR, Tohen M, Bowden CL (2010) Bipolar depression: overview and commentary. Harv Rev Psychiatry 18:143–157

Barbosa IG, Huguet RB, Sousa LP, Abreu MN, Rocha NP, Bauer ME, Carvalho LA, Teixeira AL (2011) Circulating levels of GDNF in bipolar disorder. Neurosci Lett 502:103–106

Barley K, Dracheva S, Byne W (2009) Subcortical oligodendrocyte-and astrocyte-associated gene expression in subjects with schizophrenia, major depression and bipolar disorder. Schizophr Res 112:54–64

Bauer D, Gupta D, Harotunian V, Meador-Woodruff JH, McCullumsmith RE (2008) Abnormal expression of glutamate transporter and transporter interacting molecules in prefrontal cortex in elderly patients with schizophrenia. Schizophr Res 104:108–120

Bauer D, Haroutunian V, Meador-Woodruff JH, McCullumsmith RE (2010) Abnormal glycosylation of EAAT1 and EAAT2 in prefrontal cortex of elderly patients with schizophrenia. Schizophr Res 117:92–98

Benes FM, Davidson J, Bird ED (1986) Quantitative cytoarchitectural studies of the cerebral cortex of schizophrenics. Arch Gen Psychiatry 43:31–35

Bernard R, Kerman IA, Thompson RC, Jones EG, Bunney WE, Barchas JD, Schatzberg AF, Myers RM, Akil H, Watson SJ (2011) Altered expression of glutamate signaling, growth factor, and glia genes in the locus coeruleus of patients with major depression. Mol Psychiatry 16:634–646

Bonzano S, Crisci I, Podlesny-Drabiniok A, Rolando C, Krezel W, Studer M, De Marchis S (2018) Neuron-Astroglia cell fate decision in the adult mouse hippocampal neurogenic niche is cell-intrinsically controlled by COUP-TFI in vivo. Cell Rep 24:329–341

Bowley MP, Drevets WC, Ongur D, Price JL (2002) Low glial numbers in the amygdala in major depressive disorder. Biol Psychiatry 52:404–412

Bruton CJ, Crow T, Frith C, Johnstone E, Owens D, Roberts G (1990) Schizophrenia and the brain: a prospective clinico-neuropathological study. Psychol Med 20:285–304

Catts VS, Wong J, Fillman SG, Fung SJ, Shannon Weickert C (2014a) Increased expression of astrocyte markers in schizophrenia: association with neuroinflammation. Aust N Z J Psychiatry 48:722–734

Catts VS, Wong J, Fillman SG, Fung SJ, Shannon Weickert C (2014b) Increased expression of astrocyte markers in schizophrenia: association with neuroinflammation. Aust N Z J Psychiatry 48:722–734

Chandley MJ, Szebeni A, Szebeni K, Crawford JD, Stockmeier CA, Turecki G, Kostrzewa RM, Ordway GA (2014) Elevated gene expression of glutamate receptors in noradrenergic neurons from the locus coeruleus in major depression. Int J Neuropsychopharmacol 17:1569–1578

Choudary PV, Molnar M, Evans SJ, Tomita H, Li JZ, Vawter MP, Myers RM, Bunney WE Jr, Akil H, Watson SJ, Jones EG (2005) Altered cortical glutamatergic and GABAergic signal transmission with glial involvement in depression. Proc Natl Acad Sci U S A 102:15653–15658

Close H, Reilly J, Mason JM, Kripalani M, Wilson D, Main J, Hungin AP (2014) Renal failure in lithium-treated bipolar disorder: a retrospective cohort study. PLoS One 9:e90169

Cotter D, Mackay D, Landau S, Kerwin R, Everall I (2001a) Reduced glial cell density and neuronal size in the anterior cingulate cortex in major depressive disorder. Arch Gen Psychiatry 58:545–553

Cotter DR, Pariante CM, Everall IP (2001b) Glial cell abnormalities in major psychiatric disorders: the evidence and implications. Brain Res Bull 55:585–595

Cotter D, Mackay D, Chana G, Beasley C, Landau S, Everall IP (2002) Reduced neuronal size and glial cell density in area 9 of the dorsolateral prefrontal cortex in subjects with major depressive disorder. Cereb Cortex 12:386–394

Cunningham LA, Su C (2002) Astrocyte delivery of glial cell line-derived neurotrophic factor in a mouse model of Parkinson's disease. Exp Neurol 174:230–242

da Rosa MI, Simon C, Grande AJ, Barichello T, Oses JP, Quevedo J (2016) Serum S100B in manic bipolar disorder patients: systematic review and meta-analysis. J Affect Disord 206:210–215

Damadzic R, Bigelow LB, Krimer LS, Goldenson DA, Saunders RC, Kleinman JE, Herman MM (2001) A quantitative immunohistochemical study of astrocytes in the entorhinal cortex in schizophrenia, bipolar disorder and major depression: absence of significant astrocytosis. Brain Res Bull 55:611–618

Davalieva K, Maleva Kostovska I, Dwork AJ (2016) Proteomics research in schizophrenia. Front Cell Neurosci 10:18

Dean B, Gray L, Scarr E (2006) Regionally specific changes in levels of cortical S100beta in bipolar 1 disorder but not schizophrenia. Aust N Z J Psychiatry 40:217–224

Dietz AG, Goldman SA, Nedergaard M (2020) Glial cells in schizophrenia: a unified hypothesis. Lancet Psychiatry 7:272–281

Edwards MA, Yamamoto M, Caviness VS Jr (1990) Organization of radial glia and related cells in the developing murine CNS. An analysis based upon a new monoclonal antibody marker. Neuroscience 36:121–144

English JA, Dicker P, Föcking M, Dunn MJ, Cotter DR (2009) 2-D DIGE analysis implicates cytoskeletal abnormalities in psychiatric disease. Proteomics 9:3368–3382

Falk S, Gotz M (2017) Glial control of neurogenesis. Curr Opin Neurobiol 47:188–195

Fatemi SH, Laurence JA, Araghi-Niknam M, Stary JM, Schulz SC, Lee S, Gottesman II (2004) Glial fibrillary acidic protein is reduced in cerebellum of subjects with major depression, but not schizophrenia. Schizophr Res 69:317–323

Feig SL, Haberly LB (2011) Surface-associated astrocytes, not endfeet, form the glia limitans in posterior piriform cortex and have a spatially distributed, not a domain, organization. J Comp Neurol 519:1952–1969

Feresten AH, Barakauskas V, Ypsilanti A, Barr AM, Beasley CL (2013a) Increased expression of glial fibrillary acidic protein in prefrontal cortex in psychotic illness. Schizophr Res 150:252–257

Feresten AH, Barakauskas V, Ypsilanti A, Barr AM, Beasley CL (2013b) Increased expression of glial fibrillary acidic protein in prefrontal cortex in psychotic illness. Schizophr Res 150:252–257

Fillman SG, Cloonan N, Catts VS, Miller LC, Wong J, McCrossin T, Cairns M, Weickert CS (2013) Increased inflammatory markers identified in the dorsolateral prefrontal cortex of individuals with schizophrenia. Mol Psychiatry 18:206–214

Fillman SG, Sinclair D, Fung SJ, Webster MJ, Shannon Weickert C (2014) Markers of inflammation and stress distinguish subsets of individuals with schizophrenia and bipolar disorder. Transl Psychiatry 4:e365

Garcia-Segura LM, McCarthy MM (2004) Minireview: role of glia in neuroendocrine function. Endocrinology 145:1082–1086

Gos T, Schroeter ML, Lessel W, Bernstein HG, Dobrowolny H, Schiltz K, Bogerts B, Steiner J (2013) S100B-immunopositive astrocytes and oligodendrocytes in the hippocampus are differentially afflicted in unipolar and bipolar depression: a postmortem study. J Psychiatr Res 47:1694–1699

Gosselin R-D, Gibney S, O'Malley D, Dinan T, Cryan J (2009) Region specific decrease in glial fibrillary acidic protein immunoreactivity in the brain of a rat model of depression. Neuroscience 159:915–925
Hatten ME, Mason CA (1990) Mechanisms of glial-guided neuronal migration in vitro and in vivo. Experientia 46:907–916
Hercher C, Turecki G, Mechawar N (2009) Through the looking glass: examining neuroanatomical evidence for cellular alterations in major depression. J Psychiatr Res 43:947–961
Hercher C, Chopra V, Beasley CL (2014) Evidence for morphological alterations in prefrontal white matter glia in schizophrenia and bipolar disorder. J Psychiatry Neurosci 39:376–385
Hirschfeld RM (2014) Differential diagnosis of bipolar disorder and major depressive disorder. J Affect Disord 169(Suppl 1):S12–S16
Ho B-C, Andreasen NC, Ziebell S, Pierson R, Magnotta V (2011) Long-term antipsychotic treatment and brain volumes: a longitudinal study of first-episode schizophrenia. Arch Gen Psychiatry 68:128–137
Hu W, MacDonald ML, Elswick DE, Sweet RA (2015) The glutamate hypothesis of schizophrenia: evidence from human brain tissue studies. Ann N Y Acad Sci 1338:38–57
Hwang Y, Kim J, Shin J, Kim J-I, Seo J, Webster M, Lee D, Kim S (2013) Gene expression profiling by mRNA sequencing reveals increased expression of immune/inflammation-related genes in the hippocampus of individuals with schizophrenia. Transl Psychiatry 3:e321–e321
Johnston-Wilson N, Sims C, Hofmann J, Anderson L, Shore A, Torrey E, Yolken RH (2000a) Disease-specific alterations in frontal cortex brain proteins in schizophrenia, bipolar disorder, and major depressive disorder. Mol Psychiatry 5:142–149
Johnston-Wilson NL, Sims CD, Hofmann JP, Anderson L, Shore AD, Torrey EF, Yolken RH (2000b) Disease-specific alterations in frontal cortex brain proteins in schizophrenia, bipolar disorder, and major depressive disorder. The Stanley Neuropathology Consortium. Mol Psychiatry 5:142–149
Kastin AJ, Akerstrom V, Pan W (2003) Glial cell line-derived neurotrophic factor does not enter normal mouse brain. Neurosci Lett 340:239–241
Katsel P, Byne W, Roussos P, Tan W, Siever L, Haroutunian V (2011a) Astrocyte and glutamate markers in the superficial, deep, and white matter layers of the anterior cingulate gyrus in schizophrenia. Neuropsychopharmacology 36:1171–1177
Katsel P, Byne W, Roussos P, Tan W, Siever L, Haroutunian V (2011b) Astrocyte and glutamate markers in the superficial, deep, and white matter layers of the anterior cingulate gyrus in schizophrenia. Neuropsychopharmacology 36:1171–1177
Kettenmann H, Ransom BR (2004) The concept of neuroglia: a historical perspective. In: Helmut Kettenmann BRR (ed) Neuroglia, 2nd edn. Oxford University Press
Liddelow SA, Barres BA (2017) Reactive astrocytes: production, function, and therapeutic potential. Immunity 46:957–967
Lieberman JA, Tollefson GD, Charles C, Zipursky R, Sharma T, Kahn RS, Keefe RS, Green AI, Gur RE, McEvoy J (2005) Antipsychotic drug effects on brain morphology in first-episode psychosis. Arch Gen Psychiatry 62:361–370
Maes M, Yirmyia R, Noraberg J, Brene S, Hibbeln J, Perini G, Kubera M, Bob P, Lerer B, Maj M (2009) The inflammatory & neurodegenerative (I&ND) hypothesis of depression: leads for future research and new drug developments in depression. Metab Brain Dis 24:27–53
Marin-Padilla M (1995) Prenatal development of fibrous (white matter), protoplasmic (gray matter), and layer I astrocytes in the human cerebral cortex: a Golgi study. J Comp Neurol 357:554–572
Markova E, Markov I, Revishchin A, Okhotin V, Sulimov G (2000) 3-D Golgi and image analysis of the olfactory tubercle in schizophrenia. Anal Quant Cytol Histol 22:178
Martins-de-Souza D (2010) Proteome and transcriptome analysis suggests oligodendrocyte dysfunction in schizophrenia. J Psychiatr Res 44:149–156
Matute C, Melone M, Vallejo-Illarramendi A, Conti F (2005) Increased expression of the astrocytic glutamate transporter GLT-1 in the prefrontal cortex of schizophrenics. Glia 49:451–455

McCullumsmith RE, Meador-Woodruff JH (2002) Striatal excitatory amino acid transporter transcript expression in schizophrenia, bipolar disorder, and major depressive disorder. Neuropsychopharmacology 26:368–375

McCullumsmith RE, O'Donovan SM, Drummond JB, Benesh FS, Simmons M, Roberts R, Lauriat T, Haroutunian V, Meador-Woodruff JH (2015) Cell-specific abnormalities of glutamate transporters in schizophrenia: sick astrocytes and compensating relay neurons? Mol Psychiatry 21(6):823–830

McCullumsmith RE, O'Donovan SM, Drummond JB, Benesh FS, Simmons M, Roberts R, Lauriat T, Haroutunian V, Meador-Woodruff JH (2016) Cell-specific abnormalities of glutamate transporters in schizophrenia: sick astrocytes and compensating relay neurons? Mol Psychiatry 21:823–830

Medina A, Burke S, Thompson RC, Bunney W Jr, Myers RM, Schatzberg A, Akil H, Watson SJ (2013) Glutamate transporters: a key piece in the glutamate puzzle of major depressive disorder. J Psychiatr Res 47:1150–1156

Medina A, Watson SJ, Bunney W Jr, Myers RM, Schatzberg A, Barchas J, Akil H, Thompson RC (2016) Evidence for alterations of the glial syncytial function in major depressive disorder. J Psychiatr Res 72:15–21

Miguel-Hidalgo JJ, Waltzer R, Whittom AA, Austin MC, Rajkowska G, Stockmeier CA (2010) Glial and glutamatergic markers in depression, alcoholism, and their comorbidity. J Affect Disord 127:230–240

Miguel-Hidalgo JJ, Wilson BA, Hussain S, Meshram A, Rajkowska G, Stockmeier CA (2014) Reduced connexin 43 immunolabeling in the orbitofrontal cortex in alcohol dependence and depression. J Psychiatr Res 55:101–109

Miura T, Noma H, Furukawa TA, Mitsuyasu H, Tanaka S, Stockton S, Salanti G, Motomura K, Shimano-Katsuki S, Leucht S, Cipriani A, Geddes JR, Kanba S (2014) Comparative efficacy and tolerability of pharmacological treatments in the maintenance treatment of bipolar disorder: a systematic review and network meta-analysis. Lancet Psychiatry 1:351–359

Muller MB, Lucassen PJ, Yassouridis A, Hoogendijk WJ, Holsboer F, Swaab DF (2001) Neither major depression nor glucocorticoid treatment affects the cellular integrity of the human hippocampus. Eur J Neurosci 14:1603–1612

Nichols NR, Day JR, Laping NJ, Johnson SA, Finch CE (1993) GFAP mRNA increases with age in rat and human brain. Neurobiol Aging 14:421–429

O'Donovan SM, Hasselfeld K, Bauer D, Simmons M, Roussos P, Haroutunian V, Meador-Woodruff JH, McCullumsmith RE (2015) Glutamate transporter splice variant expression in an enriched pyramidal cell population in schizophrenia. Transl Psychiatry 5:e579

O'Donovan SM, Sullivan CR, McCullumsmith RE (2017) The role of glutamate transporters in the pathophysiology of neuropsychiatric disorders. NPJ Schizophr 3:32

Oberheim NA, Takano T, Han X, He W, Lin JH, Wang F, Xu Q, Wyatt JD, Pilcher W, Ojemann JG, Ransom BR, Goldman SA, Nedergaard M (2009) Uniquely hominid features of adult human astrocytes. J Neurosci 29:3276–3287

Oberheim NA, Goldman SA, Nedergaard M (2012) Heterogeneity of astrocytic form and function. Methods Mol Biol 814:23–45

Ohnuma T, Augood SJ, Arai H, McKenna PJ, Emson PC (1998) Expression of the human excitatory amino acid transporter 2 and metabotropic glutamate receptors 3 and 5 in the prefrontal cortex from normal individuals and patients with schizophrenia. Brain Res Mol Brain Res 56:207–217

Ongur D, Drevets WC, Price JL (1998) Glial reduction in the subgenual prefrontal cortex in mood disorders. Proc Natl Acad Sci U S A 95:13290–13295

Onodera H, Nagata T, Kanazawa M, Taguma Y, Itoyama Y (1999) Increased plasma GDNF levels in patients with chronic renal diseases. Nephrol Dial Transplant 14:1604–1605

Otsuki K, Uchida S, Watanuki T, Wakabayashi Y, Fujimoto M, Matsubara T, Funato H, Watanabe Y (2008) Altered expression of neurotrophic factors in patients with major depression. J Psychiatr Res 42:1145–1153

Pakkenberg B (1990) Pronounced reduction of total neuron number in mediodorsal thalamic nucleus and nucleus accumbens in schizophrenics. Arch Gen Psychiatry 47:1023–1028

Pantazopoulos H, Woo TU, Lim MP, Lange N, Berretta S (2010) Extracellular matrix-glial abnormalities in the amygdala and entorhinal cortex of subjects diagnosed with schizophrenia. Arch Gen Psychiatry 67:155–166

Parpura V, Verkhratsky A (2012) Astrocytes revisited: concise historic outlook on glutamate homeostasis and signaling. Croat Med J 53:518–528

Peng L, Li B, Verkhratsky A (2016) Targeting astrocytes in bipolar disorder. Expert Rev Neurother 16:649–657

Prabakaran S, Swatton J, Ryan M, Huffaker S, Huang J-J, Griffin J, Wayland M, Freeman T, Dudbridge F, Lilley K (2004) Mitochondrial dysfunction in schizophrenia: evidence for compromised brain metabolism and oxidative stress. Mol Psychiatry 9:684–697

Prevot V, Dehouck B, Sharif A, Ciofi P, Giacobini P, Clasadonte J (2018) The versatile Tanycyte: a hypothalamic integrator of reproduction and energy metabolism. Endocr Rev 39:333–368

Qi XR, Kamphuis W, Shan L (2019) Astrocyte changes in the prefrontal cortex from aged non-suicidal depressed patients. Front Cell Neurosci 13:503

Rajkowska G, Miguel-Hidalgo JJ (2007) Gliogenesis and glial pathology in depression. CNS Neurol Disord Drug Targets 6:219–233

Rajkowska G, Stockmeier CA (2013) Astrocyte pathology in major depressive disorder: insights from human postmortem brain tissue. Curr Drug Targets 14:1225–1236

Rajkowska G, Miguel-Hidalgo JJ, Wei J, Dilley G, Pittman SD, Meltzer HY, Overholser JC, Roth BL, Stockmeier CA (1999) Morphometric evidence for neuronal and glial prefrontal cell pathology in major depression. Biol Psychiatry 45:1085–1098

Rajkowska G, Halaris A, Selemon LD (2001) Reductions in neuronal and glial density characterize the dorsolateral prefrontal cortex in bipolar disorder. Biol Psychiatry 49:741–752

Rajkowska G, Miguel-Hidalgo JJ, Makkos Z, Meltzer H, Overholser J, Stockmeier C (2002a) Layer-specific reductions in GFAP-reactive astroglia in the dorsolateral prefrontal cortex in schizophrenia. Schizophr Res 57:127–138

Rajkowska G, Miguel-Hidalgo JJ, Makkos Z, Meltzer H, Overholser J, Stockmeier C (2002b) Layer-specific reductions in GFAP-reactive astroglia in the dorsolateral prefrontal cortex in schizophrenia. Schizophr Res 57:127–138

Rajkowska G, Clarke G, Mahajan G, Licht CM, van de Werd HJ, Yuan P, Stockmeier CA, Manji HK, Uylings HB (2016) Differential effect of lithium on cell number in the hippocampus and prefrontal cortex in adult mice: a stereological study. Bipolar Disord 18:41–51

Ramon y Cajal S (1913) Contribucion a1 conocimiento de la neuroglia del cerebro humano. Trab Lab Inuest Biol Uniu Madrid 11:255–315

Rao JS, Harry GJ, Rapoport SI, Kim HW (2010) Increased excitotoxicity and neuroinflammatory markers in postmortem frontal cortex from bipolar disorder patients. Mol Psychiatry 15:384–392

Rivera AD, Butt AM (2019) Astrocytes are direct cellular targets of lithium treatment: novel roles for lysyl oxidase and peroxisome-proliferator activated receptor-gamma as astroglial targets of lithium. Transl Psychiatry 9:211

Rybakowski JK, Permoda-Osip A, Skibinska M, Adamski R, Bartkowska-Sniatkowska A (2013) Single ketamine infusion in bipolar depression resistant to antidepressants: are neurotrophins involved? Hum Psychopharmacol 28:87–90

Schmitt A, Steyskal C, Bernstein H-G, Schneider-Axmann T, Parlapani E, Schaeffer EL, Gattaz WF, Bogerts B, Schmitz C, Falkai P (2009) Stereologic investigation of the posterior part of the hippocampus in schizophrenia. Acta Neuropathol 117:395–407

Shan D, Lucas EK, Drummond JB, Haroutunian V, Meador-Woodruff JH, McCullumsmith RE (2013) Abnormal expression of glutamate transporters in temporal lobe areas in elderly patients with schizophrenia. Schizophr Res 144:1–8

Shao L, Vawter MP (2008) Shared gene expression alterations in schizophrenia and bipolar disorder. Biol Psychiatry 64:89–97

Si X, Miguel-Hidalgo JJ, O'Dwyer G, Stockmeier CA, Rajkowska G (2004) Age-dependent reductions in the level of glial fibrillary acidic protein in the prefrontal cortex in major depression. Neuropsychopharmacology 29:2088–2096

Sidor MM, Macqueen GM (2011) Antidepressants for the acute treatment of bipolar depression: a systematic review and meta-analysis. J Clin Psychiatry 72:156–167

Sosunov AA, Wu X, Tsankova NM, Guilfoyle E, McKhann GM 2nd, Goldman JE (2014) Phenotypic heterogeneity and plasticity of isocortical and hippocampal astrocytes in the human brain. J Neurosci 34:2285–2298

Steffek AE, McCullumsmith RE, Haroutunian V, Meador-Woodruff JH (2008) Cortical expression of glial fibrillary acidic protein and glutamine synthetase is decreased in schizophrenia. Schizophr Res 103:71–82

Steiner J, Bernstein HG, Bielau H, Berndt A, Brisch R, Mawrin C, Keilhoff G, Bogerts B (2007) Evidence for a wide extra-astrocytic distribution of S100B in human brain. BMC Neurosci 8:2

Steiner J, Bernstein H-G, Bielau H, Farkas N, Winter J, Dobrowolny H, Brisch R, Gos T, Mawrin C, Myint AM (2008) S100B-immunopositive glia is elevated in paranoid as compared to residual schizophrenia: a morphometric study. J Psychiatr Res 42:868–876

Strakowski SM, Adler CM, Almeida J, Altshuler LL, Blumberg HP, Chang KD, DelBello MP, Frangou S, McIntosh A, Phillips ML, Sussman JE, Townsend JD (2012) The functional neuroanatomy of bipolar disorder: a consensus model. Bipolar Disord 14:313–325

Takebayashi M, Hisaoka K, Nishida A, Tsuchioka M, Miyoshi I, Kozuru T, Hikasa S, Okamoto Y, Shinno H, Morinobu S, Yamawaki S (2006) Decreased levels of whole blood glial cell line-derived neurotrophic factor (GDNF) in remitted patients with mood disorders. Int J Neuropsychopharmacol 9:607–612

Tanti A, Lutz PE, Kim J, O'Leary L, Theroux JF, Turecki G, Mechawar N (2019) Evidence of decreased gap junction coupling between astrocytes and oligodendrocytes in the anterior cingulate cortex of depressed suicides. Neuropsychopharmacology 44:2099–2111

Tarasov VV, Svistunov AA, Chubarev VN, Sologova SS, Mukhortova P, Levushkin D, Somasundaram SG, Kirkland CE, Bachurin SO, Aliev G (2019) Alterations of astrocytes in the context of schizophrenic dementia. Front Pharmacol 10:1612

Tarasov VV, Svistunov AA, Chubarev VN, Sologova SS, Mukhortova P, Levushkin D, Somasundaram SG, Kirkland CE, Bachurin SO, Aliev G (2020) Alterations of astrocytes in the context of schizophrenic dementia. Front Pharmacol 10:1612–1612

Toro CT, Hallak JE, Dunham JS, Deakin JF (2006a) Glial fibrillary acidic protein and glutamine synthetase in subregions of prefrontal cortex in schizophrenia and mood disorder. Neurosci Lett 404:276–281

Toro CT, Hallak JE, Dunham JS, Deakin JF (2006b) Glial fibrillary acidic protein and glutamine synthetase in subregions of prefrontal cortex in schizophrenia and mood disorder. Neurosci Lett 404:276–281

Torres-Platas SG, Hercher C, Davoli MA, Maussion G, Labonte B, Turecki G, Mechawar N (2011) Astrocytic hypertrophy in anterior cingulate white matter of depressed suicides. Neuropsychopharmacology 36:2650–2658

Torres-Platas SG, Nagy C, Wakid M, Turecki G, Mechawar N (2016) Glial fibrillary acidic protein is differentially expressed across cortical and subcortical regions in healthy brains and downregulated in the thalamus and caudate nucleus of depressed suicides. Mol Psychiatry 21:509–515

Trépanier MO, Hopperton KE, Mizrahi R, Mechawar N, Bazinet RP (2016) Postmortem evidence of cerebral inflammation in schizophrenia: a systematic review. Mol Psychiatry 21:1009–1026

Tunca Z, Ozerdem A, Ceylan D, Yalcin Y, Can G, Resmi H, Akan P, Ergor G, Aydemir O, Cengisiz C, Kerim D (2014) Alterations in BDNF (brain derived neurotrophic factor) and GDNF (glial cell line-derived neurotrophic factor) serum levels in bipolar disorder: the role of lithium. J Affect Disord 166:193–200

Verkhratsky A, Nedergaard M (2018) Physiology of Astroglia. Physiol Rev 98:239–389

Verkhratsky A, Zorec R, Parpura V (2017) Stratification of astrocytes in healthy and diseased brain. Brain Pathol 27:629–644

Verkhratsky A, Rodrigues JJ, Pivoriunas A, Zorec R, Semyanov A (2019) Astroglial atrophy in Alzheimer's disease. Pflugers Arch 471:1247–1261

Vernadakis A (1996) Glia-neuron intercommunications and synaptic plasticity. Prog Neurobiol 49:185–214

Vieta E, Locklear J, Gunther O, Ekman M, Miltenburger C, Chatterton ML, Astrom M, Paulsson B (2010) Treatment options for bipolar depression: a systematic review of randomized, controlled trials. J Clin Psychopharmacol 30:579–590

von Lenhossék M (1895) Der feinere Bau des Nervensystems im Lichte neuester Forschung, 2nd edn. Fischer's Medicinische Buchhandlung H Kornfield, Berlin

Webster MJ, Knable MB, Johnston-Wilson N, Nagata K, Inagaki M, Yolken RH (2001) Immunohistochemical localization of phosphorylated glial fibrillary acidic protein in the prefrontal cortex and hippocampus from patients with schizophrenia, bipolar disorder, and depression. Brain Behav Immun 15:388–400

Wesseloo R, Kamperman AM, Munk-Olsen T, Pop VJ, Kushner SA, Bergink V (2016) Risk of postpartum relapse in bipolar disorder and postpartum psychosis: a systematic review and meta-analysis. Am J Psychiatry 173:117–127

Williams MR, Hampton T, Pearce RK, Hirsch SR, Ansorge O, Thom M, Maier M (2013) Astrocyte decrease in the subgenual cingulate and callosal genu in schizophrenia. Eur Arch Psychiatry Clin Neurosci 263:41–52

Williams M, Pearce RKB, Hirsch SR, Ansorge O, Thom M, Maier M (2014a) Fibrillary astrocytes are decreased in the subgenual cingulate in schizophrenia. Eur Arch Psychiatry Clin Neurosci 264:357–362

Williams M, Pearce RK, Hirsch SR, Ansorge O, Thom M, Maier M (2014b) Fibrillary astrocytes are decreased in the subgenual cingulate in schizophrenia. Eur Arch Psychiatry Clin Neurosci 264:357–362

Zhang X, Zhang Z, Sha W, Xie C, Xi G, Zhou H, Zhang Y (2010) Effect of treatment on serum glial cell line-derived neurotrophic factor in bipolar patients. J Affect Disord 126:326–329

Zhang Y, Sloan SA, Clarke LE, Caneda C, Plaza CA, Blumenthal PD, Vogel H, Steinberg GK, Edwards MS, Li G, Duncan JA 3rd, Cheshier SH, Shuer LM, Chang EF, Grant GA, Gephart MG, Barres BA (2016) Purification and characterization of progenitor and mature human astrocytes reveals transcriptional and functional differences with mouse. Neuron 89:37–53

Zhao J, Verwer RW, van Wamelen DJ, Qi XR, Gao SF, Lucassen PJ, Swaab DF (2016) Prefrontal changes in the glutamate-glutamine cycle and neuronal/glial glutamate transporters in depression with and without suicide. J Psychiatr Res 82:8–15

Astrocyte Bioenergetics and Major Psychiatric Disorders

Ivan V. Maly, Michael J. Morales, and Mikhail V. Pletnikov

Introduction

Recent research has substantially advanced our understanding of the contribution of energy metabolism in neuropathophysiology of major psychiatric disorders and the role of astrocytes in normal and pathological bioenergetics of the brain. Particularly, an appreciation of the essential function of astrocyte metabolism and metabolic regulation in sustaining cognitive activity has emerged from this work. Complementary lines of study continue to document the neurobehavioral consequences of insufficiency of these pathways, which offer valuable models of pathophysiological conditions shared by different mental diseases.

Recent reviews have highlighted the role of astrocytes in synaptic maturation (Augusto-Oliveira et al. 2020), synaptic plasticity and neurotransmission (Augusto-Oliveira et al. 2020; Kim et al. 2018; Roman et al. 2020), neuroinflammation and neurodegeneration (Sofroniew 2020), and interactions with microglia (Liddelow et al. 2020). The present chapter attempts to build on the conceptual foundation laid by the recently published reviews of astrocyte energy metabolism and contributions of astrocyte bioenergetics to neuropsychiatric conditions. While a comprehensive summary of this literature is not possible here, the reader can nevertheless be alerted to the excellent overviews and perspectives on these topics, e.g., by Zuccoli et al. (2017), Sullivan et al. (2018), Takahashi (2020), and Jha and Morrison (2018), as well as those that explore the links to other aspects of brain physiology, such as tissue pH regulation (Deitmer et al. 2019), aging (Morita et al. 2019), and pathological transformation of other glial cell types (Afridi et al. 2020).

I. V. Maly · M. J. Morales · M. V. Pletnikov (✉)
Department of Physiology and Biophysics, Jacobs School of Medicine and Biomedical Sciences, State University of New York, Buffalo, NY, USA
e-mail: ivanmaly@buffalo.edu; moralesm@buffalo.edu; mvpletni@buffalo.edu

B. Li et al. (eds.), *Astrocytes in Psychiatric Disorders*, Advances in Neurobiology 26, https://doi.org/10.1007/978-3-030-77375-5_9

Research into the metabolic role of astrocytes in the brain was presaged by the trophic hypothesis of astrocyte function that, as outlined by Bouzier-Sore and Pellerin (2013), ascends to the views of Camillo Golgi. As originally described, glia was differentiated from neurons by the absence of axons and exclusive presence of dendrite-like protrusions, to which a nutritive function was ascribed. While Santiago Ramon y Cajal did not subscribe to the trophic view, he envisioned (see Mishra 2017) astrocytes as capable of regulating blood vessel diameters. Metabolic cooperation between neurons and astrocytes, suggested by the early microanatomical concepts, found empirical support in the pioneering experiments conducted by Hamberger and Hyden (1963). When stimulated, isolated neurons exhibited an increase in oxidative capacity at the same time as astrocytes elevated their glycolytic activity, implying a division of metabolic labor between the two major cell types in the brain. The present-day picture of neuron-astrocyte cooperation has acquired a significant level of molecular and cellular details that remain broadly compatible with the early insights and conjectures. Similarly, while major psychiatric disorders are now recognized as complex disorders of the multifactorial etiology (Freund and Juckel 2019; Ohi et al. 2020; Zamanpoor 2020), attention to energy metabolism as a contributing factor dates back to the earliest days of modern biochemical research into mental disease (reviewed by Kety 1959). Considering the present-day controversies in a historical, hundred-year-old context provides a useful perspective on the recent studies.

Our review opens with an outline of the astrocytes' participation in the brain metabolism on the organism, organ, and cellular level. We briefly cover the astrocytes' place in the brain-liver circuit of glucose homeostasis and proceed to a discussion of the current model of vascular regulation mediated by calcium signaling and fatty acid metabolites in astrocytes responding to synaptic activity. These organ-level and interorgan mechanisms provide the physiological context for an overview of the neuron-astrocyte cooperation in the neuropil. An attempt is made to offer a description based on the pathway-centric, molecular-biochemical notion of metabolic fluxes in the cell as well as across cell types. We then overview the fundamental energy metabolism processes in relation to the current genetic, epigenetic, and metabolomic data concerning the molecular basis of the major mental disorders and attempt to propose new directions for research and potential therapeutic intervention.

Metabolite-Mediated Interorgan Signaling

The place of brain metabolism in interorgan regulatory axes is an active area of research that has already yielded insights into the pathophysiological mechanisms contributing to major psychiatric disorders. Ongoing research is clarifying the processes whereby energy metabolism in astrocytes is positioned at the crux of blood glucose regulation during normal cognitive activity and in organism-level dysregulation that can impact the course of mental disease. Although it is likely that astrocytes from other brain structures may contribute to these processes through regional and systemic effects, hypothalamic astrocytes are uniquely positioned to directly influence the brain sensing of blood glucose levels in an interorgan feedback loop (Camandola 2018; Fuente-Martin et al. 2019). Astrocyte-specific uptake of glucose that is mediated by glucose transporter 1 (GLUT1) is critical to glucose-sensing that takes place in the hypothalamic glia (Chari et al. 2011) and complements the glucose-sensing function of hypothalamic neurons (see Fuente-Martin et al. 2019). Elevated glucose causes activation of astrocyte-specific lactate production in the hypothalamus, which fuels adenosine triphosphate (ATP) generation in neurons of the arcuate nucleus (Lam et al. 2005). Activation of hypothalamic ATP-sensitive potassium channels, through the brain-stem projections of the arcuate nucleus and vagal efferents, suppresses gluconeogenesis and glycogenolysis in the liver (Pocai et al. 2005).

A similar process affecting glucagon production appears to be operating through glucose transporter 2 (GLUT2) (Marty et al. 2005), while insulin release by the pancreas in response to elevated glucose exhibits dependence on connexin-mediated cytoplasmic communication between the arcuate nucleus astrocytes (Allard et al. 2014). There is also evidence that astrocytes' ability to catabolize lipids and produce ketone bodies, uniquely among the brain cell types, allows them to modulate the activity of glucose- and lipid-sensing neurons, affecting food intake, particularly in animals fed with a high-fat diet. In these experiments (Le Foll 2019; Le Foll et al. 2014), firing rate of arcuate nucleus and ventromedial hypothalamus neurons was directly affected by ketone bodies characteristically produced by hypothalamic astrocytes. A pathway mediated by lipoprotein lipase appears to play a complementary role (Gao et al. 2017), as a loss of this enzyme in hypothalamic astrocytes leads to suppressed synthesis of lipid droplets in these cells and is accompanied by development of glucose intolerance and weight gain on high-fat diet. Taking into account the association of organism-level metabolic disorder, including diabetes and obesity, with the course and severity of schizophrenia, bipolar syndrome, major depression, and anxiety (Annamalai et al. 2017; Chouinard et al. 2016; Lundqvist et al. 2019), further delineating the role of astrocyte metabolism in organism-level glucose regulation and food consumption behavior represents a potentially fruitful direction.

In addition to the astrocyte-derived metabolites affecting efferent signals to the liver and pancreas, there is an intriguing potential connection between astrocyte metabolism and signaling to the brain by a major gut microflora metabolite, D-lactate. D-lactate can compete with L-lactate produced by astrocytes and transferred to neurons during cognitive tasks, impairing memory formation and synaptic plasticity (Baker and Edwards 2007; Gibbs and Hertz 2008; Murphy-Royal et al. 2020). These observations have been seen as implicating astrocytes in the pathology manifested as D-lactate encephalopathy, a rare condition which can be accompanied by delirium along with other neurological abnormalities. Comporting with this view, diet conducive to D-lactate production in the hind gut of rats and D-lactic acid concentration in the plasma of these animals have been correlated with memory impairment in behavioral studies (Hanstock et al. 2010). Among the potential mechanisms behind the link between gut dysbiosis and mental illness (Bruce-Keller et al.

2018; Rogers et al. 2016), the D-lactate and astrocytes connection deserves further scrutiny in light of the particularly strong association of schizophrenia and bipolar disorder with the *Lactobacillus* species that produce D-lactate, such as *Lactobacillus gasseri* (Dickerson et al. 2017; Nguyen et al. 2018).

Regulation of Vascular Energy Supply

Regulation of energy metabolism on the cell and tissue level in the brain is inseparable from coordination of vascular glucose and oxygen supply that occurs in response to cognitive activity and becomes impaired in neuropsychiatric conditions. Specific cortical regions display dysregulated blood flow in major depressive and bipolar disorder, as well as in schizophrenia (Ota et al. 2014; Sukumar et al. 2020; Wang et al. 2014; Wang et al. 2018; Zhu et al. 2017). These changes are also reflected in the expression of genes involved in vascular regulation and ischemia response (Moises et al. 2015), which are also prominent among genes independently associated with schizophrenia (Schmidt-Kastner et al. 2020). Recent work has begun to unravel the molecular mechanisms whereby local energy supply is regulated by vasoactive compounds released by astrocytes in coordination with the metabolic demands of synaptic transmission. Enveloping both the neuronal synapses and the capillaries, and thus encompassing what is known as neurovascular units, astrocytes are a natural conduit for regulatory information. Future studies of this function of astrocytes in the energy metabolism of the neurovascular units will likely benefit from applications of in vivo imaging of intracellular calcium signaling in astrocytes and adjacent cells using fluorescent sensor proteins, potentially in combination with in vivo functional magnetic resonance imaging (fMRI) now available to rodents (Tsurugizawa et al. 2020).

Astrocytes detect synaptic activity through a palette of transmembrane receptors positioned in immediate proximity of the neuronal synaptic cleft (Haydon 2001; Heller and Rusakov 2017; Stobart and Anderson 2013). Intracellular calcium signals in astrocytes are activated by γ-aminobutyric acid B (Ishibashi et al. 2019; Kang et al. 1998; Meier et al. 2008), muscarinic acetylcholine (Takata et al. 2011), purinergic P2Y (Simard et al. 2003), endocannabinoid (Navarrete and Araque 2010), glutamate (mGluR5), and α-adrenergic and histamine receptors (Bekar et al. 2008; Ding et al. 2013; Shelton and McCarthy 2000; Zonta et al. 2003). Calcium signaling in astrocytes is highly differentiated between regions such as soma, juxtasynaptic protrusions, and end-feet that envelop blood vessels, with intra- and intercellular calcium waves observed upon stimulation and spikes in the end-feet linked directly to the vasomotor effects (Arizono et al. 2020; Chuquet et al. 2007; Howarth 2014; Lind et al. 2013a; Lind et al. 2018; Otsu et al. 2015).

Intercellular propagation of calcium signals may be important for integrated regional responses to neural activity. They are mediated through connexin gap junctions between astrocytes (Gosejacob et al. 2011; Scemes et al. 1998) and the gliotransmitters glutamate, D-serine, and ATP (Bezzi et al. 2004; Harada et al. 2015;

Lalo et al. 2014; Mothet et al. 2005; Volterra and Meldolesi 2005). While mitochondria have been implicated in regulation of the gliotransmitter release (Parnis et al. 2013; Reyes and Parpura 2008; Vardjan et al. 2019), it has been assumed that the calcium-scavenging function of the mitochondrial lumen may be responsible for regulating this process by disrupting the intracellular calcium signal that causes the release. Given that the synthesis of gliotransmitters involved in the intercellular calcium waves is coupled to the energy metabolism, it is possible that their role in wave propagation represents another layer of regulation of this process by the metabolic state of the tissue, whereby a fully mobilized local cell metabolism makes possible a wider regional upregulation of oxygen and glucose supply.

Transmission of signal to the level of intracellular calcium is achieved through inositol trisphosphate release and action on the receptors of the endoplasmic reticulum (InsPR) but also through InsPR-independent action on the intracellular calcium store, as well as calcium-induced calcium release (Okubo et al. 2019; Rodriguez-Prados et al. 2020; Verkhratsky et al. 2012; Volterra et al. 2014). The experiments show that transfer of calcium ions to mitochondria is a particularly robust component of the endoplasmic reticulum-mediated response and can occur even during minor, local calcium release events. The mitochondrial import engages the voltage-dependent anion channel in the outer mitochondrial membrane, which displays calcium ion conductance, and the calcium uniporter in the inner membrane. Absorption of calcium endows mitochondria with an ability to attenuate calcium-mediated signaling in astrocytes (Boitier et al. 1999; Jackson and Robinson 2015; Jackson and Robinson 2018; O'Donnell et al. 2016). This ability is linked to the electrochemical effect of the transmembrane electric potential in mitochondria driving the calcium import and also to the normally occurring repositioning of mitochondria toward the intracellular sites of high energy demand. Conversely, rapid oxidative phosphorylation and the attendant generation of reactive oxygen species propagate the intracellular calcium signal through the induced opening of the permeability transition pore (Agarwal et al. 2017). Thus, recent research suggests that the transmission of the energy-demand signal from the activity in the neuronal synaptic cleft can be subject to modulation by the astrocyte's own metabolic state.

Elevation of intracellular calcium has been linked to the release of vasoactive compounds that include metabolites of arachidonic acid, nitric oxide, potassium, and carbon monoxide (Mahan 2019; Stobart and Anderson 2013). The effects mediated by arachidonic acid are regulated through the calcium dependence of phospholipase A_2 (Sun et al. 2005). Direct action of arachidonic acid on smooth muscle cells is vasoconstrictive and occurs via its conversion, in the target cells, into 20-hydroxyeicosatetraenoic acid (Mulligan and MacVicar 2004). However, elevated activity in the neural tissue causes the parallel pathway to predominate, which are mediated by conversion of arachidonic acid in the astrocytes into prostaglandin E2. The switch is effected by lactate sensitivity of prostaglandin elimination from the extracellular space, which occurs via the prostaglandin-lactate antiporter, and an increased production of lactate by astrocytes in response to tissue energy demands, to be reviewed in detail in the next section. Accumulation of extracellular prostaglandin E2 is believed to drive vasodilation that is observed under these conditions (Gordon et al. 2008).

By comparison, carbon monoxide (CO) production in astrocytes is stimulated by calmodulin-dependent activation of heme oxygenase (Xi et al. 2011) and can complement the direct action of glutamate on CO generation in the endothelium and vascular smooth muscle (Leffler et al. 2003). It is interesting to note that in addition to its vasomotor effect, CO stimulates mitochondrial metabolism in astrocytes, as well as in neurons (Almeida et al. 2012; Almeida et al. 2016). It is conceivable that this dual function serves to feed back on the energy consumption in the complex tissue when its supply via the local vasculature is ensured. Such a mechanism would comport with the hemo-neural hypothesis (Moore and Cao 2008) and also with the observation that increased arteriole blood pressure triggers calcium signaling in astrocytes, a process mediated by transient receptor potential vanilloid 4 channels (Diaz et al. 2019; Kim et al. 2015b). Perturbation of this signaling pathway in hypertension has been hypothesized to contribute to cognitive decline. Indeed, since hypertension is a frequent comorbid condition in patients with bipolar disorder and schizophrenia (Ayerbe et al. 2018), this potential connection merits further exploration. Overall, the present state of knowledge concerning the astrocytes' place in regulating the vascular energy supply to specific brain regions suggests multiple inputs reading out the state of local metabolism and integrating the signals into the vasomotor response.

Metabolic Fluxes in Gray Matter

Different aspects of the metabolic flux distributions among cell types and pathways in the gray matter of the brain have been subject of comprehensive recent reviews (see, in particular, Barros et al. 2018; Jha and Morrison 2018; Deitmer et al. 2019; Dienel 2019). In addition to qualitative conceptualizations, formal flux balance analysis has been carried out to pinpoint the main features of the energy metabolism in this complex tissue and the role played by different cell types (DiNuzzo et al. 2017; Rothman and Dienel 2019). The chief energy consumer in the brain is the neuronal electrogenic Na^+/K^+ ATPase. The electric current associated with each synaptic event is on the order of nanoampere, yet considerable energy expenditure is required to sustain a frequency of such events that may reach 100 Hz per synapse, when the total number of synapses in the human brain is estimated at 10^{15}. The work associated with synaptic activity, naturally, is added to any basal metabolic demands, in which maintenance of ionic gradients is prominently represented in all cell types. Brain energy metabolism, and particularly that in gray matter, is characterized by its reliance on glucose supply and astrocyte glycolysis to meet rising energy demand that occurs during cognitive tasks.

Pathways of Nutrient Utilization

Discussion of issues pertaining specifically to the current research in astrocyte metabolism and neuropsychiatric conditions can be aided by first revisiting the well-established general features of energy metabolism in gray matter. Glucose in the brain is transported out of the bloodstream by GLUT1 uniporter across the endothelial membranes. The same GLUT isoform is responsible for its uptake into astrocytes (Simpson et al. 2007), whereas neurons express kinetically different GLUT3 and GLUT4, whose presentation on the plasma membrane is sensitive to the energy demand (Ashrafi et al. 2017; Pearson-Leary and McNay 2016). Once in the cytosol, glucose is sequestered by conversion to glucose-6-phosphate by hexokinase. This can be regarded as the first step in glycolysis but also supplies the substrate for the pentose phosphate pathway and glycogen shunts. Glycogen synthesis and degradation cycle is a distinguishing biochemical characteristic of astrocytes, and its role in energy mobilization will be reviewed in detail below. The pentose phosphate pathway is important in the brain for generation of reducing equivalents in the form of nicotinamide adenine dinucleotide phosphate (NADPH). The two molecules of NADPH per glucose molecule shunted to this pathway can be utilized by glutathione reductase to regenerate glutathione consumed in peroxidation that eliminates hydrogen peroxide, the product of both oxygen radical generated during respiration, and monoamine oxidase-catalyzed degradation of neurotransmitters. As both processes are accelerated during cognitive tasks, and in view of the specific link of cognitive decline in schizophrenia to redox imbalance (Maas et al. 2017), a protective role of astrocyte metabolism represents a promising direction of study (Tarasov et al. 2019).

The downstream steps in glycolysis are at the core of astrocyte energy metabolism and also supply precursors for synthesis of neurotransmitters such as glycine, L-serine, D-serine, and others. One major branch of glucose metabolism in astrocytes is the reduction of pyruvate to lactate catalyzed by lactate dehydrogenase, which produces two moles of NADH per mole of glucose. However, the importance of astrocyte lactate generation in different physiological and pathological states continues to be debated, as will be reviewed below (see also Deitmer et al. 2019; Dienel 2019). Fatty acids represent an alternative energy source for astrocytes (see Lee et al. 2020). Their uptake is mediated by fatty acid binding proteins (FABP) that facilitate the transit across cell membranes (Pan et al. 2015; Rapoport et al. 2001; Smith and Nagura 2001). β-Oxidation and entry into the citric acid cycle with acetyl-CoA provide the pathway for fatty acid utilization in the energy metabolism of astrocytes and can substitute for glucose utilization pathways under conditions of glucose starvation (Weightman Potter et al. 2019). During general nutrient deprivation, astrocytes metabolize their lipid reserves contained within cytoplasmic droplets (Cabodevilla et al. 2013). However, as recent research shows, astrocyte fatty acid oxidation also operates in tandem with glycolysis (Eraso-Pichot et al. 2018; Panov et al. 2014). The significance of this process in the brain function is underscored by experiments with the astrocyte-specific FABP7 (Sharifi et al. 2011; Sharifi et al. 2013). Deletion of the *Fab7* gene in mice affects neurogenesis, neuronal morphology, and synaptic function, as well as pre-pulse inhibition of the acoustic startle that can be impaired in patients with several neuropsychiatric diseases (Ebrahimi et al. 2016; Watanabe et al. 2007). Further work is needed to more fully uncover the contribution of lipid metabolism in astrocytes to etiology of major psychiatric disorders.

Intercellular Transport of Metabolites

The energy metabolism pathways display a marked differentiation between neurons and astrocytes, leading to establishment of intercellular fluxes and metabolic cycles. There are three major metabolite exchange mechanisms mediated by lactate, glutamate, and γ-aminobutyric acid (GABA). The latter two specifically engage glutamatergic and GABAergic neurons and are directly linked to synaptic neurotransmission. By comparison, lactate exchange is believed to involve all neuronal cell types. In order to set the stage for discussion of these critical issues that may impact the interpretation of genetic associations of metabolic factors with neuropsychiatric disease, it is worth describing the current understanding of how the intercellular fluxes are organized and integrated into the intracellular pathways characterizing neurons and astrocytes.

The turnover of glutamate and its role in schizophrenia is reviewed in the preceding chapter. Certain metabolic connections of the fate of glutamate that is released in neurotransmission, nonetheless, deserve mentioning here. Although neurons express neuronal glutamate symporters, the majority of glutamate is recovered from the extracellular space surrounding the synaptic cleft by astrocytic symporters, glutamate aspartate transporter 1 (GLAST-1), and glutamate transporter 1 (GLT-1), a.k.a. excitatory amino acid transporter 1 and 2 (Pellerin and Magistretti 1994; Pines et al. 1992; Shashidharan et al. 1994; Storck et al. 1992). Inside astrocytes, glutamate is oxidized to α-ketoglutarate by either glutamate dehydrogenase or aspartate aminotransferase (Faff-Michalak and Albrecht 1993; McKenna et al. 2006a), which diverts 20–30% of the synaptic glutamate to the Krebs cycle (Dienel 2019; Stobart and Anderson 2013). Glutamate can also by synthesized de novo from α-ketoglutarate. Regardless of its source, glutamate in astrocytes is subject to conversion to glutamine (Martinez-Hernandez et al. 1977) before it can be exported back into the extracellular space. After its uptake by the neuron, glutamine undergoes deamidation inside the mitochondria to produce glutamate that is then accumulated in presynaptic vesicles. Oxidation of glutamate and entry the citric acid cycle in astrocytes serve to recover the energy expended by in its import, which dissipates the electrochemical gradient of sodium across the astrocytic plasma membrane. An interesting recent finding concerning lipid utilization in astrocytes is its downregulation by glutamate (Eraso-Pichot et al. 2018), which suggests a potential for substitution of the energy derived from fatty acid oxidation by the entry of glutamate into the citric acid cycle. The GABA shuttle operates similarly and overlaps with the glutamate-glutamine cycle due to the production of this inhibitory neurotransmitter from glutamine in GABA-ergic neurons and its entry into the astrocytic Krebs cycle via succinic semialdehyde and succinate (Cooper and Jeitner 2016; Dienel 2019).

Recent research has focused on the exchange of lactate and glutamate between neurons and astrocytes and on the role of these processes in coordination of energy metabolism between the cell types in gray matter. The role of the extracellular lactate in the astrocyte metabolism was highlighted by the observation that astrocytes in culture secrete lactate when grown in glucose-containing medium and then consume the lactate from the medium as the glucose is exhausted (Lind et al. 2013b). In contrast to neurons, which import lactate via monocarboxylate transporter 2 (MCT2), astrocytes utilize MCT1. Export of lactate from astrocytes has been associated with MCT4 (Bergersen et al. 2001; Chiry et al. 2008; Mazuel et al. 2017; Pierre and Pellerin 2005). Up to one-half of the glucose is converted to extracellular lactate in astrocytes, as well as in cortical and cerebellar glutamatergic neurons under both resting and depolarizing conditions (Gebril et al. 2016; Jekabsons et al. 2017; Lopes-Cardozo et al. 1986; Waagepetersen et al. 2000). Computational kinetic analysis suggests that the net transport between astrocytes and neurons in vivo accounts for only 2% of the taken-up glucose equivalents and for 11% of accelerated glucose consumption (Mangia et al. 2011). Thus, there may be a large difference between the intensity of the lactate exchange and the small magnitude of its directional flux, considered in relation to the overall rate of metabolism in the astrocyte-neuron system. These kinetic observations about the magnitude of flux, however, do not preclude lactate from performing an important function; rather, they are consistent with the emerging view that lactate is part of a redox buffer that equilibrates NAD^+/NADH ratio among cells in the tissue (Rabinowitz and Enerback 2020).

Flux Redistribution in Response to Demand

Redistribution of metabolic fluxes among the cells of gray matter that sustains a rise in neuronal activity is an active area of research. Several themes are emerging, which highlight, in particular, the role of astrocytic glycogen, malate-aspartate shuttle regulation, and the relative weight of the flux through the pentose phosphate shunt. Notable features that differentiate metabolism of neurons from astrocytes are low-level glycogenic enzymes and the usual absence of glycogen in neurons (cf. Saez et al. 2014; Sotelo and Palay 1968). Glycogen turnover in the brain is tightly regulated, and in the extreme case of pathology of this process, its dysregulation in Lafora disease can be fatal (Obel et al. 2012; Roach 2015). Supporting the central role of this energy depot in astrocytes, experimental knockout of glycogen synthase causes impairment of memory formation (Duran et al. 2013). The energy flux from glycogen is also crucial for maintaining the astrocytes' ability to scavenge potassium and accumulate calcium ions in their intracellular stores (Muller et al. 2014;

Xu et al. 2013). It has been suggested that glycogenolysis in astrocytes is subject to monoamine regulation (Dienel and Cruz 2015; DiNuzzo et al. 2015), which would account for accumulation of glycogen in the absence of stress (Cruz and Dienel 2002; Kong et al. 2002b; Oe et al. 2016), and its rapid utilization, amounting to half of the blood glucose uptake, which can be triggered by sensory stimulation (Dienel 2019; Dienel et al. 2002). Similarly, access to glucose following a period of utilization of glycogen in astrocytes leads to establishment of a significant glucose uptake rate as the glycogen store is replenished (Walls et al. 2009). Overall, while the studies continue, it is clear that the glycogen energy buffer in astrocytes makes a large contribution to the metabolic fluxes that precede, accompany, and follow functional activation of gray matter, and it is critical to performance in cognitive tasks.

Scavenging of NADH through the malate-aspartate shuttle is required for continued operation of the glycolysis pathway, when pyruvate enters the citric acid cycle. Interestingly, there is evidence that the shuttle is inhibited by Ca^{2+} fluctuations that exceed the threshold concentration characterizing the mitochondrial calcium uniporter in neurons, redirecting the pyruvate flux to lactate production in the cytosol (Satrustegui and Bak 2015). The lactate may then become redistributed across the activated region of gray matter through gap junctions that connect the region's astrocytes in a functional syncytium (Hertz et al. 2014). Loss of any excess lactate into the bloodstream would account for the submaximal oxygen-glucose index, whose fall in activated brain regions is well documented, reflecting incomplete oxidation of carbon taken up with glucose (Juaristi et al. 2019). Compared to neurons, expression of the malate-aspartate transporter AGC1 is lower in astrocytes (Pardo et al. 2011) and does not appear to be critical for glucose oxidation (Juaristi et al. 2017). Instead, the transport of reducing equivalents to mitochondria in stimulated astrocytes appears to be mediated largely by the glycerol phosphate shuttle, whereby the cytoplasmic glycerol-3-phosphate dehydrogenase oxidizes NADH to reduce dihydroacetone phosphate and the mitochondrial isoform oxidizes this substrate to produce $FADH_2$ (Juaristi et al. 2017). These findings are consistent with the results of the pharmacological approach to differentiating between the transporter systems (Kohler et al. 2018; McKenna et al. 2006b). Mitochondrial glycerol-3-dehydrogenase is activated by calcium from the outer side of the inner mitochondrial membrane and generates large amounts of hydrogen peroxide, which distinguishes mitochondrial respiration in astrocytes (Tretter and Adam-Vizi 2012). The demonstrated involvement of different calcium-regulated transport systems in the neuronal and astrocytic mitochondria suggests that the neuronal-astrocytic metabolic cooperation may be orchestrated by coordinated calcium signals in the two cell types. Microanatomical characterization of the astrocyte-neuron contacts, whereby a single branch of an astrocyte can engage with hundreds of synapses, along with the most recent discovery of local calcium signals in subcellular domains of astrocytes, has led to a hypothesis that the local calcium signals read out the local (subcellular) activity levels in groups of synapses, causing a localized metabolic response (Oheim et al. 2018). Advances in optical registration of calcium signals in vivo and simultaneous visualization of such subcellular parameters as NAD redox status and mitochondrial motility have the potential to clarify these issues.

Most recent work demonstrates existence of a lipid flux from neurons to astrocytes induced by neuronal activity. It has been known for some time that activated neurons generate excess fatty acids that may be peroxidized by reactive oxygen species (Reynolds and Hastings 1995; Sultana et al. 2013). Recent data show that peroxidized fatty acids in neurons follow a pathway that takes them from autophagocytosed membranes to lipid droplets (Ioannou et al. 2019). In agreement with the previously demonstrated low capacity of neurons for fatty acid utilization (Schonfeld and Reiser 2017), excess fatty acids are transported to neighboring astrocytes in dense carriers containing apolipoprotein E (Ioannou et al. 2019). Furthermore, glutamate released by the activated neurons stimulates both lipolysis and reactive oxygen species neutralization in the astrocytes. These findings elaborate on the earlier demonstration that oxidative stress in neurons is accompanied by accumulation of lipid droplets in adjacent astrocytes in a manner that is dependent on apolipoproteins (Bailey et al. 2015; Liu et al. 2015; Liu et al. 2017). More work is needed to elucidate the possible connections of this metabolic flux with psychiatric disorders, but given the well-documented involvement of neuronal oxidative stress in their etiology and progression (e.g., Flatow et al. 2013; Fraguas et al. 2017), this represents a promising direction for future research.

In culture, glutamatergic cerebellar neurons preferentially metabolize glucose over lactate when subjected to stimulation (Bak et al. 2006). The implications of this preference for the in vivo situation, where NADPH generated in the pentose phosphate pathway participates in amelioration of oxidative stress, have been disputed in view of the competition of glycolysis with the pentose phosphate shunt for glucose-6-phosphate. A key finding supporting upregulation of lactate utilization instead of glycolysis is the inability of neurons to increase the activity of phosphofructokinase 1, the gatekeeper of the specifically glycolytic energy-producing flux (Bolanos 2016; Herrero-Mendez et al. 2009). The regulation depends on production of fructose-2,6-bisphosphate, a critical phosphofructokinase 1 activator, which is catalyzed by 6-phosphofructo-2-kinase. This enzyme was found at low levels in the cultured and ex vivo neurons, compared with astrocytes, and its experimental overexpression directed the glucose flux into the glycolytic pathway at the expense of the pentose phosphate shunt. In neurons, this intervention led to lethal oxidative stress. However, other conditions, in which cultured neurons and isolated synaptic regions have been studied, demonstrated glycolysis upregulation, increased glucose use, and elevated surface expression of glucose transporters (see Dienel 2012; Patel et al. 2014). Moreover, stimulation of brain slices and the whisker barrel cortex in vivo leads to upregulation of neuronal glycolysis accompanied by lactate production, instead of uptake (Diaz-Garcia et al. 2017). These disparate observations—or, rather, interpretations that emphasize the glycolysis and pentose pathway activation—can be reconciled by recalling that fructose-6-phosphate produced in the pentose shunt can be recycled through its conversion to glucose-6-phosphate, while glyceraldehyde-3-phosphate can feed the downstream, energy-producing steps of glycolysis (Dienel 2019). This leads to a picture whereby levels of brain matter activation that do not fully engage glycogenolysis may be accompanied by a more bidirectional energy flow between neurons and astrocytes. Whereas the details of metabolic flux regulation attendant to performance of cognitive tasks continue to emerge from ongoing fundamental research, the clear importance of glycogen turnover, calcium signaling regulation of lactate production, and the pentose phosphate shunt in astrocytes can inform the interpretation of data from molecular studies of pathology in major psychiatric conditions.

Metabolic Disorders and Neuropsychiatric Disorders

As reviewed above, astrocytes represent an indispensable link in brain metabolism and organism-level regulation of energy supply to the brain. The studies on normal tissue demonstrate that metabolite-mediated interaction of astrocytes and neurons and signal propagation across the cell types in the brain are central to the exquisitely orchestrated physiological adaptation to increased neuronal activity and synaptic transmission. This physiological and biochemical information allows us to interpret the genomic, metabolomic, and expression data that document the involvement of metabolism in neuropsychiatric disorders.

Our present understanding of the molecular mechanisms underlying the core symptoms of schizophrenia and other major psychiatric disorders focuses on abnormalities in synapse development and synaptic communication (Duman et al. 2016; Frankle et al. 2003; Keshavan et al. 2020; Lee et al. 2018; Mirnics et al. 2001; Stephan et al. 2009). Comporting with this primarily developmental view, mice produced with glial progenitors derived from childhood-onset schizophrenia displayed, in particular, delayed astrocytic differentiation (Windrem et al. 2017). Notably, inhibition of SMAD signaling can both rescue the differentiation process and lead to normalization of potassium transport in the affected glia (Liu et al. 2019). These new data are consistent with the involvement of potassium transport both on the level of small conductance calcium-activated potassium channels ($K_{Ca}2.3$), encoded by the *KCNN3* gene that is associated with the 1q21-q22 locus linked to familial schizophrenia (Brzustowicz et al. 2000), and the sodium/potassium-transporting ATPase subunit α-3 encoded by *ATP1A3*, which has been linked to childhood-onset schizophrenia (Smedemark-Margulies et al. 2016).

Postmortem Evidence

Similarly, studies in human postmortem prefrontal cortex of patients with schizophrenia have detected abnormalities in the composition and distribution of the astrocytic metabolon (Genda et al. 2011) that involves glutamate transporters, hexokinase-1, and sodium/potassium-transporting ATPase, including apparent spatial uncoupling of the molecules from mitochondria (McCullumsmith et al. 2016; Shan et al. 2014). As altered glutamate turnover is implicated in schizophrenia and could represent a vital link between the neuronal and astrocytic metabolism (Section

"Intercellular Transport of Metabolites"), with the maintenance of membrane potential the leading energy expenditure, these findings suggest that it will be particularly important to investigate further the role of astrocyte energy metabolism in the pathogenesis and/or pathophysiology of this disease (Sullivan et al. 2018). Of special note is decreased expression of glial fibrillary acidic protein (GFAP), a major astrocyte marker, on astrocytes in close contact with blood vessels, suggesting changes in uncoupling of glucose energy supply within the affected neurovascular units (Webster et al. 2001). Furthermore, major depression is accompanied by downregulation of glutamate transporters along with glutamine synthase and connexins, the characteristic astrocyte proteins involved in maintaining the integrity of the glutamate-scavenging functional syncytium (Bernard et al. 2011; Blomstrand et al. 1999; Choudary et al. 2005; Sequeira et al. 2009; Zhou et al. 2019). Recent postmortem evidence also suggests impairment of glucose metabolism in the striatum of certain patients with schizophrenia (Dean et al. 2016), while both glycogen phosphorylase and its regulatory kinase are downregulated in dorsolateral prefrontal cortex astroglia of chronic patients (Pinacho et al. 2016). Thus, although our understanding of the involvement of astroglia in the developmental pathology of major psychiatric disorders can be characterized as nascent, it is already beginning to supply mechanistic hypotheses for how dysregulation of energy metabolism could affect these processes.

Genetic Evidence

Schizophrenia is a condition notable for its heritability, and genome-wide association studies have identified variations in over 100 loci linked with the disease (Pardinas et al. 2018; Schizophrenia Working Group of the Psychiatric Genomics 2014). Large and small de novo copy number variations have high penetrance and affect functionally diverse proteins (Fromer et al. 2014; Tansey et al. 2016). Enzymes of the glycolytic pathway, including the regulatory 6-phosphofructo-2-kinase/fructose-2,6-bisphosphatase 2, the flux-controlling hexokinase 3, and pyruvate kinase 3, were found to be specifically associated with development of schizophrenia (Stone et al. 2004). Glycolysis and citric acid cycle genes are also prominent among the single-nucleotide variants associated with bipolar disorder, a similarly heritable condition (Ament et al. 2015). Although some of these associations are tentative and dependent on other conditions such as the patient's race, due to their inherent nature, genetic data pointing to involvement of metabolic alterations in major psychiatric disorders have the advantage of not being impacted by the complicating developmental and environmental factors such as diet or side effects of antipsychotic medication, whose role will be reviewed below. Although existing evidence suggests that genetic risk variants may be responsible for metabolic disturbances in many psychiatric disorders, what remains unknown is whether these risk variants affect metabolic pathways or functions predominantly in neurons or also in non-neuronal brain cells and/or other tissues.

Gene Expression and Functional Data

On the gene expression level, recent work has identified functional gene sets that exhibit dysregulated levels in autism, bipolar disorder, and schizophrenia (Guan et al. 2019). The enriched sets include amino acid transport, neurotransmitter release, oxidative stress response and immune response, and also mitochondrial proteins and glycolysis. Meta-analysis of proteomic data from post-mortem brain material shows in excess of 90 differentially expressed energy metabolism genes in bipolar disorder and schizophrenia and 40 in major depressive disorder (Zuccoli et al. 2017). All three conditions implicate aldolase C, the central enzyme in glycolysis, citrate synthase and malate dehydrogenase of the citric acid cycle, cytochrome b-c1 subunit 1, and the β-subunit of ATP synthase. Intramitochondrial pyruvate dehydrogenase, aconitase, isocitrate dehydrogenase, and 2-oxoglutarate dehydrogenase of the citric acid cycle and several additional components of the electron transport chain are also affected, as are pyruvate kinase, enolase, glyceraldehyde-3-phosphate dehydrogenase, and phosphofructokinase 1 of the glycolytic pathway. These recent analyses are broadly in agreement with the targeted proteomic, transcriptomic, and functional approaches that detected alterations in glycolysis, citric acid cycle, and the respiratory chain in specific brain regions of patients with schizophrenia and other major psychiatric disorders (Beasley et al. 2006; Beasley et al. 2009; Bubber et al. 2011; Cavelier et al. 1995; Du et al. 2014; Martins-De-Souza et al. 2009a; Martins-De-Souza et al. 2009b; Maurer et al. 2001; Middleton et al. 2002; Pennington et al. 2008; Prabakaran et al. 2004; Stork and Renshaw 2005).

Similar to the genetic insights, the bulk tissue expression data is unable to identify cell-type specific metabolic alterations. Moreover, any compensatory changes across cell types would not be reflected in the bulk tissue results. Laser capture and microdissection have been used to assess changes specific to neurons, particularly in granule cells of the dentate gyrus and prefrontal pyramidal cells (Altar et al. 2005; Arion et al. 2015; Arion et al. 2017). In the hippocampus, schizophrenia patients were distinguished from bipolar and major depression cohorts by decreased expression of isocitrate, lactate, malate, NADH, and succinate dehydrogenases, as well as cytochrome C oxidase and ATP synthase. The pyramidal cells showed downregulation of mitochondrial energy metabolism genes in the schizophrenia samples with no alterations being observed in samples from patients with the bipolar disorder, major depressive disorder, or schizoaffective disorder. The technique has also been applied to study alteration of expression of astrocytic genes in specific layers of the anterior cingulate gyrus and revealed downregulation of glutaminase and excitatory amino acid transporter 2 in postmortem material from patients with schizophrenia (Katsel et al. 2011a). More cell type-specific data are necessary to form a clear picture of the metabolic alterations at the cellular level in mental illnesses.

Effects of Antipsychotic Medications

Studies on drug-naïve patients with psychiatric illnesses support the notion that the metabolic alterations are inherent in these conditions. For example, impaired glucose tolerance and metabolic syndrome are associated with first-episode schizophrenia or psychosis (Chen et al. 2016a; Perry et al. 2016; Petrikis et al. 2015; Pillinger et al. 2017; Ryan et al. 2003; Spelman et al. 2007; Zhang et al. 2015b). In the case of first-episode psychosis, however, negative findings have also been reported (Sengupta et al. 2008). By contrast, in a different study on drug-naïve patients, it was found that 8 out of 18 differentially expressed genes in peripheral monocytes of antipsychotic-naïve patients were from the glycolytic pathway (Herberth et al. 2011). Positron emission tomography demonstrates that drug-naïve patients with schizophrenia display an increased variation of glucose metabolic rates about the mean, compared with the healthy controls. This increased variation was observed in a number of different brain regions (Wiesel et al. 1987). At the same time, chronic patients show a decreased level of glucose metabolic rates. Region-specific reduction in glucose metabolism is seen in the hippocampus and anterior cingulate cortex (Tamminga et al. 1992). Despite the availability of data from drug-naïve patients, widespread and chronic use of antipsychotic drugs confounds the efforts to determine whether metabolic disturbances are the primary contributing factors or consequences of treatments that could influence the course and severity of the disease. Positron emission tomography, on the other hand, showed that haloperidol enhanced glucose utilization in the caudate and putamen, while lowering it in the anterior cingulate and frontal cortex (Holcomb et al. 1996). A more recent study has addressed the effect of a combined aripiprazole-amisulpride therapy and found glucose metabolism to be accelerated in the right superior frontal and frontal precentral gyri (Park et al. 2019).

The impact of antipsychotic medication on cell metabolism is well documented in vitro. Haloperidol, clozapine, chlorpromazine, and fluphenazine all suppress glucose uptake in experiments on PC12 cells following their administration but cause what appears to be a compensatory upregulation of GLUT1 and GLUT3 after a longer exposure (Dwyer et al. 1999). In the same neuronal model cell line, as well as in the putamen and caudate nucleus of mice, clozapine acted by reducing signaling downstream of insulin receptor, particularly rendering the phosphorylation of Akt insulin-independent (Panariello et al. 2012). The action of antipsychotic phenothiazines is not limited to their receptor- and cell type-specific interactions but can also be mediated by incorporation into the plasma membrane, as has been demonstrated in experiments on yeast, where they similarly caused inhibition of hexose transporters (Uesono et al. 2016). Further demonstrating that these effects may not be cell-type specific are the similar results obtained following in vitro exposure of a human leukemia cell line (Heiser et al. 2006). Haloperidol treatment of oligodendrocytes led to a higher concentration of glucose and a lower concentration of lactate in the extracellular medium, whereas the atypical antipsychotic clozapine had the opposite effects and, in addition, enhanced oxidative phosphorylation (Steiner et al. 2014). At the same time, the complexity of psychotropic drug action on the organ level is underscored by the outcome detected in a comparison of the clozapine and haloperidol effects after chronic administration to mice, which showed that mRNA expression related to the carbohydrate metabolism was more affected by clozapine treatment than by haloperidol (Rizig et al. 2012). A number of recent studies suggest that clozapine has a profound effect on mitochondrial functions (Contreras-Shannon et al. 2013; Scaini et al. 2018; Viana et al. 2020). Determining the place of astrocytes in the modulation of brain metabolism brought about by the chronic use of antipsychotic medications is an important direction for future work. Thus, existing data indicate that the action of antipsychotics is highly region-specific and therefore likely to be specific to the regional subtypes of neuronal and glial cells. Further deciphering this complexity of cell-specific action may be necessary for the progress in determining the contribution of astrocyte metabolism to the etiology and progression of the major neuropsychiatric conditions.

Insights from Animal Models

Pharmacological Models

Animal and cell models hold out a promise to permit disentanglement of the genetic, functional, and medication-induced contributions to the etiology of major psychiatric disorders, as well as supply material for the study of cell type-specific effects in these processes. In this section, we will cover what is known about metabolic alterations arising in different types of animal models and attempt to identify the gaps where such knowledge is lacking but where other related data make research advances likely.

Metabolic alterations are consistently observed in *N*-methyl-D-aspartate receptor (NMDAR) antagonist models of schizophrenia. Metabolomic assessment of the cortex and hippocampus in rats treated with MK-801 showed downregulation of glutamate metabolism and the citric acid cycle and, particularly, alterations in the content of citrate and succinate in addition to glutamate and glutamine (Sun et al. 2013). At the same time, a combination of methods revealed a pronounced reduction in glycolysis, "which" was constrained to the parietal and temporal cortex, and highly region-dependent reductions in fumarate and malate turnover (Eyjolfsson et al. 2011). The chronic phencyclidine rat model, by comparison, displays decreased pyruvate and pyruvate kinase levels in the animals' frontal cortex, along with alterations in the glutamate calcium signaling pathway, including calcium-calmodulin-dependent kinase II and calcineurin (Wesseling et al. 2013). Administration of ketamine similarly leads to acute alterations in glycolysis, the pentose phosphate pathway, and citric acid cycle in the hippocampus of mice (Weckmann et al. 2014).

Interestingly, in vitro investigation on neuronal, oligodendrocyte, and astrocyte cell lines demonstrates that the glycolysis inhibition effect of MK-801 is seen in all three cell types, with the oligodendrocyte glia exhibiting the largest reduction in the glycolytic enzyme expression (Guest et al. 2015). Thus, although it is clear that astrocyte metabolism contributes to the changes seen in vivo—at least with MK-801—its exact contribution to the pathology modeled with the NMDAR inhibitors remains to be determined. It can be noted that NMDAR knockdown in mice also leads to significant changes in key enzyme and calcium signaling molecule expression, including mitochondrial pyruvate kinase, aspartate aminotransferase, and calcium-calmodulin-dependent protein kinase II in the frontal cortex and hippocampus (Wesseling et al. 2014). Together with the effects triggered by administration of NMDAR inhibitors, effects that arise in this genetic model of synaptic disfunction in schizophrenia and autism spectrum disorders argue in favor of the notion that metabolic changes may be triggered downstream of the disrupted NMDAR signal transmission (Sullivan et al. 2018).

Use of dopamine enhancers such as amphetamine and apomorphine remains an informative approach in animal models of schizophrenia (Brisch et al. 2014; Lindenmayer et al. 2013; Winship et al. 2019). Whether astrocyte energy metabolism is involved in the linkage between dopamine dysregulation and its behavioral effects, however, it still unknown. Mitochondrial localization of monoamine oxidase, the dopamine-metabolizing enzyme (Edmondson et al. 2009; Schnaitman et al. 1967), including the one expressed in brain astrocytes (Levitt et al. 1982; Mallajosyula et al. 2008), is suggestive of a possible existence of connection to metabolism. Indeed, in most recent experiments on neurons, it has been demonstrated that dopamine oxidation can supply electrons directly to the mitochondrial transport chain (Graves et al. 2020). It is not clear how these new findings should be reconciled with earlier results obtained on isolated brain mitochondria, where dopamine suppressed pyruvate- and succinate-dependent oxidative phosphorylation and respiration, as well as with inhibition of complex I activity that accompanies elevated dopamine levels induced by chronic administration of *L*-DOPA and *D*-methamphetamine (see Ben-Shachar 2002). Directly addressing the possible role of similar mechanisms in astroglia may be of interest in further studies utilizing the dopamine pharmacological models.

Links to astrocyte energy metabolism have been established more firmly in 5-hydroxytryptophan (5-HT) pharmacological models, although new research is needed to clarify the connecting mechanisms and their influences on behavior. Decrease of astrocytic expression of 5-HT_{2B} receptor is observed in parallel with the progression of the depressive phenotype in a murine model of Parkinson's disease (Zhang et al. 2015a). Because of the frequent presentation with depressed mood among schizophrenia patients (Fortunati et al. 2015), the result just mentioned, which implicates specifically 5-HT signaling in astrocytes, can be directly relevant to modeling schizophrenia in the mouse using deletion of the 5-HT receptor gene or, alternatively, administration of this receptor's inhibitors (Chen et al. 2016b; Pitychoutis et al. 2015). Adding to the evidence for the importance of astrocyte 5-HT signaling are the data concerning administration of 5-HT receptor stimulants and serotonin-specific reuptake inhibitors used in alleviating major depression, such as fluoxetine (Diaz et al. 2012; Li et al. 2008; Qiao et al. 2016; Zhang et al. 2010).

Astrocytic but not neuronal 5-HT$_{2B}$ receptor becomes upregulated under the conditions of chronic treatment with fluoxetine (Hertz et al. 2015; Li et al. 2012). It has been known for some time that fluoxetine, acting via 5-HT$_{2B}$ and intracellular calcium, can stimulate glycogenolysis in astrocytes (Allaman et al. 2011; Chen et al. 1995; Kong et al. 2002a). More recently, it has been established that blocking glycogenolysis prevents serotonin- or fluoxetine-induced memory enhancement (Gibbs and Hertz 2014). The results obtained in culture or on brain slices are in concordance with the earlier observations made on the rat brain, where activators of 5-HT(2) receptor led to a significant downregulation of glycogen and elevation of extracellular glucose (Darvesh and Gudelsky 2003). Evidently, the involvement of astrocyte energy metabolism in mechanisms of major psychiatric disorders will benefit from further studies in 5-HT pharmacological models.

Neuroinflammatory Models

Maternal immune activation models have shown great potential in the study of etiological factors of psychiatric disorders (Careaga et al. 2017; Winship et al. 2019). Infection and other environmental exposures during pregnancy have been associated with schizophrenia, autism spectrum disorders, and bipolar disorder in the offspring, while data from animal models have helped connect the action of cytokines on the fetal brain with behavioral abnormalities in subsequent development. In the animal model approach, immunogens such as polyinosinic-polycytidylic acid, poly(I:C), and lipopolysaccharide (LPS) are employed for triggering maternal immune system response (Conway and Brown 2019).

Numerous lines of evidence implicate astrocytes in the pathological mechanisms triggered by maternal immune activation. Rat astrocytes, but not neurons, secrete tumor necrosis factor α (TNFα) in response to toll-like receptor (TLR) stimulation by LPS, which acts through TLR4, and poly(I:C), which acts through TLR3 (Chistyakov et al. 2018). Poly(I:C), in addition, triggers lysosomal ATP release and intracellular calcium elevation by activating TRPML1 channels in the affected astrocytes (Beckel et al. 2018). A modulating role of the astrocyte energy metabolism in these responses is suggested by most recent experiments that evaluated the activation of p38, JNK, and ERK signaling pathways in astrocytes under conditions of different glucose availability (Bellaver et al. 2015; Chistyakov et al. 2019). The release of TNFα in response to either poly(I:C) or LPS was enhanced in high glucose (25 mM), whereas the kinases were modulated reciprocally and in a time-dependent manner. These intriguing findings are in agreement with the general notion of connection of metabolism to inflammation in the brain (Ristow 2004) and with other recent investigations of modulation of astrocyte signaling by the metabolic conditions (Gandhi et al. 2010; Quincozes-Santos et al. 2017; Wang et al. 2012). Regulation of astrocyte morphology and connexin expression by LPS (Debarba et al. 2019) may also be of relevance to brain metabolism, considering the role of astrocyte connectivity in the spatial organization of neurovascular units

(Section "Regulation of Vascular Energy Supply"). Furthermore, data are beginning to accumulate that describe changes in astrocyte mitochondrial morphology and function under the conditions of inflammatory challenge, for example, with LPS. Impairment of the mitochondrial membrane potential, ATP production, and mitochondrial transcription are among the observed effects (Peng et al. 2019; Zhao et al. 2017). At the same time, the cited data point to a protective role of uncoupling protein 2. More work is needed to elucidate the role of astrocyte energy metabolism in regulating the responses triggered in the organism by maternal immune activation.

The possibility that inflammation signaling in astrocytes may be modulated by energy metabolism is suggested by recent work on the mitochondrial antiviral pathway (Koshiba 2013; Refolo et al. 2020). This pathway is triggered by retinoic acid inducible gene 1-like receptor (RLR), which, alongside the toll-like receptor, has the capacity to recognize double-stranded RNA (see Chattopadhyay and Sen 2014) and signals to membrane-associated mitochondrial antiviral signaling (MAVS) protein. *O*-linked *N*-acetylglucosamination of MAVS in myeloid cells regulates this protein's function (Song et al. 2019) in a manner that depends on the activity of the hexosamine biosynthesis pathway, providing a link between glucose metabolism and antiviral innate immunity (Li et al. 2018). Indeed, in experiments on various cell types, MAVS has been demonstrated to directly sense lactate and associate physically with hexokinase, mediating interferon production-suppressing effect of lactate (Zhang et al. 2019). Conversely, RLR activation suppresses production of glucose metabolites, including lactate, and upregulates MAVS-mediated production of type-1 interferon. This mechanism is compatible with the earlier demonstrated impairment of the MAVS-dependent interferon production, when mitochondrial oxidative phosphorylation is suppressed (Yoshizumi et al. 2017). In light of the recent findings that astrocytes are an important source of interferon produced by the MAVS-mediated pathway (Clarke et al. 2019; Giovannoni and Quintana 2020; Vaccari et al. 2012), it is conceivable that regulatory pathways similar to the ones described in other cells may operate in astrocytes as well and play a role in the neuroinflammatory etiology of autism and schizophrenia.

A new direction of research benefiting from the use of maternal immune activation models targets the role of microbiome in the etiology of major psychiatric disorders. Microbiome imbalance has been implicated in both behavioral abnormalities and immune dysfunction and, particularly, in the abnormal immune function that is associated with autism spectrum disorders (Conway and Brown 2019; Cryan and Dinan 2015; Hsiao 2013). It has been established that the relationship between gut microbiota and neuroinflammation is bidirectional, with the microbiota composition affecting regional immune responses in the brain and being itself regulated by neuroinflammation (Cryan and Dinan 2015; Erny et al. 2015). Remarkably, maternal immune activation in the animal models impacts the immune responses and microbiome composition in the offspring (Hsiao et al. 2013; Kim et al. 2017; Mandal et al. 2013). Among the important questions that remain to be investigated is how astrocyte metabolism may be affected in the enteric nervous system and what role it may play in the lasting changes that develop in the gut microbiota, neuroinflammation, and behavior following maternal immune activation. The connections of enteric glia and specifically enteric astrocytes with both neurons and immune cells of the intestine are attracting attention in other lines of study, for example, in pain research (Grubisic and Gulbransen 2017; Morales-Soto and Gulbransen 2019). There is evidence for involvement of both gut microbiota and immune system in postnatal development of the enteric nervous system (Bistoletti et al. 2019; De Vedder et al. 2018; Kabouridis and Pachnis 2015), and applications of imaging and optogenetics to target specific cell types of the enteric nervous system in situ are being developed (Boesmans et al. 2018). These advances make it both feasible and imperative to address the question of the possible involvement of enteric nervous system astrocyte energy metabolism in the pathogenesis of behavioral changes triggered by maternal immune activation.

Clues from Models of Substance Use Disorders

Substance use disorders present certain behavioral pathology that is shared with other major psychiatric conditions, and their animal models, as well as in vitro experiments motivated by their study, have great potential to inform our general understanding of the role of astrocyte energy metabolism. Much relevant data also come from research into the effects of specific substances of abuse on primary astrocytic cultures. The chapters that follow in this volume cover the involvement of astroglia in the mechanisms of addiction and alcohol dependence in detail; here, we will maintain our focus on bioenergetics and highlight the data that point at commonality of its perturbation with that in other major psychiatric disorders.

Investigations on mouse brain slices revealed that activation of endocannabinoid CB1 receptors on hippocampal astrocytes leads to calcium release from the intracellular stores, followed by release of glutamate that can act on neuronal NMDAR and potentiate synaptic transmission (Navarrete and Araque 2008, 2010). This insight has led to CB1 signaling being implicated in the disruption of working memory formation by exogenous cannabinoids in vivo (Han et al. 2012). At the same time, it is noteworthy that in both cortical and hypothalamic astrocytes, CB1 signaling sustains the expression of leptin receptors and thus may be linked to regulation of glycogen storage (Bosier et al. 2013). This process may be crucial to address the apparent imbalance between glucose uptake, which tetrahydrocannabinol inhibits in rat brain (Miederer et al. 2017), and glucose metabolism, which this compound upregulates in cultured astrocytes (Sanchez et al. 1998). Furthermore, preadolescent exposure to tetrahydrocannabinol in rats reduces the expression of glutamate synthase (Suarez et al. 2002), which has the potential to perturb the metabolic fluxes associated with the neuron-astrocyte glutamate-glutamine cycle. The specific mechanisms behind the involvement of the astrocyte energy metabolism in the etiology and clinical course of cannabinoid addiction represent a promising direction for future work.

Glutamate transporter 1 (GLT-1) expression is downregulated in the nucleus accumbens of chronically exposed, alcohol-preferring rats (Das et al. 2015), in contrast to a lack of modulation in intermittently exposed animals (Griffin et al. 2015; Madayag et al. 2017; Pati et al. 2016), suggesting that perturbation of the glutamate-glutamine cycle may be involved in the formation of alcohol addiction. This hypothesis is borne out by extensive experimentation that shows reduction in alcohol preference and consumption that can be elicited by administration of compounds, for example, ceftriaxone (a β-lactam antibiotic), which upregulate GLT-1 (Alasmari et al. 2016; Das et al. 2015; Sari et al. 2016). Additional evidence comes from experiments with designer receptors activated by designer drugs (DREADD), where stimulation of glutamate release in astrocytes of the core of the nucleus accumbens was sufficient to inhibit ethanol self-administration (Bull et al. 2014). On the whole, the findings concerning glutamate transport in the opiate abuse disorder models to date echo the results from alcohol abuse investigations.

It has been established that morphine reduces expression of GLT-1 in astrocytes in culture through its activation of protein kinase C (Wang et al. 2017), and similar results have been obtained in the murine cerebrum (Wu et al. 2008). However, the effect seen in animals is sex-specific, being limited to male mice. Illuminating results were obtained in the study of heroin self-administration and withdrawal in rats (Shen et al. 2014), where heroin-seeking behavior was found to be dependent on glutamate accumulation in the core compartment of the nucleus accumbens. The "spillover" could be linked to GLT-1 downregulation, while the attenuation of the heroin-seeking behavior by ceftriaxone was mediated by GLT-1 re-synthesis. Interestingly, the specific light chain of cystine-glutamate exchanger (xCT) becomes overexpressed under the extinction conditions. This complexity underscores the importance of further inquiry into how the varied and parallel energy metabolism fluxes involving astrocytes may be involved in substance abuse disorders.

Of the numerous brain regions studied, none exhibited any alteration of GLT-1 expression associated with amphetamine exposure in rats (Sidiropoulou et al. 2001). Effects of methamphetamine, in comparison, depend on the exact behavioral model but include upregulation of GLT-1 in the striatum and medial prefrontal cortex (Han et al. 2014; Shirai et al. 1996). Change of xCT expression is documented with short-term amphetamine administration, where a reduction in the thalamus and frontal cortex is observed (Pang et al. 2013). Thus, regulation of xCT that is seen with the amphetamines resembles the one in the models of opiate abuse in that it is reciprocal with the direction of any effects on GLT-1.

In cocaine use disorder models, both GLT-1 (Knackstedt et al. 2010; LaCrosse et al. 2016; Reissner et al. 2014; Reissner et al. 2015; Sondheimer and Knackstedt 2011) and xCT (LaCrosse et al. 2017; Sondheimer and Knackstedt 2011; Trantham-Davidson et al. 2012) are found to be downregulated in the nucleus accumbens, but not in the dorsal striatum or prefrontal cortex (Knackstedt et al. 2010; LaCrosse et al. 2016; Parikh et al. 2014; Reissner et al. 2014). Both changes, and the attendant impairment of glutamate turnover, have been linked to cocaine-seeking behavior. Interestingly, the GLT-1 downregulation is progressive with the length of cocaine administration as well as the length of the abstinence period (Fischer-Smith et al. 2012). The main features of glutamate transport dysregulation triggered by cocaine use, therefore, contrast with the pattern seen in the other reviewed substance abuse models: xCT and GLT-1 transport systems appear to be engaged in parallel pathways that drive the dependence behavior. Arriving at a unified concept of glutamate turnover dysregulation and the likely associated energy metabolism perturbations in substance use disorders will require a quantitative systems approach.

Genetic Models

Recent work has begun to address the role of metabolism and glial cells in animal models of major psychiatric disorders that incorporate genetic disruptions known to be associated with these diseases. Significant advances have been made, in particular, using the *DISC1* (disrupted in schizophrenia 1) model. Mutations in the *DISC1* gene are a rare but highly penetrant risk factor. The gene was originally identified as affected in a familial chromosomal translocation. In addition to schizophrenia, it is associated with major depression and bipolar disorder (Brandon and Sawa 2011; Millar et al. 2000; St Clair et al. 1990). The balanced (1:11) (q42.1; q14.3) translocation segregated with conditions that included schizophrenia, bipolar syndrome, and depression, in the affected family (Millar et al. 2001). Remarkably, *DISC1* was not found among the large number of loci linked to schizophrenia in a multistage genome-wide association study (Schizophrenia Working Group of the Psychiatric Genomics 2014). The totality of these data, therefore, suggests that *DISC1* may be a general psychiatric risk factor. Accordingly, mechanistic investigations of this factor could shed light on the pathophysiology of more than one condition (Farrell et al. 2015).

The translocation-related disruption results in truncated transcripts of the gene, and expression of various truncated forms of DISC1 can perturb cellular functions and brain development (Brandon and Sawa 2011; Kamiya et al. 2005). Animals expressing mutated DISC1 show behavioral phenotypes that can serve as models of schizophrenia, depression, and other major mental diseases (e.g., Abazyan et al. 2010; Abazyan et al. 2014; Ayhan et al. 2011; Clapcote et al. 2007; Lipina et al. 2013; Pletnikov et al. 2008; Shen et al. 2008; Sultana and Lee 2020). The study of the mechanisms underlying these abnormalities has highlighted a number of processes that are affected by DISC1 disruption in neurons (see, e.g., Pletnikov et al. 2007; Pletnikov et al. 2008; Jaaro-Peled et al. 2009; Brandon and Sawa 2011; Ramsey et al. 2011; Ayhan et al. 2011; Holley et al. 2013; Kim et al. 2015a; Shevelkin et al. 2017). The gene, however, is also expressed in glial cells including oligodendrocytes and astrocytes and can alter their development if disrupted (Hattori et al. 2014; Katsel et al. 2011b; Katsel et al. 2018; Kuroda et al. 2011; Seshadri et al. 2010). In addition, expression of mutant DISC1 in astrocytes can affect neuronal development, with associated behavioral effects (Terrillion et al. 2017). Thus, it is becoming clear that among the functions regulated by this pleiotropic gene, the ones mediated by its expression in astrocytes may be at the crossroads of pathways leading to major mental disease.

Recent work has begun to uncover the connections between the molecular role of the DISC1 risk factor and the role of the astrocyte energy metabolism in mental illness. Localization of *DISC1* products to mitochondria and their involvement in such functions as mitochondria trafficking and mitochondrial ATP production, oxidative phosphorylation, and calcium buffering were uncovered in the course of characterization of this gene in neurons and neuronal models (Eykelenboom et al. 2012; James et al. 2004; Ji et al. 2014; Millar et al. 2005; Norkett et al. 2016; Ogawa et al. 2014; Pinero-Martos et al. 2016), while proteomic data pinpointed involvement of DISC1 in regulation of expression of mitochondrial glucose-3-phosphate dehydrogenase in astrocytes (Xia et al. 2016). Importantly for the progress made in deciphering *DISC1* functions, astrocyte-specific expression of mutant DISC1 has the capacity to downregulate the endogenous wild-type gene in this cell type (Ma et al. 2013). The experiments also provided evidence that DISC1 binds serine racemase, while the mutant protein lacks this interaction, causing ubiquitin-mediated degradation of serine racemase in astrocytes. This is accompanied by behavioral changes characteristic of the NMDAR hypofunction models of mental disease, which could be explained by the well-documented role of D-serine as a co-ligand of NMDAR and astrocytes being its major source (e.g., Mothet et al. 2000; Panatier et al. 2006). Furthermore, when expressed in cultured astrocytes, a dominant-negative human variant of DISC1 suppresses the levels of glucose transporter 4; the same outcome can be obtained by knocking down the gene in mouse astrocytes (Jouroukhin et al. 2018). The induced changes lead to a pronounced inhibition of glucose uptake into the cells, as well as of glycolysis and oxidative phosphorylation. Both basal and maximal respiration are reduced, along with lactate production. The latter effect is not mediated by any downregulation of monocarboxylate transporter 4, indicating a purely intracellular metabolic-flux effect. The data point to downregulation of a series of steps in the glucose catabolism pathways, and a similar reduction of the lactate content could be observed in the hippocampus. In vivo, the effect on lactate is not accompanied by any change in the blood lactate concentration, supporting the notion of a local, astrocyte-dependent regulation by the dominant-negative DISC1. Strikingly, the behavioral alterations documented for astrocyte-restricted dominant negative DISC1 expression (Terrillion et al. 2017) can be rescued by administration of lactate to the mice (Jouroukhin et al. 2018). The results strongly suggest that DISC1 alters the energy metabolism in astrocytes to affect metabolite exchange with neurons, triggering the behavioral changes associated with mental disease etiology.

The implications of these findings for the role of the astrocyte energy metabolism are supported by our most recent study (Shevelkin et al. 2020). We showed that trace fear conditioning is impaired by downregulation of DISC1 in hippocampus, while no effect is observed in animals with an astrocyte-specific knockdown in the prefrontal cortex. Both spatial density of astrocytes and morphology of their protrusions are perturbed by the knockdown, as remarkably are the expression and localization of numerous proteins associated with energy metabolism. We demonstrated that pyruvate dehydrogenase and translocase of the mitochondrial inner membrane exhibit a more localized distribution in the knockdown mice. These results are consistent with DISC1 belonging to the mammalian mitochondrial contact site and cristae organizing system and being essential for the processes of oxidative phosphorylation (Park et al. 2010; Pinero-Martos et al. 2016). Additionally, GLAST becomes positioned closer to several mitochondrial markers (Shevelkin et al. 2020), which is suggestive in light of other recent findings (Jackson et al. 2014) that showed that GLAST proximity to mitochondria regulates glutamate uptake. Furthermore, nonapoptotic caspase 3 activation was detected in neurons within the territories of affected astrocytes, and these changes were associated with perturbed glutamatergic and GABAergic markers. Thus, in addition to arguing in favor of the notion that astrocyte energy metabolism is crucial to the etiology of major psychiatric disorders, these findings suggest specific new mechanisms of astrocyte-neuron cooperation, whose perturbations may underlie development of schizophrenia and other mental illnesses. At the same time, the picture remains far from complete, and new work will be required to fill such obvious gaps as the potential role of mitochondria redistribution in the cell, which may impact the metabolic flux locally without gross changes of enzyme expression. Another topic that deserves attention is how the regulation of the overall energy flux may be affected by intracellular calcium dynamics, which is likely to act in a feedback or feed-forward manner, being modulated by the state of polarization of the mitochondrial membrane while controlling both glucose and oxygen supply.

The involvement of DISC1 in neurobehavioral abnormalities observed in substance use disorder models is starting to provide additional clues about the mechanism of its action in the affected brain. In connection with the involvement of the serine oxidation metabolism in the nucleus accumbens (reviewed in the previous subsection), it is remarkable that self-administration of cocaine in the rat model upregulates expression of DISC1 in this brain region (Gancarz et al. 2016). This finding is especially suggestive when juxtaposed with the long-known accumulation of a nuclear isoform of DISC1 in the orbitofrontal cortex of patients suffering not only from schizophrenia or major depression but also from substance abuse (Sawamura et al. 2005). Conversely, mutant DISC1 expression attenuates methamphetamine-induced behavioral sensitization in mice (Pogorelov et al. 2012) while mimicking and potentiating the cannabinoid receptor 1 downregulating effect of tetrahydrocannabinol treatment (Ballinger et al. 2015). Collectively, these findings open the possibility of involvement of the astrocyte metabolism in the molecular factors contributing to the etiology of a spectrum of substance use disorders, given the intersection of serine and energy metabolism. It is also hypothetically possible that the perturbations of glycogen synthase kinase 3β activity in the nucleus accumbens, which has been detected in the cited studies, have more direct implications for astrocyte energy metabolism. Additional work is needed to probe the potential connections between expression of the DISC1 forms, metabolism, and the mechanisms driving substance abuse, as well as any generalizations between them and the findings that have recently implicated the DISC1 risk factor in the astrocyte energy metabolism contribution to other major psychiatric disorders.

Less is known, by comparison, about the role played by metabolic alterations in other genetic models. However, the recent data are indicative of their possible involvement and should motivate further targeted investigation of these questions. So, for example, knockout of STOP, a microtubule-stabilizing protein known for its involvement in glutamate turnover and association with cognitive defects, is accompanied by well-defined metabolic changes (Hanaya et al. 2008; Volle et al. 2013). While glucose utilization in this model is increased in the olfactory cortex, ventromedial and anterolateral hypothalamus, ventral tegmental area, and substantia nigra pars compacta, both total glutamine and glutamine derived from the preferred astrocyte substrate, acetate, are decreased (Brenner et al. 2007). These findings are strongly suggestive of involvement of astrocyte metabolism in hyperdopaminergia characterizing this mouse model and its attendant major behavioral changes. Further work is needed to clarify this apparent relationship. This need is only more urgent in light of the recent demonstration of sleep-wake cycle disruptions in the STOP model, which correlate with those seen in schizophrenia patients (Pocivavsek and Rowland 2018; Profitt et al. 2016).

Recent results concerning the neuregulin 1 model (Buonanno 2010; Law et al. 2004; Shi and Bergson 2020; Skirzewski et al. 2020) similarly raise the possibility that astrocyte energy metabolism may be involved. Indeed, neuregulin family ligands, including neuregulin 1, are potent modulators of glucose and lipid metabolism in various tissues (see Zhang et al. 2018). Although the best-studied molecular mechanism in the neuregulin 1 model of schizophrenia is the neuronal plasticity pathway, it can be noted that astrocytes express both neuregulin 1 and its receptors, including ErB4 receptor widely implicated in metabolic regulation in other tissues (Kronenberg et al. 2019; Lacroix-Fralish et al. 2006). Elucidation of the metabolic impact of neuregulin signaling in astrocytes, and its connection to the characteristic symptoms modeled by neuregulin 1 insufficiency, can represent a promising direction of research.

An additional line of investigation may be suggested by findings related to the reelin (*Reln*) model of schizophrenia (Sawahata et al. 2020; Sobue et al. 2018). In connection with this model, it is of interest that region- and layer-specific, bidirectional alteration of cytochrome oxidase activity has been observed in the brain of *reeler* (*Reln*$^{-/-}$) mice (Hayzoun et al. 2007; Lalonde et al. 2004; Strazielle et al. 2006). Supporting this picture, similar results have been reported in mice mutant for disabled-1 (DAB1) protein, a downstream partner of *Reln* (Jacquelin et al. 2013). The possibility that some of these changes are mediated by metabolic effects in astroglia appears plausible, given the recently documented expression of reelin in astrocytes (Zeisel et al. 2018) and fine but specific morphological alterations displayed by astrocytes of *reeler* mice (Lanjakornsiripan et al. 2018). Whether astrocyte energy metabolism is a contributing factor to the behavioral abnormalities of the *reeler* phenotype remains to be determined.

Therapeutic Possibilities

While it is probably correct to characterize the current state of developing treatments that would target astrocyte energy metabolism as nascent, at least two therapeutic avenues are being pursued. We have noted that obesity and the metabolic syndrome are frequent comorbidities with neuropsychiatric conditions. In this regard, a regimen of ketogenic diet or exogenous ketone supplementation to alleviate the symptoms of schizophrenia and other major psychiatric disorders is a noteworthy effort (Kovacs et al. 2019; Sarnyai and Palmer 2020). Ketone bodies are produced by the liver and in the brain also by astrocytes (Guzman and Blazquez 2004; Le Foll and Levin 2016). In addition to this direct involvement in ketogenesis, astrocytes' metabolism is changed in response to ketosis, as demonstrated by the finding that β-hydroxybutyrate downregulates the expression of both astrocytic GABA transporter GAT-1 and GABA transaminase, which can lead to an increased GABA concentration in the brain (Suzuki et al. 2009). Interestingly, another study found that in addition to GABA, the turnover of glutamate and glutamine, as well as aspartate in astrocytes, was increased in rats on a ketogenic diet (Melo et al. 2006). Astrocyte morphology is also significantly altered (Gzielo et al. 2019). Demonstration of clinical efficacy has been limited to an uncontrolled study and a number of individual cases, which, perhaps tellingly, include a complete remission of severe schizophrenia accompanied by reversal of severe obesity (Sarnyai et al. 2019; Sarnyai and Palmer 2020). The promise of these cases is supported by reversal of behavioral deficits in the MK-801 mouse model after 3 weeks of ketogenic diet (Kraeuter et al. 2019). A broadly analogous picture is presented by the existing experience with ketogenic diet in autism spectrum disorder patients (Evangeliou et al. 2003; Gogou and Kolios 2018; Spilioti et al. 2013) and rodent models (Ahn et al. 2014; Castro et al. 2017; Ruskin et al. 2013; Ruskin et al. 2017; Verpeut et al. 2016), where it can reduce stereotypic behavior and improve sociability. Evidence is growing that depression may respond to ketogenic diet similarly (Brietzke et al. 2018; Sussman et al. 2015), while limited data concerning bipolar disorder are mixed (Bostock et al. 2017). Additional studies are clearly warranted that would clarify the benefits of a tailored diet regimen and the role played by the astrocyte energy metabolism in any beneficial effects.

A promising medication-based direction utilizing the insights into brain metabolism to counter neuropsychiatric symptoms involves the use of pioglitazone and related thiazolidinedione compounds. Although pioglitazone is known as an agonist of peroxisome proliferator-activated receptor γ (PPARγ), there is pharmacological evidence that its mechanism of action, and that of the closely related compounds, in brain metabolism may not be mediated by PPAR (Dello Russo et al. 2003). The effect of pioglitazone on astrocytes involves cAMP and protein kinase A and comprises hyperpolarization of mitochondria as well as an increased glucose uptake through GLUT1 in the absence of this transporter's overexpression. Recent studies have found pioglitazone effective as adjuvant therapy with regard to negative symptoms in schizophrenia (Cakici et al. 2019; Iranpour et al. 2016). Detailed evaluation showed that depression, but not cognitive score, could be improved, along with organism-level metabolic parameters such as blood glucose in patients with the common comorbidity affecting this parameter (Smith et al. 2013). Consistent with these results, pioglitazone and another PPARγ agonist, rosiglitazone, were found effective in a number of trials involving patients with major depression (Colle et al. 2017; Kashani et al. 2013; Kemp et al. 2012). Furthermore, depressive and anxiety symptoms in patients with bipolar disorder who exhibited treatment-resistant depression in addition to metabolic syndrome could be relieved by adjunctive administration of pioglitazone (Kemp et al. 2014). However, the results of pioglitazone treatment in cohorts not exhibiting or not selected for the presence of metabolic syndrome—a frequent co-morbidity—have been mixed (Aftab et al. 2019; Zeinoddini et al. 2015). Also, irritability, stereotypical behavior, and hyperactivity symptoms were improved in studies that tested pioglitazone in children with autism (Boris et al. 2007; Ghaleiha et al. 2015), although it is possible that these effects are mediated by anti-inflammatory action of the thiazolidinediones (Hafizi et al. 2019). Collectively, these clinical findings have led to the current concept that PPAR agonists hold out great promise in treatment of a number of neuropsychiatric disorders (Tufano and Pinna 2020). In addition to representing an application of the studies of astrocyte metabolism to the clinic, future work on PPAR agonist-based therapy may yield further insight into the role of energy metabolism in determining the clinical course of these psychiatric conditions.

Outlook

The study of the role played by the astrocyte energy metabolism in etiology of major psychiatric conditions is experiencing an acceleration brought about by accumulation of a critical mass of results. The pathways of neuron-astrocyte metabolic cooperation are being disentangled with the use of new research technologies, and the place of the astrocytes in metabolic signaling to and from the brain is being assessed. Among the outstanding fundamental questions is not only the clarification of the intensity but also the directionality of metabolite exchange in the brain tissue under the conditions of upregulated metabolism in sustained cognitive tasks and how their disruption accompanies cognitive abnormalities characteristic of mental disease. It will be necessary to more fully integrate our understanding of genetic risk factors in major psychiatric disorders with the picture being painted by metabolomic and functional assessment of the affected brain and individual cell types, particularly glia and astrocytes. Animal models and in vitro research have the potential to uncover the missing links between genotype and the pathological behavioral phenotype. As these studies are beginning to indicate promising avenues for therapeutic treatment of the most severe symptoms, much more work is needed to arrive at a system-level understanding of the energy metabolic processes that implicate astrocytes in major mental disorders.

References

Abazyan B, Dziedzic J, Hua KG, Abazyan S, Yang CX, Mori S, Pletnikov MV, Guilarte TR (2014) Chronic exposure of mutant DISC1 mice to lead produces sex-dependent abnormalities consistent with schizophrenia and related mental disorders: a gene-environment interaction study. Schizophr Bull 40(3):575–584. https://doi.org/10.1093/schbul/sbt071

Abazyan B, Nomura J, Kannan G, Ishizuka K, Tamashiro KL, Nucifora F, Pogorelov V, Ladenheim B, Yang CX, Krasnova IN, Cadet JL, Pardo C, Mori S, Kamiya A, Vogel MW, Sawa A, Ross CA, Pletnikov MV (2010) Prenatal interaction of mutant DISC1 and immune activation produces adult psychopathology. Biol Psychiatry 68(12):1172–1181. https://doi.org/10.1016/j.biopsych.2010.09.022

Afridi R, Kim JH, Rahman MH, Suk K (2020) Metabolic regulation of glial phenotypes: implications in neuron-glia interactions and neurological disorders. Front Cell Neurosci 14:20. https://doi.org/10.3389/fncel.2020.00020

Aftab A, Kemp DE, Ganocy SJ, Schinagle M, Conroy C, Brownrigg B, D'Arcangelo N, Goto T, Woods N, Serrano MB, Han H, Calabrese JR, Gao K (2019) Double-blind, placebo-controlled trial of pioglitazone for bipolar depression. J Affect Disord 245:957–964. https://doi.org/10.1016/j.jad.2018.11.090

Agarwal A, Wu PH, Hughes EG, Fukaya M, Tischfield MA, Langseth AJ, Wirtz D, Bergles DE (2017) Transient opening of the mitochondrial permeability transition pore induces microdomain calcium transients in astrocyte processes. Neuron 93(3):587–605. e587. https://doi.org/10.1016/j.neuron.2016.12.034

Ahn Y, Narous M, Tobias R, Rho JM, Mychasiuk R (2014) The ketogenic diet modifies social and metabolic alterations identified in the prenatal valproic acid model of autism spectrum disorder. Dev Neurosci 36(5):371–380. https://doi.org/10.1159/000362645

Alasmari F, Rao PSS, Sari Y (2016) Effects of cefazolin and cefoperazone on glutamate transporter 1 isoforms and cystine/glutamate exchanger as well as alcohol drinking behavior in male alcohol-preferring rats. Brain Res 1634:150–157. https://doi.org/10.1016/j.brainres.2016.01.007

Allaman I, Fiumelli H, Magistretti PJ, Martin JL (2011) Fluoxetine regulates the expression of neurotrophic/growth factors and glucose metabolism in astrocytes. Psychopharmacology 216(1):75–84. https://doi.org/10.1007/s00213-011-2190-y

Allard C, Carneiro L, Grall S, Cline BH, Fioramonti X, Chretien C, Baba-Aissa F, Giaume C, Penicaud L, Leloup C (2014) Hypothalamic astroglial connexins are required for brain glucose sensing-induced insulin secretion. J Cereb Blood Flow Metab 34(2):339–346. https://doi.org/10.1038/jcbfm.2013.206

Almeida AS, Queiroga CS, Sousa MF, Alves PM, Vieira HL (2012) Carbon monoxide modulates apoptosis by reinforcing oxidative metabolism in astrocytes: role of Bcl-2. J Biol Chem 287(14):10761–10770. https://doi.org/10.1074/jbc.M111.306738

Almeida AS, Sonnewald U, Alves PM, Vieira HL (2016) Carbon monoxide improves neuronal differentiation and yield by increasing the functioning and number of mitochondria. J Neurochem 138(3):423–435. https://doi.org/10.1111/jnc.13653

Altar CA, Jurata LW, Charles V, Lemire A, Liu P, Bukhman Y, Young TA, Bullard J, Yokoe H, Webster MJ, Knable MB, Brockman JA (2005) Deficient hippocampal neuron expression of proteasome, ubiquitin, and mitochondrial genes in multiple schizophrenia cohorts. Biol Psychiatry 58(2):85–96. https://doi.org/10.1016/j.biopsych.2005.03.031

Ament SA, Szelinger S, Glusman G, Ashworth J, Hou LP, Akula N, Shekhtman T, Badner JA, Brunkow ME, Mauldin DE, Stittrich AB, Rouleau K, Detera-Wadleigh SD, Nurnberger JI, Edenberg HJ, Gershon ES, Schork N, Price ND, Gelinas R, Hood L, Craig D, McMahon FJ, Kelsoe JR, Roach JC, Study BG (2015) Rare variants in neuronal excitability genes influence risk for bipolar disorder. Proc Natl Acad Sci U S A 112(11):3576–3581. https://doi.org/10.1073/pnas.1424958112

Annamalai A, Kosir U, Tek C (2017) Prevalence of obesity and diabetes in patients with schizophrenia. World J Diabetes 8(8):390–396. https://doi.org/10.4239/wjd.v8.i8.390

Arion D, Corradi JP, Tang S, Datta D, Boothe F, He A, Cacace AM, Zaczek R, Albright CF, Tseng G, Lewis DA (2015) Distinctive transcriptome alterations of prefrontal pyramidal neurons in schizophrenia and schizoaffective disorder. Mol Psychiatry 20(11):1397–1405. https://doi.org/10.1038/mp.2014.171

Arion D, Huo Z, Enwright JF, Corradi JP, Tseng G, Lewis DA (2017) Transcriptome alterations in prefrontal pyramidal cells distinguish schizophrenia from bipolar and major depressive disorders. Biol Psychiatry 82(8):594–600. https://doi.org/10.1016/j.biopsych.2017.03.018

Arizono M, Inavalli V, Panatier A, Pfeiffer T, Angibaud J, Levet F, Ter Veer MJT, Stobart J, Bellocchio L, Mikoshiba K, Marsicano G, Weber B, Oliet SHR, Nagerl UV (2020) Structural basis of astrocytic ca(2+) signals at tripartite synapses. Nat Commun 11(1):1906. https://doi.org/10.1038/s41467-020-15648-4

Ashrafi G, Wu Z, Farrell RJ, Ryan TA (2017) GLUT4 mobilization supports energetic demands of active synapses. Neuron 93(3):606–615. e603. https://doi.org/10.1016/j.neuron.2016.12.020

Augusto-Oliveira M, Arrifano GP, Takeda PY, Lopes-Araujo A, Santos-Sacramento L, Anthony DC, Verkhratsky A, Crespo-Lopez ME (2020) Astroglia-specific contributions to the regulation of synapses, cognition and behaviour. Neurosci Biobehav Rev 118:331–357. https://doi.org/10.1016/j.neubiorev.2020.07.039

Ayerbe L, Forgnone I, Addo J, Siguero A, Gelati S, Ayis S (2018) Hypertension risk and clinical care in patients with bipolar disorder or schizophrenia; a systematic review and meta-analysis. J Affect Disord 225:665–670. https://doi.org/10.1016/j.jad.2017.09.002

Ayhan Y, Abazyan B, Nomura J, Kim R, Ladenheim B, Krasnova IN, Sawa A, Margolis RL, Cadet JL, Mori S, Vogel MW, Ross CA, Pletnikov MV (2011) Differential effects of prenatal and postnatal expressions of mutant human DISC1 on neurobehavioral phenotypes in transgenic mice: evidence for neurodevelopmental origin of major psychiatric disorders. Mol Psychiatry 16(3):293–306. https://doi.org/10.1038/mp.2009.144

Bailey AP, Koster G, Guillermier C, Hirst EM, MacRae JI, Lechene CP, Postle AD, Gould AP (2015) Antioxidant role for lipid droplets in a stem cell niche of drosophila. Cell 163(2):340–353. https://doi.org/10.1016/j.cell.2015.09.020

Bak LK, Schousboe A, Sonnewald U, Waagepetersen HS (2006) Glucose is necessary to maintain neurotransmitter homeostasis during synaptic activity in cultured glutamatergic neurons. J Cereb Blood Flow Metab 26(10):1285–1297. https://doi.org/10.1038/sj.jcbfm.9600281

Baker KD, Edwards TM (2007) D-Lactate inhibition of memory in a single trial discrimination avoidance task in the young chick. Neurobiol Learn Mem 88(3):269–276. https://doi.org/10.1016/j.nlm.2007.06.004

Ballinger MD, Saito A, Abazyan B, Taniguchi Y, Huang CH, Ito K, Zhu XL, Segal H, Jaaro-Peled H, Sawa A, Mackie K, Pletnikov MV, Kamiya A (2015) Adolescent cannabis exposure interacts with mutant DISC1 to produce impaired adult emotional memory. Neurobiol Dis 82:176–184. https://doi.org/10.1016/j.nbd.2015.06.006

Barros LF, Brown A, Swanson RA (2018) Glia in brain energy metabolism: a perspective. Glia 66(6):1134–1137. https://doi.org/10.1002/glia.23316

Beasley CL, Dwork AJ, Rosoklija G, Mann JJ, Mancevski B, Jakovski Z, Davceva N, Tait AR, Straus SK, Honer WG (2009) Metabolic abnormalities in fronto-striatal-thalamic white matter tracts in schizophrenia. Schizophr Res 109(1–3):159–166. https://doi.org/10.1016/j.schres.2009.01.017

Beasley CL, Pennington K, Behan A, Wait R, Dunn MJ, Cotter D (2006) Proteomic analysis of the anterior cingulate cortex in the major psychiatric disorders: evidence for disease-associated changes. Proteomics 6(11):3414–3425. https://doi.org/10.1002/pmic.200500069

Beckel JM, Gomez NM, Lu WN, Campagno KE, Nabet B, Albalawi F, Lim JC, Boesze-Battaglia K, Mitchell CH (2018) Stimulation of TLR3 triggers release of lysosomal ATP in astrocytes and epithelial cells that requires TRPML1 channels. Sci Rep 8:5726. https://doi.org/10.1038/s41598-018-23877-3

Bekar LK, He W, Nedergaard M (2008) Locus coeruleus alpha-adrenergic-mediated activation of cortical astrocytes in vivo. Cereb Cortex 18(12):2789–2795. https://doi.org/10.1093/cercor/bhn040

Bellaver B, Souza DG, Bobermin LD, Souza DO, Goncalves CA, Quincozes-Santos A (2015) Resveratrol protects hippocampal astrocytes against LPS-induced neurotoxicity through HO-1, p38 and ERK pathways. Neurochem Res 40(8):1600–1608. https://doi.org/10.1007/s11064-015-1636-8

Ben-Shachar D (2002) Mitochondrial dysfunction in schizophrenia: a possible linkage to dopamine. J Neurochem 83(6):1241–1251. https://doi.org/10.1046/j.1471-4159.2002.01263.x

Bergersen L, Waerhaug O, Helm J, Thomas M, Laake P, Davies AJ, Wilson MC, Halestrap AP, Ottersen OP (2001) A novel postsynaptic density protein: the monocarboxylate transporter MCT2 is co-localized with delta-glutamate receptors in postsynaptic densities of parallel fiber-Purkinje cell synapses. Exp Brain Res 136(4):523–534. https://doi.org/10.1007/s002210000600

Bernard R, Kerman IA, Thompson RC, Jones EG, Bunney WE, Barchas JD, Schatzberg AF, Myers RM, Akil H, Watson SJ (2011) Altered expression of glutamate signaling, growth factor, and glia genes in the locus coeruleus of patients with major depression. Mol Psychiatry 16(6):634–646. https://doi.org/10.1038/mp.2010.44

Bezzi P, Gundersen V, Galbete JL, Seifert G, Steinhauser C, Pilati E, Volterra A (2004) Astrocytes contain a vesicular compartment that is competent for regulated exocytosis of glutamate. Nat Neurosci 7(6):613–620. https://doi.org/10.1038/nn1246

Bistoletti M, Caputi V, Baranzini N, Marchesi N, Filpa V, Marsilio I, Cerantola S, Terova G, Baj A, Grimaldi A, Pascale A, Frigo G, Crema F, Giron MC, Giaroni C (2019) Antibiotic treatment-induced dysbiosis differently affects BDNF and TrkB expression in the brain and in the gut of juvenile mice. Plos One 14(2):e0212856. https://doi.org/10.1371/journal.pone.0212856

Blomstrand F, Aberg ND, Eriksson PS, Hansson E, Ronnback L (1999) Extent of intercellular calcium wave propagation is related to gap junction permeability and level of connexin-43 expression in astrocytes in primary cultures from four brain regions. Neuroscience 92(1):255–265. https://doi.org/10.1016/s0306-4522(98)00738-6

Boesmans W, Hao MM, Berghe PV (2018) Optogenetic and chemogenetic techniques for neurogastroenterology. Nat Rev Gastroenterol Hepatol 15(1):21–38. https://doi.org/10.1038/nrgastro.2017.151

Boitier E, Rea R, Duchen MR (1999) Mitochondria exert a negative feedback on the propagation of intracellular Ca2+ waves in rat cortical astrocytes. J Cell Biol 145(4):795–808. https://doi.org/10.1083/jcb.145.4.795

Bolanos JP (2016) Bioenergetics and redox adaptations of astrocytes to neuronal activity. J Neurochem 139(Suppl 2):115–125. https://doi.org/10.1111/jnc.13486

Boris M, Kaiser CC, Goldblatt A, Elice MW, Edelson SM, Adams JB, Feinstein DL (2007) Effect of pioglitazone treatment on behavioral symptoms in autistic children. J Neuroinflamm 4:3. https://doi.org/10.1186/1742-2094-4-3

Bosier B, Bellocchio L, Metna-Laurent M, Soria-Gomez E, Matias I, Hebert-Chatelain E, Cannich A, Maitre M, Leste-Lasserre T, Cardinal P, Mendizabal-Zubiaga J, Canduela MJ, Reguero L, Hermans E, Grandes P, Cota D, Marsicano G (2013) Astroglial CB1 cannabinoid receptors regulate leptin signaling in mouse brain astrocytes. Mol Metab 2(4):393–404. https://doi.org/10.1016/j.molmet.2013.08.001

Bostock ECS, Kirkby KC, Taylor BVM (2017) The current status of the ketogenic diet in psychiatry. Front Psych 8:43. https://doi.org/10.3389/fpsyt.2017.00043

Bouzier-Sore AK, Pellerin L (2013) Unraveling the complex metabolic nature of astrocytes. Front Cell Neurosci 7:179. https://doi.org/10.3389/fncel.2013.00179

Brandon NJ, Sawa A (2011) Linking neurodevelopmental and synaptic theories of mental illness through DISC1. Nat Rev Neurosci 12(12):707–722. https://doi.org/10.1038/nrn3120

Brenner E, Sonnewald U, Schweitzer A, Andrieux A, Nehlig A (2007) Hypoglutamatergic activity in the STOP knockout mouse: a potential model for chronic untreated schizophrenia. J Neurosci Res 85(15):3487–3493. https://doi.org/10.1002/jnr.21200

Brietzke E, Mansur RB, Subramaniapillai M, Balanza-Martinez V, Vinberg M, Gonzalez-Pinto A, Rosenblat JD, Ho R, McIntyre RS (2018) Ketogenic diet as a metabolic therapy for mood disorders: evidence and developments. Neurosci Biobehav Rev 94:11–16. https://doi.org/10.1016/j.neubiorev.2018.07.020

Brisch R, Saniotis A, Wolf R, Bielau H, Bernstein HG, Steiner J, Bogerts B, Braun K, Jankowski Z, Kumaratilake J, Henneberg M, Gos T (2014) The role of dopamine in schizophrenia from a neurobiological and evolutionary perspective: old fashioned, but still in vogue. Front Psych 5:47. https://doi.org/10.3389/fpsyt.2014.00047

Bruce-Keller AJ, Salbaum JM, Berthoud HR (2018) Harnessing gut microbes for mental health: getting from here to there. Biol Psychiatry 83(3):214–223. https://doi.org/10.1016/j.biopsych.2017.08.014

Brzustowicz LM, Hodgkinson KA, Chow EW, Honer WG, Bassett AS (2000) Location of a major susceptibility locus for familial schizophrenia on chromosome 1q21-q22. Science 288(5466):678–682. https://doi.org/10.1126/science.288.5466.678

Bubber P, Hartounian V, Gibson GE, Blass JP (2011) Abnormalities in the tricarboxylic acid (TCA) cycle in the brains of schizophrenia patients. Eur Neuropsychopharmacol 21(3):254–260. https://doi.org/10.1016/j.euroneuro.2010.10.007

Bull C, Freitas KCC, Zou SP, Poland RS, Syed WA, Urban DJ, Minter SC, Shelton KL, Hauser KF, Negus SS, Knapp PE, Bowers MS (2014) Rat nucleus accumbens core astrocytes modulate reward and the motivation to self-administer ethanol after abstinence. Neuropsychopharmacology 39(12):2835–2845. https://doi.org/10.1038/npp.2014.135

Buonanno A (2010) The neuregulin signaling pathway and schizophrenia: from genes to synapses and neural circuits. Brain Res Bull 83(3–4):122–131. https://doi.org/10.1016/j.brainresbull.2010.07.012

Cabodevilla AG, Sanchez-Caballero L, Nintou E, Boiadjieva VG, Picatoste F, Gubern A, Claro E (2013) Cell survival during complete nutrient deprivation depends on lipid droplet-fueled beta-oxidation of fatty acids. J Biol Chem 288(39):27777–27788. https://doi.org/10.1074/jbc.M113.466656

Cakici N, van Beveren NJM, Judge-Hundal G, Koola MM, Sommer IEC (2019) An update on the efficacy of anti-inflammatory agents for patients with schizophrenia: a meta-analysis. Psychol Med 49(14):2307–2319. https://doi.org/10.1017/S0033291719001995

Camandola S (2018) Astrocytes, emerging stars of energy homeostasis. Cell Stress 2(10):246–252. https://doi.org/10.15698/cst2018.10.157

Careaga M, Murai T, Bauman MD (2017) Maternal immune activation and autism spectrum disorder: from rodents to nonhuman and human primates. Biol Psychiatry 81(5):391–401. https://doi.org/10.1016/j.biopsych.2016.10.020

Castro K, Baronio D, Perry IS, Riesgo RD, Gottfried C (2017) The effect of ketogenic diet in an animal model of autism induced by prenatal exposure to valproic acid. Nutr Neurosci 20(6):343–350. https://doi.org/10.1080/1028415x.2015.1133029

Cavelier L, Jazin EE, Eriksson I, Prince J, Bave U, Oreland L, Gyllensten U (1995) Decreased cytochrome-C-oxidase activity and lack of age-related accumulation of mitochondrial-DNA deletions in the brains of schizophrenics. Genomics 29(1):217–224. https://doi.org/10.1006/geno.1995.1234

Chari M, Yang CS, Lam CK, Lee K, Mighiu P, Kokorovic A, Cheung GW, Lai TY, Wang PY, Lam TK (2011) Glucose transporter-1 in the hypothalamic glial cells mediates glucose sensing to regulate glucose production in vivo. Diabetes 60(7):1901–1906. https://doi.org/10.2337/db11-0120

Chattopadhyay S, Sen GC (2014) dsRNA-activation of TLR3 and RLR signaling: gene induction-dependent and independent effects. J Interferon Cytokine Res 34(6):427–436. https://doi.org/10.1089/jir.2014.0034

Chen DC, Du XD, Yin GZ, Yang KB, Nie Y, Wang N, Li YL, Xiu MH, He SC, Yang FD, Cho RY, Kosten TR, Soares JC, Zhao JP, Zhang XY (2016a) Impaired glucose tolerance in first-episode drug-naive patients with schizophrenia: relationships with clinical phenotypes and cognitive deficits. Psychol Med 46(15):3219–3230. https://doi.org/10.1017/S0033291716001902

Chen Y, Du T, Peng L, Gibbs ME, Hertz L (2016b) Sequential astrocytic 5-HT2B receptor stimulation, [Ca2+](i) regulation, glycogenolysis, glutamate synthesis, and K+ homeostasis are similar but not identical in learning and mood regulation. Front Integr Neurosc 9:67. https://doi.org/10.3389/fnint.2015.00067

Chen Y, Peng LA, Zhang XH, Stolzenburg JU, Hertz L (1995) Further evidence that fluoxetine interacts with a 5-Ht2c receptor in glial-cells. Brain Res Bull 38(2):153–159. https://doi.org/10.1016/0361-9230(95)00082-P

Chiry O, Fishbein WN, Merezhinskaya N, Clarke S, Galuske R, Magistretti PJ, Pellerin L (2008) Distribution of the monocarboxylate transporter MCT2 in human cerebral cortex: an immunohistochemical study. Brain Res 1226:61–69. https://doi.org/10.1016/j.brainres.2008.06.025

Chistyakov DV, Azbukina NV, Astakhova AA, Polozhintsev AI, Sergeeva MG, Reiser G (2019) Toll-like receptors control p38 and JNK MAPK signaling pathways in rat astrocytes differently, when cultured in normal or high glucose concentrations. Neurochem Int 131:104513. https://doi.org/10.1016/j.neuint.2019.104513

Chistyakov DV, Azbukina NV, Lopachev AV, Kulichenkova KN, Astakhova AA, Sergeeva MG (2018) Rosiglitazone as a modulator of TLR4 and TLR3 signaling pathways in rat primary neurons and astrocytes. Int J Mol Sci 19(1):113. https://doi.org/10.3390/ijms19010113

Choudary PV, Molnar M, Evans SJ, Tomita H, Li JZ, Vawter MP, Myers RM, Bunney WE Jr, Akil H, Watson SJ, Jones EG (2005) Altered cortical glutamatergic and GABAergic signal transmission with glial involvement in depression. Proc Natl Acad Sci U S A 102(43):15653–15658. https://doi.org/10.1073/pnas.0507901102

Chouinard VA, Pingali SM, Chouinard G, Henderson DC, Mallya SG, Cypess AM, Cohen BM, Ongur D (2016) Factors associated with overweight and obesity in schizophrenia, schizoaffective and bipolar disorders. Psychiatry Res 237:304–310. https://doi.org/10.1016/j.psychres.2016.01.024

Chuquet J, Hollender L, Nimchinsky EA (2007) High-resolution in vivo imaging of the neurovascular unit during spreading depression. J Neurosci 27(15):4036–4044. https://doi.org/10.1523/JNEUROSCI.0721-07.2007

Clapcote SJ, Lipina TV, Millar JK, Mackie S, Christie S, Ogawa F, Lerch JP, Trimble K, Uchiyama M, Sakuraba Y, Kaneda H, Shiroishi T, Houslay MD, Henkelman RM, Sled JG, Gondo Y, Porteous DJ, Roder JC (2007) Behavioral phenotypes of Disc1 missense mutations in mice. Neuron 54(3):387–402. https://doi.org/10.1016/j.neuron.2007.04.015

Clarke P, Zhuang YH, Berens HM, Leser JS, Tyler KL (2019) Interferon beta contributes to astrocyte activation in the brain following reovirus infection. J Virol 93(10):e02027–e02018. https://doi.org/10.1128/JVI.02027-18

Colle R, de Larminat D, Rotenberg S, Hozer F, Hardy P, Verstuyft C, Feve B, Corruble E (2017) PPAR-gamma agonists for the treatment of major depression: a review. Pharmacopsychiatry 50(2):49–55. https://doi.org/10.1055/s-0042-120120

Contreras-Shannon V, Heart DL, Paredes RM, Navaira E, Catano G, Maffi SK, Walss-Bass C (2013) Clozapine-induced mitochondria alterations and inflammation in brain and insulin-responsive cells. PLoS One 8(3):e59012. https://doi.org/10.1371/journal.pone.0059012

Conway F, Brown AS (2019) Maternal immune activation and related factors in the risk of offspring psychiatric disorders. Front Psych 10:430. https://doi.org/10.3389/fpsyt.2019.00430

Cooper AJ, Jeitner TM (2016) Central role of glutamate metabolism in the maintenance of nitrogen homeostasis in normal and hyperammonemic brain. Biomol Ther 6(2):16. https://doi.org/10.3390/biom6020016

Cruz NF, Dienel GA (2002) High glycogen levels in brains of rats with minimal environmental stimuli: implications for metabolic contributions of working astrocytes. J Cereb Blood Flow Metab 22(12):1476–1489. https://doi.org/10.1097/01.WCB.0000034362.37277.C0

Cryan JF, Dinan TG (2015) GUT MICROBIOTA Microbiota and neuroimmune signalling-Metchnikoff to microglia. Nat Rev Gastroenterol Hepatol 12(9):495–496. https://doi.org/10.1038/nrgastro.2015.127

Darvesh AS, Gudelsky GA (2003) Activation of 5-HT2 receptors induces glycogenolysis in the rat brain. Eur J Pharmacol 464(2–3):135–140. https://doi.org/10.1016/S0014-2999(03)01432-8

Das SC, Yamamoto BK, Hristov AM, Sari Y (2015) Ceftriaxone attenuates ethanol drinking and restores extracellular glutamate concentration through normalization of GLT-1 in nucleus accumbens of male alcohol-preferring rats. Neuropharmacology 97:67–74. https://doi.org/10.1016/j.neuropharm.2015.05.009

De Vedder F, Grasset E, Holm LM, Karsenty G, Macpherson AJ, Olofsson LE, Backhed F (2018) Gut microbiota regulates maturation of the adult enteric nervous system via enteric serotonin networks. Proc Natl Acad Sci U S A 115(25):6458–6463. https://doi.org/10.1073/pnas.1720017115

Dean B, Thomas N, Scarr E, Udawela M (2016) Evidence for impaired glucose metabolism in the striatum, obtained postmortem, from some subjects with schizophrenia. Transl Psychiat 6(11):e949. https://doi.org/10.1038/tp.2016.226

Debarba LK, Vechiato FMV, Veida-Silva H, Borges BC, Jamur MC, Antunes-Rodrigues J, Elias LLK (2019) The role of TCPTP on leptin effects on astrocyte morphology. Mol Cell Endocrinol 482:62–69. https://doi.org/10.1016/j.mce.2018.12.010

Deitmer JW, Theparambil SM, Ruminot I, Noor SI, Becker HM (2019) Energy dynamics in the brain: contributions of astrocytes to metabolism and pH homeostasis. Front Neurosci 13:1301. https://doi.org/10.3389/fnins.2019.01301

Dello Russo C, Gavrilyuk V, Weinberg G, Almeida A, Bolanos JP, Palmer J, Pelligrino D, Galea E, Feinstein DL (2003) Peroxisome proliferator-activated receptor gamma thiazolidinedione agonists increase glucose metabolism in astrocytes. J Biol Chem 278(8):5828–5836. https://doi.org/10.1074/jbc.M208132200

Diaz-Garcia CM, Mongeon R, Lahmann C, Koveal D, Zucker H, Yellen G (2017) Neuronal stimulation triggers neuronal glycolysis and not lactate uptake. Cell Metab 26(2):361–374. e364. https://doi.org/10.1016/j.cmet.2017.06.021

Diaz JR, Kim KJ, Brands MW, Filosa JA (2019) Augmented astrocyte microdomain Ca(2+) dynamics and parenchymal arteriole tone in angiotensin II-infused hypertensive mice. Glia 67(3):551–565. https://doi.org/10.1002/glia.23564

Diaz SL, Doly S, Narboux-Neme N, Fernandez S, Mazot P, Banas SM, Boutourlinsky K, Moutkine I, Belmer A, Roumier A, Maroteaux L (2012) 5-HT(2B) receptors are required for serotonin-selective antidepressant actions. Mol Psychiatry 17(2):154–163. https://doi.org/10.1038/mp.2011.159

Dickerson F, Severance E, Yolken R (2017) The microbiome, immunity, and schizophrenia and bipolar disorder. Brain Behav Immun 62:46–52. https://doi.org/10.1016/j.bbi.2016.12.010

Dienel GA (2012) Brain lactate metabolism: the discoveries and the controversies. J Cereb Blood Flow Metab 32(7):1107–1138. https://doi.org/10.1038/jcbfm.2011.175

Dienel GA (2019) Brain glucose metabolism: integration of energetics with function. Physiol Rev 99(1):949–1045. https://doi.org/10.1152/physrev.00062.2017

Dienel GA, Cruz NF (2015) Contributions of glycogen to astrocytic energetics during brain activation. Metab Brain Dis 30(1):281–298. https://doi.org/10.1007/s11011-014-9493-8

Dienel GA, Wang RY, Cruz NF (2002) Generalized sensory stimulation of conscious rats increases labeling of oxidative pathways of glucose metabolism when the brain glucose-oxygen uptake ratio rises. J Cereb Blood Flow Metab 22(12):1490–1502. https://doi.org/10.1097/01.WCB.0000034363.37277.89

Ding F, O'Donnell J, Thrane AS, Zeppenfeld D, Kang H, Xie L, Wang F, Nedergaard M (2013) alpha1-Adrenergic receptors mediate coordinated Ca2+ signaling of cortical astrocytes in awake, behaving mice. Cell Calcium 54(6):387–394. https://doi.org/10.1016/j.ceca.2013.09.001

DiNuzzo M, Giove F, Maraviglia B, Mangia S (2015) Monoaminergic control of cellular glucose utilization by glycogenolysis in neocortex and hippocampus. Neurochem Res 40(12):2493–2504. https://doi.org/10.1007/s11064-015-1656-4

DiNuzzo M, Giove F, Maraviglia B, Mangia S (2017) Computational flux balance analysis predicts that stimulation of energy metabolism in astrocytes and their metabolic interactions with neurons depend on uptake of K(+) rather than glutamate. Neurochem Res 42(1):202–216. https://doi.org/10.1007/s11064-016-2048-0

Du F, Cooper AJ, Thida T, Sehovic S, Lukas SE, Cohen BM, Zhang XL, Ongur D (2014) In vivo evidence for cerebral bioenergetic abnormalities in schizophrenia measured using P-31 magnetization transfer spectroscopy. JAMA Psychiat 71(1):19–27. https://doi.org/10.1001/jamapsychiatry.2013.2287

Duman RS, Aghajanian GK, Sanacora G, Krystal JH (2016) Synaptic plasticity and depression: new insights from stress and rapid-acting antidepressants. Nat Med 22(3):238–249. https://doi.org/10.1038/nm.4050

Duran J, Saez I, Gruart A, Guinovart JJ, Delgado-Garcia JM (2013) Impairment in long-term memory formation and learning-dependent synaptic plasticity in mice lacking glycogen synthase in the brain. J Cereb Blood Flow Metab 33(4):550–556. https://doi.org/10.1038/jcbfm.2012.200

Dwyer DS, Pinkofsky HB, Liu Y, Bradley RJ (1999) Antipsychotic drugs affect glucose uptake and the expression of glucose transporters in PC12 cells. Prog Neuro-Psychopharmacol Biol Psychiatry 23(1):69–80. https://doi.org/10.1016/s0278-5846(98)00092-x

Ebrahimi M, Yamamoto Y, Sharifi K, Kida H, Kagawa Y, Yasumoto Y, Islam A, Miyazaki H, Shimamoto C, Maekawa M, Mitsushima D, Yoshikawa T, Owada Y (2016) Astrocyte-expressed FABP7 regulates dendritic morphology and excitatory synaptic function of cortical neurons. Glia 64(1):48–62. https://doi.org/10.1002/glia.22902

Edmondson DE, Binda C, Wang J, Upadhyay AK, Mattevi A (2009) Molecular and mechanistic properties of the membrane-bound mitochondrial monoamine oxidases. Biochemistry 48(20):4220–4230. https://doi.org/10.1021/bi900413g

Eraso-Pichot A, Braso-Vives M, Golbano A, Menacho C, Claro E, Galea E, Masgrau R (2018) GSEA of mouse and human mitochondriomes reveals fatty acid oxidation in astrocytes. Glia 66(8):1724–1735. https://doi.org/10.1002/glia.23330

Erny D, de Angelis ALH, Jaitin D, Wieghofer P, Staszewski O, David E, Keren-Shaul H, Mahlakoiv T, Jakobshagen K, Buch T, Schwierzeck V, Utermohlen O, Chun E, Garrett WS, Mccoy KD, Diefenbach A, Staeheli P, Stecher B, Amit I, Prinz M (2015) Host microbiota constantly control maturation and function of microglia in the CNS. Nature Neuroscience 18(7):965. https://doi.org/10.1038/nn.4030

Evangeliou A, Vlachonikolis I, Mihailidou H, Spilioti M, Skarpalezou A, Makaronas N, Prokopiou A, Christodoulou P, Liapi-Adamidou G, Helidonis E, Sbyrakis S, Smeitink J (2003) Application of a ketogenic diet in children with autistic behavior: pilot study. J Child Neurol 18(2):113–118. https://doi.org/10.1177/08830738030180020501

Eyjolfsson EM, Nilsen LH, Kondziella D, Brenner E, Haberg A, Sonnewald U (2011) Altered 13C glucose metabolism in the cortico-striato-thalamo-cortical loop in the MK-801 rat model of schizophrenia. J Cereb Blood Flow Metab 31(3):976–985. https://doi.org/10.1038/jcbfm.2010.193

Eykelenboom JE, Briggs GJ, Bradshaw NJ, Soares DC, Ogawa F, Christie S, Malavasi ELV, Makedonopoulou P, Mackie S, Malloy MP, Wear MA, Blackburn EA, Bramham J, McIntosh AM, Blackwood DH, Muir WJ, Porteous DJ, Millar JK (2012) A t(1;11) translocation linked to schizophrenia and affective disorders gives rise to aberrant chimeric DISC1 transcripts that encode structurally altered, deleterious mitochondrial proteins. Hum Mol Genet 21(15):3374–3386. https://doi.org/10.1093/hmg/dds169

Faff-Michalak L, Albrecht J (1993) Hyperammonemia and hepatic encephalopathy stimulate rat cerebral synaptic mitochondrial glutamate dehydrogenase activity specifically in the direction of glutamate oxidation. Brain Res 618(2):299–302. https://doi.org/10.1016/0006-8993(93)91279-2

Farrell MS, Werge T, Sklar P, Owen MJ, Ophoff RA, O'Donovan MC, Corvin A, Cichon S, Sullivan PF (2015) Evaluating historical candidate genes for schizophrenia. Mol Psychiatry 20(5):555–562. https://doi.org/10.1038/mp.2015.16

Fischer-Smith KD, Houston ACW, Rebec GV (2012) Differential effects of cocaine access and withdrawal on glutamate type 1 transporter expression in rat nucleus accumbens core and shell. Neuroscience 210:333–339. https://doi.org/10.1016/j.neuroscience.2012.02.049

Flatow J, Buckley P, Miller BJ (2013) Meta-analysis of oxidative stress in schizophrenia. Biol Psychiatry 74(6):400–409. https://doi.org/10.1016/j.biopsych.2013.03.018

Fortunati R, Ossola P, Camerlengo A, Bettini E, De Panfilis C, Maggini C, Marchesi C (2015) Anhedonia in schizophrenia: the role of subjective experiences in the emotion paradox. Eur Psychiat 30(S1):1. https://doi.org/10.1016/S0924-9338(15)30177-2

Fraguas D, Diaz-Caneja CM, Rodriguez-Quiroga A, Arango C (2017) Oxidative stress and inflammation in early onset first episode psychosis: a systematic review and meta-analysis. Int J Neuropsychopharmacol 20(6):435–444. https://doi.org/10.1093/ijnp/pyx015

Frankle WG, Lerma J, Laruelle M (2003) The synaptic hypothesis of schizophrenia. Neuron 39(2):205–216. https://doi.org/10.1016/s0896-6273(03)00423-9

Freund N, Juckel G (2019) Bipolar disorder: its etiology and how to model in rodents. Methods Mol Biol 2011:61–77. https://doi.org/10.1007/978-1-4939-9554-7_4

Fromer M, Pocklington AJ, Kavanagh DH, Williams HJ, Dwyer S, Gormley P, Georgieva L, Rees E, Palta P, Ruderfer DM, Carrera N, Humphreys I, Johnson JS, Roussos P, Barker DD, Banks E, Milanova V, Grant SG, Hannon E, Rose SA, Chambert K, Mahajan M, Scolnick EM, Moran JL, Kirov G, Palotie A, McCarroll SA, Holmans P, Sklar P, Owen MJ, Purcell SM, O'Donovan MC (2014) De novo mutations in schizophrenia implicate synaptic networks. Nature 506(7487):179–184. https://doi.org/10.1038/nature12929

Fuente-Martin E, Mellado-Gil JM, Cobo-Vuilleumier N, Martin-Montalvo A, Romero-Zerbo SY, Diaz Contreras I, Hmadcha A, Soria B, Martin Bermudo F, Reyes JC, Bermudez-Silva FJ, Lorenzo PI, Gauthier BR (2019) Dissecting the brain/islet axis in metabesity. Genes (Basel) 10(5):350. https://doi.org/10.3390/genes10050350

Gancarz A, Jouroukhin Y, Saito A, Shevelkin A, Mueller LE, Kamiya A, Dietz DM, Pletnikov MV (2016) DISC1 signaling in cocaine addiction: towards molecular mechanisms of co-morbidity. Neurosci Res 105:70–74. https://doi.org/10.1016/j.neures.2015.09.001

Gandhi GK, Ball KK, Cruz NF, Dienel GA (2010) Hyperglycaemia and diabetes impair gap junctional communication among astrocytes. Asn Neuro 2(2):e00030. https://doi.org/10.1042/AN20090048

Gao Y, Layritz C, Legutko B, Eichmann TO, Laperrousaz E, Moulle VS, Cruciani-Guglielmacci C, Magnan C, Luquet S, Woods SC, Eckel RH, Yi CX, Garcia-Caceres C, Tschop MH (2017) Disruption of lipid uptake in astroglia exacerbates diet-induced obesity. Diabetes 66(10):2555–2563. https://doi.org/10.2337/db16-1278

Gebril HM, Avula B, Wang YH, Khan IA, Jekabsons MB (2016) (13)C metabolic flux analysis in neurons utilizing a model that accounts for hexose phosphate recycling within the pentose phosphate pathway. Neurochem Int 93:26–39. https://doi.org/10.1016/j.neuint.2015.12.008

Genda EN, Jackson JG, Sheldon AL, Locke SF, Greco TM, O'Donnell JC, Spruce LA, Xiao R, Guo W, Putt M, Seeholzer S, Ischiropoulos H, Robinson MB (2011) Co-compartmentalization of the astroglial glutamate transporter, GLT-1, with glycolytic enzymes and mitochondria. J Neurosci 31(50):18275–18288. https://doi.org/10.1523/JNEUROSCI.3305-11.2011

Ghaleiha A, Rasa SM, Nikoo M, Farokhnia M, Mohammadi MR, Akhondzadeh S (2015) A pilot double-blind placebo-controlled trial of pioglitazone as adjunctive treatment to risperidone: effects on aberrant behavior in children with autism. Psychiatry Res 229(1–2):181–187. https://doi.org/10.1016/j.psychres.2015.07.043

Gibbs ME, Hertz L (2008) Inhibition of astrocytic energy metabolism by D-lactate exposure impairs memory. Neurochem Int 52(6):1012–1018. https://doi.org/10.1016/j.neuint.2007.10.014

Gibbs ME, Hertz L (2014) Serotonin mediation of early memory formation via 5-HT2B receptor-induced glycogenolysis in the day-old chick. Front Pharmacol 5:54. https://doi.org/10.3389/fphar.2014.00054

Giovannoni F, Quintana FJ (2020) The role of astrocytes in CNS inflammation. Trends Immunol 41(9):805–819. https://doi.org/10.1016/j.it.2020.07.007

Gogou M, Kolios G (2018) Are therapeutic diets an emerging additional choice in autism spectrum disorder management? World J Pediatr 14(3):215–223. https://doi.org/10.1007/s12519-018-0164-4

Gordon GR, Choi HB, Rungta RL, Ellis-Davies GC, MacVicar BA (2008) Brain metabolism dictates the polarity of astrocyte control over arterioles. Nature 456(7223):745–749. https://doi.org/10.1038/nature07525

Gosejacob D, Dublin P, Bedner P, Huttmann K, Zhang J, Tress O, Willecke K, Pfrieger F, Steinhauser C, Theis M (2011) Role of astroglial connexin30 in hippocampal gap junction coupling. Glia 59(3):511–519. https://doi.org/10.1002/glia.21120

Graves SM, Xie Z, Stout KA, Zampese E, Burbulla LF, Shih JC, Kondapalli J, Patriarchi T, Tian L, Brichta L, Greengard P, Krainc D, Schumacker PT, Surmeier DJ (2020) Dopamine metabolism by a monoamine oxidase mitochondrial shuttle activates the electron transport chain. Nat Neurosci 23(1):15–20. https://doi.org/10.1038/s41593-019-0556-3

Griffin WC, Ramachandra VS, Knackstedt LA, Becker HC (2015) Repeated cycles of chronic intermittent ethanol exposure increases basal glutamate in the nucleus accumbens of mice without affecting glutamate transport. Front Pharmacol 6:27. https://doi.org/10.3389/fphar.2015.00027

Grubisic V, Gulbransen BD (2017) Enteric glia: the most alimentary of all glia. J Physiol 595(2):542–555. https://doi.org/10.1113/Jp271021

Guan JT, Cai JJ, Ji GL, Sham PC (2019) Commonality in dysregulated expression of gene sets in cortical brains of individuals with autism, schizophrenia, and bipolar disorder. Transl Psychiat 9:152. https://doi.org/10.1038/s41398-019-0488-4

Guest PC, Iwata K, Kato TA, Steiner J, Schmitt A, Turck CW, Martins-de-Souza D (2015) MK-801 treatment affects glycolysis in oligodendrocytes more than in astrocytes and neuronal cells: insights for schizophrenia. Front Cell Neurosci 9:180. https://doi.org/10.3389/fncel.2015.00180

Guzman M, Blazquez C (2004) Ketone body synthesis in the brain: possible neuroprotective effects. Prostaglandins Leukot Essent Fatty Acids 70(3):287–292. https://doi.org/10.1016/j.plefa.2003.05.001

Gzielo K, Soltys Z, Rajfur Z, Setkowicz ZK (2019) The impact of the ketogenic diet on glial cells morphology. A quantitative morphological analysis. Neuroscience 413:239–251. https://doi.org/10.1016/j.neuroscience.2019.06.009

Hafizi S, Tabatabaei D, Lai MC (2019) Review of clinical studies targeting inflammatory pathways for individuals with autism. Front Psych 10:849. https://doi.org/10.3389/fpsyt.2019.00849

Hamberger A, Hyden H (1963) Inverse enzymatic changes in neurons and glia during increased function and hypoxia. J Cell Biol 16:521–525. https://doi.org/10.1083/jcb.16.3.521

Han J, Kesner P, Metna-Laurent M, Duan TT, Xu L, Georges F, Koehl M, Abrous DN, Mendizabal-Zubiaga J, Grandes P, Liu QS, Bai G, Wang W, Xiong LZ, Ren W, Marsicano G, Zhang X (2012) Acute cannabinoids impair working memory through astroglial CB1 receptor modulation of hippocampal LTD. Cell 148(5):1039–1050. https://doi.org/10.1016/j.cell.2012.01.037

Han WY, Du P, Fu SY, Wang F, Song M, Wu CF, Yang JY (2014) Oxytocin via its receptor affects restraint stress-induced methamphetamine CPP reinstatement in mice: involvement of the medial prefrontal cortex and dorsal hippocampus glutamatergic system. Pharmacol Biochem Behav 119:80–87. https://doi.org/10.1016/j.pbb.2013.11.014

Hanaya R, Koning E, Ferrandon A, Schweitzer A, Andrieux A, Nehlig A (2008) Deletion of the STOP gene, a microtubule stabilizing factor, leads only to discrete cerebral metabolic changes in mice. J Neurosci Res 86(4):813–820. https://doi.org/10.1002/jnr.21550

Hanstock TL, Mallet PE, Clayton EH (2010) Increased plasma d-lactic acid associated with impaired memory in rats. Physiol Behav 101(5):653–659. https://doi.org/10.1016/j.physbeh.2010.09.018

Harada K, Kamiya T, Tsuboi T (2015) Gliotransmitter release from astrocytes: functional, developmental, and pathological implications in the brain. Front Neurosci 9:499. https://doi.org/10.3389/fnins.2015.00499

Hattori T, Shimizu S, Koyama Y, Emoto H, Matsumoto Y, Kumamoto N, Yamada K, Takamura H, Matsuzaki S, Katayama T, Tohyama M, Ito A (2014) DISC1 (disrupted-in-Schizophrenia-1) regulates differentiation of oligodendrocytes. PLoS One 9(2):e88506. https://doi.org/10.1371/journal.pone.0088506

Haydon PG (2001) GLIA: listening and talking to the synapse. Nat Rev Neurosci 2(3):185–193. https://doi.org/10.1038/35058528

Hayzoun K, Lalonde R, Mariani J, Strazielle C (2007) Regional variations of cytochrome oxidase activity in the central auditory system of Relnrl-Orl (reeler) mutant mice. Neurosci Res 58(4):378–385. https://doi.org/10.1016/j.neures.2007.04.008

Heiser P, Singh S, Krieg JC, Vedder H (2006) Effects of different antipsychotics and the antidepressant mirtazapine on glucose transporter mRNA levels in human blood cells. J Psychiatr Res 40(4):374–379. https://doi.org/10.1016/j.jpsychires.2005.04.016

Heller JP, Rusakov DA (2017) The nanoworld of the tripartite synapse: insights from super-resolution microscopy. Front Cell Neurosci 11:374. https://doi.org/10.3389/fncel.2017.00374

Herberth M, Koethe D, Cheng TM, Krzyszton ND, Schoeffmann S, Guest PC, Rahmoune H, Harris LW, Kranaster L, Leweke FM, Bahn S (2011) Impaired glycolytic response in peripheral blood mononuclear cells of first-onset antipsychotic-naive schizophrenia patients. Mol Psychiatry 16(8):848–859. https://doi.org/10.1038/mp.2010.71

Herrero-Mendez A, Almeida A, Fernandez E, Maestre C, Moncada S, Bolanos JP (2009) The bioenergetic and antioxidant status of neurons is controlled by continuous degradation of a key glycolytic enzyme by APC/C-Cdh1. Nat Cell Biol 11(6):747–752. https://doi.org/10.1038/ncb1881

Hertz L, Gibbs ME, Dienel GA (2014) Fluxes of lactate into, from, and among gap junction-coupled astrocytes and their interaction with noradrenaline. Front Neurosci 8:261. https://doi.org/10.3389/fnins.2014.00261

Hertz L, Rothman DL, Li B, Peng L (2015) Chronic SSRI stimulation of astrocytic 5-HT2B receptors change multiple gene expressions/editings and metabolism of glutamate, glucose and glycogen: a potential paradigm shift. Front Behav Neurosci 9:25. https://doi.org/10.3389/fnbeh.2015.00025

Holcomb HH, Cascella NG, Thaker GK, Medoff DR, Dannals RF, Tamminga CA (1996) Functional sites of neuroleptic drug action in the human brain: PET/FDG studies with and without haloperidol. Am J Psychiatry 153(1):41–49. https://doi.org/10.1176/ajp.153.1.41

Holley SM, Wang EA, Cepeda C, Jentsch JD, Ross CA, Pletnikov MV, Levine MS (2013) Frontal cortical synaptic communication is abnormal in Disc1 genetic mouse models of schizophrenia. Schizophr Res 146(1–3):264–272. https://doi.org/10.1016/j.schres.2013.02.007

Howarth C (2014) The contribution of astrocytes to the regulation of cerebral blood flow. Front Neurosci 8:103. https://doi.org/10.3389/fnins.2014.00103

Hsiao EY (2013) Immune dysregulation in autism spectrum disorder. Int Rev Neurobiol 113:269–302. https://doi.org/10.1016/B978-0-12-418700-9.00009-5

Hsiao EY, McBride SW, Hsien S, Sharon G, Hyde ER, McCue T, Codelli JA, Chow J, Reisman SE, Petrosino JF, Patterson PH, Mazmanian SK (2013) Microbiota modulate behavioral and physiological abnormalities associated with neurodevelopmental disorders. Cell 155(7):1451–1463. https://doi.org/10.1016/j.cell.2013.11.024

Ioannou MS, Jackson J, Sheu SH, Chang CL, Weigel AV, Liu H, Pasolli HA, Xu CS, Pang S, Matthies D, Hess HF, Lippincott-Schwartz J, Liu Z (2019) Neuron-astrocyte metabolic coupling protects against activity-induced fatty acid toxicity. Cell 177(6):1522–1535. e1514. https://doi.org/10.1016/j.cell.2019.04.001

Iranpour N, Zandifar A, Farokhnia M, Goguol A, Yekehtaz H, Khodaie-Ardakani MR, Salehi B, Esalatmanesh S, Zeionoddini A, Mohammadinejad P, Zeinoddini A, Akhondzadeh S (2016) The effects of pioglitazone adjuvant therapy on negative symptoms of patients with

chronic schizophrenia: a double-blind and placebo-controlled trial. Hum Psychopharmacol 31(2):103–112. https://doi.org/10.1002/hup.2517
Ishibashi M, Egawa K, Fukuda A (2019) Diverse actions of astrocytes in GABAergic signaling. Int J Mol Sci 20(12):2964. https://doi.org/10.3390/ijms20122964
Jaaro-Peled H, Hayashi-Takagi A, Seshadri S, Kamiya A, Brandon NJ, Sawa A (2009) Neurodevelopmental mechanisms of schizophrenia: understanding disturbed postnatal brain maturation through neuregulin-1-ErbB4 and DISC1. Trends Neurosci 32(9):485–495. https://doi.org/10.1016/j.tins.2009.05.007
Jackson JG, O'Donnell JC, Takano H, Coulter DA, Robinson MB (2014) Neuronal activity and glutamate uptake decrease mitochondrial mobility in astrocytes and position mitochondria near glutamate transporters. J Neurosci 34(5):1613–1624. https://doi.org/10.1523/Jneurosci.3510-13.2014
Jackson JG, Robinson MB (2015) Reciprocal regulation of mitochondrial dynamics and calcium signaling in astrocyte processes. J Neurosci 35(45):15199–15213. https://doi.org/10.1523/JNEUROSCI.2049-15.2015
Jackson JG, Robinson MB (2018) Regulation of mitochondrial dynamics in astrocytes: mechanisms, consequences, and unknowns. Glia 66(6):1213–1234. https://doi.org/10.1002/glia.23252
Jacquelin C, Lalonde R, Jantzen-Ossola C, Strazielle C (2013) Neurobehavioral performances and brain regional metabolism in Dab1(scm) (scrambler) mutant mice. Behav Brain Res 252:92–100. https://doi.org/10.1016/j.bbr.2013.05.024
James R, Adams RR, Christie S, Buchanan SR, Porteous DJ, Millar JK (2004) Disrupted in schizophrenia 1 (DISC1) is a multicompartmentalized protein that predominantly localizes to mitochondria. Mol Cell Neurosci 26(1):112–122. https://doi.org/10.1016/j.mcn.2004.01.013
Jekabsons MB, Gebril HM, Wang YH, Avula B, Khan IA (2017) Updates to a (13)C metabolic flux analysis model for evaluating energy metabolism in cultured cerebellar granule neurons from neonatal rats. Neurochem Int 109:54–67. https://doi.org/10.1016/j.neuint.2017.03.020
Jha MK, Morrison BM (2018) Glia-neuron energy metabolism in health and diseases: new insights into the role of nervous system metabolic transporters. Exp Neurol 309:23–31. https://doi.org/10.1016/j.expneurol.2018.07.009
Ji B, Higa KK, Kim M, Zhou L, Young JW, Geyer MA, Zhou X (2014) Inhibition of protein translation by the DISC1-Boymaw fusion gene from a Scottish family with major psychiatric disorders. Hum Mol Genet 23(21):5683–5705. https://doi.org/10.1093/hmg/ddu285
Jouroukhin Y, Kageyama Y, Misheneva V, Shevelkin A, Andrabi S, Prandovszky E, Yolken RH, Dawson VL, Dawson TM, Aja S, Sesaki H, Pletnikov MV (2018) DISC1 regulates lactate metabolism in astrocytes: implications for psychiatric disorders. Transl Psychiatry 8(1):76. https://doi.org/10.1038/s41398-018-0123-9
Juaristi I, Contreras L, Gonzalez-Sanchez P, Perez-Liebana I, Gonzalez-Moreno L, Pardo B, Del Arco A, Satrustegui J (2019) The response to stimulation in neurons and astrocytes. Neurochem Res 44(10):2385–2391. https://doi.org/10.1007/s11064-019-02803-7
Juaristi I, Garcia-Martin ML, Rodrigues TB, Satrustegui J, Llorente-Folch I, Pardo B (2017) ARALAR/AGC1 deficiency, a neurodevelopmental disorder with severe impairment of neuronal mitochondrial respiration, does not produce a primary increase in brain lactate. J Neurochem 142(1):132–139. https://doi.org/10.1111/jnc.14047
Kabouridis PS, Pachnis V (2015) Emerging roles of gut microbiota and the immune system in the development of the enteric nervous system. J Clin Investig 125(3):956–964. https://doi.org/10.1172/Jci76308
Kamiya A, Kubo K, Tomoda T, Takaki M, Youn R, Ozeki Y, Sawamura N, Park U, Kudo C, Okawa M, Ross CA, Hatten ME, Nakajima K, Sawa A (2005) A schizophrenia-associated mutation of DISC1 perturbs cerebral cortex development. Nat Cell Biol 7(12):1167–1178. https://doi.org/10.1038/ncb1328
Kang J, Jiang L, Goldman SA, Nedergaard M (1998) Astrocyte-mediated potentiation of inhibitory synaptic transmission. Nat Neurosci 1(8):683–692. https://doi.org/10.1038/3684

Kashani L, Omidvar T, Farazmand B, Modabbernia A, Ramzanzadeh F, Tehraninejad ES, Ashrafi M, Tabrizi M, Akhondzadeh S (2013) Does pioglitazone improve depression through insulin-sensitization? Results of a randomized double-blind metformin-controlled trial in patients with polycystic ovarian syndrome and comorbid depression. Psychoneuroendocrinology 38(6):767–776. https://doi.org/10.1016/j.psyneuen.2012.08.010

Katsel P, Byne W, Roussos P, Tan W, Siever L, Haroutunian V (2011a) Astrocyte and glutamate markers in the superficial, deep, and white matter layers of the anterior cingulate gyrus in schizophrenia. Neuropsychopharmacology 36(6):1171–1177. https://doi.org/10.1038/npp.2010.252

Katsel P, Fam P, Tan WL, Khan S, Yang CX, Jouroukhin Y, Rudchenko S, Pletnikov MV, Haroutunian V (2018) Overexpression of truncated human DISC1 induces appearance of hindbrain oligodendroglia in the forebrain during development. Schizophr Bull 44(3):515–524. https://doi.org/10.1093/schbul/sbx106

Katsel P, Tan WL, Abazyan B, Davis KL, Ross C, Pletnikov MV, Haroutunian V (2011b) Expression of mutant human DISC1 in mice supports abnormalities in differentiation of oligodendrocytes. Schizophr Res 130(1–3):238–249. https://doi.org/10.1016/j.schres.2011.04.021

Kemp DE, Ismail-Beigi F, Ganocy SJ, Conroy C, Gao K, Obral S, Fein E, Findling RL, Calabrese JR (2012) Use of insulin sensitizers for the treatment of major depressive disorder: a pilot study of pioglitazone for major depression accompanied by abdominal obesity. J Affect Disord 136(3):1164–1173. https://doi.org/10.1016/j.jad.2011.06.033

Kemp DE, Schinagle M, Gao K, Conroy C, Ganocy SJ, Ismail-Beigi F, Calabrese JR (2014) PPAR-gamma agonism as a modulator of mood: proof-of-concept for pioglitazone in bipolar depression. CNS Drugs 28(6):571–581. https://doi.org/10.1007/s40263-014-0158-2

Keshavan M, Lizano P, Prasad K (2020) The synaptic pruning hypothesis of schizophrenia: promises and challenges. World Psychiatry 19(1):110–111. https://doi.org/10.1002/wps.20725

Kety SS (1959) Biochemical theories of schizophrenia. I. Science 129(3362):1528–1532. https://doi.org/10.1126/science.129.3362.1528

Kim J, Horti AG, Mathews WB, Pogorelov V, Valentine H, Brasic JR, Holt DP, Ravert HT, Dannals RF, Zhou LW, Jedynak B, Kamiya A, Pletnikov MV, Wong DF (2015a) Quantitative multimodal brain autoradiography of glutamatergic, dopaminergic, cannabinoid, and nicotinic receptors in mutant disrupted-in-schizophrenia-1 (DISC1) mice. Mol Imaging Biol 17(3):355–363. https://doi.org/10.1007/s11307-014-0786-4

Kim KJ, Iddings JA, Stern JE, Blanco VM, Croom D, Kirov SA, Filosa JA (2015b) Astrocyte contributions to flow/pressure-evoked parenchymal arteriole vasoconstriction. J Neurosci 35(21):8245–8257. https://doi.org/10.1523/JNEUROSCI.4486-14.2015

Kim R, Healey KL, Sepulveda-Orengo MT, Reissner KJ (2018) Astroglial correlates of neuropsychiatric disease: from astrocytopathy to astrogliosis. Prog Neuropsychopharmacol Biol Psychiatry 87(Pt A):126–146. https://doi.org/10.1016/j.pnpbp.2017.10.002

Kim S, Kim H, Yim YS, Ha S, Atarashi K, Tan TG, Longman RS, Honda K, Littman DR, Choi GB, Huh JR (2017) Maternal gut bacteria promote neurodevelopmental abnormalities in mouse offspring. Nature 549(7673):528. https://doi.org/10.1038/nature23910

Knackstedt LA, Melendez RI, Kalivas PW (2010) Ceftriaxone restores glutamate homeostasis and prevents relapse to cocaine seeking. Biol Psychiatry 67(1):81–84. https://doi.org/10.1016/j.biopsych.2009.07.018

Kohler S, Winkler U, Sicker M, Hirrlinger J (2018) NBCe1 mediates the regulation of the NADH/NAD(+) redox state in cortical astrocytes by neuronal signals. Glia 66(10):2233–2245. https://doi.org/10.1002/glia.23504

Kong EKC, Peng L, Chen Y, Yu ACH, Hertz L (2002a) Up-regulation of 5-HT2B receptor density and receptor-mediated glycogenolysis in mouse astrocytes by long-term fluoxetine administration. Neurochem Res 27(1–2):113–120. https://doi.org/10.1023/A:1014862808126

Kong J, Shepel PN, Holden CP, Mackiewicz M, Pack AI, Geiger JD (2002b) Brain glycogen decreases with increased periods of wakefulness: implications for homeostatic drive to sleep. J Neurosci 22(13):5581–5587

Koshiba T (2013) Mitochondrial-mediated antiviral immunity. BBA-Mol Cell Res 1833(1):225–232. https://doi.org/10.1016/j.bbamcr.2012.03.005

Kovacs Z, D'Agostino DP, Diamond D, Kindy MS, Rogers C, Ari C (2019) Therapeutic potential of exogenous ketone supplement induced ketosis in the treatment of psychiatric disorders: review of current literature. Front Psych 10:363. https://doi.org/10.3389/fpsyt.2019.00363

Kraeuter AK, van den Buuse M, Sarnyai Z (2019) Ketogenic diet prevents impaired prepulse inhibition of startle in an acute NMDA receptor hypofunction model of schizophrenia. Schizophr Res 206:244–250. https://doi.org/10.1016/j.schres.2018.11.011

Kronenberg J, Merkel L, Heckers S, Gudi V, Schwab HM, Stangel M (2019) Investigation of neuregulin-1 and glial cell-derived neurotrophic factor in rodent astrocytes and microglia. J Mol Neurosci 67(3):484–493. https://doi.org/10.1007/s12031-019-1258-8

Kuroda K, Yamada S, Tanaka M, Iizuka M, Yano H, Mori D, Tsuboi D, Nishioka T, Namba T, Iizuka Y, Kubota S, Nagai T, Ibi D, Wang R, Enomoto A, Isotani-Sakakibara M, Asai N, Kimura K, Kiyonari H, Abe T, Mizoguchi A, Sokabe M, Takahashi M, Yamada K, Kaibuchi K (2011) Behavioral alterations associated with targeted disruption of exons 2 and 3 of the Disc1 gene in the mouse. Hum Mol Genet 20(23):4666–4683. https://doi.org/10.1093/hmg/ddr400

Lacroix-Fralish ML, Tawfik VL, Nutile-McMenemy N, Harris BT, DeLeo JA (2006) Differential regulation of neuregulin 1 expression by progesterone in astrocytes and neurons. Neuron Glia Biol 2:227–234. https://doi.org/10.1017/S1740925x07000385

LaCrosse AL, Hill K, Knackstedt LA (2016) Ceftriaxone attenuates cocaine relapse after abstinence through modulation of nucleus accumbens AMPA subunit expression. Eur Neuropsychopharmacol 26(2):186–194. https://doi.org/10.1016/j.euroneuro.2015.12.022

LaCrosse AL, O'Donovan SM, Sepulveda-Orengo MT, McCullumsmith RE, Reissner KJ, Schwendt M, Knackstedt LA (2017) Contrasting the role of xCT and GLT-1 upregulation in the ability of ceftriaxone to attenuate the Cue-induced reinstatement of cocaine seeking and normalize AMPA receptor subunit expression. J Neurosci 37(24):5809–5821. https://doi.org/10.1523/Jneurosci.3717-16.2017

Lalo U, Palygin O, Rasooli-Nejad S, Andrew J, Haydon PG, Pankratov Y (2014) Exocytosis of ATP from astrocytes modulates phasic and tonic inhibition in the neocortex. PLoS Biol 12(1):e1001747. https://doi.org/10.1371/journal.pbio.1001747

Lalonde R, Hayzoun K, Derer M, Mariani J, Strazielle C (2004) Neurobehavioral evaluation of Reln-rl-orl mutant mice and correlations with cytochrome oxidase activity. Neurosci Res 49(3):297–305. https://doi.org/10.1016/j.neures.2004.03.012

Lam TK, Gutierrez-Juarez R, Pocai A, Rossetti L (2005) Regulation of blood glucose by hypothalamic pyruvate metabolism. Science 309(5736):943–947. https://doi.org/10.1126/science.1112085

Lanjakornsiripan D, Pior BJ, Kawaguchi D, Furutachi S, Tahara T, Katsuyama Y, Suzuki Y, Fukazawa Y, Gotoh Y (2018) Layer-specific morphological and molecular differences in neocortical astrocytes and their dependence on neuronal layers. Nat Commun 9:1623. https://doi.org/10.1038/s41467-018-03940-3

Law AJ, Weickert CS, Hyde TM, Kleinman JE, Harrison PJ (2004) Neuregulin-1 (NRG-1) mRNA and protein in the adult human brain. Neuroscience 127(1):125–136. https://doi.org/10.1016/j.neuroscience.2004.04.026

Le Foll C (2019) Hypothalamic fatty acids and ketone bodies sensing and role of FAT/CD36 in the regulation of food intake. Front Physiol 10:1036. https://doi.org/10.3389/fphys.2019.01036

Le Foll C, Dunn-Meynell AA, Miziorko HM, Levin BE (2014) Regulation of hypothalamic neuronal sensing and food intake by ketone bodies and fatty acids. Diabetes 63(4):1259–1269. https://doi.org/10.2337/db13-1090

Le Foll C, Levin BE (2016) Fatty acid-induced astrocyte ketone production and the control of food intake. Am J Physiol Regul Integr Comp Physiol 310(11):R1186–R1192. https://doi.org/10.1152/ajpregu.00113.2016

Lee JA, Hall B, Allsop J, Alqarni R, Allen SP (2020) Lipid metabolism in astrocytic structure and function. Semin Cell Dev Biol 112:123–136. https://doi.org/10.1016/j.semcdb.2020.07.017

Lee Y, Zhang Y, Kim S, Han K (2018) Excitatory and inhibitory synaptic dysfunction in mania: an emerging hypothesis from animal model studies. Exp Mol Med 50(4):12. https://doi.org/10.1038/s12276-018-0028-y

Leffler CW, Balabanova L, Fedinec AL, Waters CM, Parfenova H (2003) Mechanism of glutamate stimulation of CO production in cerebral microvessels. Am J Physiol Heart Circ Physiol 285(1):H74–H80. https://doi.org/10.1152/ajpheart.01081.2002

Levitt P, Pintar JE, Breakefield XO (1982) Immunocytochemical demonstration of monoamine oxidase B in brain astrocytes and serotonergic neurons. Proc Natl Acad Sci U S A 79(20):6385–6389. https://doi.org/10.1073/pnas.79.20.6385

Li B, Zhang S, Zhang H, Nu W, Cai L, Hertz L, Peng L (2008) Fluoxetine-mediated 5-HT2B receptor stimulation in astrocytes causes EGF receptor transactivation and ERK phosphorylation. Psychopharmacology 201(3):443–458. https://doi.org/10.1007/s00213-008-1306-5

Li BM, Dong L, Wang B, Cai LP, Jiang N, Peng L (2012) Cell type-specific gene expression and editing responses to chronic fluoxetine treatment in the in vivo mouse brain and their relevance for stress-induced anhedonia. Neurochem Res 37(11):2480–2495. https://doi.org/10.1007/s11064-012-0814-1

Li TL, Li XH, Attri KS, Liu CH, Li LP, Herring LE, Asara JM, Lei YL, Singh PK, Gao CJ, Wen HT (2018) O-GlcNAc transferase links glucose metabolism to MAVS-mediated antiviral innate immunity. Cell Host Microbe 24(6):791. https://doi.org/10.1016/j.chom.2018.11.001

Liddelow SA, Marsh SE, Stevens B (2020) Microglia and astrocytes in disease: dynamic duo or partners in crime? Trends Immunol 41(9):820–835. https://doi.org/10.1016/j.it.2020.07.006

Lind BL, Brazhe AR, Jessen SB, Tan FC, Lauritzen MJ (2013a) Rapid stimulus-evoked astrocyte Ca2+ elevations and hemodynamic responses in mouse somatosensory cortex in vivo. Proc Natl Acad Sci U S A 110(48):E4678–E4687. https://doi.org/10.1073/pnas.1310065110

Lind BL, Jessen SB, Lonstrup M, Josephine C, Bonvento G, Lauritzen M (2018) Fast Ca(2+) responses in astrocyte end-feet and neurovascular coupling in mice. Glia 66(2):348–358. https://doi.org/10.1002/glia.23246

Lind KR, Ball KK, Cruz NF, Dienel GA (2013b) The unfolded protein response to endoplasmic reticulum stress in cultured astrocytes and rat brain during experimental diabetes. Neurochem Int 62(5):784–795. https://doi.org/10.1016/j.neuint.2013.02.009

Lindenmayer JP, Nasrallah H, Pucci M, James S, Citrome L (2013) A systematic review of psychostimulant treatment of negative symptoms of schizophrenia: challenges and therapeutic opportunities. Schizophr Res 147(2–3):241–252. https://doi.org/10.1016/j.schres.2013.03.019

Lipina TV, Fletcher PJ, Lee FH, Wong AHC, Roder JC (2013) Disrupted-in-schizophrenia-1 Gln31Leu polymorphism results in social anhedonia associated with monoaminergic imbalance and reduction of CREB and beta-arrestin-1,2 in the nucleus accumbens in a mouse model of depression. Neuropsychopharmacology 38(3):423–436. https://doi.org/10.1038/npp.2012.197

Liu L, MacKenzie KR, Putluri N, Maletic-Savatic M, Bellen HJ (2017) The glia-neuron lactate shuttle and elevated ROS promote lipid synthesis in neurons and lipid droplet accumulation in glia via APOE/D. Cell Metab 26(5):719–737. e716. https://doi.org/10.1016/j.cmet.2017.08.024

Liu L, Zhang K, Sandoval H, Yamamoto S, Jaiswal M, Sanz E, Li Z, Hui J, Graham BH, Quintana A, Bellen HJ (2015) Glial lipid droplets and ROS induced by mitochondrial defects promote neurodegeneration. Cell 160(1–2):177–190. https://doi.org/10.1016/j.cell.2014.12.019

Liu Z, Osipovitch M, Benraiss A, Huynh NPT, Foti R, Bates J, Chandler-Militello D, Findling RL, Tesar PJ, Nedergaard M, Windrem MS, Goldman SA (2019) Dysregulated glial differentiation in schizophrenia may be relieved by suppression of SMAD4- and REST-dependent signaling. Cell Rep 27(13):3832–3843. e3836. https://doi.org/10.1016/j.celrep.2019.05.088

Lopes-Cardozo M, Larsson OM, Schousboe A (1986) Acetoacetate and glucose as lipid precursors and energy substrates in primary cultures of astrocytes and neurons from mouse cerebral cortex. J Neurochem 46(3):773–778. https://doi.org/10.1111/j.1471-4159.1986.tb13039.x

Lundqvist MH, Almby K, Abrahamsson N, Eriksson JW (2019) Is the brain a key player in glucose regulation and development of type 2 diabetes? Front Physiol 10:457. https://doi.org/10.3389/fphys.2019.00457

Ma TM, Abazyan S, Abazyan B, Nomura J, Yang C, Seshadri S, Sawa A, Snyder SH, Pletnikov MV (2013) Pathogenic disruption of DISC1-serine racemase binding elicits schizophrenia-like behavior via D-serine depletion. Mol Psychiatry 18(5):557–567. https://doi.org/10.1038/mp.2012.97

Maas DA, Valles A, Martens GJM (2017) Oxidative stress, prefrontal cortex hypomyelination and cognitive symptoms in schizophrenia. Transl Psychiatry 7(7):e1171. https://doi.org/10.1038/tp.2017.138

Madayag AC, Stringfield SJ, Reissner KJ, Boettiger CA, Robinson DL (2017) Sex and adolescent ethanol exposure influence pavlovian conditioned approach. Alcohol Clin Exp Res 41(4):846–856. https://doi.org/10.1111/acer.13354

Mahan VL (2019) Neurointegrity and neurophysiology: astrocyte, glutamate, and carbon monoxide interactions. Med Gas Res 9(1):24–45. https://doi.org/10.4103/2045-9912.254639

Mallajosyula JK, Kaur D, Chinta SJ, Rajagopalan S, Rane A, Nicholls DG, Di Monte DA, Macarthur H, Andersen JK (2008) MAO-B elevation in mouse brain astrocytes results in Parkinson's pathology. PLoS One 3(2):e1616. https://doi.org/10.1371/journal.pone.0001616

Mandal M, Donnelly R, Elkabes S, Zhang P, Davini D, David BT, Ponzio NM (2013) Maternal immune stimulation during pregnancy shapes the immunological phenotype of offspring. Brain Behav Immun 33:33–45. https://doi.org/10.1016/j.bbi.2013.04.012

Mangia S, DiNuzzo M, Giove F, Carruthers A, Simpson IA, Vannucci SJ (2011) Response to 'comment on recent modeling studies of astrocyte-neuron metabolic interactions': much ado about nothing. J Cereb Blood Flow Metab 31(6):1346–1353. https://doi.org/10.1038/jcbfm.2011.29

Martinez-Hernandez A, Bell KP, Norenberg MD (1977) Glutamine synthetase: glial localization in brain. Science 195(4284):1356–1358. https://doi.org/10.1126/science.14400

Martins-De-Souza D, Gattaz WF, Schmitt A, Maccarrone G, Hunyadi-Gulyas E, Eberlin MN, Souza GHMF, Marangoni S, Novello JC, Turck CW, Dias-Neto E (2009a) Proteomic analysis of dorsolateral prefrontal cortex indicates the involvement of cytoskeleton, oligodendrocyte, energy metabolism and new potential markers in schizophrenia. J Psychiatr Res 43(11):978–986. https://doi.org/10.1016/j.jpsychires.2008.11.006

Martins-De-Souza D, Gattaz WF, Schmitt A, Novello JC, Marangoni S, Turck CW, Dias-Neto E (2009b) Proteome analysis of schizophrenia patients Wernicke's area reveals an energy metabolism dysregulation. BMC Psychiatry 9:17. https://doi.org/10.1186/1471-244x-9-17

Marty N, Dallaporta M, Foretz M, Emery M, Tarussio D, Bady I, Binnert C, Beermann F, Thorens B (2005) Regulation of glucagon secretion by glucose transporter type 2 (glut2) and astrocyte-dependent glucose sensors. J Clin Invest 115(12):3545–3553. https://doi.org/10.1172/JCI26309

Maurer I, Zierz S, Moller HJ (2001) Evidence for a mitochondrial oxidative phosphorylation defect in brains from patients with schizophrenia. Schizophr Res 48(1):125–136. https://doi.org/10.1016/S0920-9964(00)00075-X

Mazuel L, Blanc J, Repond C, Bouchaud V, Raffard G, Deglon N, Bonvento G, Pellerin L, Bouzier-Sore AK (2017) A neuronal MCT2 knockdown in the rat somatosensory cortex reduces both the NMR lactate signal and the BOLD response during whisker stimulation. PLoS One 12(4):e0174990. https://doi.org/10.1371/journal.pone.0174990

McCullumsmith RE, O'Donovan SM, Drummond JB, Benesh FS, Simmons M, Roberts R, Lauriat T, Haroutunian V, Meador-Woodruff JH (2016) Cell-specific abnormalities of glutamate transporters in schizophrenia: sick astrocytes and compensating relay neurons? Mol Psychiatry 21(6):823–830. https://doi.org/10.1038/mp.2015.148

McKenna MC, Hopkins IB, Lindauer SL, Bamford P (2006a) Aspartate aminotransferase in synaptic and nonsynaptic mitochondria: differential effect of compounds that influence transient hetero-enzyme complex (metabolon) formation. Neurochem Int 48(6–7):629–636. https://doi.org/10.1016/j.neuint.2005.11.018

McKenna MC, Waagepetersen HS, Schousboe A, Sonnewald U (2006b) Neuronal and astrocytic shuttle mechanisms for cytosolic-mitochondrial transfer of reducing equivalents: current evidence and pharmacological tools. Biochem Pharmacol 71(4):399–407. https://doi.org/10.1016/j.bcp.2005.10.011

Meier SD, Kafitz KW, Rose CR (2008) Developmental profile and mechanisms of GABA-induced calcium signaling in hippocampal astrocytes. Glia 56(10):1127–1137. https://doi.org/10.1002/glia.20684

Melo TM, Nehlig A, Sonnewald U (2006) Neuronal-glial interactions in rats fed a ketogenic diet. Neurochem Int 48(6–7):498–507. https://doi.org/10.1016/j.neuint.2005.12.037

Middleton FA, Mirnics K, Pierri JN, Lewis DA, Levitt P (2002) Gene expression profiling reveals alterations of specific metabolic pathways in schizophrenia. J Neurosci 22(7):2718–2729. https://doi.org/10.1523/Jneurosci.22-07-02718.2002

Miederer I, Uebbing K, Rohrich J, Maus S, Bausbacher N, Krauter K, Weyer-Elberich V, Lutz B, Schreckenberger M, Urban R (2017) Effects of tetrahydrocannabinol on glucose uptake in the rat brain. Neuropharmacology 117:273–281. https://doi.org/10.1016/j.neuropharm.2017.02.011

Millar JK, Christie S, Anderson S, Lawson D, Loh DHW, Devon RS, Arveiler B, Muir WJ, Blackwood DHR, Porteous DJ (2001) Genomic structure and localisation within a linkage hotspot of disrupted in schizophrenia 1, a gene disrupted by a translocation segregating with schizophrenia. Mol Psychiatr 6(2):173–178. https://doi.org/10.1038/sj.mp.4000784

Millar JK, James R, Christie S, Porteous DJ (2005) Disrupted in schizophrenia 1 (DISC1): subcellular targeting and induction of ring mitochondria. Mol Cell Neurosci 30(4):477–484. https://doi.org/10.1016/j.mcn.2005.08.021

Millar JK, Wilson-Annan JC, Anderson S, Christie S, Taylor MS, Semple CA, Devon RS, St Clair DM, Muir WJ, Blackwood DH, Porteous DJ (2000) Disruption of two novel genes by a translocation co-segregating with schizophrenia. Hum Mol Genet 9(9):1415–1423. https://doi.org/10.1093/hmg/9.9.1415

Mirnics K, Middleton FA, Lewis DA, Levitt P (2001) Analysis of complex brain disorders with gene expression microarrays: schizophrenia as a disease of the synapse. Trends Neurosci 24(8):479–486. https://doi.org/10.1016/s0166-2236(00)01862-2

Mishra A (2017) Binaural blood flow control by astrocytes: listening to synapses and the vasculature. J Physiol 595(6):1885–1902. https://doi.org/10.1113/JP270979

Moises HW, Wollschlager D, Binder H (2015) Functional genomics indicate that schizophrenia may be an adult vascular-ischemic disorder. Transl Psychiatry 5:e616. https://doi.org/10.1038/tp.2015.103

Moore CI, Cao R (2008) The hemo-neural hypothesis: on the role of blood flow in information processing. J Neurophysiol 99(5):2035–2047. https://doi.org/10.1152/jn.01366.2006

Morales-Soto W, Gulbransen BD (2019) Enteric glia: a new player in abdominal pain. Cell Mol Gastroenterol Hepatol 7(2):433–445. https://doi.org/10.1016/j.jcmgh.2018.11.005

Morita M, Ikeshima-Kataoka H, Kreft M, Vardjan N, Zorec R, Noda M (2019) Metabolic plasticity of astrocytes and aging of the brain. Int J Mol Sci 20(4):941. https://doi.org/10.3390/ijms20040941

Mothet JP, Parent AT, Wolosker H, Brady RO, Linden DJ, Ferris CD, Rogawski MA, Snyder SH (2000) D-serine is an endogenous ligand for the glycine site of the N-methyl-D-aspartate receptor. Proc Natl Acad Sci U S A 97(9):4926–4931. https://doi.org/10.1073/pnas.97.9.4926

Mothet JP, Pollegioni L, Ouanounou G, Martineau M, Fossier P, Baux G (2005) Glutamate receptor activation triggers a calcium-dependent and SNARE protein-dependent release of the gliotransmitter D-serine. Proc Natl Acad Sci U S A 102(15):5606–5611. https://doi.org/10.1073/pnas.0408483102

Muller MS, Fox R, Schousboe A, Waagepetersen HS, Bak LK (2014) Astrocyte glycogenolysis is triggered by store-operated calcium entry and provides metabolic energy for cellular calcium homeostasis. Glia 62(4):526–534. https://doi.org/10.1002/glia.22623

Mulligan SJ, MacVicar BA (2004) Calcium transients in astrocyte endfeet cause cerebrovascular constrictions. Nature 431(7005):195–199. https://doi.org/10.1038/nature02827

Murphy-Royal C, Johnston AD, Boyce AKJ, Diaz-Castro B, Institoris A, Peringod G, Zhang O, Stout RF, Spray DC, Thompson RJ, Khakh BS, Bains JS, Gordon GR (2020) Stress gates an astrocytic energy reservoir to impair synaptic plasticity. Nat Commun 11(1):2014. https://doi.org/10.1038/s41467-020-15778-9

Navarrete M, Araque A (2008) Endocannabinoids mediate neuron-astrocyte communication. Neuron 57(6):883–893. https://doi.org/10.1016/j.neuron.2008.01.029

Navarrete M, Araque A (2010) Endocannabinoids potentiate synaptic transmission through stimulation of astrocytes. Neuron 68(1):113–126. https://doi.org/10.1016/j.neuron.2010.08.043

Nguyen TT, Kosciolek T, Eyler LT, Knight R, Jeste DV (2018) Overview and systematic review of studies of microbiome in schizophrenia and bipolar disorder. J Psychiatr Res 99:50–61. https://doi.org/10.1016/j.jpsychires.2018.01.013

Norkett R, Modi S, Birsa N, Atkin TA, Ivankovic D, Pathania M, Trossbach SV, Korth C, Hirst WD, Kittler JT (2016) DISC1-dependent regulation of mitochondrial dynamics controls the morphogenesis of complex neuronal dendrites. J Biol Chem 291(2):613–629. https://doi.org/10.1074/jbc.M115.699447

O'Donnell JC, Jackson JG, Robinson MB (2016) Transient oxygen/glucose deprivation causes a delayed loss of mitochondria and increases spontaneous calcium signaling in astrocytic processes. J Neurosci 36(27):7109–7127. https://doi.org/10.1523/JNEUROSCI.4518-15.2016

Obel LF, Muller MS, Walls AB, Sickmann HM, Bak LK, Waagepetersen HS, Schousboe A (2012) Brain glycogen-new perspectives on its metabolic function and regulation at the subcellular level. Front Neuroenerg 4:3. https://doi.org/10.3389/fnene.2012.00003

Oe Y, Baba O, Ashida H, Nakamura KC, Hirase H (2016) Glycogen distribution in the microwave-fixed mouse brain reveals heterogeneous astrocytic patterns. Glia 64(9):1532–1545. https://doi.org/10.1002/glia.23020

Ogawa F, Malavasi ELV, Crummie DK, Eykelenboom JE, Soares DC, Mackie S, Porteous DJ, Millar JK (2014) DISC1 complexes with TRAK1 and Miro1 to modulate anterograde axonal mitochondrial trafficking. Hum Mol Genet 23(4):906–919. https://doi.org/10.1093/hmg/ddt485

Oheim M, Schmidt E, Hirrlinger J (2018) Local energy on demand: are 'spontaneous' astrocytic Ca(2+)-microdomains the regulatory unit for astrocyte-neuron metabolic cooperation? Brain Res Bull 136:54–64. https://doi.org/10.1016/j.brainresbull.2017.04.011

Ohi K, Otowa T, Shimada M, Sasaki T, Tanii H (2020) Shared genetic etiology between anxiety disorders and psychiatric and related intermediate phenotypes. Psychol Med 50(4):692–704. https://doi.org/10.1017/S003329171900059X

Okubo Y, Kanemaru K, Suzuki J, Kobayashi K, Hirose K, Iino M (2019) Inositol 1,4,5-trisphosphate receptor type 2-independent Ca(2+) release from the endoplasmic reticulum in astrocytes. Glia 67(1):113–124. https://doi.org/10.1002/glia.23531

Ota M, Noda T, Sato N, Hattori K, Teraishi T, Hori H, Nagashima A, Shimoji K, Higuchi T, Kunugi H (2014) Characteristic distributions of regional cerebral blood flow changes in major depressive disorder patients: a pseudo-continuous arterial spin labeling (pCASL) study. J Affect Disord 165:59–63. https://doi.org/10.1016/j.jad.2014.04.032

Otsu Y, Couchman K, Lyons DG, Collot M, Agarwal A, Mallet JM, Pfrieger FW, Bergles DE, Charpak S (2015) Calcium dynamics in astrocyte processes during neurovascular coupling. Nat Neurosci 18(2):210–218. https://doi.org/10.1038/nn.3906

Pan Y, Scanlon MJ, Owada Y, Yamamoto Y, Porter CJ, Nicolazzo JA (2015) Fatty acid-binding protein 5 facilitates the blood-brain barrier transport of docosahexaenoic acid. Mol Pharm 12(12):4375–4385. https://doi.org/10.1021/acs.molpharmaceut.5b00580

Panariello F, Perruolo G, Cassese A, Giacco F, Botta G, Barbagallo AP, Muscettola G, Beguinot F, Formisano P, de Bartolomeis A (2012) Clozapine impairs insulin action by up-regulating Akt phosphorylation and Ped/Pea-15 protein abundance. J Cell Physiol 227(4):1485–1492. https://doi.org/10.1002/jcp.22864

Panatier A, Theodosis DT, Mothet JP, Touquet B, Pollegioni L, Poulain DA, Oliet SHR (2006) Glia-derived D-serine controls NMDA receptor activity and synaptic memory. Cell 125(4):775–784. https://doi.org/10.1016/j.cell.2006.02.051

Pang XS, Panee J, Liu XQ, Berry MJ, Chang SLL, Chang LD (2013) Regional variations of antioxidant capacity and oxidative stress responses in HIV-1 transgenic rats with and without methamphetamine administration. J Neuroimmune Pharmacol 8(3):691–704. https://doi.org/10.1007/s11481-013-9454-8

Panov A, Orynbayeva Z, Vavilin V, Lyakhovich V (2014) Fatty acids in energy metabolism of the central nervous system. Biomed Res Int 2014:472459. https://doi.org/10.1155/2014/472459

Pardinas AF, Holmans P, Pocklington AJ, Escott-Price V, Ripke S, Carrera N, Legge SE, Bishop S, Cameron D, Hamshere ML, Han J, Hubbard L, Lynham A, Mantripragada K, Rees E, MacCabe JH, McCarroll SA, Baune BT, Breen G, Byrne EM, Dannlowski U, Eley TC, Hayward C, Martin NG, McIntosh AM, Plomin R, Porteous DJ, Wray NR, Caballero A, Geschwind DH, Huckins LM, Ruderfer DM, Santiago E, Sklar P, Stahl EA, Won H, Agerbo E, Als TD, Andreassen OA, Baekvad-Hansen M, Mortensen PB, Pedersen CB, Borglum AD, Bybjerg-Grauholm J, Djurovic S, Durmishi N, Pedersen MG, Golimbet V, Grove J, Hougaard DM, Mattheisen M, Molden E, Mors O, Nordentoft M, Pejovic-Milovancevic M, Sigurdsson E, Silagadze T, Hansen CS, Stefansson K, Stefansson H, Steinberg S, Tosato S, Werge T, Consortium G, Consortium C, Collier DA, Rujescu D, Kirov G, Owen MJ, O'Donovan MC, Walters JTR (2018) Common schizophrenia alleles are enriched in mutation-intolerant genes and in regions under strong background selection. Nat Genet 50(3):381–389. https://doi.org/10.1038/s41588-018-0059-2

Pardo B, Rodrigues TB, Contreras L, Garzon M, Llorente-Folch I, Kobayashi K, Saheki T, Cerdan S, Satrustegui J (2011) Brain glutamine synthesis requires neuronal-born aspartate as amino donor for glial glutamate formation. J Cereb Blood Flow Metab 31(1):90–101. https://doi.org/10.1038/jcbfm.2010.146

Parikh V, Naughton SX, Shi XD, Kelley LK, Yegla B, Tallarida CS, Rawls SM, Unterwald EM (2014) Cocaine-induced neuroadaptations in the dorsal striatum: glutamate dynamics and behavioral sensitization. Neurochem Int 75:54–65. https://doi.org/10.1016/j.neuint.2014.05.016

Park JH, Hong JS, Kim SM, Min KJ, Chung US, Han DH (2019) Effects of amisulpride adjunctive therapy on working memory and brain metabolism in the frontal cortex of patients with schizophrenia: a preliminary positron emission tomography/computerized tomography investigation. Clin Psychopharmacol Neurosci 17(2):250–260. https://doi.org/10.9758/cpn.2019.17.2.250

Park YU, Jeong J, Lee H, Mun JY, Kim JH, Lee JS, Nguyen MD, Han SS, Suh PG, Park SK (2010) Disrupted-in-schizophrenia 1 (DISC1) plays essential roles in mitochondria in collaboration with Mitofilin. Proc Natl Acad Sci U S A 107(41):17785–17790. https://doi.org/10.1073/pnas.1004361107

Parnis J, Montana V, Delgado-Martinez I, Matyash V, Parpura V, Kettenmann H, Sekler I, Nolte C (2013) Mitochondrial exchanger NCLX plays a major role in the intracellular Ca2+ signaling, gliotransmission, and proliferation of astrocytes. J Neurosci 33(17):7206–7219. https://doi.org/10.1523/JNEUROSCI.5721-12.2013

Patel AB, Lai JC, Chowdhury GM, Hyder F, Rothman DL, Shulman RG, Behar KL (2014) Direct evidence for activity-dependent glucose phosphorylation in neurons with implications for the astrocyte-to-neuron lactate shuttle. Proc Natl Acad Sci U S A 111(14):5385–5390. https://doi.org/10.1073/pnas.1403576111

Pati D, Kelly K, Stennett B, Frazier CJ, Knackstedt LA (2016) Alcohol consumption increases basal extracellular glutamate in the nucleus accumbens core of Sprague-Dawley rats without increasing spontaneous glutamate release. Eur J Neurosci 44(2):1896–1905. https://doi.org/10.1111/ejn.13284

Pearson-Leary J, McNay EC (2016) Novel roles for the insulin-regulated glucose transporter-4 in hippocampally dependent memory. J Neurosci 36(47):11851–11864. https://doi.org/10.1523/JNEUROSCI.1700-16.2016

Pellerin L, Magistretti PJ (1994) Glutamate uptake into astrocytes stimulates aerobic glycolysis: a mechanism coupling neuronal activity to glucose utilization. Proc Natl Acad Sci U S A 91(22):10625–10629. https://doi.org/10.1073/pnas.91.22.10625

Peng WW, Huang JD, Zheng YJ, Ding Y, Li ST, Zhang JL, Lyu JJ, Zeng QY (2019) UCP2 silencing aggravates mitochondrial dysfunction in astrocytes under septic conditions. Mol Med Rep 20(5):4459–4466. https://doi.org/10.3892/mmr.2019.10721

Pennington K, Beasley CL, Dicker P, Fagan A, English J, Pariante CM, Wait R, Dunn MJ, Cotter DR (2008) Prominent synaptic and metabolic abnormalities revealed by proteomic analysis of the dorsolateral prefrontal cortex in schizophrenia and bipolar disorder. Mol Psychiatry 13(12):1102–1117. https://doi.org/10.1038/sj.mp.4002098

Perry BI, McIntosh G, Weich S, Singh S, Rees K (2016) The association between first-episode psychosis and abnormal glycaemic control: systematic review and meta-analysis. Lancet Psychiatry 3(11):1049–1058. https://doi.org/10.1016/S2215-0366(16)30262-0

Petrikis P, Tigas S, Tzallas AT, Papadopoulos I, Skapinakis P, Mavreas V (2015) Parameters of glucose and lipid metabolism at the fasted state in drug-naive first-episode patients with psychosis: evidence for insulin resistance. Psychiatry Res 229(3):901–904. https://doi.org/10.1016/j.psychres.2015.07.041

Pierre K, Pellerin L (2005) Monocarboxylate transporters in the central nervous system: distribution, regulation and function. J Neurochem 94(1):1–14. https://doi.org/10.1111/j.1471-4159.2005.03168.x

Pillinger T, Beck K, Gobjila C, Donocik JG, Jauhar S, Howes OD (2017) Impaired glucose homeostasis in first-episode schizophrenia a systematic review and meta-analysis. JAMA Psychiat 74(3):261–269. https://doi.org/10.1001/jamapsychiatry.2016.3803

Pinacho R, Vila E, Prades R, Tarrago T, Castro E, Ferrer I, Ramos B (2016) The glial phosphorylase of glycogen isoform is reduced in the dorsolateral prefrontal cortex in chronic schizophrenia. Schizophr Res 177(1–3):37–43. https://doi.org/10.1016/j.schres.2016.04.024

Pinero-Martos E, Ortega-Vila B, Pol-Fuster J, Cisneros-Barroso E, Ruiz-Guerra L, Medina-Dols A, Heine-Suner D, Llado J, Olmos G, Vives-Bauza C (2016) Disrupted in schizophrenia 1 (DISC1) is a constituent of the mammalian mitochondrial contact site and cristae organizing system (MICOS) complex, and is essential for oxidative phosphorylation. Hum Mol Genet 25(19):4157–4169. https://doi.org/10.1093/hmg/ddw250

Pines G, Danbolt NC, Bjoras M, Zhang Y, Bendahan A, Eide L, Koepsell H, Storm-Mathisen J, Seeberg E, Kanner BI (1992) Cloning and expression of a rat brain L-glutamate transporter. Nature 360(6403):464–467. https://doi.org/10.1038/360464a0

Pitychoutis PM, Belmer A, Moutkine I, Adrien J, Maroteaux L (2015) Mice lacking the serotonin Htr2B receptor gene present an antipsychotic-sensitive schizophrenic-like phenotype. Neuropsychopharmacology 40(12):2764–2773. https://doi.org/10.1038/npp.2015.126

Pletnikov MV, Ayhan Y, Nikolskaia O, Xu Y, Ovanesov MV, Huang H, Mori S, Moran TH, Ross CA (2008) Inducible expression of mutant human DISC1 in mice is associated with brain and behavioral abnormalities reminiscent of schizophrenia. Mol Psychiatry 13(2):173–186. https://doi.org/10.1038/sj.mp.4002079

Pletnikov MV, Xu Y, Ovanesov MV, Kamiya A, Sawa A, Ross CA (2007) PC12 cell model of inducible expression of mutant DISC1: new evidence for a dominant-negative mechanism of abnormal neuronal differentiation. Neurosci Res 58(3):234–244. https://doi.org/10.1016/j.neures.2007.03.003

Pocai A, Lam TK, Gutierrez-Juarez R, Obici S, Schwartz GJ, Bryan J, Aguilar-Bryan L, Rossetti L (2005) Hypothalamic K(ATP) channels control hepatic glucose production. Nature 434(7036):1026–1031. https://doi.org/10.1038/nature03439

Pocivavsek A, Rowland LM (2018) Basic neuroscience illuminates causal relationship between sleep and memory: translating to schizophrenia. Schizophr Bull 44(1):7–14. https://doi.org/10.1093/schbul/sbx151

Pogorelov VM, Nomura J, Kim J, Kannan G, Ayhan Y, Yang CX, Taniguchi Y, Abazyan B, Valentine H, Krasnova IN, Kamiya A, Cadet JL, Wong DF, Pletnikov MV (2012) Mutant DISC1 affects methamphetamine-induced sensitization and conditioned place preference: a comorbidity model. Neuropharmacology 62(3):1242–1251. https://doi.org/10.1016/j.neuropharm.2011.02.003

Prabakaran S, Swatton JE, Ryan MM, Huffaker SJ, Huang JTJ, Griffin JL, Wayland M, Freeman T, Dudbridge F, Lilley KS, Karp NA, Hester S, Tkachev D, Mimmack ML, Yolken RH, Webster MJ, Torrey EF, Bahn S (2004) Mitochondrial dysfunction in schizophrenia: evidence for compromised brain metabolism and oxidative stress. Mol Psychiatry 9(7):684–697. https://doi.org/10.1038/sj.mp.4001511

Profitt MF, Deurveilher S, Robertson GS, Rusak B, Semba K (2016) Disruptions of sleep/wake patterns in the stable tubule only polypeptide (STOP) null mouse model of schizophrenia. Schizophr Bull 42(5):1207–1215. https://doi.org/10.1093/schbul/sbw017

Qiao J, Wang J, Wang H, Zhang Y, Zhu S, Adilijiang A, Guo H, Zhang R, Guo W, Luo G, Qiu Y, Xu H, Kong J, Huang Q, Li XM (2016) Regulation of astrocyte pathology by fluoxetine prevents the deterioration of Alzheimer phenotypes in an APP/PS1 mouse model. Glia 64(2):240–254. https://doi.org/10.1002/glia.22926

Quincozes-Santos A, Bobermin LD, de Assis AM, Goncalves CA, Souza DO (2017) Fluctuations in glucose levels induce glial toxicity with glutamatergic, oxidative and inflammatory implications. BBA-Mol Basis Dis 1863(1):1–14. https://doi.org/10.1016/j.bbadis.2016.09.013

Rabinowitz JD, Enerback S (2020) Lactate: the ugly duckling of energy metabolism. Nat Metab 2(7):566–571. https://doi.org/10.1038/s42255-020-0243-4

Ramsey AJ, Milenkovic M, Oliveira AF, Escobedo-Lozoya Y, Seshadri S, Salahpour A, Sawa A, Yasuda R, Caron MG (2011) Impaired NMDA receptor transmission alters striatal synapses and DISC1 protein in an age-dependent manner. Proc Natl Acad Sci U S A 108(14):5795–5800. https://doi.org/10.1073/pnas.1012621108

Rapoport SI, Chang MC, Spector AA (2001) Delivery and turnover of plasma-derived essential PUFAs in mammalian brain. J Lipid Res 42(5):678–685

Refolo G, Vescovo T, Piacentini M, Fimia GM, Ciccosanti F (2020) Mitochondrial interactome: a focus on antiviral signaling pathways. Front Cell Dev Biol 8:8. https://doi.org/10.3389/fcell.2020.00008

Reissner KJ, Brown RM, Spencer S, Tran PK, Thomas CA, Kalivas PW (2014) Chronic administration of the methylxanthine propentofylline impairs reinstatement to cocaine by a GLT-1-dependent mechanism. Neuropsychopharmacology 39(2):499–506. https://doi.org/10.1038/npp.2013.223

Reissner KJ, Gipson CD, Tran PK, Knackstedt LA, Scofield MD, Kalivas PW (2015) Glutamate transporter GLT-1 mediates N-acetylcysteine inhibition of cocaine reinstatement. Addict Biol 20(2):316–323. https://doi.org/10.1111/adb.12127

Reyes RC, Parpura V (2008) Mitochondria modulate Ca2+-dependent glutamate release from rat cortical astrocytes. J Neurosci 28(39):9682–9691. https://doi.org/10.1523/JNEUROSCI.3484-08.2008

Reynolds IJ, Hastings TG (1995) Glutamate induces the production of reactive oxygen species in cultured forebrain neurons following NMDA receptor activation. J Neurosci 15(5 Pt 1):3318–3327

Ristow M (2004) Neurodegenerative disorders associated with diabetes mellitus. J Mol Med (Berl) 82(8):510–529. https://doi.org/10.1007/s00109-004-0552-1

Rizig MA, McQuillin A, Ng A, Robinson M, Harrison A, Zvelebil M, Hunt SP, Gurling HM (2012) A gene expression and systems pathway analysis of the effects of clozapine compared to haloperidol in the mouse brain implicates susceptibility genes for schizophrenia. J Psychopharmacol 26(9):1218–1230. https://doi.org/10.1177/0269881112450780

Roach PJ (2015) Glycogen phosphorylation and Lafora disease. Mol Asp Med 46:78–84. https://doi.org/10.1016/j.mam.2015.08.003

Rodriguez-Prados M, Rojo-Ruiz J, Garcia-Sancho J, Alonso MT (2020) Direct monitoring of ER Ca(2+) dynamics reveals that Ca(2+) entry induces ER-Ca(2+) release in astrocytes. Pflugers Arch 472(4):439–448. https://doi.org/10.1007/s00424-020-02364-7

Rogers GB, Keating DJ, Young RL, Wong ML, Licinio J, Wesselingh S (2016) From gut dysbiosis to altered brain function and mental illness: mechanisms and pathways. Mol Psychiatry 21(6):738–748. https://doi.org/10.1038/mp.2016.50

Roman C, Egert L, Di Benedetto B (2020) Astrocytic-neuronal crosstalk gets jammed: alternative perspectives on the onset of neuropsychiatric disorders. Eur J Neurosci:e14900. https://doi.org/10.1111/ejn.14900

Rothman DL, Dienel GA (2019) Development of a model to test whether glycogenolysis can support astrocytic energy demands of Na(+), K(+)-ATPase and glutamate-glutamine cycling, sparing an equivalent amount of glucose for neurons. Adv Neurobiol 23:385–433. https://doi.org/10.1007/978-3-030-27480-1_14

Ruskin DN, Fortin JA, Bisnauth SN, Masino SA (2017) Ketogenic diets improve behaviors associated with autism spectrum disorder in a sex-specific manner in the EL mouse. Physiol Behav 168:138–145. https://doi.org/10.1016/j.physbeh.2016.10.023

Ruskin DN, Svedova J, Cote JL, Sandau U, Rho JM, Kawamura M, Boison D, Masino SA (2013) Ketogenic diet improves core symptoms of autism in BTBR mice. Plos One 8(6):e65021. https://doi.org/10.1371/journal.pone.0065021

Ryan MC, Collins P, Thakore JH (2003) Impaired fasting glucose tolerance in first-episode, drug-naive patients with schizophrenia. Am J Psychiatry 160(2):284–289. https://doi.org/10.1176/appi.ajp.160.2.284

Saez I, Duran J, Sinadinos C, Beltran A, Yanes O, Tevy MF, Martinez-Pons C, Milan M, Guinovart JJ (2014) Neurons have an active glycogen metabolism that contributes to tolerance to hypoxia. J Cereb Blood Flow Metab 34(6):945–955. https://doi.org/10.1038/jcbfm.2014.33

Sanchez C, Galve-Roperh I, Rueda D, Guzman M (1998) Involvement of sphingomyelin hydrolysis and the mitogen-activated protein kinase cascade in the Delta(9)-tetrahydrocannabinol-induced stimulation of glucose metabolism in primary astrocytes. Mol Pharmacol 54(5):834–843

Sari Y, Toalston JE, Rao PSS, Bell RL (2016) Effects of ceftriaxone on ethanol, nicotine or sucrose intake by alcohol-preferring (P) rats and its association with Glt-1 expression. Neuroscience 326:117–125. https://doi.org/10.1016/j.neuroscience.2016.04.004

Sarnyai Z, Kraeuter AK, Palmer CM (2019) Ketogenic diet for schizophrenia: clinical implication. Curr Opin Psychiatry 32(5):394–401. https://doi.org/10.1097/Yco.0000000000000535

Sarnyai Z, Palmer CM (2020) Ketogenic therapy in serious mental illness: emerging evidence. Int J Neuropsychopharmacol 23(7):434–439. https://doi.org/10.1093/ijnp/pyaa036

Satrustegui J, Bak LK (2015) Fluctuations in cytosolic calcium regulate the neuronal malate-aspartate NADH shuttle: implications for neuronal energy metabolism. Neurochem Res 40(12):2425–2430. https://doi.org/10.1007/s11064-015-1652-8

Sawahata M, Mori D, Arioka Y, Kubo H, Kushima I, Kitagawa K, Sobue A, Shishido E, Sekiguchi M, Kodama A, Ikeda R, Aleksic B, Kimura H, Ishizuka K, Nagai T, Kaibuchi K, Nabeshima T, Yamada K, Ozaki N (2020) Generation and analysis of novel Reln-deleted mouse model corresponding to exonic Reln deletion in schizophrenia. Psychiatry Clin Neurosci 74(5):318–327. https://doi.org/10.1111/pcn.12993

Sawamura N, Sawamura-Yamamoto T, Ozeki Y, Ross CA, Sawa A (2005) A form of DISC1 enriched in nucleus: altered subcellular distribution in orbitofrontal cortex in psychosis and substance/alcohol abuse. Proc Natl Acad Sci U S A 102(4):1187–1192. https://doi.org/10.1073/pnas.0406543102

Scaini G, Quevedo J, Velligan D, Roberts DL, Raventos H, Walss-Bass C (2018) Second generation antipsychotic-induced mitochondrial alterations: implications for increased risk of metabolic syndrome in patients with schizophrenia. Eur Neuropsychopharmacol 28(3):369–380. https://doi.org/10.1016/j.euroneuro.2018.01.004

Scemes E, Dermietzel R, Spray DC (1998) Calcium waves between astrocytes from Cx43 knockout mice. Glia 24(1):65–73. https://doi.org/10.1002/(sici)1098-1136(199809)24:1<65::aid-glia7>3.0.co;2-#

Schizophrenia Working Group of the Psychiatric Genomics Consortium (2014) Biological insights from 108 schizophrenia-associated genetic loci. Nature 511(7510):421–427. https://doi.org/10.1038/nature13595

Schmidt-Kastner R, Guloksuz S, Kietzmann T, van Os J, Rutten BPF (2020) Analysis of GWAS-derived schizophrenia genes for links to ischemia-hypoxia response of the brain. Front Psych 11:393. https://doi.org/10.3389/fpsyt.2020.00393

Schnaitman C, Erwin VG, Greenawalt JW (1967) The submitochondrial localization of monoamine oxidase. An enzymatic marker for the outer membrane of rat liver mitochondria. J Cell Biol 32(3):719–735. https://doi.org/10.1083/jcb.32.3.719

Schonfeld P, Reiser G (2017) Brain energy metabolism spurns fatty acids as fuel due to their inherent mitotoxicity and potential capacity to unleash neurodegeneration. Neurochem Int 109:68–77. https://doi.org/10.1016/j.neuint.2017.03.018

Sengupta S, Parrilla-Escobar MA, Klink R, Fathalli F, Ng YK, Stip E, Baptista T, Malla A, Joober R (2008) Are metabolic indices different between drug-naive first-episode psychosis patients and healthy controls? Schizophr Res 102(1–3):329–336. https://doi.org/10.1016/j.schres.2008.02.013

Sequeira A, Mamdani F, Ernst C, Vawter MP, Bunney WE, Lebel V, Rehal S, Klempan T, Gratton A, Benkelfat C, Rouleau GA, Mechawar N, Turecki G (2009) Global brain gene expression analysis links glutamatergic and GABAergic alterations to suicide and major depression. PLoS One 4(8):e6585. https://doi.org/10.1371/journal.pone.0006585

Seshadri S, Kamiya A, Yokota Y, Prikulis I, Kano S, Hayashi-Takagi A, Stanco A, Eom TY, Rao S, Ishizuka K, Wong P, Korth C, Anton ES, Sawa A (2010) Disrupted-in-Schizophrenia-1 expression is regulated by beta-site amyloid precursor protein cleaving enzyme-1-neuregulin cascade. Proc Natl Acad Sci U S A 107(12):5622–5627. https://doi.org/10.1073/pnas.0909284107

Shan D, Mount D, Moore S, Haroutunian V, Meador-Woodruff JH, McCullumsmith RE (2014) Abnormal partitioning of hexokinase 1 suggests disruption of a glutamate transport protein complex in schizophrenia. Schizophr Res 154(1–3):1–13. https://doi.org/10.1016/j.schres.2014.01.028

Sharifi K, Ebrahimi M, Kagawa Y, Islam A, Tuerxun T, Yasumoto Y, Hara T, Yamamoto Y, Miyazaki H, Tokuda N, Yoshikawa T, Owada Y (2013) Differential expression and regulatory roles of FABP5 and FABP7 in oligodendrocyte lineage cells. Cell Tissue Res 354(3):683–695. https://doi.org/10.1007/s00441-013-1730-7

Sharifi K, Morihiro Y, Maekawa M, Yasumoto Y, Hoshi H, Adachi Y, Sawada T, Tokuda N, Kondo H, Yoshikawa T, Suzuki M, Owada Y (2011) FABP7 expression in normal and stab-injured brain cortex and its role in astrocyte proliferation. Histochem Cell Biol 136(5):501–513. https://doi.org/10.1007/s00418-011-0865-4

Shashidharan P, Wittenberg I, Plaitakis A (1994) Molecular cloning of human brain glutamate/aspartate transporter II. Biochim Biophys Acta 1191(2):393–396. https://doi.org/10.1016/0005-2736(94)90192-9

Shelton MK, McCarthy KD (2000) Hippocampal astrocytes exhibit Ca2+-elevating muscarinic cholinergic and histaminergic receptors in situ. J Neurochem 74(2):555–563. https://doi.org/10.1046/j.1471-4159.2000.740555.x

Shen HW, Scofield MD, Boger H, Hensley M, Kalivas PW (2014) Synaptic glutamate spillover due to impaired glutamate uptake mediates heroin relapse. J Neurosci 34(16):5649–5657. https://doi.org/10.1523/Jneurosci.4564-13.2014

Shen SB, Lang B, Nakamoto C, Zhang F, Pu J, Kuan SL, Chatzi C, He S, Mackie I, Brandon NJ, Marquis KL, Day M, Hurko O, McCaig CD, Riedel G, St Clair D (2008) Schizophrenia-related neural and behavioral phenotypes in transgenic mice expressing truncated Disc1. J Neurosci 28(43):10893–10904. https://doi.org/10.1523/Jneurosci.3299-08.2008

Shevelkin AV, Terrillion CE, Abazyan BN, Kajstura TJ, Jouroukhin YA, Rudow GL, Troncoso JC, Linden DJ, Pletnikov MV (2017) Expression of mutant DISC1 in Purkinje cells increases their spontaneous activity and impairs cognitive and social behaviors in mice. Neurobiol Dis 103:144–153. https://doi.org/10.1016/j.nbd.2017.04.008

Shevelkin AV, Terrillion CE, Hasegawa Y, Mychko OA, Jouroukhin Y, Sawa A, Kamiya A, Pletnikov MV (2020) Astrocyte DISC1 contributes to cognitive function in a brain region-dependent manner. Hum Mol Genet 29(17):2936–2950. https://doi.org/10.1093/hmg/ddaa180

Shi L, Bergson CM (2020) Neuregulin 1: an intriguing therapeutic target for neurodevelopmental disorders. Transl Psychiat 10(1):190. https://doi.org/10.1038/s41398-020-00868-5

Shirai Y, Shirakawa O, Nishino N, Saito N, Nakai H (1996) Increased striatal glutamate transporter by repeated intermittent administration of methamphetamine. Psychiatry Clin Neurosci 50(3):161–164. https://doi.org/10.1111/j.1440-1819.1996.tb01682.x

Sidiropoulou K, Chao S, Lu W, Wolf ME (2001) Amphetamine administration does not alter protein levels of the GLT-1 and EAAC1 glutamate transporter subtypes in rat midbrain, nucleus accumbens, striatum, or prefrontal cortex. Brain Res Mol Brain Res 90(2):187–192. https://doi.org/10.1016/s0169-328x(01)00110-3

Simard M, Arcuino G, Takano T, Liu QS, Nedergaard M (2003) Signaling at the gliovascular interface. J Neurosci 23(27):9254–9262

Simpson IA, Carruthers A, Vannucci SJ (2007) Supply and demand in cerebral energy metabolism: the role of nutrient transporters. J Cereb Blood Flow Metab 27(11):1766–1791. https://doi.org/10.1038/sj.jcbfm.9600521

Skirzewski M, Cronin ME, Murphy R, Fobbs W, Kravitz AV, Buonanno A (2020) ErbB4 null mice display altered mesocorticolimbic and nigrostriatal dopamine levels as well as deficits in cognitive and motivational behaviors. Eneuro 7(3):0395-19.2020. https://doi.org/10.1523/Eneuro.0395-19.2020

Smedemark-Margulies N, Brownstein CA, Vargas S, Tembulkar SK, Towne MC, Shi J, Gonzalez-Cuevas E, Liu KX, Bilguvar K, Kleiman RJ, Han MJ, Torres A, Berry GT, Yu TW, Beggs AH, Agrawal PB, Gonzalez-Heydrich J (2016) A novel de novo mutation in ATP1A3 and childhood-onset schizophrenia. Cold Spring Harb Mol Case Stud 2(5):a001008. https://doi.org/10.1101/mcs.a001008

Smith QR, Nagura H (2001) Fatty acid uptake and incorporation in brain: studies with the perfusion model. J Mol Neurosci 16(2–3):167–172.; discussion 215-121. https://doi.org/10.1385/JMN:16:2-3:167

Smith RC, Jin H, Li C, Bark N, Shekhar A, Dwivedi S, Mortiere C, Lohr J, Hu Q, Davis JM (2013) Effects of pioglitazone on metabolic abnormalities, psychopathology, and cognitive function in schizophrenic patients treated with antipsychotic medication: a randomized double-blind study. Schizophr Res 143(1):18–24. https://doi.org/10.1016/j.schres.2012.10.023

Sobue A, Kushima I, Nagai T, Shan W, Kohno T, Aleksic B, Aoyama Y, Mori D, Arioka Y, Kawano N, Yamamoto M, Hattori M, Nabeshima T, Yamada K, Ozaki N (2018) Genetic and animal model analyses reveal the pathogenic role of a novel deletion of RELN in schizophrenia. Sci Rep 8:13046. https://doi.org/10.1038/s41598-018-31390-w

Sofroniew MV (2020) Astrocyte reactivity: subtypes, states, and functions in CNS innate immunity. Trends Immunol 41(9):758–770. https://doi.org/10.1016/j.it.2020.07.004

Sondheimer I, Knackstedt LA (2011) Ceftriaxone prevents the induction of cocaine sensitization and produces enduring attenuation of cue- and cocaine-primed reinstatement of cocaine-seeking. Behav Brain Res 225(1):252–258. https://doi.org/10.1016/j.bbr.2011.07.041

Song N, Qi Q, Cao RY, Qin BJ, Wang B, Wang YX, Zhao L, Li W, Du XL, Liu F, Yan YZ, Yi W, Jiang HL, Li T, Zhou T, Li HY, Xia Q, Zhang XM, Zhong W, Li AL, Duan XT (2019) MAVS O-GlcNAcylation is essential for host antiviral immunity against lethal RNA viruses. Cell Rep 28(9):2386. https://doi.org/10.1016/j.celrep.2019.07.085

Sotelo C, Palay SL (1968) The fine structure of the lateral vestibular nucleus in the rat. I. Neurons and neuroglial cells. J Cell Biol 36(1):151–179

Spelman LM, Walsh PI, Sharifi N, Collins P, Thakore JH (2007) Impaired glucose tolerance in first-episode drug-naive patients with schizophrenia. Diabet Med 24(5):481–485. https://doi.org/10.1111/j.1464-5491.2007.02092.x

Spilioti M, Evangeliou AE, Tramma D, Theodoridou Z, Metaxas S, Michailidi E, Bonti E, Frysira H, Haidopoulou A, Asprangathou D, Tsalkidis AJ, Kardaras P, Wevers RA, Jakobs C, Gibson KM (2013) Evidence for treatable inborn errors of metabolism in a cohort of 187 Greek patients with autism spectrum disorder (ASD). Front Hum Neurosci 7:858. https://doi.org/10.3389/fnhum.2013.00858

St Clair D, Blackwood D, Muir W, Carothers A, Walker M, Spowart G, Gosden C, Evans HJ (1990) Association within a family of a balanced autosomal translocation with major mental illness. Lancet 336(8706):13–16. https://doi.org/10.1016/0140-6736(90)91520-k

Steiner J, Martins-de-Souza D, Schiltz K, Sarnyai Z, Westphal S, Isermann B, Dobrowolny H, Turck CW, Bogerts B, Bernstein HG, Horvath TL, Schild L, Keilhoff G (2014) Clozapine promotes glycolysis and myelin lipid synthesis in cultured oligodendrocytes. Front Cell Neurosci 8:384. https://doi.org/10.3389/fncel.2014.00384

Stephan KE, Friston KJ, Frith CD (2009) Dysconnection in schizophrenia: from abnormal synaptic plasticity to failures of self-monitoring. Schizophr Bull 35(3):509–527. https://doi.org/10.1093/schbul/sbn176

Stobart JL, Anderson CM (2013) Multifunctional role of astrocytes as gatekeepers of neuronal energy supply. Front Cell Neurosci 7:38. https://doi.org/10.3389/fncel.2013.00038

Stone WS, Faraone SV, Su J, Tarbox SI, Van Eerdewegh P, Tsuang MT (2004) Evidence for linkage between regulatory enzymes in glycolysis and schizophrenia in a multiplex sample. Am J Med Genet B 127b(1):5–10. https://doi.org/10.1002/ajmg.b.20132

Storck T, Schulte S, Hofmann K, Stoffel W (1992) Structure, expression, and functional analysis of a Na(+)-dependent glutamate/aspartate transporter from rat brain. Proc Natl Acad Sci U S A 89(22):10955–10959. https://doi.org/10.1073/pnas.89.22.10955

Stork C, Renshaw PF (2005) Mitochondrial dysfunction in bipolar disorder: evidence from magnetic resonance spectroscopy research. Mol Psychiatry 10(10):900–919. https://doi.org/10.1038/sj.mp.4001711

Strazielle C, Hayzoun K, Derer M, Mariani J, Lalonde R (2006) Regional brain variations of cytochrome oxidase activity in Relnrl-orl mutant mice. J Neurosci Res 83(5):821–831. https://doi.org/10.1002/jnr.20772

Suarez I, Bodega G, Fernandez-Ruiz JJ, Ramos JA, Rubio M, Fernandez B (2002) Reduced glial fibrillary acidic protein and glutamine synthetase expression in astrocytes and Bergmann glial cells in the rat cerebellum caused by Delta(9)-tetrahydrocannabinol administration during development. Dev Neurosci 24(4):300–312. https://doi.org/10.1159/000066744

Sukumar N, Sabesan P, Anazodo U, Palaniyappan L (2020) Neurovascular uncoupling in schizophrenia: a bimodal meta-analysis of brain perfusion and glucose metabolism. Front Psych 11:754. https://doi.org/10.3389/fpsyt.2020.00754

Sullivan CR, O'Donovan SM, McCullumsmith RE, Ramsey A (2018) Defects in bioenergetic coupling in schizophrenia. Biol Psychiatry 83(9):739–750. https://doi.org/10.1016/j.biopsych.2017.10.014

Sultana R, Lee CLC (2020) Expression of behavioral phenotypes in genetic and environmental mouse models of schizophrenia. Front Behav Neurosci 14:29. https://doi.org/10.3389/fnbeh.2020.00029

Sultana R, Perluigi M, Butterfield DA (2013) Lipid peroxidation triggers neurodegeneration: a redox proteomics view into the Alzheimer disease brain. Free Radic Biol Med 62:157–169. https://doi.org/10.1016/j.freeradbiomed.2012.09.027

Sun GY, Xu J, Jensen MD, Yu S, Wood WG, Gonzalez FA, Simonyi A, Sun AY, Weisman GA (2005) Phospholipase A2 in astrocytes: responses to oxidative stress, inflammation, and G protein-coupled receptor agonists. Mol Neurobiol 31(1–3):27–41. https://doi.org/10.1385/MN:31:1-3:027

Sun L, Li J, Zhou K, Zhang M, Yang J, Li Y, Ji B, Zhang Z, Zhu H, Yang L, He G, Gao L, Wei Z, Wang K, Han X, Liu W, Tan L, Yu Y, He L, Wan C (2013) Metabolomic analysis reveals metabolic disturbance in the cortex and hippocampus of subchronic MK-801 treated rats. PLoS One 8(4):e60598. https://doi.org/10.1371/journal.pone.0060598

Sussman D, Germann J, Henkelman M (2015) Gestational ketogenic diet programs brain structure and susceptibility to depression & anxiety in the adult mouse offspring. Brain Behav 5(2):e00300. https://doi.org/10.1002/brb3.300

Suzuki Y, Takahashi H, Fukuda M, Hino H, Kobayashi K, Tanaka J, Ishii E (2009) Beta-hydroxybutyrate alters GABA-transaminase activity in cultured astrocytes. Brain Res 1268:17–23. https://doi.org/10.1016/j.brainres.2009.02.074

Takahashi S (2020) Metabolic compartmentalization between astroglia and neurons in physiological and pathophysiological conditions of the neurovascular unit. Neuropathology 40(2):121–137. https://doi.org/10.1111/neup.12639

Takata N, Mishima T, Hisatsune C, Nagai T, Ebisui E, Mikoshiba K, Hirase H (2011) Astrocyte calcium signaling transforms cholinergic modulation to cortical plasticity in vivo. J Neurosci 31(49):18155–18165. https://doi.org/10.1523/JNEUROSCI.5289-11.2011

Tamminga CA, Thaker GK, Buchanan R, Kirkpatrick B, Alphs LD, Chase TN, Carpenter WT (1992) Limbic system abnormalities identified in schizophrenia using positron emission tomography with fluorodeoxyglucose and neocortical alterations with deficit syndrome. Arch Gen Psychiatry 49(7):522–530. https://doi.org/10.1001/archpsyc.1992.01820070016003

Tansey KE, Rees E, Linden DE, Ripke S, Chambert KD, Moran JL, McCarroll SA, Holmans P, Kirov G, Walters J, Owen MJ, O'Donovan MC (2016) Common alleles contribute to schizophrenia in CNV carriers. Mol Psychiatry 21(8):1085–1089. https://doi.org/10.1038/mp.2015.143

Tarasov VV, Svistunov AA, Chubarev VN, Sologova SS, Mukhortova P, Levushkin D, Somasundaram SG, Kirkland CE, Bachurin SO, Aliev G (2019) Alterations of astrocytes in the context of schizophrenic dementia. Front Pharmacol 10:1612. https://doi.org/10.3389/fphar.2019.01612

Terrillion CE, Abazyan B, Yang ZX, Crawford J, Shevelkin AV, Jouroukhin Y, Yoo KH, Cho CH, Roychaudhuri R, Snyder SH, Jang MH, Pletnikov MV (2017) DISC1 in astrocytes influences adult neurogenesis and hippocampus-dependent behaviors in mice. Neuropsychopharmacology 42(11):2242–2251. https://doi.org/10.1038/npp.2017.129

Trantham-Davidson H, LaLumiere RT, Reissner KJ, Kalivas PW, Knackstedt LA (2012) Ceftriaxone normalizes nucleus accumbens synaptic transmission, glutamate transport, and export following cocaine self-administration and extinction training. J Neurosci 32(36):12406–12410. https://doi.org/10.1523/Jneurosci.1976-12.2012

Tretter L, Adam-Vizi V (2012) High Ca2+ load promotes hydrogen peroxide generation via activation of alpha-glycerophosphate dehydrogenase in brain mitochondria. Free Radic Biol Med 53(11):2119–2130. https://doi.org/10.1016/j.freeradbiomed.2012.09.029

Tsurugizawa T, Tamada K, Ono N, Karakawa S, Kodama Y, Debacker C, Hata J, Okano H, Kitamura A, Zalesky A, Takumi T (2020) Awake functional MRI detects neural circuit dysfunction in a mouse model of autism. Sci Adv 6(6):eaav4520. https://doi.org/10.1126/sciadv.aav4520

Tufano M, Pinna G (2020) Is there a future for PPARs in the treatment of neuropsychiatric disorders? Molecules 25(5):1062. https://doi.org/10.3390/molecules25051062

Uesono Y, Toh-e A, Kikuchi Y, Araki T, Hachiya T, Watanabe CK, Noguchi K, Terashima I (2016) Local anesthetics and antipsychotic phenothiazines interact nonspecifically with membranes and inhibit hexose transporters in yeast. Genetics 202(3):997–1012. https://doi.org/10.1534/genetics.115.183806

Vaccari JPD, Minkiewicz J, Wang XL, Vaccari JCD, German R, Marcillo AE, Dietrich WD, Keane RW (2012) Astrogliosis involves activation of retinoic acid-inducible gene-like signaling in the innate immune response after spinal cord injury. Glia 60(3):414–421. https://doi.org/10.1002/glia.22275

Vardjan N, Parpura V, Verkhratsky A, Zorec R (2019) Gliocrine system: astroglia as secretory cells of the CNS. Adv Exp Med Biol 1175:93–115. https://doi.org/10.1007/978-981-13-9913-8_4

Verkhratsky A, Rodriguez JJ, Parpura V (2012) Calcium signalling in astroglia. Mol Cell Endocrinol 353(1–2):45–56. https://doi.org/10.1016/j.mce.2011.08.039

Verpeut JL, DiCicco-Bloom E, Bello NT (2016) Ketogenic diet exposure during the juvenile period increases social behaviors and forebrain neural activation in adult Engrailed 2 null mice. Physiol Behav 161:90–98. https://doi.org/10.1016/j.physbeh.2016.04.001

Viana J, Wildman N, Hannon E, Farbos A, Neill PO, Moore K, van Aerle R, Paull G, Santos E, Mill J (2020) Clozapine-induced transcriptional changes in the zebrafish brain. NPJ Schizophr 6(1):3. https://doi.org/10.1038/s41537-019-0092-x

Volle J, Brocard J, Saoud M, Gory-Faure S, Brunelin J, Andrieux A, Suaud-Chagny MF (2013) Reduced expression of STOP/MAP6 in mice leads to cognitive deficits. Schizophr Bull 39(5):969–978. https://doi.org/10.1093/schbul/sbs113

Volterra A, Liaudet N, Savtchouk I (2014) Astrocyte Ca(2)(+) signalling: an unexpected complexity. Nat Rev Neurosci 15(5):327–335. https://doi.org/10.1038/nrn3725

Volterra A, Meldolesi J (2005) Astrocytes, from brain glue to communication elements: the revolution continues. Nat Rev Neurosci 6(8):626–640. https://doi.org/10.1038/nrn1722

Waagepetersen HS, Sonnewald U, Larsson OM, Schousboe A (2000) A possible role of alanine for ammonia transfer between astrocytes and glutamatergic neurons. J Neurochem 75(2):471–479. https://doi.org/10.1046/j.1471-4159.2000.0750471.x

Walls AB, Heimburger CM, Bouman SD, Schousboe A, Waagepetersen HS (2009) Robust glycogen shunt activity in astrocytes: effects of glutamatergic and adrenergic agents. Neuroscience 158(1):284–292. https://doi.org/10.1016/j.neuroscience.2008.09.058

Wang H, Wang SQ, Zhang K, Wang H, Lan LT, Ma XY, Liu XY, Zhang SZ, Zheng JQ, Wei XL, Yan HT (2017) Aquaporin 4 forms a macromolecular complex with glutamate transporter 1 and mu opioid receptor in astrocytes and participates in morphine dependence. J Mol Neurosci 62(1):17–27. https://doi.org/10.1007/s12031-017-0905-1

Wang J, Li G, Wang Z, Zhang X, Yao L, Wang F, Liu S, Yin J, Ling EA, Wang L, Hao A (2012) High glucose-induced expression of inflammatory cytokines and reactive oxygen species in cultured astrocytes. Neuroscience 202:58–68. https://doi.org/10.1016/j.neuroscience.2011.11.062

Wang Y, Liu X, Li P, Zhou H, Yang L, Zheng L, Xie P, Li L, Liao DJ, Liu Q, Fang D (2018) Regional cerebral blood flow in mania: assessment using 320-slice computed tomography. Front Psych 9:296. https://doi.org/10.3389/fpsyt.2018.00296

Wang Y, Zhang H, Tang S, Liu X, O'Neil A, Turner A, Chai F, Chen F, Berk M (2014) Assessing regional cerebral blood flow in depression using 320-slice computed tomography. PLoS One 9(9):e107735. https://doi.org/10.1371/journal.pone.0107735

Watanabe A, Toyota T, Owada Y, Hayashi T, Iwayama Y, Matsumata M, Ishitsuka Y, Nakaya A, Maekawa M, Ohnishi T, Arai R, Sakurai K, Yamada K, Kondo H, Hashimoto K, Osumi N, Yoshikawa T (2007) Fabp7 maps to a quantitative trait locus for a schizophrenia endophenotype. PLoS Biol 5(11):e297. https://doi.org/10.1371/journal.pbio.0050297

Webster MJ, Knable MB, Johnston-Wilson N, Nagata K, Inagaki M, Yolken RH (2001) Immunohistochemical localization of phosphorylated glial fibrillary acidic protein in the prefrontal cortex and hippocampus from patients with schizophrenia, bipolar disorder, and depression. Brain Behav Immun 15(4):388–400. https://doi.org/10.1006/brbi.2001.0646

Weckmann K, Labermaier C, Asara JM, Muller MB, Turck CW (2014) Time-dependent metabolomic profiling of Ketamine drug action reveals hippocampal pathway alterations and biomarker candidates. Transl Psychiatry 4:e481. https://doi.org/10.1038/tp.2014.119

Weightman Potter PG, Vlachaki Walker JM, Robb JL, Chilton JK, Williamson R, Randall AD, Ellacott KLJ, Beall C (2019) Basal fatty acid oxidation increases after recurrent low glucose in human primary astrocytes. Diabetologia 62(1):187–198. https://doi.org/10.1007/s00125-018-4744-6

Wesseling H, Chan MK, Tsang TM, Ernst A, Peters F, Guest PC, Holmes E, Bahn S (2013) A combined metabonomic and proteomic approach identifies frontal cortex changes in a chronic phencyclidine rat model in relation to human schizophrenia brain pathology. Neuropsychopharmacology 38(12):2532–2544. https://doi.org/10.1038/npp.2013.160

Wesseling H, Guest PC, Lee CM, Wong EH, Rahmoune H, Bahn S (2014) Integrative proteomic analysis of the NMDA NR1 knockdown mouse model reveals effects on central and peripheral pathways associated with schizophrenia and autism spectrum disorders. Mol Autism 5:38. https://doi.org/10.1186/2040-2392-5-38

Wiesel FA, Wik G, Sjogren I, Blomqvist G, Greitz T, Stone-Elander S (1987) Regional brain glucose metabolism in drug free schizophrenic patients and clinical correlates. Acta Psychiatr Scand 76(6):628–641. https://doi.org/10.1111/j.1600-0447.1987.tb02933.x

Windrem MS, Osipovitch M, Liu Z, Bates J, Chandler-Militello D, Zou L, Munir J, Schanz S, McCoy K, Miller RH, Wang S, Nedergaard M, Findling RL, Tesar PJ, Goldman SA (2017) Human iPSC glial mouse chimeras reveal glial contributions to schizophrenia. Cell Stem Cell 21(2):195–208. e196. https://doi.org/10.1016/j.stem.2017.06.012

Winship IR, Dursun SM, Baker GB, Balista PA, Kandratavicius L, Maia-de-Oliveira JP, Hallak J, Howland JG (2019) An overview of animal models related to schizophrenia. Can J Psychiatr 64(1):5–17. https://doi.org/10.1177/0706743718773728

Wu N, Lu XQ, Yan HT, Su RB, Wang JF, Liu Y, Hu G, Li J (2008) Aquaporin 4 deficiency modulates morphine pharmacological actions. Neurosci Lett 448(2):221–225. https://doi.org/10.1016/j.neulet.2008.10.065

Xi Q, Tcheranova D, Basuroy S, Parfenova H, Jaggar JH, Leffler CW (2011) Glutamate-induced calcium signals stimulate CO production in piglet astrocytes. Am J Physiol Heart Circ Physiol 301(2):H428–H433. https://doi.org/10.1152/ajpheart.01277.2010

Xia M, Broek JA, Jouroukhin Y, Schoenfelder J, Abazyan S, Jaaro-Peled H, Sawa A, Bahn S, Pletnikov M (2016) Cell type-specific effects of mutant DISC1: a proteomics study. Mol Neuropsychiatry 2(1):28–36. https://doi.org/10.1159/000444587

Xu J, Song D, Xue Z, Gu L, Hertz L, Peng L (2013) Requirement of glycogenolysis for uptake of increased extracellular K+ in astrocytes: potential implications for K+ homeostasis and glycogen usage in brain. Neurochem Res 38(3):472–485. https://doi.org/10.1007/s11064-012-0938-3

Yoshizumi T, Imamura H, Taku T, Kuroki T, Kawaguchi A, Ishikawa K, Nakada K, Koshiba T (2017) RLR-mediated antiviral innate immunity requires oxidative phosphorylation activity. Sci Rep 7:5379. https://doi.org/10.1038/s41598-017-05808-w

Zamanpoor M (2020) Schizophrenia in a genomic era: a review from the pathogenesis, genetic and environmental etiology to diagnosis and treatment insights. Psychiatr Genet 30(1):1–9. https://doi.org/10.1097/YPG.0000000000000245

Zeinoddini A, Sorayani M, Hassanzadeh E, Arbabi M, Farokhnia M, Salimi S, Ghaleiha A, Akhondzadeh S (2015) Pioglitazone adjunctive therapy for depressive episode of bipolar disorder: a randomized, double-blind, placebo-controlled trial. Depress Anxiety 32(3):167–173. https://doi.org/10.1002/da.22340

Zeisel A, Hochgerner H, Lonnerberg P, Johnsson A, Memic F, van der Zwan J, Haring M, Braun E, Borm LE, La Manno G, Codeluppi S, Furlan A, Lee K, Skene N, Harris KD, Hjerling-Leffler J, Arenas E, Ernfors P, Marklund U, Linnarsson S (2018) Molecular architecture of the mouse nervous system. Cell 174(4):999. https://doi.org/10.1016/j.cell.2018.06.021

Zhang P, Kuang H, He YL, Idiga SO, Li SM, Chen ZM, Yang Z, Cai X, Zhang KZ, Potthoff MJ, Xu Y, Lin JD (2018) NRG1-Fc improves metabolic health via dual hepatic and central action. JCI Insight 3(5):e98522. https://doi.org/10.1172/jci.insight.98522

Zhang SQ, Li BM, Lovatt D, Xu JN, Song D, Goldman SA, Nedergaard M, Hertz L, Peng LA (2010) 5-HT2B receptors are expressed on astrocytes from brain and in culture and are a chronic target for all five conventional 'serotonin-specific reuptake inhibitors'. Neuron Glia Biol 6(2):113–125. https://doi.org/10.1017/S1740925x10000141

Zhang WN, Wang GH, Xu ZG, Tu HQ, Hu FQ, Dai J, Chang Y, Chen YQ, Lu YJ, Zeng HL, Cai Z, Han F, Xu C, Jin GX, Sun L, Pan BS, Lai SW, Hsu CC, Xu J, Chen ZZ, Li HY, Seth P, Hu JB, Zhang XM, Li HY, Lin HK (2019) Lactate is a natural suppressor of RLR signaling by targeting MAVS. Cell 178(1):176. https://doi.org/10.1016/j.cell.2019.05.003

Zhang XQ, Song D, Gu L, Ren Y, Verkhratsky A, Peng L (2015a) Decrease of gene expression of astrocytic 5-HT2B receptors parallels development of depressive phenotype in a mouse model of Parkinson's disease. Frontiers in cellular. Neuroscience 9:388. https://doi.org/10.3389/fncel.2015.00388

Zhang XY, Chen DC, Tan YL, An HM, Zunta-Soares GB, Huang XF, Soares JC (2015b) Glucose disturbances in first-episode drug-naive schizophrenia: relationship to psychopathology. Psychoneuroendocrinology 62:376–380. https://doi.org/10.1016/j.psyneuen.2015.09.005

Zhao YZ, Gao ZY, Ma LQ, Zhuang YY, Guan FL (2017) Research on biogenesis of mitochondria in astrocytes in sepsis-associated encephalopathy models. Eur Rev Med Pharmacol Sci 21(17):3924–3934

Zhou X, Xiao Q, Xie L, Yang F, Wang L, Tu J (2019) Astrocyte, a promising target for mood disorder interventions. Front Mol Neurosci 12:136. https://doi.org/10.3389/fnmol.2019.00136

Zhu J, Zhuo C, Xu L, Liu F, Qin W, Yu C (2017) Altered coupling between resting-state cerebral blood flow and functional connectivity in schizophrenia. Schizophr Bull 43(6):1363–1374. https://doi.org/10.1093/schbul/sbx051

Zonta M, Angulo MC, Gobbo S, Rosengarten B, Hossmann KA, Pozzan T, Carmignoto G (2003) Neuron-to-astrocyte signaling is central to the dynamic control of brain microcirculation. Nat Neurosci 6(1):43–50. https://doi.org/10.1038/nn980

Zuccoli GS, Saia-Cereda VM, Nascimento JM, Martins-de-Souza D (2017) The energy metabolism dysfunction in psychiatric disorders postmortem brains: focus on proteomic evidence. Front Neurosci 11:493. https://doi.org/10.3389/fnins.2017.00493

Part IV
Astroglia in Addictive Disorders

Astrocytes in Addictive Disorders

Anna Kruyer and Michael D. Scofield

Overview of Addiction Biology

Circuitry and Transmitters

The elevation of extracellular dopamine in the ventral striatum in response to salient and rewarding stimuli underlies several reinforcement and addiction-related behaviors extensively studied in preclinical models of substance use and relapse (Scofield et al. 2016a). Anatomically, the ventral striatum is divided into two subregions, the nucleus accumbens core (NAcore) and shell (NAshell), which both receive dopaminergic innervation from the midbrain nucleus the ventral tegmental area (VTA) (Ikemoto 2007) and that are both involved in encoding reward and motivation. Increased levels of extracellular dopamine in the ventral striatum are evoked by administration of addictive drugs (reviewed in (Willuhn et al. 2010)) including alcohol (Howard et al. 2008), as well as by sucrose intake (Bassareo et al. 2017), and a substantial body of lesion, microdialysis, voltammetry, and pharmacology studies has led to the general conception that the NAshell is involved in encoding the rewarding effects of addictive drugs, while the NAcore coordinates behavioral outputs in response to salient stimuli (Scofield et al. 2016a; Sellings and Clarke 2003; Salgado and Kaplitt 2015).

A. Kruyer
Department of Neurosciences, Medical University of South Carolina, Charleston, SC, USA

M. D. Scofield (✉)
Department of Neurosciences, Medical University of South Carolina, Charleston, SC, USA

Department of Anesthesia and Perioperative Medicine, Medical University of South Carolina, Charleston, SC, USA
e-mail: Scofield@musc.edu

B. Li et al. (eds.), *Astrocytes in Psychiatric Disorders*, Advances in Neurobiology 26, https://doi.org/10.1007/978-3-030-77375-5_10

Though natural (i.e., sucrose) as well as pathological (i.e., drugs of abuse) rewards elevate dopamine in the ventral striatum, dopaminergic signaling within this circuit is modulated by glutamatergic transmission, and glutamatergic signaling is uniquely disrupted following repeated intake of addictive substances (Kalivas et al. 1989, 2009; Kalivas 2009). In keeping with important but separable roles for these two transmitters in addiction-related behaviors are recent studies demonstrating that subpopulations of dopaminergic VTA neurons that project to the NAshell co-release glutamate along with dopamine (Stuber et al. 2010; Mongia et al. 2019; Mingote et al. 2019). Glutamate release from dopamine neurons in the NAshell is linked to intake of sucrose and psychostimulants and locomotor responses to acute and repeated psychostimulant administration (Hnasko et al. 2010; Alsio et al. 2011; Birgner et al. 2010; Bimpisidis and Wallen-Mackenzie 2019). Glutamate modulates dopamine neurotransmission through its action on glutamate receptors expressed on dopaminergic terminals (Floresco et al. 1998; Howland et al. 2002) and through facilitation of dopamine loading in synaptic vesicles (Hnasko et al. 2010; Hnasko and Edwards 2012). Though VTA inputs to the NAcore contribute to drug seeking induced by drug prime (Shen et al. 2014a), the impact of dopamine transmission on seeking behavior can be blocked by glutamate receptor antagonists even in the presence of dopamine (Cornish and Kalivas 2000), consistent with a critical role for glutamatergic transmission in drug seeking behavior.

The ventral striatum receives excitatory input from a number of brain regions, in addition to VTA. Cortical regions provide glutamatergic input to the NAcore and NAshell, with prelimbic (PL) cortical afferents in the NAcore or infralimbic (IL) afferents in the NAshell contributing to drug seeking or refraining behavior, respectively. Glutamate transmission in the NAcore is necessary for initial learning of drug-reward associations (Kelley et al. 1997; Smith-Roe and Kelley 2000) and later for retrieval of drug-contingency information during reinstatement of drug seeking following extinction of behavioral responding (Kalivas 2009; Kalivas and Volkow 2005; Koob and Volkow 2010). PL afferents in the NAcore are engaged during cue-, drug-, context- and stress-reinstated seeking of different classes of addictive substances, including opioids, psychostimulants, and alcohol (Rogers et al. 2008; Ball and Slane 2012; Rocha and Kalivas 2010; Doncheck et al. 2020; McFarland et al. 2003; Willcocks and McNally 2013; Chaudhri et al. 2008). The IL input to the NAshell, on the other hand, is recruited during extinction training, whereby operant responding for drug reinforcers is gradually reduced over time when the drug is no longer available (Peters et al. 2008).

The NAcore also receives significant glutamatergic input from basolateral amygdala (BLA) that contributes to both acquisition of self-administration and reinstatement initiated by cues (Carelli et al. 2003; Whitelaw et al. 1996; Di Ciano and Everitt 2004; Puaud et al. 2020), through direct glutamatergic projections to the NAcore and/or projections to the PL (Stefanik and Kalivas 2013). BLA glutamate may also be engaged to a certain extent by operant training for natural reinforcers, since optogenetically stimulating BLA projections to the NAcore can initiate responding for sucrose (Stuber et al. 2011). A number of other brain regions have been strongly implicated in promoting addiction-related behaviors, including dorsal and ventral hippocampus (Fuchs et al. 2005; Lasseter et al. 2010; Atkins et al. 2008), ventral subiculum (Bossert and Stern 2014), and lateral septum (McGlinchey and Aston-Jones 2018), though not necessarily via direct NAcore projections. Interestingly, excitatory projections to the NAcore can also suppress seeking behavior, since glutamate release from the paraventricular thalamus in the NAcore serves to suppress seeking behavior in the absence of a reinforcer (Do-Monte et al. 2017), arguing that excitatory transmission in the NAcore is necessary, but not sufficient for reward seeking, and that unique innervation of postsynaptic targets may be an important factor in ultimate behavioral outputs.

Astrocytes are likely mediators of the long-lasting excitatory plasticity that arises within the corticostriatal circuitry in response to chronic drug intake. Astroglial processes are positioned proximal to excitatory synapses, and their expression of neurotransmitter transporters regulates stimulation of pre- and postsynaptic receptors. Moreover, astroglia are capable of signaling directly to synapses through transmitter release and play critical roles in homeostatic regulation of synapses. Astroglial adaptations following exposure to addictive drugs are being actively investigated within the circuits outlined above. Given that vast brain region-dependent heterogeneity in astroglial structure and transcriptomic profiles has recently been described (Batiuk et al. 2020; Matias et al. 2019; Cuevas-Diaz Duran et al. 2019; Kohler et al. 2019), astrocyte function as it pertains to addiction is likely unique in each region. Below we explore what is known regarding astroglial contributions to synaptic physiology at baseline and after use of addictive substances.

Transmitter Uptake

Glutamate

The glutamate transporter GLT-1 (EAAT2 in humans) conducts the majority of glutamate uptake and is mostly expressed by astroglia in the adult brain (Danbolt 2001), despite reports of low-level expression in some neurons (Rimmele and Rosenberg 2016). GLT-1 has long been thought to contribute to addiction-related glutamatergic dysregulation in the ventral striatum since its expression and/or function is disrupted after long-term use of psychostimulants, opioids, and alcohol (Roberts-Wolfe and Kalivas 2015). GLT-1 knockdown in astroglia during the early postnatal period leads to repetitive behaviors in drug-naïve animals and enhanced neuronal excitability in the dorsal striatum (Aida et al. 2015). In rats trained to self-administer cocaine, GLT-1 downregulation in the NAcore is dependent upon withdrawal duration as well as cocaine intake, with increasing amounts of either exacerbating the severity of GLT-1 downregulation (Fischer-Smith et al. 2012). Further, extended withdrawal from long-access cocaine self-administration (i.e., daily 6-hour sessions) reduces mRNA levels of the predominant of three GLT-1 isoforms, GLT-1a (Holmseth et al. 2009), in the NAcore, coincident with enhanced methylation of *Slc1a2*, the gene that encodes GLT-1 (Kim et al. 2018). Since the authors of this study found no changes in GLT-1a or b mRNA levels after withdrawal or extinction from short access cocaine self-administration, but many other studies show GLT-1 downregulation in these paradigms (Roberts-Wolfe and Kalivas 2015), it may be that posttranslational rather than transcriptional mechanisms are responsible for its reduced expression and/or function in short-access models. Interestingly, up- or downregulation of GLT-1 in the striatum disrupts spike-timing dependent plasticity, a type of long-term potentiation that requires a tight temporal relationship between pre- and postsynaptic activity (Valtcheva and Venance 2016). When the action of GLT-1 is blocked, long-term potentiation requires signaling through extrasynaptic GluN2B-containing NMDA receptors. Similarly, signaling through GluN2B is required for cue- or heroin prime-induced reinstatement of heroin seeking (Shen et al. 2011). These data are consistent with diminished function of GLT-1 after extinction from heroin self-administration (Shen et al. 2014b). Further, disrupted forms of synaptic plasticity involving downregulated GLT-1 after drug use and/or withdrawal may conceivably impair mechanisms required for extinction learning, which is somewhat impaired in drug-trained compared with sucrose-trained animals (Martin-Fardon and Weiss 2017). Notably, operant sucrose training does not produce downregulation of GLT-1 (Kruyer and Kalivas 2020a), which may account for these behavioral differences.

A significant body of work has established that compounds that restore GLT-1 expression reduce drug craving and relapse, both in animal models and in human clinical trials (Roberts-Wolfe and Kalivas 2015). For example, the antioxidant N-acetylcysteine, the ß-lactam antibiotic ceftriaxone, and the xanthine derivative propentofylline have all demonstrated promise in restoring GLT-1 expression after drug withdrawal and in blunting reinstated drug seeking in preclinical models (summarized in (Scofield et al. 2016a)). N-acetylcysteine is extremely safe for use in humans but has proven only marginally efficacious in reducing craving and drug use in patients (summarized in (Roberts-Wolfe and Kalivas 2015)), potentially due to limited bioavailability. Given that recent studies indicate that GLT-1 upregulation alone is not sufficient to blunt reinstated cocaine seeking (Logan et al. 2018), it may very well be that additional manipulations are required to adequately suppress relapse vulnerability. Moreover, it may be the case that upregulation rather than homeostatic restoration of GLT-1 expression may not entirely restore mechanisms of synaptic plasticity, as both GLT-1 blockade and overexpression do not properly permit spike timing-dependent plasticity at striatal synapses (Valtcheva and Venance 2016).

Although drug-induced changes in GLT-1 expression after chronic use of addictive drugs have been less well documented in the midbrain when compared to the ventral striatum, astroglial glutamate transporters in the VTA also play an important role in modulating avoidance behaviors (Gomez et al. 2019). As mentioned above, the VTA sends dopaminergic, glutamatergic, as well as GABAergic projections to the nucleus accumbens, and these projections encode the reinforcing and aversive aspects of drug use (Morales and Margolis 2017). Additionally, GABAergic interneurons within the VTA regulate dopaminergic projection neurons that signal avoidance vs. motivated approach behaviors (McCullough et al. 1993; Tan et al. 2012). This was demonstrated with optogenetic stimulation of VTA astrocytes, which permits excitation of GABAergic interneurons that inhibit dopaminergic projections and promote avoidance behavior (Gomez et al. 2019). It was subsequently shown that astrocyte-dependent facilitation of GABAergic activation depends on GLT-1, as conditional knockout of GLT-1 in VTA astrocytes impairs the optogenetic stimulation-mediated GABAergic excitation (Gomez et al. 2019). While the authors do not clarify the mechanism by which GLT-1 contributes to GABAergic excitation, they illustrate that manipulation of astroglial GLT-1 interferes with conditioned place avoidance, but does not impact conditioned place preference (Gomez et al. 2019), suggesting divergence of these circuits and a contribution of astrocytes through GLT-1 to avoidance, but not approach behavior (Gomez et al. 2019). Although changes in GLT-1 expression have not been as clearly demonstrated in the VTA after drug use and withdrawal compared with the NAcore (Knackstedt et al. 2009), analysis of GLT-1 function may be warranted given these surprising findings. These data also highlight the importance of analyzing transporter function in a pathway-specific manner, rather than in whole tissue, since functional changes that impact discrete subcircuits may not be discernable when protein levels are assessed in tissue extracts.

GABA

While uptake of glutamate release from cortical terminals in the striatum is thought to directly regulate relapse-like behaviors in animal models (Kalivas 2008), cortical stimulation triggers uptake of both GABA and glutamate by striatal astrocytes and clearance of both transmitters contributes to postsynaptic excitation of MSNs (Goubard et al. 2011). These findings illustrate the relevance of local GABAergic signaling on corticostriatal synaptic transmission and the contribution of transporter uptake in regulating striatal outputs. In support of this concept, studies where Ca^{2+} signaling in astrocytes is inhibited in vivo through viral delivery of a genetically encoded plasma membrane Ca^{2+} pump that expels cytosolic Ca^{2+} from astrocytes (CalEx) (Yu et al. 2018) demonstrate the behavioral involvement of GABA transporters expressed on striatal astrocytes in compulsive-like behaviors. Importantly, the CalEx vector reduces astroglial intracellular Ca^{2+} levels at baseline and also reduces the amplitude and duration of Ca^{2+} elevations. When delivered to astroglia in the dorsolateral striatum, CalEx expression results in upregulation of the largely astroglial GABA transporter GAT-3 and produces excessive self-grooming behavior, reminiscent of features of obsessive-compulsive disorder (Yu et al. 2018). Further, GAT-3 upregulation reduces tonic inhibition, a known role for GAT-3 that is expressed most densely in extrasynaptic zones (Melone et al. 2015), selectively at D1 receptor expressing MSNs (D1-MSNs) (Yu et al. 2018). Interestingly, when a different strategy was used to upregulate astroglial GAT-3 in a pathway-nonspecific manner, the authors found no impact on grooming behavior, suggesting that perhaps the cell selectivity of the effects were necessary for the disrupted behavior observed by the authors.

The dorsal striatum is involved in habit learning (Yin et al. 2004) and is uniquely recruited in cases of substance use disorder characterized by compulsive seeking. For example, in humans who drink socially, but not compulsively, alcohol-associated cues stimulate activation of the ventral striatum. Instead, the same cues stimulate activation of the dorsal rather than the ventral striatum in heavy drinkers, and ventral striatal activity in response to alcohol-associated cues is negatively associated with measures of compulsive craving in these individuals (Vollstadt-Klein et al. 2010). Interestingly, compounds effective in suppressing compulsive behaviors, like N-acetylcysteine, which reduces symptoms of trichotillomania (Farhat et al. 2020) and pathological gambling (Grant et al. 2007), also reduce reinstated drug seeking in animal models (Kalivas and Kalivas 2016). As mentioned above, N-acetylcysteine was most studied in the context of addiction for its ability to upregulate GLT-1 in the ventral striatum, but its impact on astroglial expression of GABA transporters like GAT-3 in the ventral or dorsal striatum has not been assessed.

Dopamine

Although there is in vitro evidence that astrocytes can recover extracellular dopamine through expression of the dopamine transporter (DAT), norepinephrine transporter (NET), and/or plasma membrane monoamine transporter (PMAT) (Pelton 2nd et al. 1981; Takeda et al. 2002; Naganuma et al. 2014), evidence is less clear-cut in vivo. The organic cation transporter 3 (OCT3), a low-affinity monoamine transporter, has been reported on neurons and astroglia in the rodent substantia nigra and striatum (Cui et al. 2009), and electron microscopy studies revealed OCT3 on the plasma membrane of perisynaptic astroglial processes in the rodent amygdala, consistent with its role in monoamine uptake near synaptic sites (Gasser et al. 2017). OCT3 was recently found to function in reverse in the presence of amphetamine, releasing dopamine into the extracellular space (Mayer et al. 2018). The authors also confirmed that while OCT3 was expressed by dopaminergic neurons, the majority of OCT3-expressing cells were likely glia as well as non-dopaminergic neurons (Mayer et al. 2018). Importantly, while astroglia express OCT3 and therefore may also theoretically employ this mechanism, releasing dopamine in the presence of amphetamine, amphetamine enters into dopaminergic neurons through DAT and not through OCT3 (Mayer et al. 2018). Thus, in the absence of a mechanism for amphetamine uptake by astroglia, astroglial OCT3 is not likely to be involved in this process. Because studies on the contribution of OCT3 to dopamine transport in the striatum and elsewhere have emerged relatively recently (Holleran et al. 2020), there are no studies investigating changes in OCT3 expression in astrocytes following drug use and withdrawal in brain regions that contribute to addiction-related behaviors, though it remains a fundamental research question.

Gliotransmission

Glutamate

Gliotransmission by astrocytes is most often described as involving intracellular Ca^{2+} flux and vesicular exocytosis of transmitters, which signal to nearby neurons (Parpura et al. 1994; Araque et al. 2014; Papouin et al. 2017; Scofield 2018). Though electron microscopy and expression studies support the existence of vesicular release machinery in astrocytes, the relevance of this modality for gliotransmission has been contested by studies indicating a lack of glutamate receptor expression in astrocytes in the adult brain in vivo (Sun et al. 2013) as well as studies where deletion of key mediators of intracellular Ca^{2+} flux in astrocytes had little impact on neuronal function (Petravicz et al. 2014; Bazargani and Attwell 2016). Despite this, there are a number of well-established mechanisms by which astroglia signal to neurons through non-exocytotic mechanisms and without associated changes in intracellular Ca^{2+} (Agulhon et al. 2010; Gomez-Gonzalo et al. 2018). Here we will discuss evidence for exocytotic and non-exocytotic mechanisms of transmitter release given the important role for excitatory signaling in formation of drug-cue associations and in reinstatement of drug seeking.

Perhaps the least controversial mechanism for glutamate transmission by astroglia is through the cystine-glutamate antiporter, system xc-, which contributes to 60% of extracellular glutamate in the NAcore in drug-naïve animals (Baker et al. 2002). Extracellular glutamate derived from system xc- tonically stimulates presynaptic autoinhibitory mGluRs on glutamatergic and dopaminergic terminals, regulating transmitter release during neural firing (Baker et al. 2002). System xc-, which is expressed to the greatest extent by astroglia (Pow 2001; Sagara et al. 1993), is downregulated after chronic intake of psychostimulants in the NAcore, and this downregulation is thought to disrupt autoinhibitory tone at presynaptic mGluR2/3 on glutamatergic terminals, leading to enhanced glutamate release in response to drug-associated cues (Baker et al. 2002). It should be noted that system xc- is not similarly downregulated after chronic intake of opioids (Shen et al. 2014b), so this molecular adaptation may not generalize across all substances of abuse. However, the treatments described above that have been tested preclinically for their ability to attenuate reinstated seeking through restoration of GLT-1, including N-acetylcysteine and ceftriaxone, also increase expression of system xc- (Knackstedt et al. 2010), and N-aceytylcysteine may even increase extracellular cystine levels, promoting cystine-glutamate exchange. The contribution of GLT-1 and system xc- upregulation to relapse suppression has been tested by pairing these pharmacological treatments with morpholino knockdown of either GLT-1 or xCT, a subunit of system xc-, in the NAcore prior to reinstatement. In these studies, the authors found that GLT-1, not xCT, had the greatest impact on reinstated drug seeking (Reissner et al. 2015).

Studies exploring the involvement of exocytotic transmitter release by astrocytes in addiction-related behaviors has made use of transgenic mice that express a dominant negative SNARE protein (dnSNARE) in astroglia. These mice lose the capacity to exocytose transmitter-containing vesicles through SNARE mechanisms in astroglia but retain normal mechanisms for neuronal exocytosis (Pascual et al. 2005). dnSNARE mice exhibit similar operant responding for food compared to wild-type animals but appear to take less cocaine during self-administration, despite similar rates of operant acquisition (Turner et al. 2013). dnSNARE mice also reinstate less robustly to cocaine-paired cues (Turner et al. 2013). Whether this reinstatement deficit derives in part from changes in cocaine intake is an important question, since dnSNARE animals appear to receive fewer cocaine infusions and consequently fewer cocaine-cue pairings during training. However, reinstatement of conditioned place preference (CPP) is also abolished in these mice, supporting a role for vesicular gliotransmission in reinstatement behaviors after exposure to psychostimulants (Turner et al. 2013). One limitation of these studies is that global dnSNARE expression in astroglia does not provide information regarding which brain region(s) orchestrate gliotransmission-dependent modulation of reinstatement. Interestingly, it was found that chemogenetic stimulation of Gq signaling in NAcore astrocytes, a manipulation that engages flux of internal Ca^{2+} and activates SNARE-dependent vesicular release of glutamate, reduces reinstated cocaine seeking initiated by cocaine-associated cues (Scofield et al. 2015). Similarly, astroglial Gq stimulation reduces methamphetamine (Siemsen et al. 2019) and ethanol seeking (Bull et al. 2014). It was demonstrated that the suppression of cocaine seeking using this technique was a consequence of astroglial-glutamate release that stimulated presynaptic mGluR2/3 and ultimately reduced transmitter release from presynaptic terminals (Scofield et al. 2015).

In another study, engaging Gq signaling in cultured astrocytes evoked non-vesicular glutamate release through the glutamate-permeable anion channel Bestrophin 1 (Best1) (Woo et al. 2012). Importantly, Best1 is situated near synapses in vivo and would be expected to raise extracellular glutamate levels less robustly than vesicular release. Consequently, release of glutamate through this modality is more likely to impact high-affinity NMDA receptors, compared with vesicular release which might also engage lower-affinity receptors. Accordingly, it might be expected that separate modalities of glutamate release from astrocytes engage functionally distinct intracellular signaling cascades in nearby neurons, despite both modalities being linked to Gq-coupled receptor activation. Together, the findings described above lead to important questions regarding how these cellular mechanisms are employed within the reward circuitry of drug-naïve animals and how their deployment may be impacted by chronic intake of and withdrawal from addictive substances.

One valid critique of studies where either optogenetic or chemogenetic manipulation of astrocytes is employed to understand astroglial biology is that the abundance of the exogenously expressed proteins, as well as their proximity to synapses, may not reflect normal physiological features of astroglia. Nonetheless, studies in the dorsal striatum and elsewhere clearly illustrate that astrocytes signal to local MSNs using endogenous mechanisms of glutamate release. Dorsal striatal astrocytes are uniquely tuned to respond functionally to electrical stimulation of D1- or D2-MSN subtypes in response to neuronally released endocannabinoids (eCBs), with Ca^{2+} flux and ultimately glutamate release (Martin et al. 2015). Astrocyte glutamate release produces slow inward currents (SICs) in adjacent neurons of the same subtype as the stimulated neuron, presumably through stimulation of extrasynaptic NMDA receptors (D'Ascenzo et al. 2007), but SICs are rare in heterotypic pairs. These findings are particularly relevant given that stimulation of striatal D1-MSN projections or inhibition of D2-MSN projections trigger reinstated seeking (Heinsbroek et al. 2017; Pardo-Garcia et al. 2019). There is not yet definitive evidence for how astrocytes distinguish between neural subcircuits. It is abundantly clear that striatal astrocytes encompass both D1- and D2-MSN somata within their cellular territories (Octeau et al. 2018), so spatial segregation does not seem a likely possibility. Further, studies using dye-filling strategies illustrate that astrocytes are extensively coupled to one another through gap junctions (Octeau et al. 2018), forming large densely interconnected syncytia. Thus, it does not appear that segregated astrocyte subpopulations are joined into networks through gap junctions, at least not in drug-naïve animals (but see "Homeostatic Functions" for a discussion of drug-induced changes in astrocyte gap junctions). Still, there are a number of hypothetical ways that astrocytes might coordinate signaling within, but not between circuits, when the same receptors and transmitters are employed in both pathways. One possibility is that additional signaling molecules accompany transmitter release and astrocytes selectively express receptors for molecules released by one or the other pathway. As an example, in the ventral pallidum, D1- and D2-MSN terminals co-release unique neuropeptides, with D1-MSNs co-releasing substance P and dynorphin and D2-MSNs co-releasing enkephalin and neurotensin (Kupchik et al. 2014). The same could occur postsynaptically, with the release of different eCBs following postsynaptic excitation of either D1- or D2-MSNs or unique expression patterns of cannabinoid receptors on astroglial subtypes. Whether release of unique pre- or postsynaptic signaling molecules selectively recruits astrocytes to signal to adjacent MSNs in a homotypic manner remains to be shown. Another possibility is that astrocytes are selective in their synaptic proximity, with each astrocyte exhibiting unique proximity to D1- or D2-MSNs in order to receive and transmit signals selectively. While it does not appear that D1 or D2-MSN-containing synapses receive different degrees of astrocyte insulation overall, there is tremendous variability in synaptic insulation by astroglia in the striatum (Octeau et al. 2018; Chai et al. 2017). Thus, it remains possible that each astrocyte may exhibit some selectivity in which synapses it approaches most closely. This concept will be discussed in more depth in the subsequent section on astrocyte morphological plasticity.

D-Serine

D-serine is the R-enantiomer of the amino acid serine and functions as an NMDA receptor co-agonist, with selective affinity for the glycine binding site (Mothet et al. 2000; MacKay et al. 2019). D-serine has been studied for its improvement of positive and cognitive symptoms in schizophrenia patients when co-administered with antipsychotics (Tsai et al. 1998), based on evidence of hypofunctional NMDA receptor activity in the disorder (Coyle 2012). Whether D-serine is released by astrocytes has been a somewhat controversial topic since serine racemase, the enzyme that converts L-serine to D-serine in cells, is mostly expressed by neurons (Yoshikawa et al. 2007; Benneyworth et al. 2012). It appears that instead, D-serine is produced by serine racemase in neurons and is shuttled between neurons and astroglia for vesicular release (Wolosker 2011; Martineau et al. 2013). Regardless of its cellular origin, D-serine has been shown to play an important role in signaling via NMDA receptors in the accumbens. D-serine is required for NMDA-dependent synaptic potentiation and depression in the NAcore (Curcio et al. 2013), both of which are disrupted after withdrawal from cocaine self-administration (Moussawi et al. 2009). Linking these findings, researchers found that cocaine reduces D-serine concentration in the NAcore through increased expression of D-amino acid oxidase, an enzyme that degrades D-serine (Curcio et al. 2013; D'Ascenzo et al. 2014). The reduction in D-serine levels increases the relative proportion of AMPA/NMDA receptors in neurons postsynaptically and permits locomotor sensitization and CPP (Curcio et al. 2013; D'Ascenzo et al. 2014; Yang et al. 2013). Subsequent studies found that morphine also decreases D-serine in the nucleus accumbens and does so by reducing surface expression of AMPA receptors on astrocytes that normally trigger Ca^{2+} flux and vesicular D-serine release (Wu et al. 2017). Despite these promising findings, additional experimentation is required to determine the precise role of D-serine gliotransmission in mediating addiction-relevant synaptic plasticity.

ATP

Astrocyte-derived adenosine has been shown to contribute to synaptic modulation in various brain regions and in response to synaptically released glutamate as well as dopamine (Pascual et al. 2005; Zhang et al. 2003; Quon et al. 2018; Corkrum et al. 2020). In the dorsolateral striatum, high-frequency stimulation of cortical terminals stimulates astroglial mGluR5, a Gq-coupled receptor, to produce Ca^{2+} increases, ATP release, and adenosine receptor-mediated long-term depression of postsynaptic cells (Cavaccini et al. 2020). In a similar study in the dorsomedial striatum, the authors show that astroglial expression of the adenosine transporter equilibrative nucleoside transporter 1, ENT1 is required for astroglial Gq stimulation to impact neural activity. Interestingly, triggering this signaling cascade alters firing properties of both D1- and D2-MSNs, reducing spontaneous EPSCs in D1- and increasing them in D2-MSNs, promoting goal-directed rather than habitual behaviors (Kang et al. 2020).

Altogether, Gq-dependent signaling in striatal astrocytes has been linked to vesicular glutamate release (Scofield et al. 2015), slow and low-volume glutamate transmission through Best1 (Woo et al. 2012), and adenosine transport through ENT1 (Kang et al. 2020), and activation of astroglial mGluR5, a Gq-coupled receptor, is linked to astrocyte release of ATP (Cavaccini et al. 2020). Perhaps it is not problematic that such a diverse range of outcomes can be recruited using the same tools or by engaging the same signaling cascade in astroglia. In most cases, it cannot be ruled out that multiple forms of gliotransmission occur following Gq receptor stimulation, as exhaustive tests are often not feasible. Also, gliotransmission in vivo would be expected to be more refined, since local neural signals would be detected by receptors on astrocyte processes to tune local responses. Indeed, it has been demonstrated previously that the same astrocyte is capable of signaling in distinct ways using different gliotransmitters depending on the dynamics of incoming signals (Covelo and Araque 2018), supporting the tremendous plasticity of this cell type.

Synaptic Proximity

Morphological Plasticity of Astrocytes

A central aspect underlying efficacy of transporter uptake and gliotransmission is the physical proximity of astroglial processes with synapses (see Fig. 1). The fidelity of this spatial relationship is crucial for normal neurobiology since perisynaptic astroglial processes (PAPs) express machinery for uptake and release of transmitters and maintenance of homeostatic neuronal function. Interestingly, PAPs display morphological plasticity more dynamic than what is observed in dendritic spines (Haber et al. 2006), and astrocyte morphological plasticity has been reported in the presence of extracellular glutamate (Genoud et al. 2006; Bernardinelli et al. 2014a; Verbich et al. 2012; Perez-Alvarez et al. 2014) and dopamine (Galloway et al. 2018). In the hippocampus, astroglial processes exhibit enhanced physical interaction with

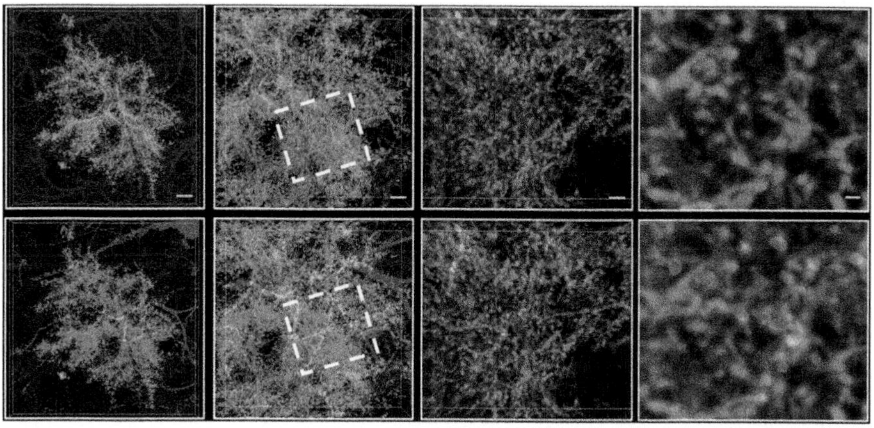

Fig. 1 Super-resolution confocal imaging of an astroglial plasma membrane, GFAP arbor and neighboring dendritic segment on an NAcore-projecting PL neuron. *Top*, plasma membrane labeling of an astrocyte in the PL using the AAV-GFAP-LcK-GFP viral vector from the Khakh laboratory (green) (Shigetomi et al. 2013). Immunohistochemical labeling of the cytoskeletal protein GFAP is shown in blue. *Bottom*, association of the astrocyte plasma membrane (green) with a retrogradely labeled dendritic segment from a neighboring PL neuron that projects to the NAcore (red). Here the neuron was labeled with retrograde delivery of Cre in the NAcore combined with Cre-dependent AAV-DIO-mCherry expression in the PL. For both the top and bottom series, panels transition to higher magnification to show detail from left to right. The dashed box depicts the inset region, scale bars are 10, 5, 2 and 0.5 microns

the postsynaptic compared with the presynaptic compartment (Lehre and Rusakov 2002). Further, astroglial interaction with neurons in this region tends to be biased toward dendritic spines as opposed to the larger dendritic shaft (Gavrilov et al. 2018). The general consensus regarding this conformation is that PAPs are positioned to perform glutamate transport upon transmitter release while also permitting a degree of transmitter spillover presynaptically, allowing for stimulation of presynaptic autoreceptors as a negative feedback loop that maintains elegant regulation of transmitter release probability.

In the striatum, there is a tremendous degree of variability in synaptic insulation by astroglial processes, perhaps due to the different input types (dopaminergic vs. glutamatergic), with astrocytes more closely abutting glutamatergic vs. dopaminergic terminals (Octeau et al. 2018). Regarding postsynaptic selectivity, high-resolution techniques have been applied to astrocytes and different synapse types in the striatum to demonstrate that astrocytes encompass similar numbers of D1- and D2-MSNs, with a slight bias toward D1-MSNs (Octeau et al. 2018). Nevertheless, both neuronal types receive similar degrees of synaptic insulation by astrocytes (Octeau et al. 2018).

Impact of Synaptic Insulation on Synapse Function

Synaptic proximity of astrocyte processes promotes synapse stability and maturation (Bernardinelli et al. 2014b; Nishida and Okabe 2007; Blanco-Suarez et al. 2018), and it is taken for granted that transmitter uptake and release by astrocytes requires a high degree of synaptic proximity to effectively regulate synaptic physiology. Not only would loss of synaptic apposition impact efficacy of these functions, but synaptic retraction of astrocyte processes has been shown to favor synaptic recruitment at excitatory synapses, with transmitter spillover potentiating nearby synapses when astrocyte processes have retracted (Henneberger et al. 2020). Retraction of astroglial processes may also permit stimulation of extrasynaptic glutamate receptors (Pal 2018; Kruyer and Kalivas 2020b) and impact neuronal excitability and plasticity. Extrasynaptic mGluRs pertinent to excitatory signaling underlying relapse include the aforementioned presynaptic mGluR2/3 that serves as a brake on glutamate and dopamine release (Xi et al. 2002) and mGluR5 expressed on nNOS interneurons that initiate degradation of the extracellular matrix and facilitate postsynaptic potentiation (Smith et al. 2017; Kruyer et al. 2019a).

Glutamate spillover permitted by retraction of astrocyte processes also recruits high-affinity NMDA receptors, which play an important role during drug relapse. The GluN2b (NR2b or NMDAR2b) subunit is largely extrasynaptic, as opposed to the mostly synaptic GluN2a (D'Ascenzo et al. 2007). Stimulation of GluN2b-containing NMDA receptors contributes to postsynaptic potentiation during reinstated heroin seeking, and GluN2b knockdown or blockade prevents increases in AMPA/NMDA and spine head diameter induced by a priming heroin injection and reduces reinstated seeking induced by cues or heroin prime (Shen et al. 2011).

Drug-Induced Morphological Plasticity

Astrocytes exhibit profound morphological adaptations after exposure to substances of abuse. For example, astrocyte volume is reduced in the NAcore by exposure to both opioids and psychostimulants (Siemsen et al. 2019; Scofield et al. 2016b; Kruyer et al. 2019b). NAcore astrocytes are also more densely packed after alcohol and astrocyte density correlate positively with breakpoint for ethanol, a measure of motivation to acquire alcohol (Bull et al. 2014). Analysis of astrocyte plasticity is complicated given that both immunological and synaptic events can impact their morphology, and substances of abuse can trigger both types of CNS adaptations. For example, chronic cocaine or morphine exposure causes reactive astrogliosis characterized by changes in GFAP expression, a type of neuroinflammation that induces an altered functional and morphological state in astroglia (Beitner-Johnson et al. 1993; Sil et al. 2018; Bowers and Kalivas 2003). Whether changes in astrocyte proximity to synapses is a by-product of an immunoreactive state and whether synaptic functions of astroglia are interrupted by immunological processes is an ongoing question that necessitates further investigation.

Recent studies show that cues that stimulate drug seeking trigger morphological plasticity in NAcore astrocytes in animals trained to self-administer heroin, but not sucrose (Kruyer et al. 2019b). This plasticity appears homeostatic in nature, resulting from cue-induced neuronal activity and glutamate release. Moreover, interrupting the re-association of astroglial processes with synapse during reinstatement elevates responding for cues that signal heroin availability. Changes in synaptic proximity of astrocyte processes in this case are linked to phosphorylation of ezrin, an actin-blinding protein selectively expressed in astroglial processes (Derouiche and Frotscher 2001). Currently, neither the signaling cascade that drives ezrin phosphorylation nor the mechanism by which astrocyte process motility suppresses reinstated seeking has been uncovered. Additionally, whether NAcore D1- or D2-MSN synapses retain different degrees of astrocyte insulation during withdrawal from drug use or during reinstated seeking is a fundamental remaining question.

Homeostatic Functions

Synaptogenesis

Astrocytes play key roles in both synapse formation and elimination. Astrocytes express and release several signaling factors that contribute to formation of dendritic spines and spine density (Walker et al. 2020; Ikeda et al. 2010; Wang et al. 2020a; Chung et al. 2015), a measure that is aberrantly altered in the striatum after drug use and during reinstated drug seeking (Shen et al. 2011; Dos Santos et al. 2017; Anderson and Self 2017). During development, thrombospondins released by astrocytes promote genesis of new synapses that are postsynaptically silent

(Christopherson et al. 2005). In adult animals, thrombospondin expression is elevated in response to cocaine exposure, and astrocyte thrombospondins in the NAshell contribute to generation of silent synapses (Wang et al. 2020b). In both cases, it is expected that presence of silent synapses facilitates induction of functional synapses rapidly upon further signaling that induces postsynaptic insertion of AMPA receptors. Indeed blocking thrombospondin release by astroglia during exposure to cocaine cues blunts cued reinstatement of seeking (Wang et al. 2020b), which is strongly linked to measures of postsynaptic potentiation (Gipson et al. 2013). Astrocytes also participate in synapse elimination, through phagocytosis directly, and through coordinated signaling with microglia (Chung et al. 2013, 2015; Wilton et al. 2019). While astrocytes prune synapses to refine neural circuits during development, there is evidence that astrocytes express the machinery for synapse engulfment into adulthood (Chung et al. 2015), and recruitment of this process during acquisition of drug taking behavior or extinction learning is a possibility.

BDNF, a growth factor expressed and released by both astrocytes and neurons (Zafra et al. 1991; Ohno et al. 2018; Bergami et al. 2008), confers bidirectional effects on addiction-related behaviors (McGinty et al. 2010). For instance, animals that receive BDNF infusions in the VTA 2 h after a cocaine self-administration session exhibit enhanced reinstated seeking after withdrawal (Lu et al. 2004). Instead, BDNF infusion in the PL attenuates reinstated seeking as well as reinstatement-associated increases in accumbens glutamate according to the same timeline (McGinty et al. 2010). Generally BDNF undergoes activity-dependent upregulation and contributes to synapse stability and plasticity (Gomez-Palacio-Schjetnan and Escobar 2013). Interestingly, BDNF also increases astrocyte morphological complexity (Holt et al. 2019). Whether BDNF expression is altered selectively in astrocytes within corticostriatal or other circuitries pertinent to addiction and relapse or whether neuronal BDNF exerts its effects in part through astroglial signaling has not been determined.

Network Homeostasis

The coupling of astroglia through gap junctions provides them with the unique capacity to scale network activity. Although some connexin proteins are expressed at relatively high levels in the nucleus accumbens of naïve animals, connexin protein expression is reduced up to 21 days after cocaine self-administration (Bennett et al. 1999). Consistent with this finding, methamphetamine reduces gap junction coupling when delivered to cultured astroglia directly, indicated by lack of dye transfer to nearby astroglia that are normally extensively coupled (Castellano et al. 2016). The gap junction protein connexin 30 regulates excitatory synaptic strength broadly in the hippocampus by coordinating widespread synaptic insertion leading to efficient glutamate uptake through increased synaptic proximity, but not increased transporter expression (Pannasch et al. 2014). Perhaps relevant to these findings is the discovery that inhibiting astrocytes in the medial prefrontal cortex impairs cognitive flexibility, involving coordinated neural oscillations through astroglial transmission (Brockett et al. 2018). How this may relate with flexibility in drug-related learned behaviors, such as extinction, is a relevant question given the involvement of corticostriatal glutamate in measures of drug seeking and refraining. Interestingly, blockade of gap junction proteins in astrocytes in the PL prevents extinction and reinstatement of cocaine CPP (Fitzgerald 2016). Research momentum on the involvement of gap junction proteins in addiction-related behaviors is slowly growing, and there is enthusiasm for the concept of lateral regulation by astroglia, where given their coupling, plasticity, and the multitude of mechanisms by which they impact synaptic function, astrocytes are poised to coordinate synapses that are neither directly nor indirectly linked otherwise (Covelo and Araque 2016).

Conclusions

Emerging research highlights the critical contribution of astrocytes to synaptic function, and astroglial adaptations across a number of brain regions have been shown to contribute to the encoding and expression of motivated behaviors relevant to drug addiction. A majority of early literature on astroglial function was generated in vitro and ex vivo, providing information on general astroglial responses to neural activity or pharmacological manipulation. An accumulating body of work links these early findings with in vivo behavioral measures, highlighting new avenues for research and experimentation. Here, we have highlighted research avenues poised to promote future discovery. Generation of new tools, including CalEx and viruses for astroglial labeling and functional manipulation of gene expression, and improved imaging methodologies are facilitating studies of astroglial function in rodent models of drug addiction and relapse. We expect that advances made in the coming years using these tools will drastically expand our understanding of ways in which astrocytes impact synaptic physiology during normal motivated behavior and in psychiatric disorders such as drug addiction.

References

Agulhon C, Fiacco TA, McCarthy KD (2010) Hippocampal short- and long-term plasticity are not modulated by astrocyte Ca2+ signaling. Science 327:1250–1254. https://doi.org/10.1126/science.1184821

Aida T et al (2015) Astroglial glutamate transporter deficiency increases synaptic excitability and leads to pathological repetitive behaviors in mice. Neuropsychopharmacology 40:1569–1579. https://doi.org/10.1038/npp.2015.26

Alsio J et al (2011) Enhanced sucrose and cocaine self-administration and cue-induced drug seeking after loss of VGLUT2 in midbrain dopamine neurons in mice. J Neurosci 31:12593–12603. https://doi.org/10.1523/JNEUROSCI.2397-11.2011

Anderson EM, Self DW (2017) It's only a matter of time: longevity of cocaine-induced changes in dendritic spine density in the nucleus accumbens. Curr Opin Behav Sci 13:117–123. https://doi.org/10.1016/j.cobeha.2016.11.013

Araque A et al (2014) Gliotransmitters travel in time and space. Neuron 81:728–739. https://doi.org/10.1016/j.neuron.2014.02.007

Atkins AL, Mashhoon Y, Kantak KM (2008) Hippocampal regulation of contextual cue-induced reinstatement of cocaine-seeking behavior. Pharmacol Biochem Behav 90:481–491. https://doi.org/10.1016/j.pbb.2008.04.007

Baker DA, Xi ZX, Shen H, Swanson CJ, Kalivas PW (2002) The origin and neuronal function of in vivo nonsynaptic glutamate. J Neurosci 22:9134–9141

Ball KT, Slane M (2012) Differential involvement of prelimbic and infralimbic medial prefrontal cortex in discrete cue-induced reinstatement of 3,4-methylenedioxymethamphetamine (MDMA; ecstasy) seeking in rats. Psychopharmacology 224:377–385. https://doi.org/10.1007/s00213-012-2762-5

Bassareo V, Cucca F, Frau R, Di Chiara G (2017) Changes in dopamine transmission in the nucleus Accumbens Shell and Core during ethanol and sucrose self-administration. Front Behav Neurosci 11:71. https://doi.org/10.3389/fnbeh.2017.00071

Batiuk MY et al (2020) Identification of region-specific astrocyte subtypes at single cell resolution. Nat Commun 11:1220. https://doi.org/10.1038/s41467-019-14198-8

Bazargani N, Attwell D (2016) Astrocyte calcium signaling: the third wave. Nat Neurosci 19:182–189. https://doi.org/10.1038/nn.4201

Beitner-Johnson D, Guitart X, Nestler EJ (1993) Glial fibrillary acidic protein and the mesolimbic dopamine system: regulation by chronic morphine and Lewis-Fischer strain differences in the rat ventral tegmental area. J Neurochem 61:1766–1773. https://doi.org/10.1111/j.1471-4159.1993.tb09814.x

Bennett SA et al (1999) Long-term changes in connexin32 gap junction protein and mRNA expression following cocaine self-administration in rats. Eur J Neurosci 11:3329–3338. https://doi.org/10.1046/j.1460-9568.1999.00752.x

Benneyworth MA, Li Y, Basu AC, Bolshakov VY, Coyle JT (2012) Cell selective conditional null mutations of serine racemase demonstrate a predominate localization in cortical glutamatergic neurons. Cell Mol Neurobiol 32:613–624. https://doi.org/10.1007/s10571-012-9808-4

Bergami M et al (2008) Uptake and recycling of pro-BDNF for transmitter-induced secretion by cortical astrocytes. J Cell Biol 183:213–221. https://doi.org/10.1083/jcb.200806137

Bernardinelli Y et al (2014a) Activity-dependent structural plasticity of perisynaptic astrocytic domains promotes excitatory synapse stability. Curr Biol 24:1679–1688. https://doi.org/10.1016/j.cub.2014.06.025

Bernardinelli Y, Muller D, Nikonenko I (2014b) Astrocyte-synapse structural plasticity. Neural Plast 2014:232105. https://doi.org/10.1155/2014/232105

Bimpisidis Z, Wallen-Mackenzie A (2019) Neurocircuitry of reward and addiction: potential impact of dopamine-glutamate co-release as future target in substance use disorder. J Clin Med 8. https://doi.org/10.3390/jcm8111887

Birgner C et al (2010) VGLUT2 in dopamine neurons is required for psychostimulant-induced behavioral activation. Proc Natl Acad Sci U S A 107:389–394. https://doi.org/10.1073/pnas.0910986107

Blanco-Suarez E, Liu TF, Kopelevich A, Allen NJ (2018) Astrocyte-secreted chordin-like 1 drives synapse maturation and limits plasticity by increasing synaptic GluA2 AMPA receptors. Neuron 100:1116–1132.e1113. https://doi.org/10.1016/j.neuron.2018.09.043

Bossert JM, Stern AL (2014) Role of ventral subiculum in context-induced reinstatement of heroin seeking in rats. Addict Biol 19:338–342. https://doi.org/10.1111/adb.12015

Bowers MS, Kalivas PW (2003) Forebrain astroglial plasticity is induced following withdrawal from repeated cocaine administration. Eur J Neurosci 17:1273–1278. https://doi.org/10.1046/j.1460-9568.2003.02537.x

Brockett AT et al (2018) Evidence supporting a role for astrocytes in the regulation of cognitive flexibility and neuronal oscillations through the Ca2+ binding protein S100beta. PLoS One 13:e0195726. https://doi.org/10.1371/journal.pone.0195726

Bull C et al (2014) Rat nucleus accumbens core astrocytes modulate reward and the motivation to self-administer ethanol after abstinence. Neuropsychopharmacology 39:2835–2845. https://doi.org/10.1038/npp.2014.135

Carelli RM, Williams JG, Hollander JA (2003) Basolateral amygdala neurons encode cocaine self-administration and cocaine-associated cues. J Neurosci 23:8204–8211

Castellano P, Nwagbo C, Martinez LR, Eugenin EA (2016) Methamphetamine compromises gap junctional communication in astrocytes and neurons. J Neurochem 137:561–575. https://doi.org/10.1111/jnc.13603

Cavaccini A, Durkee C, Kofuji P, Tonini R, Araque A (2020) Astrocyte signaling gates long-term depression at Corticostriatal synapses of the direct pathway. J Neurosci 40:5757–5768. https://doi.org/10.1523/JNEUROSCI.2369-19.2020

Chai H et al (2017) Neural circuit-specialized astrocytes: transcriptomic, proteomic, morphological, and functional evidence. Neuron 95:531–549.e539. https://doi.org/10.1016/j.neuron.2017.06.029

Chaudhri N, Sahuque LL, Cone JJ, Janak PH (2008) Reinstated ethanol-seeking in rats is modulated by environmental context and requires the nucleus accumbens core. Eur J Neurosci 28:2288–2298. https://doi.org/10.1111/j.1460-9568.2008.06517.x

Christopherson KS et al (2005) Thrombospondins are astrocyte-secreted proteins that promote CNS synaptogenesis. Cell 120:421–433. https://doi.org/10.1016/j.cell.2004.12.020

Chung WS et al (2013) Astrocytes mediate synapse elimination through MEGF10 and MERTK pathways. Nature 504:394–400. https://doi.org/10.1038/nature12776

Chung WS, Allen NJ, Eroglu C (2015) Astrocytes control synapse formation, function, and elimination. Cold Spring Harb Perspect Biol 7:a020370. https://doi.org/10.1101/cshperspect.a020370

Corkrum M et al (2020) Dopamine-evoked synaptic regulation in the nucleus Accumbens requires astrocyte activity. Neuron 105:1036–1047.e1035. https://doi.org/10.1016/j.neuron.2019.12.026

Cornish JL, Kalivas PW (2000) Glutamate transmission in the nucleus accumbens mediates relapse in cocaine addiction. J Neurosci 20:RC89

Covelo A, Araque A (2016) Lateral regulation of synaptic transmission by astrocytes. Neuroscience 323:62–66. https://doi.org/10.1016/j.neuroscience.2015.02.036

Covelo A, Araque A (2018) Neuronal activity determines distinct gliotransmitter release from a single astrocyte. elife 7. https://doi.org/10.7554/eLife.32237

Coyle JT (2012) NMDA receptor and schizophrenia: a brief history. Schizophr Bull 38:920–926. https://doi.org/10.1093/schbul/sbs076

Cuevas-Diaz Duran R, Wang CY, Zheng H, Deneen B, Wu JQ (2019) Brain region-specific gene signatures revealed by distinct astrocyte subpopulations unveil links to glioma and neurodegenerative diseases. eNeuro 6. https://doi.org/10.1523/ENEURO.0288-18.2019

Cui M et al (2009) The organic cation transporter-3 is a pivotal modulator of neurodegeneration in the nigrostriatal dopaminergic pathway. Proc Natl Acad Sci U S A 106:8043–8048. https://doi.org/10.1073/pnas.0900358106

Curcio L et al (2013) Reduced D-serine levels in the nucleus accumbens of cocaine-treated rats hinder the induction of NMDA receptor-dependent synaptic plasticity. Brain 136:1216–1230. https://doi.org/10.1093/brain/awt036

Danbolt NC (2001) Glutamate uptake. Prog Neurobiol 65:1–105

D'Ascenzo M et al (2007) mGluR5 stimulates gliotransmission in the nucleus accumbens. Proc Natl Acad Sci U S A 104:1995–2000. https://doi.org/10.1073/pnas.0609408104

D'Ascenzo M, Podda MV, Grassi C (2014) The role of D-serine as co-agonist of NMDA receptors in the nucleus accumbens: relevance to cocaine addiction. Front Synaptic Neurosci 6:16. https://doi.org/10.3389/fnsyn.2014.00016

Derouiche A, Frotscher M (2001) Peripheral astrocyte processes: monitoring by selective immunostaining for the actin-binding ERM proteins. Glia 36:330–341

Di Ciano P, Everitt BJ (2004) Direct interactions between the basolateral amygdala and nucleus accumbens core underlie cocaine-seeking behavior by rats. J Neurosci 24:7167–7173. https://doi.org/10.1523/JNEUROSCI.1581-04.2004

Do-Monte FH, Minier-Toribio A, Quinones-Laracuente K, Medina-Colon EM, Quirk GJ (2017) Thalamic regulation of sucrose seeking during unexpected reward omission. Neuron 94:388–400.e384. https://doi.org/10.1016/j.neuron.2017.03.036

Doncheck EM et al (2020) Sex, stress, and prefrontal cortex: influence of biological sex on stress-promoted cocaine seeking. Neuropsychopharmacology. https://doi.org/10.1038/s41386-020-0674-3

Dos Santos M et al (2017) Rapid synaptogenesis in the nucleus Accumbens is induced by a single cocaine administration and stabilized by mitogen-activated protein kinase interacting Kinase-1 activity. Biol Psychiatry 82:806–818. https://doi.org/10.1016/j.biopsych.2017.03.014

Farhat LC et al (2020) Pharmacological and behavioral treatment for trichotillomania: an updated systematic review with meta-analysis. Depress Anxiety 37:715–727. https://doi.org/10.1002/da.23028

Fischer-Smith KD, Houston AC, Rebec GV (2012) Differential effects of cocaine access and withdrawal on glutamate type 1 transporter expression in rat nucleus accumbens core and shell. Neuroscience 210:333–339. https://doi.org/10.1016/j.neuroscience.2012.02.049

Fitzgerald M (2016) Gap junction communication in memory retrieval and extinction of cocaine seeking. Theses Dissertations 1140:39–40

Floresco SB, Yang CR, Phillips AG, Blaha CD (1998) Basolateral amygdala stimulation evokes glutamate receptor-dependent dopamine efflux in the nucleus accumbens of the anaesthetized rat. Eur J Neurosci 10:1241–1251. https://doi.org/10.1046/j.1460-9568.1998.00133.x

Fuchs RA et al (2005) The role of the dorsomedial prefrontal cortex, basolateral amygdala, and dorsal hippocampus in contextual reinstatement of cocaine seeking in rats. Neuropsychopharmacology 30:296–309. https://doi.org/10.1038/sj.npp.1300579

Galloway A et al (2018) Dopamine triggers CTCF-dependent morphological and genomic remodeling of astrocytes. J Neurosci 38:4846–4858. https://doi.org/10.1523/JNEUROSCI.3349-17.2018

Gasser PJ, Hurley MM, Chan J, Pickel VM (2017) Organic cation transporter 3 (OCT3) is localized to intracellular and surface membranes in select glial and neuronal cells within the basolateral amygdaloid complex of both rats and mice. Brain Struct Funct 222:1913–1928. https://doi.org/10.1007/s00429-016-1315-9

Gavrilov N et al (2018) Astrocytic coverage of dendritic spines, dendritic shafts, and axonal boutons in hippocampal neuropil. Front Cell Neurosci 12:248. https://doi.org/10.3389/fncel.2018.00248

Genoud C et al (2006) Plasticity of astrocytic coverage and glutamate transporter expression in adult mouse cortex. PLoS Biol 4:e343. https://doi.org/10.1371/journal.pbio.0040343

Gipson CD et al (2013) Relapse induced by cues predicting cocaine depends on rapid, transient synaptic potentiation. Neuron 77:867–872. https://doi.org/10.1016/j.neuron.2013.01.005

Gomez JA et al (2019) Ventral tegmental area astrocytes orchestrate avoidance and approach behavior. Nat Commun 10:1455. https://doi.org/10.1038/s41467-019-09131-y

Gomez-Gonzalo M, Zehnder T, Requie LM, Bezzi P, Carmignoto G (2018) Insights into the release mechanism of astrocytic glutamate evoking in neurons NMDA receptor-mediated slow depolarizing inward currents. Glia 66:2188–2199. https://doi.org/10.1002/glia.23473

Gomez-Palacio-Schjetnan A, Escobar ML (2013) Neurotrophins and synaptic plasticity. Curr Top Behav Neurosci 15:117–136. https://doi.org/10.1007/7854_2012_231

Goubard V, Fino E, Venance L (2011) Contribution of astrocytic glutamate and GABA uptake to corticostriatal information processing. J Physiol 589:2301–2319. https://doi.org/10.1113/jphysiol.2010.203125

Grant JE, Kim SW, Odlaug BL (2007) N-acetyl cysteine, a glutamate-modulating agent, in the treatment of pathological gambling: a pilot study. Biol Psychiatry 62:652–657. https://doi.org/10.1016/j.biopsych.2006.11.021

Haber M, Zhou L, Murai KK (2006) Cooperative astrocyte and dendritic spine dynamics at hippocampal excitatory synapses. J Neurosci 26:8881–8891. https://doi.org/10.1523/JNEUROSCI.1302-06.2006

Heinsbroek JA et al (2017) Loss of plasticity in the D2-Accumbens Pallidal pathway promotes cocaine seeking. J Neurosci 37:757–767. https://doi.org/10.1523/JNEUROSCI.2659-16.2016

Henneberger C et al (2020) LTP induction boosts glutamate spillover by driving withdrawal of Perisynaptic Astroglia. Neuron. https://doi.org/10.1016/j.neuron.2020.08.030

Hnasko TS, Edwards RH (2012) Neurotransmitter corelease: mechanism and physiological role. Annu Rev Physiol 74:225–243. https://doi.org/10.1146/annurev-physiol-020911-153315

Hnasko TS et al (2010) Vesicular glutamate transport promotes dopamine storage and glutamate corelease in vivo. Neuron 65:643–656. https://doi.org/10.1016/j.neuron.2010.02.012

Holleran KM et al (2020) Organic cation transporter 3 and the dopamine transporter differentially regulate catecholamine uptake in the basolateral amygdala and nucleus accumbens. Eur J Neurosci. https://doi.org/10.1111/ejn.14927

Holmseth S et al (2009) The concentrations and distributions of three C-terminal variants of the GLT1 (EAAT2; slc1a2) glutamate transporter protein in rat brain tissue suggest differential regulation. Neuroscience 162:1055–1071. https://doi.org/10.1016/j.neuroscience.2009.03.048

Holt LM et al (2019) Astrocyte morphogenesis is dependent on BDNF signaling via astrocytic TrkB.T1. elife 8. https://doi.org/10.7554/eLife.44667

Howard EC, Schier CJ, Wetzel JS, Duvauchelle CL, Gonzales RA (2008) The shell of the nucleus accumbens has a higher dopamine response compared with the core after non-contingent intravenous ethanol administration. Neuroscience 154:1042–1053. https://doi.org/10.1016/j.neuroscience.2008.04.014

Howland JG, Taepavarapruk P, Phillips AG (2002) Glutamate receptor-dependent modulation of dopamine efflux in the nucleus accumbens by basolateral, but not central, nucleus of the amygdala in rats. J Neurosci 22:1137–1145

Ikeda H et al (2010) Morphine modulation of thrombospondin levels in astrocytes and its implications for neurite outgrowth and synapse formation. J Biol Chem 285:38415–38427. https://doi.org/10.1074/jbc.M110.109827

Ikemoto S (2007) Dopamine reward circuitry: two projection systems from the ventral midbrain to the nucleus accumbens-olfactory tubercle complex. Brain Res Rev 56:27–78. https://doi.org/10.1016/j.brainresrev.2007.05.004

Kalivas PW (2008) Addiction as a pathology in prefrontal cortical regulation of corticostriatal habit circuitry. Neurotox Res 14:185–189. https://doi.org/10.1007/BF03033809

Kalivas PW (2009) The glutamate homeostasis hypothesis of addiction. Nat Rev Neurosci 10:561–572. https://doi.org/10.1038/nrn2515

Kalivas BC, Kalivas PW (2016) Corticostriatal circuitry in regulating diseases characterized by intrusive thinking. Dialogues Clin Neurosci 18:65–76

Kalivas PW, Volkow ND (2005) The neural basis of addiction: a pathology of motivation and choice. Am J Psychiatry 162:1403–1413. https://doi.org/10.1176/appi.ajp.162.8.1403

Kalivas PW, Duffy P, Barrow J (1989) Regulation of the mesocorticolimbic dopamine system by glutamic acid receptor subtypes. J Pharmacol Exp Ther 251:378–387

Kalivas PW, Lalumiere RT, Knackstedt L, Shen H (2009) Glutamate transmission in addiction. Neuropharmacology 56(Suppl 1):169–173. https://doi.org/10.1016/j.neuropharm.2008.07.011

Kang S et al (2020) Activation of astrocytes in the dorsomedial striatum facilitates transition from habitual to goal-directed reward-seeking behavior. Biol Psychiatry. https://doi.org/10.1016/j.biopsych.2020.04.023

Kelley AE, Smith-Roe SL, Holahan MR (1997) Response-reinforcement learning is dependent on N-methyl-D-aspartate receptor activation in the nucleus accumbens core. Proc Natl Acad Sci U S A 94:12174–12179. https://doi.org/10.1073/pnas.94.22.12174

Kim R, Sepulveda-Orengo MT, Healey KL, Williams EA, Reissner KJ (2018) Regulation of glutamate transporter 1 (GLT-1) gene expression by cocaine self-administration and withdrawal. Neuropharmacology 128:1–10. https://doi.org/10.1016/j.neuropharm.2017.09.019

Knackstedt LA et al (2009) The role of cystine-glutamate exchange in nicotine dependence in rats and humans. Biol Psychiatry 65:841–845. https://doi.org/10.1016/j.biopsych.2008.10.040
Knackstedt LA, Melendez RI, Kalivas PW (2010) Ceftriaxone restores glutamate homeostasis and prevents relapse to cocaine seeking. Biol Psychiatry 67:81–84. https://doi.org/10.1016/j.biopsych.2009.07.018
Kohler S, Winkler U, Hirrlinger J (2019) Heterogeneity of astrocytes in Grey and white matter. Neurochem Res. https://doi.org/10.1007/s11064-019-02926-x
Koob GF, Volkow ND (2010) Neurocircuitry of addiction. Neuropsychopharmacology 35:217–238. https://doi.org/10.1038/npp.2009.110
Kruyer AK, Kalivas PW (2020a) Heroin cues reveal astroglial heterogeneity in the nucleus Accumbens core. bioRxiv. https://doi.org/10.1101/2020.07.22.216036
Kruyer A, Kalivas PW (2020b) Astrocytes as cellular mediators of cue reactivity in addiction. Curr Opin Pharmacol 56:1–6. https://doi.org/10.1016/j.coph.2020.07.009
Kruyer A, Chioma VC, Kalivas PW (2019a) The opioid-addicted Tetrapartite synapse. Biol Psychiatry. https://doi.org/10.1016/j.biopsych.2019.05.025
Kruyer A, Scofield MD, Wood D, Reissner KJ, Kalivas PW (2019b) Heroin Cue-evoked astrocytic structural plasticity at nucleus Accumbens synapses inhibits heroin seeking. Biol Psychiatry. https://doi.org/10.1016/j.biopsych.2019.06.026
Kupchik YM et al (2014) Cocaine dysregulates opioid gating of GABA neurotransmission in the ventral pallidum. J Neurosci 34:1057–1066. https://doi.org/10.1523/JNEUROSCI.4336-13.2014
Lasseter HC, Xie X, Ramirez DR, Fuchs RA (2010) Sub-region specific contribution of the ventral hippocampus to drug context-induced reinstatement of cocaine-seeking behavior in rats. Neuroscience 171:830–839. https://doi.org/10.1016/j.neuroscience.2010.09.032
Lehre KP, Rusakov DA (2002) Asymmetry of glia near central synapses favors presynaptically directed glutamate escape. Biophys J 83:125–134. https://doi.org/10.1016/S0006-3495(02)75154-0
Logan CN, LaCrosse AL, Knackstedt LA (2018) Nucleus accumbens GLT-1a overexpression reduces glutamate efflux during reinstatement of cocaine-seeking but is not sufficient to attenuate reinstatement. Neuropharmacology 135:297–307. https://doi.org/10.1016/j.neuropharm.2018.03.022
Lu L, Dempsey J, Liu SY, Bossert JM, Shaham Y (2004) A single infusion of brain-derived neurotrophic factor into the ventral tegmental area induces long-lasting potentiation of cocaine seeking after withdrawal. J Neurosci 24:1604–1611. https://doi.org/10.1523/JNEUROSCI.5124-03.2004
MacKay MB et al (2019) D-serine: potential therapeutic agent and/or biomarker in schizophrenia and depression? Front Psych 10:25. https://doi.org/10.3389/fpsyt.2019.00025
Martin R, Bajo-Graneras R, Moratalla R, Perea G, Araque A (2015) Circuit-specific signaling in astrocyte-neuron networks in basal ganglia pathways. Science 349:730–734. https://doi.org/10.1126/science.aaa7945
Martineau M et al (2013) Storage and uptake of D-serine into astrocytic synaptic-like vesicles specify gliotransmission. J Neurosci 33:3413–3423. https://doi.org/10.1523/JNEUROSCI.3497-12.2013
Martin-Fardon R, Weiss F (2017) Perseveration of craving: effects of stimuli conditioned to drugs of abuse versus conventional reinforcers differing in demand. Addict Biol 22:923–932. https://doi.org/10.1111/adb.12374
Matias I, Morgado J, Gomes FCA (2019) Astrocyte heterogeneity: impact to brain aging and disease. Front Aging Neurosci 11:59. https://doi.org/10.3389/fnagi.2019.00059
Mayer FP et al (2018) An unsuspected role for organic cation transporter 3 in the actions of amphetamine. Neuropsychopharmacology 43:2408–2417. https://doi.org/10.1038/s41386-018-0053-5
McCullough LD, Sokolowski JD, Salamone JD (1993) A neurochemical and behavioral investigation of the involvement of nucleus accumbens dopamine in instrumental avoidance. Neuroscience 52:919–925. https://doi.org/10.1016/0306-4522(93)90538-q

McFarland K, Lapish CC, Kalivas PW (2003) Prefrontal glutamate release into the core of the nucleus accumbens mediates cocaine-induced reinstatement of drug-seeking behavior. J Neurosci 23:3531–3537
McGinty JF, Whitfield TW Jr, Berglind WJ (2010) Brain-derived neurotrophic factor and cocaine addiction. Brain Res 1314:183–193. https://doi.org/10.1016/j.brainres.2009.08.078
McGlinchey EM, Aston-Jones G (2018) Dorsal Hippocampus drives context-induced cocaine seeking via inputs to lateral septum. Neuropsychopharmacology 43:987–1000. https://doi.org/10.1038/npp.2017.144
Melone M, Ciappelloni S, Conti F (2015) A quantitative analysis of cellular and synaptic localization of GAT-1 and GAT-3 in rat neocortex. Brain Struct Funct 220:885–897. https://doi.org/10.1007/s00429-013-0690-8
Mingote S, Amsellem A, Kempf A, Rayport S, Chuhma N (2019) Dopamine-glutamate neuron projections to the nucleus accumbens medial shell and behavioral switching. Neurochem Int 129:104482. https://doi.org/10.1016/j.neuint.2019.104482
Mongia S et al (2019) The ventral tegmental area has calbindin neurons with the capability to co-release glutamate and dopamine into the nucleus accumbens. Eur J Neurosci 50:3968–3984. https://doi.org/10.1111/ejn.14493
Morales M, Margolis EB (2017) Ventral tegmental area: cellular heterogeneity, connectivity and behaviour. Nat Rev Neurosci 18:73–85. https://doi.org/10.1038/nrn.2016.165
Mothet JP et al (2000) D-serine is an endogenous ligand for the glycine site of the N-methyl-D-aspartate receptor. Proc Natl Acad Sci U S A 97:4926–4931. https://doi.org/10.1073/pnas.97.9.4926
Moussawi K et al (2009) N-Acetylcysteine reverses cocaine-induced metaplasticity. Nat Neurosci 12:182–189. https://doi.org/10.1038/nn.2250
Naganuma F et al (2014) Predominant role of plasma membrane monoamine transporters in monoamine transport in 1321N1, a human astrocytoma-derived cell line. J Neurochem 129:591–601. https://doi.org/10.1111/jnc.12665
Nishida H, Okabe S (2007) Direct astrocytic contacts regulate local maturation of dendritic spines. J Neurosci 27:331–340. https://doi.org/10.1523/JNEUROSCI.4466-06.2007
Octeau JC et al (2018) An optical neuron-astrocyte proximity assay at synaptic distance scales. Neuron 98:49–66 e49. https://doi.org/10.1016/j.neuron.2018.03.003
Ohno Y, Kinboshi M, Shimizu S (2018) Inwardly rectifying Potassium Channel Kir4.1 as a novel modulator of BDNF expression in astrocytes. Int J Mol Sci 19. https://doi.org/10.3390/ijms19113313
Pal B (2018) **Involvement of extrasynaptic glutamate in physiological and pathophysiological changes of neuronal excitability. Cell Mol Life Sci 75:2917–2949. https://doi.org/10.1007/s00018-018-2837-5
Pannasch U et al (2014) Connexin 30 sets synaptic strength by controlling astroglial synapse invasion. Nat Neurosci 17:549–558. https://doi.org/10.1038/nn.3662
Papouin T, Dunphy J, Tolman M, Foley JC, Haydon PG (2017) Astrocytic control of synaptic function. Philos Trans R Soc Lond Ser B Biol Sci 372. https://doi.org/10.1098/rstb.2016.0154
Pardo-Garcia TR et al (2019) Ventral pallidum is the primary target for Accumbens D1 projections driving cocaine seeking. J Neurosci 39:2041–2051. https://doi.org/10.1523/JNEUROSCI.2822-18.2018
Parpura V et al (1994) Glutamate-mediated astrocyte-neuron signalling. Nature 369:744–747. https://doi.org/10.1038/369744a0
Pascual O et al (2005) Astrocytic purinergic signaling coordinates synaptic networks. Science 310:113–116. https://doi.org/10.1126/science.1116916
Pelton EW 2nd, Kimelberg HK, Shipherd SV, Bourke RS (1981) Dopamine and norepinephrine uptake and metabolism by astroglial cells in culture. Life Sci 28:1655–1663. https://doi.org/10.1016/0024-3205(81)90322-2

Perez-Alvarez A, Navarrete M, Covelo A, Martin ED, Araque A (2014) Structural and functional plasticity of astrocyte processes and dendritic spine interactions. J Neurosci 34:12738–12744. https://doi.org/10.1523/JNEUROSCI.2401-14.2014

Peters J, LaLumiere RT, Kalivas PW (2008) Infralimbic prefrontal cortex is responsible for inhibiting cocaine seeking in extinguished rats. J Neurosci 28:6046–6053. https://doi.org/10.1523/JNEUROSCI.1045-08.2008

Petravicz J, Boyt KM, McCarthy KD (2014) Astrocyte IP3R2-dependent Ca(2+) signaling is not a major modulator of neuronal pathways governing behavior. Front Behav Neurosci 8:384. https://doi.org/10.3389/fnbeh.2014.00384

Pow DV (2001) Visualising the activity of the cystine-glutamate antiporter in glial cells using antibodies to aminoadipic acid, a selectively transported substrate. Glia 34:27–38. https://doi.org/10.1002/glia.1037

Puaud M, Higuera-Matas A, Brunault P, Everitt BJ, Belin D (2020) The basolateral amygdala to nucleus Accumbens core circuit mediates the conditioned reinforcing effects of cocaine-paired cues on cocaine seeking. Biol Psychiatry. https://doi.org/10.1016/j.biopsych.2020.07.022

Quon EF, Wotton CA, Bekar LK (2018) Evidence for astrocyte purinergic signaling in cortical sensory adaptation and serotonin-mediated neuromodulation. Mol Cell Neurosci 88:53–61. https://doi.org/10.1016/j.mcn.2017.12.008

Reissner KJ et al (2015) Glutamate transporter GLT-1 mediates N-acetylcysteine inhibition of cocaine reinstatement. Addict Biol 20:316–323. https://doi.org/10.1111/adb.12127

Rimmele TS, Rosenberg PA (2016) GLT-1: the elusive presynaptic glutamate transporter. Neurochem Int 98:19–28. https://doi.org/10.1016/j.neuint.2016.04.010

Roberts-Wolfe DJ, Kalivas PW (2015) Glutamate transporter GLT-1 as a therapeutic target for substance use disorders. CNS Neurol Disord Drug Targets 14:745–756

Rocha A, Kalivas PW (2010) Role of the prefrontal cortex and nucleus accumbens in reinstating methamphetamine seeking. Eur J Neurosci 31:903–909. https://doi.org/10.1111/j.1460-9568.2010.07134.x

Rogers JL, Ghee S, See RE (2008) The neural circuitry underlying reinstatement of heroin-seeking behavior in an animal model of relapse. Neuroscience 151:579–588. https://doi.org/10.1016/j.neuroscience.2007.10.012

Sagara JI, Miura K, Bannai S (1993) Maintenance of neuronal glutathione by glial cells. J Neurochem 61:1672–1676. https://doi.org/10.1111/j.1471-4159.1993.tb09802.x

Salgado S, Kaplitt MG (2015) The nucleus Accumbens: a comprehensive review. Stereotact Funct Neurosurg 93:75–93. https://doi.org/10.1159/000368279

Scofield MD (2018) Exploring the role of Astroglial glutamate release and association with synapses in neuronal function and behavior. Biol Psychiatry 84:778–786. https://doi.org/10.1016/j.biopsych.2017.10.029

Scofield MD et al (2015) Gq-DREADD selectively initiates glial glutamate release and inhibits Cue-induced cocaine seeking. Biol Psychiatry 78:441–451. https://doi.org/10.1016/j.biopsych.2015.02.016

Scofield MD et al (2016a) The nucleus Accumbens: mechanisms of addiction across drug classes reflect the importance of glutamate homeostasis. Pharmacol Rev 68:816–871. https://doi.org/10.1124/pr.116.012484

Scofield MD et al (2016b) Cocaine Self-administration and extinction leads to reduced glial fibrillary acidic protein expression and morphometric features of astrocytes in the nucleus Accumbens Core. Biol Psychiatry 80:207–215. https://doi.org/10.1016/j.biopsych.2015.12.022

Sellings LH, Clarke PB (2003) Segregation of amphetamine reward and locomotor stimulation between nucleus accumbens medial shell and core. J Neurosci 23:6295–6303

Shen H, Moussawi K, Zhou W, Toda S, Kalivas PW (2011) Heroin relapse requires long-term potentiation-like plasticity mediated by NMDA2b-containing receptors. Proc Natl Acad Sci U S A 108:19407–19412. https://doi.org/10.1073/pnas.1112052108

Shen HW, Gipson CD, Huits M, Kalivas PW (2014a) Prelimbic cortex and ventral tegmental area modulate synaptic plasticity differentially in nucleus accumbens during cocaine-reinstated drug seeking. Neuropsychopharmacology 39:1169–1177. https://doi.org/10.1038/npp.2013.318

Shen HW, Scofield MD, Boger H, Hensley M, Kalivas PW (2014b) Synaptic glutamate spillover due to impaired glutamate uptake mediates heroin relapse. J Neurosci 34:5649–5657. https://doi.org/10.1523/JNEUROSCI.4564-13.2014

Shigetomi E et al (2013) Imaging calcium microdomains within entire astrocyte territories and endfeet with GCaMPs expressed using adeno-associated viruses. J Gen Physiol 141:633–647. https://doi.org/10.1085/jgp.201210949

Siemsen BM et al (2019) Effects of methamphetamine Self-administration and extinction on astrocyte structure and function in the nucleus Accumbens Core. Neuroscience 406:528–541. https://doi.org/10.1016/j.neuroscience.2019.03.040

Sil S, Periyasamy P, Guo ML, Callen S, Buch S (2018) Morphine-mediated brain region-specific Astrocytosis involves the ER stress-autophagy Axis. Mol Neurobiol 55:6713–6733. https://doi.org/10.1007/s12035-018-0878-2

Smith ACW et al (2017) Accumbens nNOS interneurons regulate cocaine relapse. J Neurosci 37:742–756. https://doi.org/10.1523/JNEUROSCI.2673-16.2016

Smith-Roe SL, Kelley AE (2000) Coincident activation of NMDA and dopamine D1 receptors within the nucleus accumbens core is required for appetitive instrumental learning. J Neurosci 20:7737–7742

Stefanik MT, Kalivas PW (2013) Optogenetic dissection of basolateral amygdala projections during cue-induced reinstatement of cocaine seeking. Front Behav Neurosci 7:213. https://doi.org/10.3389/fnbeh.2013.00213

Stuber GD, Hnasko TS, Britt JP, Edwards RH, Bonci A (2010) Dopaminergic terminals in the nucleus accumbens but not the dorsal striatum corelease glutamate. J Neurosci 30:8229–8233. https://doi.org/10.1523/JNEUROSCI.1754-10.2010

Stuber GD et al (2011) Excitatory transmission from the amygdala to nucleus accumbens facilitates reward seeking. Nature 475:377–380. https://doi.org/10.1038/nature10194

Sun W et al (2013) Glutamate-dependent neuroglial calcium signaling differs between young and adult brain. Science 339:197–200. https://doi.org/10.1126/science.1226740

Takeda H, Inazu M, Matsumiya T (2002) Astroglial dopamine transport is mediated by norepinephrine transporter. Naunyn Schmiedeberg's Arch Pharmacol 366:620–623. https://doi.org/10.1007/s00210-002-0640-0

Tan KR et al (2012) GABA neurons of the VTA drive conditioned place aversion. Neuron 73:1173–1183. https://doi.org/10.1016/j.neuron.2012.02.015

Tsai G, Yang P, Chung LC, Lange N, Coyle JT (1998) D-serine added to antipsychotics for the treatment of schizophrenia. Biol Psychiatry 44:1081–1089. https://doi.org/10.1016/s0006-3223(98)00279-0

Turner JR, Ecke LE, Briand LA, Haydon PG, Blendy JA (2013) Cocaine-related behaviors in mice with deficient gliotransmission. Psychopharmacology 226:167–176. https://doi.org/10.1007/s00213-012-2897-4

Valtcheva S, Venance L (2016) Astrocytes gate Hebbian synaptic plasticity in the striatum. Nat Commun 7:13845. https://doi.org/10.1038/ncomms13845

Verbich D, Prenosil GA, Chang PK, Murai KK, McKinney RA (2012) Glial glutamate transport modulates dendritic spine head protrusions in the hippocampus. Glia 60:1067–1077. https://doi.org/10.1002/glia.22335

Vollstadt-Klein S et al (2010) Initial, habitual and compulsive alcohol use is characterized by a shift of cue processing from ventral to dorsal striatum. Addiction 105:1741–1749. https://doi.org/10.1111/j.1360-0443.2010.03022.x

Walker CD, Risher WC, Risher ML (2020) Regulation of synaptic development by astrocyte signaling factors and their emerging roles in substance abuse. Cell 9. https://doi.org/10.3390/cells9020297

Wang W et al (2020a) TREK-1 null impairs neuronal excitability, synaptic plasticity, and cognitive function. Mol Neurobiol 57:1332–1346. https://doi.org/10.1007/s12035-019-01828-x

Wang J et al (2020b) Cocaine triggers astrocyte-mediated synaptogenesis. Biol Psychiatry. https://doi.org/10.1016/j.biopsych.2020.08.012

Whitelaw RB, Markou A, Robbins TW, Everitt BJ (1996) Excitotoxic lesions of the basolateral amygdala impair the acquisition of cocaine-seeking behaviour under a second-order schedule of reinforcement. Psychopharmacology 127:213–224

Willcocks AL, McNally GP (2013) The role of medial prefrontal cortex in extinction and reinstatement of alcohol-seeking in rats. Eur J Neurosci 37:259–268. https://doi.org/10.1111/ejn.12031

Willuhn I, Wanat MJ, Clark JJ, Phillips PE (2010) Dopamine signaling in the nucleus accumbens of animals self-administering drugs of abuse. Curr Top Behav Neurosci 3:29–71. https://doi.org/10.1007/7854_2009_27

Wilton DK, Dissing-Olesen L, Stevens B (2019) Neuron-glia signaling in synapse elimination. Annu Rev Neurosci 42:107–127. https://doi.org/10.1146/annurev-neuro-070918-050306

Wolosker H (2011) Serine racemase and the serine shuttle between neurons and astrocytes. Biochim Biophys Acta 1814:1558–1566. https://doi.org/10.1016/j.bbapap.2011.01.001

Woo DH et al (2012) TREK-1 and Best1 channels mediate fast and slow glutamate release in astrocytes upon GPCR activation. Cell 151:25–40. https://doi.org/10.1016/j.cell.2012.09.005

Wu J, Zhao R, Guo L, Zhen X (2017) Morphine-induced inhibition of Ca(2+) -dependent d-serine release from astrocytes suppresses excitability of GABAergic neurons in the nucleus accumbens. Addict Biol 22:1289–1303. https://doi.org/10.1111/adb.12417

Xi ZX, Baker DA, Shen H, Carson DS, Kalivas PW (2002) Group II metabotropic glutamate receptors modulate extracellular glutamate in the nucleus accumbens. J Pharmacol Exp Ther 300:162–171. https://doi.org/10.1124/jpet.300.1.162

Yang FY et al (2013) D-cycloserine, sarcosine and D-serine diminish the expression of cocaine-induced conditioned place preference. J Psychopharmacol 27:550–558. https://doi.org/10.1177/0269881110388333

Yin HH, Knowlton BJ, Balleine BW (2004) Lesions of dorsolateral striatum preserve outcome expectancy but disrupt habit formation in instrumental learning. Eur J Neurosci 19:181–189. https://doi.org/10.1111/j.1460-9568.2004.03095.x

Yoshikawa M et al (2007) The serine racemase mRNA is predominantly expressed in rat brain neurons. Arch Histol Cytol 70:127–134. https://doi.org/10.1679/aohc.70.127

Yu X et al (2018) Reducing astrocyte calcium signaling in vivo alters striatal microcircuits and causes repetitive behavior. Neuron 99:1170–1187.e1179. https://doi.org/10.1016/j.neuron.2018.08.015

Zafra F, Castren E, Thoenen H, Lindholm D (1991) Interplay between glutamate and gamma-aminobutyric acid transmitter systems in the physiological regulation of brain-derived neurotrophic factor and nerve growth factor synthesis in hippocampal neurons. Proc Natl Acad Sci U S A 88:10037–10041. https://doi.org/10.1073/pnas.88.22.10037

Zhang JM et al (2003) ATP released by astrocytes mediates glutamatergic activity-dependent heterosynaptic suppression. Neuron 40:971–982. https://doi.org/10.1016/s0896-6273(03)00717-7

Astroglia in the Vulnerability and Maintenance of Alcohol Use Disorders

José Javier Miguel-Hidalgo

Introduction

As it can be appreciated in previous chapters of this book, a great variety of functional and structural features of astrocytes are critical to the support of neuronal metabolism, synaptic activity, myelin maintenance, and repair of the injured nervous system (Verkhratsky and Nedergaard 2018). These functions are precisely some of the most important that are pathologically disturbed by chronic or acute alcohol abuse (de la Monte and Kril 2014; Harper 2009; Kyzar and Pandey 2015; Vargas et al. 2014). Thus, it is not surprising that a large body of research on human subjects with AUDs and animal models of AUD has focused on the direct effects of alcohol on the cellular and molecular biology of astrocytes (Adermark and Bowers 2016; Verkhratsky and Parpura 2010) as well as on the roles of astrocytes in alcohol-related pathological mechanisms leading to brain damage and the establishment and maintenance of alcohol dependence (Adermark and Bowers 2016; Miguel-Hidalgo 2018). Since earlier studies on astrocytes using Golgi techniques or immunohistochemistry of cytoskeletal markers (Babu et al. 1994; Davies and Vernadakis 1984; Mayordomo et al. 1992; Popova and Shchekalina 1980; Renau-Piqueras et al. 1989), alcohol has been known to alter their morphology upon prolonged or binge exposure in developing and mature animals, although the mechanisms involved and functional consequences of those changes were not completely understood. These alterations appear to depend on the duration of the exposure to alcohol. Binge alcohol intake (or subacute exposure in astrocyte cultures) appears to cause hypertrophy of astrocyte cell bodies and processes and increased synthesis of cytoskeletal proteins such as glial fibrillary acidic protein or vimentin, while prolonged or repetitive

J. J. Miguel-Hidalgo (✉)
Department of Psychiatry and Human Behavior, University of Mississippi Medical Center, Jackson, MS, USA
e-mail: jmiguel-hidalgo@umc.edu

B. Li et al. (eds.), *Astrocytes in Psychiatric Disorders*, Advances in Neurobiology 26, https://doi.org/10.1007/978-3-030-77375-5_11

exposure to alcohol may result in loss of astrocytes processes and the expression of their cytoskeletal proteins (Franke 1995; Kane et al. 1996; Sáez et al. 1991).

During neurodevelopment, most astrocytes are born after neuronal genesis ceases, so that alcohol exposure before astrogliogenesis tends to affect the generation and differentiation of neural precursors common to both neurons and glial cells, while exposure after the period of neurogenesis would mainly target the generation of astrocytes and other glial cells such as oligodendrocytes (Rubert et al. 2006), although, clearly, alcohol is still able to affect differentiation and plasticity of neurons and emerging astrocytes. In either case, alcohol exposure during pregnancy often results in developmental skeletal, neurological, and behavioral abnormalities that are grouped under the name of fetal alcohol syndrome spectrum disorder (FASD) (Saito et al. 2016; Wilhelm and Guizzetti 2015). Even in cases when the physical abnormalities of FASD are not obvious, alcohol exposure may result in behavioral disturbances associated with increased probability for later AUDs. In addition, other factors such as stress during infancy, preadolescence, and young adulthood are known to increase the chance of later alcohol abuse and dependence (Portero-Tresserra et al. 2018) or help to acquire excessive alcohol consumption (Kaufman et al. 2007; Meyer et al. 2013), probably mediated by permanent changes in neurohormones and the hypothalamic-pituitary-adrenal axis (Enoch 2011). Glial cells are sensitive to such changes and may participate in the onset and maintenance of AUDs, although the specific relevant glial responses and the mechanism of their exact putative contribution to the vulnerability to alcohol dependence largely remain to be fully understood. However, over the last decade, progress has been made in identifying a variety of mechanisms by which astrocyte dysfunction participates directly or indirectly in the onset, vulnerability, and maintenance of alcohol abuse and dependence. This chapter will address the pathological effects of alcohol exposure on the morphology and function astrocytes as well as the role that astrocyte disturbances caused by alcohol itself or other factors play in the maintenance and vulnerability to alcohol use disorders (Fig. 1).

Molecular Pathology of Astrocytes in Alcohol Use Disorders

Neuropathology of Astrocytes in Alcohol-Related Neurological Disorders

The size and morphology of astrocytes and their precursors in gray matter (GM) and white matter (WM) pathologically change in subjects with prolonged and repetitive histories of alcohol abuse (Miguel-Hidalgo and Rajkowska 2003). Some brain regions, such as the PFC and hippocampus, appear to be especially affected by alcohol-induced atrophic changes, including reduced staining of the GFAP cytoskeletal marker or lower numbers of GFAP+ astrocyte (Korbo 1999; Miguel-Hidalgo et al. 2002, 2006). Those alterations are accompanied by downregulation of

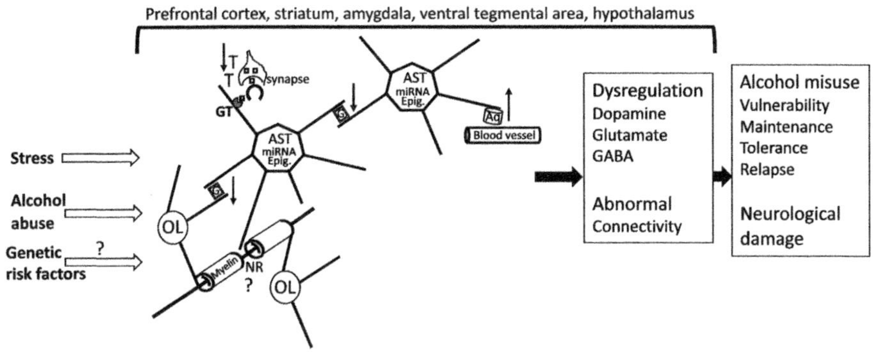

Fig. 1 Schematic illustration of several interactions of CNS astrocytes that are affected by stress, alcohol exposure, and possibly predisposing genetic factors that result not only in structural neuropathology but also in increased vulnerability to alcohol misuse by increasing tolerance and preference and favoring reinstatement of alcohol intake. The pathological interactions with astrocytes result in a variety of molecular and neurotransmitters alterations in the prefrontal cortex, ventral tegmental area, dorsal and ventral striatum, amygdala, and ventral tegmental area involving dysregulation of at least dopamine, glutamate, and GABA neurotransmission, with varying degrees of regional specificity. Alcohol exposure and other risk factors for neuropathology itself also result in pathological alterations of astrocyte-derived components at the blood-brain barrier and possibly the extracellular matrix and NRs. More recent research work has revealed epigenetic abnormalities (Epig.) or changed levels of specific miRNAs and other non-coding RNAs in alcohol use disorders that may contribute to alcohol abuse. Nevertheless, further research is needed to understand the mechanisms by which alcohol-related functional pathology at the cell and protein expression levels is linked to epigenetic and non-coding RNA markers and the involvement of astrocyte contacts at NR and oligodendrocytes that may contribute to abnormal connectivity in alcohol use disorders. Abbreviations: Aq Aquaporin 4, AST astrocytes, G gap junctions, GFAP glial fibrillary acidic protein, GT glutamate and GABA transporters, NR node of Ranvier, OL oligodendrocyte, T thrombospondin. ↑ upregulation, ↓ downregulation. Small squares at the synapse represent neurotransmitters

astrocyte-specific genes, particularly in subjects with hepatic pathology (Liu et al. 2006), indicating that indirect alcohol-related metabolic pathways must be taken into account when explaining astrocyte-related disturbances. Chronic AUDs are often concomitant with important nutritional deficiencies and hepatic pathology, causing severe neurological disorders such as Wernicke's encephalopathy, hepatic encephalopathy, or demyelinating disorders (de la Monte and Kril 2014; Verkhratsky et al. 2014). The grave neurological symptoms that accompany degeneration of GM or WM are macroanatomically detectable in the cerebral cortex, cerebellum, thalamus, or mammillary bodies (Kril and Harper 2012; Phillips et al. 1990). The cellular basis of mechanisms leading to the structural and metabolic dysfunction in those neurological disorders is not limited to oligodendrocytes or neurons but also involves importantly astrocytes (Hazell 2009), suggesting that even in subjects without major neurological signs, the alterations in emotional regulation and behavior may be dependent, at least partly on astrocyte dysfunction.

Neuropathological changes in response to alcohol abuse are not limited to alterations in number, morphology, or development of astrocytes but encompass many of their roles in the nervous system. These roles involve calcium signaling, the balance of excitatory and inhibitory neurotransmission, regulation of neuroinflammatory processes, water balance in the CNS, as well as the regulation of dopamine-dependent behavioral processes in brain reward circuits (Adermark and Bowers 2016). The interference of alcohol actions with those roles can be understood as modifying the functions of local components of neuronal circuits with repercussions for the overall integration of nervous signals. However, alcohol-related pathology of astrocytes may also significantly disturb the connections between brain regions by directly modifying the propagation of signals in the between those regions. For instance, pathology alterations of astrocyte connexins caused by alcohol may impair the communication between astrocytes and oligodendrocytes disturbing the maintenance of myelin (Hazell 2009). Likewise, pathology of astrocytes processes and the extracellular matrix they support around nodes of Ranvier may disturb the buffering of ions in the proximity of nodes Ranvier or the aggregation of voltage-gated sodium channels. Altering ion buffering and the osmotic regulation that results from astrocyte interactions with oligodendrocytes around nodes of Ranvier causes abnormal action potential propagation in WM and myelinated portions of GM (Gankam Kengne et al. 2011).

The inhibitory action of alcohol on the proliferation and turnover of astrocytes may be a major cause for the reduction in astrocyte numbers and their functional markers. Alcohol causes significant inhibition of astrocyte proliferation and DNA and protein synthesis, including a diminution of the major astrocyte marker GFAP in cultured neonatal astrocytes (Davies and Cox 1991; Guerri 1998; Guerri and Renau-Piqueras 1997) and in astrocytes from postmortem human brain tissue (Kane et al. 1996). The reduced number of astrocytes and the shrinkage of their processes may as well impair some of their critical functions. The responses of astrocytes to chronic alcohol, although mostly inhibitory, may also lead to secondary activation of gliosis-like astrocyte responses when AUDs prolong sufficiently into senescence (Miguel-Hidalgo 2009; Miguel-Hidalgo and Rajkowska 2003). Deficits in astrocyte structure and function accumulated during aging in AUD subjects may contribute to neuronal degeneration, unleashing a secondary increase of astrocyte reactivity or gliosis, which would be reflected in increased GFAP production and gliosis, although it is still unclear how other markers of astrocytes respond to aging-associated neuronal degeneration. The response to acute alcohol exposure also may involve extensive activation gene expression for more fundamental pathways linked to cellular stress such as the heat shock response (Pignataro et al. 2013).

The cycle of alcohol abuse is considered to depend not solely on the rewarding properties of alcohol and the involvement of altered dopaminergic transmission but on the negative behavioral and organic consequences upon withdrawal from alcohol intake (Koob 2013). Aversion to these consequences reinforces the behaviors leading to excessive alcohol consumption and favors relapse and the maintenance of alcohol abuse and addiction (Gilpin and Koob 2008). The aversion to the unpleasant consequences of withdrawal seems to greatly depend on the activation of neurons and the expression of glucocorticoid and CRF receptors in the amygdala (Gass et al. 2011; Gilpin and Roberto 2012; Koob 2009; Läck et al. 2005; Patkar et al. 2016;

Vendruscolo et al. 2012) after chronic alcohol abuse, although other brain regions are also recruited in changes leading to reinstatement of drinking (Schroeder et al. 2008). Disturbances of neurotransmission in the amygdala are attributed to mostly neuronal pathology; however, some postmortem studies have shown that one of the most distinctive changes associated with chronic alcoholism in the basolateral amygdala is the reduction of glutamate transporters GLT1 and GLAST, the two major transporters for glutamate expressed by astrocytes (Kryger and Wilce 2010). Gene expression changes also occur in the amygdala that involve several pathways present n astrocytes and other glia (McBride et al. 2014); for some of which, there is evidence supporting a role in the vulnerability or maintenance of behaviors of alcohol addiction (Lee et al. 2013; Nam et al. 2012; Paul and Medina 2012).

Alcohol Effects on Glutamate Receptors and Astrocyte Components of the Cycle for Release and Reuptake of Glutamate

A major tenet of the theories explaining the neurobiological underpinnings of the vulnerability to alcohol abuse is that, initially, repetitive or binge exposure to alcohol causes excessive activation of reward pathways mediated by increased release of dopamine in the nucleus accumbens and the prefrontal cortex by synaptic terminals of dopaminergic neurons with cell bodies located in the ventral tegmental area (VTA) (Boileau et al. 2003; Brodie et al. 1999; Cowen and Lawrence 1999; Heinz et al. 2009; Pascual et al. 2009; Tupala and Tiihonen 2004). VTA dopaminergic neurons are negatively regulated by synapses formed with axon terminals of interneurons that release gamma-aminobutyric acid (GABA). GABA release from these terminals in turn is inhibited by opioids acting on presynaptic GABAA receptors (Koob et al. 1998). Given the critical role of astrocytes in the synaptic reuptake of inhibitory and excitatory neurotransmitters in all the implicated structures (VAT, NACC, PFC) (Ayers-Ringler et al. 2016; Forget et al. 2006; Fouyssac and Belin 2019; Gomez et al. 2019; Haydon et al. 2009; Schwarcz et al. 1994; Vollbrecht et al. 2014; Xin et al. 2019), it seems natural to think that there is a role for astrocytes in the altered regulation of the effects and release of neurotransmitters, including dopamine, in alcoholism (Deng et al. 2009; Flatscher-Bader et al. 2007; Morikawa and Morrisett 2010; Trantham-Davidson and Chandler 2015; Volkow et al. 2007; Xiao et al. 2009) and thus in the corresponding disturbances of function and behavior linked to them. The participation of astrocytes in the vulnerability to AUD, thus, may involve direct actions of alcohol exposure on astrocytes or indirect effects by first causing neuronal, microglial, or oligodendroglial physiological disturbances that impinge on astrocyte function, which in turn further modifies neuronal function and behavioral outcomes. In addition, abnormalities of astrocyte gene expression or physiology preceding AUDs, including developmental disturbances, may exist that increase the vulnerability to alcohol abuse and may or may not involve the same astrocytic pathways affected by AUD-promoting alcohol exposure.

AUDs and the Role of NMDA-Type Glutamate Receptors

The available evidence points to an increase, at least in some brain regions, in the content of NMDA-type glutamate receptors and a decrease in GABA receptors, particularly in chronic alcoholism (Davis and Wu 2001) which would be at the root of further changes in dopamine release in acute and chronic alcohol exposure and thus importantly contribute to craving, tolerance, and dependence. Ethanol acts as an antagonist of the actions of glutamate at NMDA receptors. Chronic alcohol consumption leads to enhanced expression of NR2A, NR2B, and NR1 NMDA receptors in the neocortex and the hippocampus (Gass and Olive 2008; Nixon et al. 2004), which would account for increased neuronal excitability that occurs after alcohol withdrawal in animal models. However, more variable, human studies in chronic alcoholics demonstrate higher NMDA receptor ligand binding (Tsai 1998; Tsai et al. 1998) in the PFC, but not in other brain regions including the cingulate cortex, hippocampus, and cerebellar vermis (Freund and Anderson 1996; Freund and Anderson 1999). mRNA levels for the same NMDAR subunits are unchanged in uncomplicated alcoholics but are lower in AUD subjects with cirrhosis (Ridge et al. 2008).

AUDs and the Release of NMDA Receptor Co-agonists by Astrocytes

One of the best-known pharmacological effects of ethanol is its antagonism of the activation of NMDA receptors by the neurotransmitter glutamate. In animal models of chronic alcoholism, there is also an increase in the expression of glutamate receptors that may explain increased behavioral agitation and neuronal hyperexcitability following withdrawal form alcohol intake (Gass and Olive 2008). These withdrawal effects act as negative reinforcers to facilitate relapse into alcohol drinking. NMDA receptor activation by glutamate is also dependent on the binding of co-transmitter glycine to a specific site (different from that for glutamate) of the receptor. Astrocytes release D-serine, which specifically binds the glycine site of NMDA receptors and thus can regulate their activation state (Tsai 1998). Thus, astrocyte integrity may directly affect the effects of ethanol on neuronal excitability and consequently on the probability to engage in alcohol abuse. Interestingly, ethanol competes with D-serine for the occupancy of the glycine-site, which would further amplify the effects of astrocyte inability to provide D-serine, helping to explain the development of reduced sensitivity to ethanol effects in mice (Kiefer et al. 2003) or increased tolerance to partial agonists of the glycine site in AUD subjects (Krystal et al. 2011).

Astrocytic Glutamate and GABA Transporters and Glutamine Synthetase in the Vulnerability to Alcohol Abuse

Alcohol is known to block glutamate NMDA receptors and potentiate the activation of GABA receptors and thus contribute to the neurotropic responses to alcohol. Those actions of alcohol result in alterations of glutamate and GABA release and their concentrations in the peri-synaptic space, where astrocytes regulate extracellular GLU and GABA by uptake with specific transporters lodged in their cell membranes. In addition, the neurotransmitters taken up by astrocytes are recycled first by converting glutamate into glutamine through the action of the astrocyte-located enzyme glutamine synthetase. Thus, the ability of astrocytes to regulate the neurotransmitters around synapses may be affected by changes in the transporters or glutamine synthetase. In fact, studies of the mRNA and protein levels as well as the distribution of glutamate transporters and glutamine synthetase have been performed in the neocortex of AUD subjects (Ayers-Ringler et al. 2016; Miguel-Hidalgo et al. 2010). However, these studies have not detected significant chronic alcohol abuse-associated changes in protein levels of either EAAT1, EAAT2, or GS caused by prolonged alcohol exposure, while in vitro studies rather have shown that alcohol treatment may lead to an increase in the rate of glutamate transport per astrocyte (Smith 1997; Smith and Zsigo 1996; Zink et al. 2004), although in some rat brain areas such the nucleus accumbens, there appears to be an ethanol-induced decrease in glutamate transport that still is not linked to a reduced expression of glutamate transporters (Melendez et al. 2005). Nonetheless, in alcohol self-administering animals, we have observed that withdrawal of the access to alcohol results in increased glutamine synthetase immunoreactivity in the rat PFC, although the role of such an increase in the mechanisms of relapse or maintenance of alcohol intake remains to be ascertained.

Previously we have argued that unaltered or increased expression of glutamate receptors or transporters in several brain regions of AUD subjects may depend on the severity of the neural damage caused by alcohol exposure, because in a cohort of AUD subjects with comorbid major depression (unlike "non-complicated" alcoholics), we found reduced levels of glutamate transporters and GS, suggesting the possibility that more severe alcohol-related pathology actually dampens the expression of glutamate cycle constituents in astrocytes (Hazell et al. 2010; Miguel-Hidalgo et al. 2010). Alternatively, if initial damage is not important in "uncomplicated" AUD subjects, there might be compensatory mechanisms in astrocyte components of the glutamate cycle. For instance, astrocytic glutamate transporter levels (Wu et al. 2011) can increase even if alcohol itself blocks the activation of those transporters (Mulholland et al. 2009). Nevertheless, regulation of glutamate transporter gene expression may partake in the vulnerability to alcoholism because mice lacking the astrocyte glutamate transporter GLAST do not develop place preference for alcohol and show diminished voluntary alcohol consumption as compared to control mice (Karlsson et al. 2012). Likewise, intracerebral blockade of GLT-1 reduced binge ethanol drinking in mice (Smith et al. 2014). In other less studied parts of the brain, but gaining importance for their involvement in the regulation of alcohol intake, such as the suprachiasmatic nucleus, the contribution of astrocytic transporters to abuse behaviors may be instrumental since enhancement of GLT-1-based glutamate transport with ceftriaxone diminishes alcohol drinking (Lee et al. 2013), while reduced astrocytic GLT-1 resulting from deletion of the circadian period gene (Per2) in mice leads to augmented alcohol intake (Spanagel et al. 2005).

Gap Junctions of Astrocytes in AUDs

Astrocytes are capable of communication among themselves and with oligodendrocytes and neurons through gap junctions. These junctions include groups of channels connecting the cytoplasms of two cells. Each gap junction channel is formed by two connexons composed of six protein subunits each, with two connexons (one for each cell forming the junction) forming a gap junction. The subunits in each connexon are cell-type specific. In the CNS, astrocyte connexons are mainly formed with connexin 43 (Cx43) or connexin 30. Gap junctions allow for fast communication of calcium signals between astrocytes, the buffering of potassium, the diffusion of small molecules, and the passage of water exchanged with the blood circulation. They also allow for the exchange of intracellular signals with oligodendrocytes that are important in the maintenance of myelin.

Prolonged alcohol exposure also appears to reduce the localization of Cx43 to the cell membrane in astrocytes and other cells (Miguel-Hidalgo et al. 2014; Wentlandt et al. 2004). This reduction may have significant effects on the ability of astrocytes to spread calcium transients. In addition, Cx43 is also the major constituent of hemichannels (connexons sitting on the membrane not coupled to connexons in adjacent cells) that are involved in the release of molecules such as ATP and D-serine to regulate calcium signals between astrocytes and glutamate receptor in neurons. Alterations in the extracellular concentrations of those molecules may directly affect neurotransmission in brain regions such as the prefrontal cortex, the ventral tegmental area (VTA), and nucleus accumbens (NAcc) and thus increase the vulnerability to abuse alcohol. For instance, D-serine is a co-transmitter that greatly enhances the actions of glutamate at NMDA receptors, so that reduced release because of limited Cx43 hemichannel numbers may contribute to enhanced activation of the reward circuits, due to the inactivation of inhibitory neurons that control the activity of dopaminergic neurons in the VTA. In the prefrontal cortex of human subjects with AUDs, we have found a dramatic decrease of immunohistochemical staining and protein levels for Cx43 (Miguel-Hidalgo et al. 2014). Since the prefrontal cortex is a major site for pathology in AUD subjects, and is heavily involved in emotional and appetitive control, the alteration of Cx43 may contribute to the vulnerability to and the maintenance of AUDs. In fact, we found that localized infusion of a gap junction/hemichannel inhibitor or an astrocyte-directed toxin in the prefrontal cortex of the rat increased transiently but significantly alcohol intake in rats (Miguel-Hidalgo et al. 2009). Further involvement of an alcohol-induced reduction of astrocyte gap junction hemichannels in increasing the vulnerability to alcohol abuse is suggested by the pharmacological blockade of hemichannels, which enhances the motivation to drink alcohol after 3 weeks of abstinence from alcohol (Bennett et al. 2003; Bull et al. 2014). Nonetheless, the exact role of astrocyte gap junctional changes in regulating AUDs in adults remains to be fully elucidated since prenatal exposure to ethanol, which increases the probability of epilepsy in humans and in young mice, results in increased connexins 43 and 30 in the hippocampus and cortex of these mice (Ramani et al. 2016). Whether these early astrocyte-related changes have any bearing on the expression of AUDs later in life will require further study.

Involvement of Astrocytes in Neuroimmune Responses Elicited by Alcohol Exposure

Astrocytes are capable of releasing cytokines and other inflammatory factors involved in neuroimmune signaling and, conversely, they respond to neuroimmune signals (Warden et al. 2016). In addition, astrocytes carry toll-like like receptors in their cell membrane. These receptors mediate neuroimmune responses to a variety of signals. Exposure to excessive ethanol induces the release of inflammatory mediators from glial cells, including astrocytes, and binding of those molecules to TLRs further facilitates cytokine release and the activation of inflammatory responses (Alfonso-Loeches et al. 2010, 2012; Montesinos et al. 2016). Recent research suggests that signaling at TLR4 in the prefrontal cortex may have a direct role in the vulnerability to and maintenance of alcohol intake. While wild-type adolescent mice increase the rate of alcohol consumption over time, knockout mice lacking TLR4 do no escalate alcohol drinking, pointing to the possibility that activation of TLR4 receptors in astrocytes participates in facilitating alcohol drinking after exposure to alcohol (Montesinos et al. 2016).

Astrocyte Thrombospondin in AUD-Related Synaptic Alterations

Reduced numbers of synaptic contacts and structural alterations of pre- and presynaptic components are also caused by excessive or chronic alcohol intake. Besides the roles of peri-synaptic astrocyte processes in the recycling of neurotransmitter and the regulation of their actions, astrocytes are known to secrete factors, such as thrombospondins, that are involved in the maturation and formation of synapses. Impairments to the role of astrocytes in secreting thrombospondins or alterations of their receptors can result in deficient synaptic function and formation (Ullian et al. 2004). In animal models, alcohol exposure can result in persistent reduction of thrombospondin release in addition to impaired matching of presynaptic and postsynaptic structures (Risher et al. 2015). Liver damage in alcoholism may also result in synaptic dysfunction mediated by astrocytes, because increased ammonia levels reduce astrocyte-released thrombospondin and levels of synaptic proteins (Jayakumar et al. 2014). Alcohol exposure during early or prenatal stages of development, and maybe later too, may cause persistent changes in synapse formation involving thrombospondin (Trindade et al. 2016). Other factors secreted by astrocytes to the extracellular matrix such as laminin or heparan-sulfate proteoglycan are also reduced by alcohol exposure (Lasek 2016; Trindade et al. 2016). In summary, repeated alcohol exposure at different stages of prenatal and postnatal development may result in abnormal regulation of astrocyte-released factors that participate in synaptogenesis.

Astrocyte Processes at the Blood-Brain Barrier and the Involvement of Aquaporins in AUD-Related Neuropathology

The endings of astrocyte processes distal to their cell bodies tightly sheath the basal lamina around the endothelial cells of small blood vessels, so contributing to the maintenance of the blood-brain barrier (BBB) (Prat et al. 2001). Moreover, those processes are critically involved in the exchange of metabolic mediators with the circulating blood and to the control of blood flow (Koehler et al. 2009). Chronic alcoholism BBB components cause disruption (Haorah et al. 2005; Rubio-Araiz et al. 2017), impairing the exchange of metabolites such as glucose and trophic factors, which potentially affects the function of surrounding neurons and glial cells (Abdul Muneer et al. 2011a, b).

Effects of alcohol mediated by astrocytes at the BBB are likely to involve changes in aquaporins as well (Kong et al. 2013), in particular aquaporin 4 (AQ4), a membrane protein highly expressed in astrocytes processes abutting at the BBB that allows passage of water through the cell membrane (Badaut et al. 2002; Rajkowska et al. 2013). Repeated alcohol binging in rats results in higher aquaporin levels, swelling of astrocytes, and activation of neuroinflammatory factors (Collins et al.

2013; Collins and Neafsey 2012). Anti-inflammatory treatments can prevent the effects of the AQ4 elevation that is concomitant with increases in neuroinflammatory markers (Tajuddin et al. 2014). Aquaporin loss in astrocytes may contribute to the loss of myelin in central pontine myelinolysis, although more research seems to be needed to fully ascertain this possibility (Popescu et al. 2013).

Alcohol-induced AQ4 increase in astrocytes of the NAcc and the concomitant cell swelling may contribute to increased vulnerability to excessive alcohol intake because the upregulation of the dopamine level in the NAcc is positively correlated with the concentration of AQ4 (Kuppers et al. 2008; Lee et al. 2013), and inhibition of cell swelling results in the inhibition of ethanol-induced dopamine release (Adermark et al. 2011a).

Epigenetic Changes in Astrocytes in AUDs

mRNA transcription from DNA is heavily regulated by epigenetic modifications such as acetylation and methylation of DNA itself, the associated chromatin histones, or both (Graff et al. 2011). Translation of mRNA into proteins is further regulated by the presence of noncoding fragments of microRNAs (miRNAs) that are about 22 nucleotides long and bind to complementary sequences in mRNA to impede translation of specific proteins (Emery and Lu 2015; Li and Yao 2012; Liu and Casaccia 2010).

Various studies have revealed that the development of astrocytes and other glial cells depends on an intricate network epigenetic pathways (Bian et al. 2013; Emery and Lu 2015; MacDonald and Roskams 2009; Namihira and Nakashima 2013). The fate and differentiation of precursors into astrocytes are dependent on methylation of DNA at specific nucleotides and on specific histone modifications (Moyon et al. 2016). In fact, alcohol exposure can significantly disturb epigenetic pathways, suggesting that pathology of astrocytes and their dysfunction may depend on enduring, long-term changes in gene expression in the components of both regulatory and effector intracellular pathways (Alfonso-Loeches et al. 2012; Aspberg and Tottmar 1994; Bichenkov and Ellingson 2009; Coutts and Harrison 2015; Creeley et al. 2013; Newville et al. 2017).

There is limited knowledge on the effects of alcohol exposure on astrocyte epigenetic modifications and the expression of proteins and pathways affected by those modifications. However, available evidence raises the possibility that alcohol exposure causes persistent epigenetic and gene expression changes in astrocytes. For instance, prenatal ethanol exposure in rats produces hypermethylation of the promoter for the astrocyte GFAP gene leading to diminished GFAP expression during postnatal development (Vallés et al. 1997).

Astrocyte Epigenetic Markers

AUDs result in extensive changes of brain epigenetic markers (Farris et al. 2015; Legastelois et al. 2017; Weng et al. 2015; Zhou et al. 2011) and higher levels of miRNAs regulating the expression of several proteins (Lewohl et al. 2011). Alcohol induces epigenetic alterations of DNA methylation and in methylation and acetylation of histones in the human PFC, hippocampus, and amygdala (Farris et al. 2015; Ponomarev 2013) as well as in cultured astrocytes (Zhang et al. 2014). Some research indicates that alcohol intake causes inhibition of histone deacetylase (HDAC) in the amygdala leading to increased histone acetylation and reduced anxiety, while withdrawal, anxiety, or alcohol exposure in adolescence would be rather linked to increased HDAC activity (Pandey et al. 2017) and low acetylation (Pandey et al. 2008). Since acetylation level is inversely related to gene expression, higher HDAC would result in lower expression of genes associated with synaptic plasticity. In consequence, HDAC inhibitors have been insinuated as therapeutic agents to mitigate anxiety and alcohol intake (Pandey et al. 2017). Other research has found mainly DNA methylation reductions in humans with AUDs, although a recent study used a different methodology and found a higher proportion of hypermethylated sites in brain DNA of AUD subjects (Tulisiak et al. 2017).

DNA methylation depends on the activity of DNA methyltransferases (DNMTs). In experimental animals, prolonged alcohol intake induces high DNMT levels, but in the human brain, overall methylation changes are rather associated with lower DNMT mRNA (Tulisiak et al. 2017). Nonetheless, DNA methylation patterns in AUDs are complex and include both de- and hypermethylated promoters of particular genes (Tulisiak et al. 2017). Thus, in view of the higher DNMT level in experimental models of AUDs, some researchers have attempted to leverage DNMT inhibition as a possible therapeutic approach for AUDs. These investigators have discovered that the DNMT inhibitors 5-aza and decitabine can reduce excessive alcohol drinking in rodents (Ponomarev et al. 2017; Tulisiak et al. 2017). In mice, these behavioral changes are paralleled by increased expression of genes preferentially present in astrocytes in the VTA (Ponomarev et al. 2017). DNA methylation and other epigenetic markers in the developing CNS (Laufer et al. 2017; Mahnke et al. 2017; Ozturk et al. 2017) are also disturbed as a consequence of early alcohol exposure.

Research on the effects of ethanol on cultured astrocytes has also found increased methylation in the promoter of the tissue plasminogen activator (Tulisiak et al. 2017), which is involved in the degradation of components of the extracellular matrix and has been found upregulated in animal models of AUDs (Zhang et al. 2014). In rats, prenatal ethanol exposure leads to hypermethylation of the promoter for the astrocyte-expressed GFAP gene and to reduction in GFAP expression during early postnatal development (Vallés et al. 1997). These findings suggest that alteration of DNA or histone epigenetic markers in astrocytes may play an important role in mediating behavioral disturbances in AUDs, although it is unclear whether alterations during early development or later, as a consequence of alcohol abuse, are etiologically related to increased vulnerability to alcohol abuse and whether particular brain structures are related to that vulnerability.

Variety of Mechanisms Involving Astrocytes in the Vulnerability to AUDs

Plasticity in the morphology of astrocytes processes and in the expression of their cytoskeletal proteins has been directly linked to physiological and behavioral processes, and is maybe related to resumption of alcohol intake after a period of abstinence. In particular, increases in the number of GFAP-positive astrocytes or in the morphology of processes in the NAcc are observed after 3 weeks of abstinence from alcohol self-administration and are parallel to increases in the motivation to drink ethanol (Bull et al. 2014). Given the importance of astrocytes in the regulation of synaptic activity in the NAcc, relapse to alcohol self-administration may well depend, at least partly, on the numbers and morphology of astrocytes involved in such regulation. Interestingly, the effects of excessive alcohol exposure on NAcc astrocytes may also influence other appetitive behaviors as shown recently in female mice undergoing several days of chronic intermittent alcohol exposure. In these experiments, extinction of conditioned place preference for food rewards was reduced in alcohol-dependent animals in parallel with a reduction in the GFAP staining of astrocytes in the NAcc (Giacometti et al. 2020). At the same time, selective DREADDS-based chemogenetic activation of NAcc astrocytes restored extinction of food CPP in alcohol-dependent animals, suggesting that NAcc astrocyte activity is involved in the regulation of appetitive behaviors and in the effects of alcohol on the expression of those behaviors. Likewise, in the prefrontal cortex of rats abstinent from prolonged alcohol intake, glutamine synthetase immunoreactive (GS-I) astrocytes are increased (Miguel-Hidalgo 2006). These animals also show an increased rate of alcohol consumption (as compared to before abstinence), suggesting that an increased glutamate processing ability by newly recruited GS-I astrocytes in the PFC may contribute to a relapse into excessive alcohol drinking.

Despite the effects of early alcohol exposure on astrocyte numbers and astrocytic proteins, a form of early-life stress, maternal separation, although increasing the chance for enhanced alcohol seeking behavior in rats, does not appear to cause changes in the numbers GFAP-immunostained astrocytes or in the length of their processes in the hippocampus. This effect occurs despite the ability of maternal separation to also increase neuronal numbers in the amygdala but decrease them in the dentate gyrus of the hippocampus (Gondre-Lewis et al. 2016), suggesting that there are differential pathways of astrocyte involvement in the contribution to AUDs.

Various mechanisms with components localized at the cell membrane or intracellularly, some of which have been mentioned above, result in dramatic changes in the concentration of calcium within astrocytes. The membrane mechanisms include receptors to various neurotransmitters and growth factors, while the spread of calcium t increases to adjacent astrocytes involves gap junctions. One of the main mechanisms to increase calcium within astrocytes depends on the activation of GqGPCRs. Expression of genes for G-protein-coupled receptors and calcium regulation is decreased by sustained exposure to alcohol, but acutely, alcohol activates GqGPCR resulting in increased intracellular calcium in astrocyte and augments release of ATP, which once rapidly converted into adenosine acutely contributes to behavioral alterations such as hyperlocomotion and sedation (Erickson et al. 2018; Guerra-Gomes et al. 2017). In recent work, Erickson et al. (2020) using a chemogenetic approach have shown that stimulation of calcium signals in astrocytes increases alcohol consumption and preference but that reduction of intracellular calcium in those cells reduces alcohol intake (Erickson et al. 2020). In addition, GqGPCR activation in PFC astrocytes potentiates sensitivity to acute hyperlocomotion and the subsequent sedation caused by alcohol, but blockade of astrocytic calcium increases has the inverse effect (Erickson et al. 2020). These effects are dependent on the release of ATP, because blockade of the adenosine receptor A1R reduces the enhancement of alcohol intoxication caused by the chemogenetic activation of the Gq subunit of GPCR in astrocytes (Erickson et al. 2020; Moffat and Ron 2020). These experiments indicate that the ability of alcohol to alter calcium signaling in astrocytes of brain regions relevant to addiction, such as the prefrontal cortex, may be an important determinant of the vulnerability to alcohol abuse.

As somewhat hinted above, several of the mechanisms pathologically targeted in astrocytes by excessive (binge or chronic) alcohol intake not only affect negatively neuronal function, and thus indirectly behavior, but may be major mediators for further and sustained alcohol dependence and abuse. For instance, the increase in the density of GFAP-positive astrocytes in the nucleus accumbens shell during abstinence in rats is positively correlated with the motivation to retake alcohol self-administration (Bull et al. 2014). The effects of deficient metabolic function in astrocytes caused by ethanol exposure may also play a role in facilitating alcohol self-administration, since infusion of fluorocitrate, a metabolic inhibitor mainly taken up by astrocytes, in the prefrontal cortex temporarily increases preference for alcohol intake, which could be mediated by reduced taurine release by astrocytes, because taurine release can be inhibited by fluorocitrate (Choe et al. 2012). By contrast, swelling of astrocytes in the nucleus accumbens, another consequence of alcohol exposure (Adermark et al. 2011a; Vargova and Sykova 2014), results in the release of taurine, involved in the release of various neurotransmitters including dopamine in the NAcc, resulting in activation of dopamine D1 and D2 receptors, which are involved in the reinforcing properties of alcohol and other drugs.

Recent studies also point to an involvement of the astrocyte cystine-glutamate antiporter function in the vulnerability to alcohol dependence since reduced content of the antiporter in the medial prefrontal cortex results in resistance to alcohol-induced conditioned place preference and facilitates stress-induced reinstatement of CPP for alcohol in rats (Amaral et al. 2020).

In addition to altered astrocyte function in the prefrontal cortex and nucleus accumbens favoring reinstatement and facilitating self-administration, pathology of astrocytes in the dorsomedial striatum (DMS) may be involved in the establishment of habitual reward-seeking behavior. Experiments in a rat model have shown that calcium increases in DMS astrocytes induced with chemogenetic technology cause a change from habitual behaviors to goal-directed behaviors that is mediated by astrocyte adenosine metabolism (Kang et al. 2020). These results are relevant because alcohol exposure is associated with alterations in the capacity of astrocytes to elicit calcium transients, which are dependent on the neurotransmitters astrocytes are exposed to around synapses (Catlin et al. 2000; Kimelberg et al. 1993; Salazar et al. 2008). At the same time, intercellular communication through gap junctions that allow for the spread of Ca++ transients in astrocyte networks may also be inhibited by ethanol (Adermark et al. 2004), which could be further complicated by ethanol-associated reduction of astrocyte-expressed connexins that form gap junctions and hemichannels.

The abnormal regulation of intracellular calcium levels in astrocytes by altered activation of A1 adenosine receptors may as well contribute importantly to the regulation of alcohol intake and to the sedative effects of alcohol. In mice, activation of calcium signaling by G-protein-coupled receptors in astrocytes of the prefrontal cortex results in increased ethanol consumption in alcohol-naïve mice but not in ethanol addicted mice, which suggests that calcium regulation within astrocytes contributes to the establishment of alcohol abuse and possibly to the tolerance to alcohol effects (Erickson et al. 2020). In addition, A1 adenosine receptors in astrocytes were shown to participate significantly in the sedative effects of alcohol. Thus, the various effects of ethanol impinging on the regulation of intracellular calcium changes in astrocytes would further contribute to the maintenance of excessive alcohol intake.

Effects of Early Alcohol Exposure and Stress on Astrocyte and Their Role in AUDs

Various prenatal and postnatal disturbances of neural development increase the vulnerability to drug abuse or dependence including alcohol abuse (Campbell et al. 2009). Since those disturbances include pathological changes of glial cells, in particular astroglia (Guerri and Renau-Piqueras 1997), it is possible that the increased vulnerability depends partly on anomalous generation of astrocyte precursors and astrocytes during gliogenesis or damage to newly differentiated astrocytes that would play roles in the development of neurons and oligodendrocytes, including myelin formation by the latter. These developmental glial alterations are involved in the structural, cognitive, and behavioral abnormalities that characterize fetal spectrum disorder (Guizzetti et al. 2014; Wilhelm and Guizzetti 2015). However, the specific roles that the developmental deficits in astrocyte formation play in leading to increased probability to alcohol abuse and dependence in adolescence and adulthood are still unclear.

The grave features of fetal alcohol spectrum disorders and, in general, of prenatal and early postnatal alcohol exposure are not limited to generalized physical, cognitive, and psychiatric consequences, but also may induce dramatic increase in the probability of promoting alcohol abuse in those subjects involuntarily exposed to alcohol early in development. In fact, it seems that for many adolescents and young adults, early alcohol exposure is a better predictor of later abuse than the history of alcohol problems per se (Baer et al. 1998, 2003; Miller and Spear 2006). Given the roles of astrocytes in induction, reinstatement and maintenance of alcohol abuse mentioned earlier, and the known negative consequences of early alcohol exposure on glial cell development (Wilhelm and Guizzetti 2015), it is very likely that early alcohol-related pathology of astrocytes and the processes they support (such as myelination) may contribute to set the stage for adolescent or adult acquisition of alcohol abuse disorders. Consistent with this view are studies showing that even moderate prenatal ethanol exposure in rats can result in significant alterations of GFAP and astrocytic glutamate transport when measured during the adolescence of the exposed rats (Brolese et al. 2014, 2015).

Early stress is a factor for later engagement in behaviors of addiction including alcohol abuse (Gondre-Lewis et al. 2016; Penasco et al. 2015). A possible role for astrocytes in early stress-linked abuse liability is consistent with studies in rats subjected to maternal separation; after which, they present altered content of astrocytic CB1 receptors, which are known for their involvement in the regulation of behaviors of addiction (Lopez-Gallardo et al. 2012). These rats also develop increased alcohol drinking and reduction of endocannabinoid levels in the striatum and prefrontal cortex (Portero-Tresserra et al. 2018), thus suggesting that an abnormal role of astrocytes in endocannabinoid neurotransmission alterations caused by early stress may contribute to development of AUDs (Adermark et al., 2011b; Karlsson et al., 2012). In addition, early-life stress can also result in anomalous function of hypothalamic astrocytes, which has been linked to susceptibility to stress during adulthood, a risk factor for AUDs (Gunn et al. 2013).

A multitude of epigenetic changes have been identified in the brain as a consequence of FSAD-inducing exposure to alcohol (Lussier et al. 2017). Those changes affect many pathways involved in neuronal plasticity and survival. Some changes that affect glial cells such as DNA hypermethylation of the GFAP gene indicate that astrocytes may undergo long-lasting gene expression changes that may affect later alcohol consummatory behavior (Basavarajappa and Subbanna 2016; Vallés et al. 1997), although experiments are needed to further ascertain this possibility.

References

Abdul Muneer PM, Alikunju S, Szlachetka AM, Haorah J (2011a) Inhibitory effects of alcohol on glucose transport across the blood-brain barrier leads to neurodegeneration: preventive role of acetyl-L: -carnitine. Psychopharmacology 214:707–718

Abdul Muneer PM, Alikunju S, Szlachetka AM, Mercer AJ, Haorah J (2011b) Ethanol impairs glucose uptake by human astrocytes and neurons: protective effects of acetyl-L-carnitine. Int J Physiol Pathophysiol Pharmacol 3:48–56

Adermark L, Bowers MS (2016) Disentangling the role of astrocytes in alcohol use disorder. Alcohol Clin Exp Res 40:1802–1816

Adermark L, Olsson T, Hansson E (2004) Ethanol acutely decreases astroglial gap junction permeability in primary cultures from defined brain regions. Neurochem Int 45:971–978

Adermark L, Clarke RB, Olsson T, Hansson E, Soderpalm B, Ericson M (2011a) Implications for glycine receptors and astrocytes in ethanol-induced elevation of dopamine levels in the nucleus accumbens. Addict Biol 16:43–54

Adermark L, Jonsson S, Ericson M, Soderpalm B (2011b) Intermittent ethanol consumption depresses endocannabinoid-signaling in the dorsolateral striatum of rat. Neuropharmacology 61:1160–1165

Alfonso-Loeches S, Pascual-Lucas M, Blanco AM, Sanchez-Vera I, Guerri C (2010) Pivotal role of TLR4 receptors in alcohol-induced neuroinflammation and brain damage. J Neurosci 30:8285–8295

Alfonso-Loeches S, Pascual M, Gomez-Pinedo U, Pascual-Lucas M, Renau-Piqueras J, Guerri C (2012) Toll-like receptor 4 participates in the myelin disruptions associated with chronic alcohol abuse. Glia 60:948–964

Amaral VCS, Morais-Silva G, Laverde CF, Marin MT (2020) Susceptibility to extinction and reinstatement of ethanol-induced conditioned place preference is related to differences in astrocyte cystine-glutamate antiporter content. Neurosci Res. https://doi.org/10.1016/j.neures.2020.07.002

Aspberg A, Tottmar O (1994) Ethanol-induced increase in catalase activity in reaggregation cultures of rat brain cells is due to increased oligodendrocyte differentiation. Alcohol Clin Exp Res 18:620–624

Ayers-Ringler JR, Jia YF, Qiu YY, Choi DS (2016) Role of astrocytic glutamate transporter in alcohol use disorder. World J Psychiatry 6:31–42

Babu PP, Kumari LR, Vemuri MC (1994) Differential changes in cell morphology, macromolecular composition and membrane protein profiles of neurons and astrocytes in chronic ethanol treated rats. Mol Cell Biochem 130:29–40

Badaut J, Lasbennes F, Magistretti PJ, Regli L (2002) Aquaporins in brain: distribution, physiology, and pathophysiology. J Cereb Blood Flow Metab 22:367–378

Baer JS, Barr HM, Bookstein FL, Sampson PD, Streissguth AP (1998) Prenatal alcohol exposure and family history of alcoholism in the etiology of adolescent alcohol problems. J Stud Alcohol 59:533–543

Baer JS, Sampson PD, Barr HM, Connor PD, Streissguth AP (2003) A 21-year longitudinal analysis of the effects of prenatal alcohol exposure on young adult drinking. Arch Gen Psychiatry 60:377–385

Basavarajappa BS, Subbanna S (2016) Epigenetic mechanisms in developmental alcohol-induced neurobehavioral deficits. Brain Sci 6:12

Bennett MV, Contreras JE, Bukauskas FF, Saez JC (2003) New roles for astrocytes: gap junction hemichannels have something to communicate. Trends Neurosci 26:610–617

Bian S, Xu TL, Sun T (2013) Tuning the cell fate of neurons and glia by microRNAs. Curr Opin Neurobiol 23:928–934

Bichenkov E, Ellingson JS (2009) Ethanol alters the expressions of c-Fos and myelin basic protein in differentiating oligodendrocytes. Alcohol 43:627–634

Boileau I, Assaad JM, Pihl RO, Benkelfat C, Leyton M, Diksic M, Tremblay RE, Dagher A (2003) Alcohol promotes dopamine release in the human nucleus accumbens. Synapse 49:226–231

Brodie MS, Pesold C, Appel SB (1999) Ethanol directly excites dopaminergic ventral tegmental area reward neurons. Alcohol Clin Exp Res 23:1848–1852

Brolese G, Lunardi P, Broetto N, Engelke DS, Lirio F, Batassini C, Tramontina AC, Goncalves CA (2014) Moderate prenatal alcohol exposure alters behavior and neuroglial parameters in adolescent rats. Behav Brain Res 269:175–184

Brolese G, Lunardi P, de Souza DF, Lopes FM, Leite MC, Goncalves CA (2015) Pre- and postnatal exposure to moderate levels of ethanol can have long-lasting effects on hippocampal glutamate uptake in adolescent offspring. PLoS One 10:e0127845

Bull C, Freitas KC, Zou S, Poland RS, Syed WA, Urban DJ, Minter SC, Shelton KL et al (2014) Rat nucleus accumbens core astrocytes modulate reward and the motivation to self-administer ethanol after abstinence. Neuropsychopharmacology 39:2835–2845

Campbell JC, Szumlinski KK, Kippin TE (2009) Contribution of early environmental stress to alcoholism vulnerability. Alcohol 43:547–554

Catlin MC, Guizzetti M, Costa LG (2000) Effect of ethanol on muscarinic receptor-induced calcium responses in astroglia. J Neurosci Res 60:345–355

Choe KY, Olson JE, Bourque CW (2012) Taurine release by astrocytes modulates osmosensitive glycine receptor tone and excitability in the adult supraoptic nucleus. J Neurosci 32:12518–12527

Collins MA, Neafsey EJ (2012) Neuroinflammatory pathways in binge alcohol-induced neuronal degeneration: oxidative stress cascade involving aquaporin, brain edema, and phospholipase A2 activation. Neurotox Res 21:70–78

Collins MA, Moon KH, Tajuddin N, Neafsey EJ, Kim HY (2013) Docosahexaenoic acid (DHA) prevents binge ethanol-dependent aquaporin-4 elevations while inhibiting neurodegeneration: experiments in rat adult-age entorhino-hippocampal slice cultures. Neurotox Res 23:105–110

Coutts DJ, Harrison NL (2015) Acetaldehyde, not ethanol, impairs myelin formation and viability in primary mouse oligodendrocytes. Alcohol Clin Exp Res 39:455–462

Cowen MS, Lawrence AJ (1999) The role of opioid-dopamine interactions in the induction and maintenance of ethanol consumption. Prog Neuro-Psychopharmacol Biol Psychiatry 23:1171–1212

Creeley CE, Dikranian KT, Johnson SA, Farber NB, Olney JW (2013) Alcohol-induced apoptosis of oligodendrocytes in the fetal macaque brain. Acta Neuropathol Commun 1:23

Davies DL, Cox WE (1991) Delayed growth and maturation of astrocytic cultures following exposure to ethanol: electron microscopic observations. Brain Res 547:53–61

Davies DL, Vernadakis A (1984) Effects of ethanol on cultured glial cells: proliferation and glutamine synthetase activity. Brain Res 318:27–35

Davis KM, Wu JY (2001) Role of glutamatergic and GABAergic systems in alcoholism. J Biomed Sci 8:7–19

de la Monte SM, Kril JJ (2014) Human alcohol-related neuropathology. Acta Neuropathol 127:71–90

Deng C, Li K-Y, Zhou C, Ye J-H (2009) Ethanol enhances glutamate transmission by retrograde dopamine Signaling in a postsynaptic neuron/synaptic bouton preparation from the ventral tegmental area. Neuropsychopharmacology 34:1233–1244

Emery B, Lu QR (2015) Transcriptional and epigenetic regulation of oligodendrocyte development and myelination in the central nervous system. Cold Spring Harb Perspect Biol 7:a020461

Enoch MA (2011) The role of early life stress as a predictor for alcohol and drug dependence. Psychopharmacology 214:17–31

Erickson EK, Farris SP, Blednov YA, Mayfield RD, Harris RA (2018) Astrocyte-specific transcriptome responses to chronic ethanol consumption. Pharmacogenomics J 18:578–589

Erickson EK, DaCosta AJ, Mason SC, Blednov YA, Mayfield RD, Harris RA (2020) Cortical astrocytes regulate ethanol consumption and intoxication in mice. Neuropsychopharmacology 46:500–508

Farris SP, Harris RA, Ponomarev I (2015) Epigenetic modulation of brain gene networks for cocaine and alcohol abuse. Front Neurosci 9:176

Flatscher-Bader T, Zuvela N, Landis N, Wilce PA (2007) Smoking and alcoholism target genes associated with plasticity and glutamate transmission in the human ventral tegmental area. Hum Mol Genet 17:38–51

Forget C, Stewart J, Trudeau L-É (2006) Impact of basic FGF expression in astrocytes on dopamine neuron synaptic function and development. Eur J Neurosci 23:608–616

Fouyssac M, Belin D (2019) Beyond drug-induced alteration of glutamate homeostasis, astrocytes may contribute to dopamine-dependent intrastriatal functional shifts that underlie the development of drug addiction: a working hypothesis. Eur J Neurosci 50:3014–3027

Franke H (1995) Influence of chronic alcohol treatment on the GFAP-immunoreactivity in astrocytes of the hippocampus in rats. Acta Histochem 97:263–271
Freund G, Anderson KJ (1996) Glutamate receptors in the frontal cortex of alcoholics. Alcohol Clin Exp Res 20:1165–1172
Freund G, Anderson KJ (1999) Glutamate receptors in the cingulate cortex, hippocampus, and cerebellar vermis of alcoholics. Alcohol Clin Exp Res 23:1–6
Gankam Kengne F, Nicaise C, Soupart A, Boom A, Schiettecatte J, Pochet R, Brion JP, Decaux G (2011) Astrocytes are an early target in osmotic demyelination syndrome. J Am Soc Nephrol 22:1834–1845
Gass JT, Olive MF (2008) Glutamatergic substrates of drug addiction and alcoholism. Biochem Pharmacol 75:218–265
Gass JT, Sinclair CM, Cleva RM, Widholm JJ, Olive MF (2011) Alcohol-seeking behavior is associated with increased glutamate transmission in basolateral amygdala and nucleus accumbens as measured by glutamate-oxidase-coated biosensors. Addict Biol 16:215–228
Giacometti LL, Chandran K, Figueroa LA, Barker JM (2020) Astrocyte modulation of extinction impairments in ethanol-dependent female mice. Neuropharmacology 179:108272
Gilpin NW, Koob GF (2008) Neurobiology of alcohol dependence: focus on motivational mechanisms. Alcohol Res Health 31:185–195
Gilpin NW, Roberto M (2012) Neuropeptide modulation of central amygdala neuroplasticity is a key mediator of alcohol dependence. Neurosci Biobehav Rev 36:873–888
Gomez JA, Perkins JM, Beaudoin GM, Cook NB, Quraishi SA, Szoeke EA, Thangamani K, Tschumi CW et al (2019) Ventral tegmental area astrocytes orchestrate avoidance and approach behavior. Nat Commun 10:1455
Gondre-Lewis MC, Darius PJ, Wang H, Allard JS (2016) Stereological analyses of reward system nuclei in maternally deprived/separated alcohol drinking rats. J Chem Neuroanat 76:122–132
Graff J, Kim D, Dobbin MM, Tsai LH (2011) Epigenetic regulation of gene expression in physiological and pathological brain processes. Physiol Rev 91:603–649
Guerra-Gomes S, Sousa N, Pinto L, Oliveira JF (2017) Functional roles of astrocyte calcium elevations: from synapses to behavior. Front Cell Neurosci 11:427
Guerri C (1998) Neuroanatomical and neurophysiological mechanisms involved in central nervous system dysfunctions induced by prenatal alcohol exposure. Alcohol Clin Exp Res 22:304–312
Guerri C, Renau-Piqueras J (1997) Alcohol, astroglia, and brain development. Mol Neurobiol 15:65–81
Guizzetti M, Zhang X, Goeke C, Gavin DP (2014) Glia and neurodevelopment: focus on fetal alcohol spectrum disorders. Front Pediatr 2:123
Gunn BG, Cunningham L, Cooper MA, Corteen NL, Seifi M, Swinny JD, Lambert JJ, Belelli D (2013) Dysfunctional astrocytic and synaptic regulation of hypothalamic glutamatergic transmission in a mouse model of early-life adversity: relevance to neurosteroids and programming of the stress response. J Neurosci 33:19534–19554
Haorah J, Knipe B, Leibhart J, Ghorpade A, Persidsky Y (2005) Alcohol-induced oxidative stress in brain endothelial cells causes blood-brain barrier dysfunction. J Leukoc Biol 78:1223–1232
Harper C (2009) The neuropathology of alcohol-related brain damage. Alcohol Alcohol 44:136–140
Haydon PG, Blendy J, Moss SJ, Rob Jackson F (2009) Astrocytic control of synaptic transmission and plasticity: a target for drugs of abuse? Neuropharmacology 56(Suppl 1):83–90
Hazell AS (2009) Astrocytes are a major target in thiamine deficiency and Wernicke's encephalopathy. Neurochem Int 55:129–135
Hazell AS, Sheedy D, Oanea R, Aghourian M, Sun S, Jung JY, Wang D, Wang C (2010) Loss of astrocytic glutamate transporters in Wernicke encephalopathy. Glia 58:148–156
Heinz A, Beck A, Grusser SM, Grace AA, Wrase J (2009) Identifying the neural circuitry of alcohol craving and relapse vulnerability. Addict Biol 14:108–118
Jayakumar AR, Tong XY, Curtis KM, Ruiz-Cordero R, Shamaladevi N, Abuzamel M, Johnstone J, Gaidosh G et al (2014) Decreased astrocytic thrombospondin-1 secretion after chronic ammo-

nia treatment reduces the level of synaptic proteins: in vitro and in vivo studies. J Neurochem 131:333–347
Kane CJ, Berry A, Boop FA, Davies DL (1996) Proliferation of astroglia from the adult human cerebrum is inhibited by ethanol in vitro. Brain Res 731:39–44
Kang S, Hong SI, Lee J, Peyton L, Baker M, Choi S, Kim H, Chang SY et al (2020) Activation of astrocytes in the dorsomedial striatum facilitates transition from habitual to goal-directed reward-seeking behavior. Biol Psychiatry 88:797–808
Karlsson RM, Adermark L, Molander A, Perreau-Lenz S, Singley E, Solomon M, Holmes A, Tanaka K et al (2012) Reduced alcohol intake and reward associated with impaired endocannabinoid signaling in mice with a deletion of the glutamate transporter GLAST. Neuropharmacology 63:181–189
Kaufman J, Yang BZ, Douglas-Palumberi H, Crouse-Artus M, Lipschitz D, Krystal JH, Gelernter J (2007) Genetic and environmental predictors of early alcohol use. Biol Psychiatry 61:1228–1234
Kiefer F, Jahn H, Koester A, Montkowski A, Reinscheid RK, Wiedemann K (2003) Involvement of NMDA receptors in alcohol-mediated behavior: mice with reduced affinity of the NMDA R1 glycine binding site display an attenuated sensitivity to ethanol. Biol Psychiatry 53:345–351
Kimelberg HK, Cheema M, O 'Connor ER, Tong H, Goderie SK, Rossman PA (1993) Ethanol-induced aspartate and taurine release from primary astrocyte cultures. J Neurochem 60:1682–1689
Koehler RC, Roman RJ, Harder DR (2009) Astrocytes and the regulation of cerebral blood flow. Trends Neurosci 32:160–169
Kong L, Lian G, Zheng W, Liu H, Zhang H, Chen R (2013) Effect of alcohol on diffuse axonal injury in rat brainstem: diffusion tensor imaging and aquaporin-4 expression study. Biomed Res Int 2013:798261
Koob GF (2009) Brain stress systems in the amygdala and addiction. Brain Res 1293:61–75
Koob GF (2013) Theoretical frameworks and mechanistic aspects of alcohol addiction: alcohol addiction as a reward deficit disorder. Curr Top Behav Neurosci 13:3–30
Koob GF, Roberts AJ, Schulteis G, Parsons LH, Heyser CJ, Hyytia P, Merlo-Pich E, Weiss F (1998) Neurocircuitry targets in ethanol reward and dependence. Alcohol Clin Exp Res 22:3–9
Korbo L (1999) Glial cell loss in the hippocampus of alcoholics. Alcohol Clin Exp Res 23:164–168
Kril JJ, Harper CG (2012) Neuroanatomy and neuropathology associated with Korsakoff's syndrome. Neuropsychol Rev 22:72–80
Kryger R, Wilce PA (2010) The effects of alcoholism on the human basolateral amygdala. Neuroscience 167:361–371
Krystal JH, Petrakis IL, Limoncelli D, Nappi SK, Trevisan L, Pittman B, D'Souza DC, Suckow RF (2011) Characterization of the interactive effects of glycine and D-cycloserine in men: further evidence for enhanced NMDA receptor function associated with human alcohol dependence. Neuropsychopharmacology 36:701–710
Kuppers E, Gleiser C, Brito V, Wachter B, Pauly T, Hirt B, Grissmer S (2008) AQP4 expression in striatal primary cultures is regulated by dopamine--implications for proliferation of astrocytes. Eur J Neurosci 28:2173–2182
Kyzar EJ, Pandey SC (2015) Molecular mechanisms of synaptic remodeling in alcoholism. Neurosci Lett 601:11–19
Läck AK, Floyd DW, McCool BA (2005) Chronic ethanol ingestion modulates proanxiety factors expressed in rat central amygdala. Alcohol 36:83–90
Lasek AW (2016) Effects of ethanol on brain extracellular matrix: implications for alcohol use disorder. Alcohol Clin Exp Res 40:2030–2042
Laufer BI, Chater-Diehl EJ, Kapalanga J, Singh SM (2017) Long-term alterations to DNA methylation as a biomarker of prenatal alcohol exposure: from mouse models to human children with fetal alcohol spectrum disorders. Alcohol 60:67–75

Lee MR, Ruby CL, Hinton DJ, Choi S, Adams CA, Young Kang N, Choi D-S (2013) Striatal adenosine signaling regulates EAAT2 and astrocytic AQP4 expression and alcohol drinking in mice. Neuropsychopharmacology 38:437–445
Legastelois R, Jeanblanc J, Vilpoux C, Bourguet E, Naassila M (2017) Epigenetic mechanisms and alcohol use disorders: a potential therapeutic target. Biol Aujourdhui 211:83–91
Lewohl JM, Nunez YO, Dodd PR, Tiwari GR, Harris RA, Mayfield RD (2011) Up-regulation of microRNAs in brain of human alcoholics. Alcohol Clin Exp Res 35:1928–1937
Li JS, Yao ZX (2012) MicroRNAs: novel regulators of oligodendrocyte differentiation and potential therapeutic targets in demyelination-related diseases. Mol Neurobiol 45:200–212
Liu J, Casaccia P (2010) Epigenetic regulation of oligodendrocyte identity. Trends Neurosci 33:193–201
Liu J, Lewohl JM, Harris RA, Iyer VR, Dodd PR, Randall PK, Mayfield RD (2006) Patterns of gene expression in the frontal cortex discriminate alcoholic from nonalcoholic individuals. Neuropsychopharmacology 31:1574–1582
Lopez-Gallardo M, Lopez-Rodriguez AB, Llorente-Berzal A, Rotllant D, Mackie K, Armario A, Nadal R, Viveros MP (2012) Maternal deprivation and adolescent cannabinoid exposure impact hippocampal astrocytes, CB1 receptors and brain-derived neurotrophic factor in a sexually dimorphic fashion. Neuroscience 204:90–103
Lussier AA, Weinberg J, Kobor MS (2017) Epigenetics studies of fetal alcohol spectrum disorder: where are we now? Epigenomics 9:291–311
MacDonald JL, Roskams AJ (2009) Epigenetic regulation of nervous system development by DNA methylation and histone deacetylation. Prog Neurobiol 88:170–183
Mahnke AH, Miranda RC, Homanics GE (2017) Epigenetic mediators and consequences of excessive alcohol consumption. Alcohol 60:1–6
Mayordomo F, Renau-Piqueras J, Megias L, Guerri C, Iborra FJ, Azorin I, Ledig M (1992) Cytochemical and stereological analysis of rat cortical astrocytes during development in primary culture. Effect of prenatal exposure to ethanol. Int J Dev Biol 36:311–321
McBride WJ, Kimpel MW, McClintick JN, Ding Z-M, Edenberg HJ, Liang T, Rodd ZA, Bell RL (2014) Changes in gene expression within the extended amygdala following binge-like alcohol drinking by adolescent alcohol-preferring (P) rats. Pharmacol Biochem Behav 117:52–60
Melendez RI, Hicks MP, Cagle SS, Kalivas PW (2005) Ethanol exposure decreases glutamate uptake in the nucleus accumbens. Alcohol Clin Exp Res 29:326–333
Meyer EM, Long V, Fanselow MS, Spigelman I (2013) Stress increases voluntary alcohol intake, but does not alter established drinking habits in a rat model of posttraumatic stress disorder. Alcohol Clin Exp Res 37:566–574
Miguel-Hidalgo JJ (2006) Withdrawal from free-choice ethanol consumption results in increased packing density of glutamine synthetase-immunoreactive astrocytes in the prelimbic cortex of alcohol-preferring rats. Alcohol Alcohol 41:379–385
Miguel-Hidalgo JJ (2009) Neuropathology of depression, alcoholism, and their comorbidity in the human prefrontal cortex. In: Sher L (ed) Comorbidity of depression and alcohol use disorders, vol. Nova Science Publishers, Inc., New York, pp 171–179
Miguel-Hidalgo JJ (2018) Molecular neuropathology of astrocytes and oligodendrocytes in alcohol use disorders. Front Mol Neurosci 11:78
Miguel-Hidalgo JJ, Rajkowska G (2003) Comparison of prefrontal cell pathology between depression and alcohol dependence. J Psychiatr Res 37:411–420
Miguel-Hidalgo JJ, Wei JR, Andrew M, Overholser JC, Jurjus G, Stockmeier CA, Rajkowska G (2002) Glia pathology in the prefrontal cortex in alcohol dependence with and without depressive symptoms. Biol Psychiatry 52:1121–1133
Miguel-Hidalgo JJ, Overholser JC, Meltzer HY, Stockmeier CA, Rajkowska G (2006) Reduced glial and neuronal packing density in the orbitofrontal cortex in alcohol dependence and its relationship with suicide and duration of alcohol dependence. Alcohol Clin Exp Res 30:1845–1855
Miguel-Hidalgo J, Shoyama Y, Wanzo V (2009) Infusion of gliotoxins or a gap junction blocker in the prelimbic cortex increases alcohol preference in Wistar rats. J Psychopharmacol 23:551–558

Miguel-Hidalgo JJ, Waltzer R, Whittom AA, Austin MC, Rajkowska G, Stockmeier CA (2010) Glial and glutamatergic markers in depression, alcoholism, and their comorbidity. J Affect Disord 127:230–240
Miguel-Hidalgo JJ, Wilson BA, Hussain S, Meshram A, Rajkowska G, Stockmeier CA (2014) Reduced connexin 43 immunolabeling in the orbitofrontal cortex in alcohol dependence and depression. J Psychiatr Res 55:101–109
Miller MW, Spear LP (2006) The alcoholism generator. Alcohol Clin Exp Res 30:1466–1469
Moffat JJ, Ron D (2020) Astrocytes and alcohol: cortical astrocytes regulate alcohol consumption and intoxication. Neuropsychopharmacology 46:487–488
Montesinos J, Pascual M, Rodriguez-Arias M, Minarro J, Guerri C (2016) Involvement of TLR4 in the long-term epigenetic changes, rewarding and anxiety effects induced by intermittent ethanol treatment in adolescence. Brain Behav Immun 53:159–171
Morikawa H, Morrisett RA (2010) Ethanol action on dopaminergic neurons in the ventral tegmental area: interaction with intrinsic ion channels and neurotransmitter inputs. In: Reilly MT, Lovinger DM (eds) International review of neurobiology, vol 91. Academic Press, pp 235–288
Moyon S, Liang J, Casaccia P (2016) Epigenetics in NG2 glia cells. Brain Res 1638:183–198
Mulholland PJ, Carpenter-Hyland EP, Woodward JJ, Chandler LJ (2009) Ethanol disrupts NMDA receptor and astroglial EAAT2 modulation of Kv2.1 potassium channels in hippocampus. Alcohol 43:45–50
Nam HW, McIver SR, Hinton DJ, Thakkar MM, Sari Y, Parkinson FE, Haydon PG, Choi D-S (2012) Adenosine and glutamate signaling in neuron–glial interactions: implications in alcoholism and sleep disorders. Alcohol Clin Exp Res 36:1117–1125
Namihira M, Nakashima K (2013) Mechanisms of astrocytogenesis in the mammalian brain. Curr Opin Neurobiol 23:921–927
Newville J, Valenzuela CF, Li L, Jantzie LL, Cunningham LA (2017) Acute oligodendrocyte loss with persistent white matter injury in a third trimester equivalent mouse model of fetal alcohol spectrum disorder. Glia 65:1317–1332
Nixon K, Hughes PD, Amsel A, Leslie SW (2004) NMDA receptor subunit expression after combined prenatal and postnatal exposure to ethanol. Alcohol Clin Exp Res 28:105–112
Ozturk NC, Resendiz M, Ozturk H, Zhou FC (2017) DNA methylation program in normal and alcohol-induced thinning cortex. Alcohol 60:135–147
Pandey SC, Ugale R, Zhang H, Tang L, Prakash A (2008) Brain chromatin remodeling: a novel mechanism of alcoholism. J Neurosci 28:3729–3737
Pandey SC, Kyzar EJ, Zhang H (2017) Epigenetic basis of the dark side of alcohol addiction. Neuropharmacology 122:74–84
Pascual M, Boix J, Felipo V, Guerri C (2009) Repeated alcohol administration during adolescence causes changes in the mesolimbic dopaminergic and glutamatergic systems and promotes alcohol intake in the adult rat. J Neurochem 108:920–931
Patkar OL, Belmer A, Bartlett S (2016) Contribution of noradrenaline, serotonin, and the basolateral amygdala to alcohol addiction: implications for novel pharmacotherapies for AUDs. In: Meil WM, Ruby CL (eds) Recent advances in drug addiction research and clinical applications, vol. InTech, Croatia, pp 115–141
Paul AP, Medina AE (2012) Overexpression of serum response factor in astrocytes improves neuronal plasticity in a model of early alcohol exposure. Neuroscience 221:193–202
Penasco S, Mela V, Lopez-Moreno JA, Viveros MP, Marco EM (2015) Early maternal deprivation enhances voluntary alcohol intake induced by exposure to stressful events later in life. Neural Plast 2015:342761
Phillips SC, Harper CG, Kril JJ (1990) The contribution of Wernicke's encephalopathy to alcohol-related cerebellar damage. Drug Alcohol Rev 9:53–60
Pignataro L, Varodayan FP, Tannenholz LE, Protiva P, Harrison NL (2013) Brief alcohol exposure alters transcription in astrocytes via the heat shock pathway. Brain Behav 3:114–133
Ponomarev I (2013) Epigenetic control of gene expression in the alcoholic brain. Alcohol Res 35:69–76

Ponomarev I, Stelly CE, Morikawa H, Blednov YA, Mayfield RD, Harris RA (2017) Mechanistic insights into epigenetic modulation of ethanol consumption. Alcohol 60:95–101
Popescu BF, Bunyan RF, Guo Y, Parisi JE, Lennon VA, Lucchinetti CF (2013) Evidence of aquaporin involvement in human central pontine myelinolysis. Acta Neuropathol Commun 1:40
Popova EN, Shchekalina GA (1980) Effect of alcohol on glial cells. Zh Nevropatol Psikhiatr 80:539–544
Portero-Tresserra M, Gracia-Rubio I, Cantacorps L, Pozo OJ, Gomez-Gomez A, Pastor A, Lopez-Arnau R, de la Torre R et al (2018) Maternal separation increases alcohol-drinking behaviour and reduces endocannabinoid levels in the mouse striatum and prefrontal cortex. Eur Neuropsychopharmacol 28:499–512
Prat A, Biernacki K, Wosik K, Antel JP (2001) Glial cell influence on the human blood-brain barrier. Glia 36:145–155
Rajkowska G, Hughes J, Stockmeier CA, Javier Miguel-Hidalgo J, Maciag D (2013) Coverage of blood vessels by astrocytic endfeet is reduced in major depressive disorder. Biol Psychiatry 73:613–621
Ramani M, Mylvaganam S, Krawczyk M, Wang L, Zoidl C, Brien J, Reynolds JN, Kapur B et al (2016) Differential expression of astrocytic connexins in a mouse model of prenatal alcohol exposure. Neurobiol Dis 91:83–93
Renau-Piqueras J, Zaragoza R, De Paz P, Baguena-Cervellera R, Megias L, Guerri C (1989) Effects of prolonged ethanol exposure on the glial fibrillary acidic protein-containing intermediate filaments of astrocytes in primary culture: a quantitative immunofluorescence and immunogold electron microscopic study. J Histochem Cytochem 37:229–240
Ridge JP, Ho AM, Innes DJ, Dodd PR (2008) The expression of NMDA receptor subunit mRNA in human chronic alcoholics. Ann N Y Acad Sci 1139:10–19
Risher ML, Sexton HG, Risher WC, Wilson WA, Fleming RL, Madison RD, Moore SD, Eroglu C et al (2015) Adolescent intermittent alcohol exposure: dysregulation of thrombospondins and synapse formation are associated with decreased neuronal density in the adult hippocampus. Alcohol Clin Exp Res 39:2403–2413
Rubert G, Minana R, Pascual M, Guerri C (2006) Ethanol exposure during embryogenesis decreases the radial glial progenitor pool and affects the generation of neurons and astrocytes. J Neurosci Res 84:483–496
Rubio-Araiz A, Porcu F, Perez-Hernandez M, Garcia-Gutierrez MS, Aracil-Fernandez MA, Gutierrez-Lopez MD, Guerri C, Manzanares J et al (2017) Disruption of blood-brain barrier integrity in postmortem alcoholic brain: preclinical evidence of TLR4 involvement from a binge-like drinking model. Addict Biol 22:1103–1116
Sáez R, Burgal M, Renau-Piqueras J, Marqués A, Guerri C (1991) Evolution of several cytoskeletal proteins of astrocytes in primary culture: effect of prenatal alcohol exposure. Neurochem Res 16:737–747
Saito M, Chakraborty G, Hui M, Masiello K, Saito M (2016) Ethanol-induced neurodegeneration and glial activation in the developing brain. Brain Sci 6:31
Salazar M, Pariente JA, Salido GM, Gonzalez A (2008) Ethanol induces glutamate secretion by Ca2+ mobilization and ROS generation in rat hippocampal astrocytes. Neurochem Int 52:1061–1067
Schroeder JP, Spanos M, Stevenson JR, Besheer J, Salling M, Hodge CW (2008) Cue-induced reinstatement of alcohol-seeking behavior is associated with increased ERK1/2 phosphorylation in specific limbic brain regions: blockade by the mGluR5 antagonist MPEP. Neuropharmacology 55:546–554
Schwarcz R, Wu H-Q, Shepard PD, Roberts RC (1994) 24 – astrocytic kynurenines as modulators of dopaminergic function in the rat basal ganglia. In: Fuxe K, Agnati LF, Bjelke B, Ottoson D (eds) Trophic regulation of the basal ganglia. Pergamon, pp 365–379
Smith TL (1997) Regulation of glutamate uptake in astrocytes continuously exposed to ethanol. Life Sci 61:2499–2505

Smith TL, Zsigo A (1996) Increased Na(+)-dependent high affinity uptake of glutamate in astrocytes chronically exposed to ethanol. Neurosci Lett 218:142–144

Smith KL, John CS, Sypek EI, Ongur D, Cohen BM, Barry SM, Bechtholt AJ (2014) Exploring the role of central astrocytic glutamate uptake in ethanol reward in mice. Alcohol Clin Exp Res 38:1307–1314

Spanagel R, Pendyala G, Abarca C, Zghoul T, Sanchis-Segura C, Magnone MC, Lascorz J, Depner M et al (2005) The clock gene Per2 influences the glutamatergic system and modulates alcohol consumption. Nat Med 11:35–42

Tajuddin N, Moon KH, Marshall SA, Nixon K, Neafsey EJ, Kim HY, Collins MA (2014) Neuroinflammation and neurodegeneration in adult rat brain from binge ethanol exposure: abrogation by docosahexaenoic acid. PLoS One 9:e101223

Trantham-Davidson H, Chandler LJ (2015) Alcohol-induced alterations in dopamine modulation of prefrontal activity. Alcohol 49:773–779

Trindade P, Hampton B, Manhaes AC, Medina AE (2016) Developmental alcohol exposure leads to a persistent change on astrocyte secretome. J Neurochem 137:730–743

Tsai G (1998) Glutamatergic neurotransmission in alcoholism. J Biomed Sci 5:309–320

Tsai GE, Ragan P, Chang R, Chen S, Linnoila VM, Coyle JT (1998) Increased glutamatergic neurotransmission and oxidative stress after alcohol withdrawal. Am J Psychiatr 155:726–732

Tulisiak CT, Harris RA, Ponomarev I (2017) DNA modifications in models of alcohol use disorders. Alcohol 60:19–30

Tupala E, Tiihonen J (2004) Dopamine and alcoholism: neurobiological basis of ethanol abuse. Prog Neuro-Psychopharmacol Biol Psychiatry 28:1221–1247

Ullian EM, Christopherson KS, Barres BA (2004) Role for glia in synaptogenesis. Glia 47:209–216

Vallés S, Pitarch J, Renau-Piqueras J, Guerri C (1997) Ethanol exposure affects glial fibrillary acidic protein gene expression and transcription during rat brain development. J Neurochem 69:2484–2493

Vargas WM, Bengston L, Gilpin NW, Whitcomb BW, Richardson HN (2014) Alcohol binge drinking during adolescence or dependence during adulthood reduces prefrontal myelin in male rats. J Neurosci 34:14777–14782

Vargova L, Sykova E (2014) Astrocytes and extracellular matrix in extrasynaptic volume transmission. Philos Trans R Soc Lond Ser B Biol Sci 369:20130608

Vendruscolo LF, Barbier E, Schlosburg JE, Misra KK, Whitfield TW, Logrip ML, Rivier C, Repunte-Canonigo V et al (2012) Corticosteroid-dependent plasticity mediates compulsive alcohol drinking in rats. J Neurosci 32:7563

Verkhratsky A, Nedergaard M (2018) Physiology of astroglia. Physiol Rev 98:239–389

Verkhratsky A, Parpura V (2010) Recent advances in (patho)physiology of astroglia. Acta Pharmacol Sin 31:1044–1054

Verkhratsky A, Rodriguez JJ, Steardo L (2014) Astrogliopathology: a central element of neuropsychiatric diseases? Neuroscientist 20:576–588

Volkow ND, Wang G-J, Telang F, Fowler JS, Logan J, Jayne M, Ma Y, Pradhan K et al (2007) Profound decreases in dopamine release in striatum in detoxified alcoholics: possible orbitofrontal involvement. J Neurosci 27:12700–12706

Vollbrecht PJ, Simmler LD, Blakely RD, Deutch AY (2014) Dopamine denervation of the prefrontal cortex increases expression of the astrocytic glutamate transporter GLT-1. J Neurochem 130:109–114

Warden A, Erickson E, Robinson G, Harris RA, Mayfield RD (2016) The neuroimmune transcriptome and alcohol dependence: potential for targeted therapies. Pharmacogenomics 17:2081–2096

Weng JT, Wu LS, Lee CS, Hsu PW, Cheng AT (2015) Integrative epigenetic profiling analysis identifies DNA methylation changes associated with chronic alcohol consumption. Comput Biol Med 64:299–306

Wentlandt K, Kushnir M, Naus CC, Carlen PL (2004) Ethanol inhibits gap-junctional coupling between P19 cells. Alcohol Clin Exp Res 28:1284–1290

Wilhelm CJ, Guizzetti M (2015) Fetal alcohol spectrum disorders: an overview from the glia perspective. Front Integr Neurosci 9:65

Wu J, Lee MR, Kim T, Johng S, Rohrback S, Kang N, Choi DS (2011) Regulation of ethanol-sensitive EAAT2 expression through adenosine A1 receptor in astrocytes. Biochem Biophys Res Commun 406:47–52

Xiao C, Shao XM, Olive MF, Griffin WC, Li K-Y, Krnjević K, Zhou C, Ye J-H (2009) Ethanol facilitates glutamatergic transmission to dopamine neurons in the ventral tegmental area. Neuropsychopharmacology 34:307–318

Xin W, Schuebel KE, Jair K-W, Cimbro R, De Biase LM, Goldman D, Bonci A (2019) Ventral midbrain astrocytes display unique physiological features and sensitivity to dopamine D2 receptor signaling. Neuropsychopharmacology 44:344–355

Zhang X, Kusumo H, Sakharkar AJ, Pandey SC, Guizzetti M (2014) Regulation of DNA methylation by ethanol induces tissue plasminogen activator expression in astrocytes. J Neurochem 128:344–349

Zhou Z, Yuan Q, Mash DC, Goldman D (2011) Substance-specific and shared transcription and epigenetic changes in the human hippocampus chronically exposed to cocaine and alcohol. Proc Natl Acad Sci U S A 108:6626–6631

Zink M, Schmitt A, Vengeliene V, Henn FA, Spanagel R (2004) Ethanol induces expression of the glutamate transporters EAAT1 and EAAT2 in organotypic cortical slice cultures. Alcohol Clin Exp Res 28:1752–1757

Part V
Astroglia in Eating Disorders

Brain Volume Loss, Astrocyte Reduction, and Inflammation in Anorexia Nervosa

Jochen Seitz, Stefanie Trinh, Vanessa Kogel, and Cordian Beyer

Introduction

Anorexia nervosa (AN) is characterized by strongly reduced weight or insufficient development-adapted weight gain while feeling overweight (body image distortion), being afraid of gaining weight (weight phobia) and often accompanied by increased physical activity. It is the third most common chronic illness in adolescent patients and has a lifetime prevalence of 0.5–1% and the highest mortality of all psychiatric diseases (Gonzalez et al. 2007). Incidence appears to be on the rise, and first-onset patients become younger and younger (Smink et al. 2012; Steinhausen and Jensen 2015). Exact etiology remains unclear but includes a strong genetic component explaining about 60–70% of the variance, coupled with weight loss due to dieting or supposedly "healthy diets" or excessive sport with insufficient energy intake. Character traits like perfectionism, anxiety, insecurity, and poor self-insight as well as societal influences of slim body ideals appear to have additional effects. AN has a strong gender imbalance of 10–15 females vs 1 male (Herpertz-Dahlmann 2015), which also remains poorly understood. Patients and relatives often suffer extensively and enduringly during this often chronic disease (average illness duration is 6 years (Darby et al. 2009)), with major implications for healthcare resources. Semi-starvation as experienced by patients with AN also has many endocrine consequences: decreased thyroid function to conserve energy, increased cortisol level associated with the stress of the disease, altered hunger and satiety hormones like increased levels of ghrelin and reduced levels of leptin and amenorrhea with a reduction in sex hormones, and a delay of pubertal development and growth as well as related osteopenia and later osteoporosis due to lack of estrogen

J. Seitz (✉) · S. Trinh · V. Kogel · C. Beyer
Clinic of Child and Adolescent Psychiatry, University Hospital RWTH Aachen, Aachen, Germany
e-mail: jseitz@ukaachen.de

B. Li et al. (eds.), *Astrocytes in Psychiatric Disorders*, Advances in Neurobiology 26, https://doi.org/10.1007/978-3-030-77375-5_12

(Herpertz-Dahlmann 2015). Chronically ill patients, especially after more than 3 years of illness, increasingly show neuroprogessive changes with neuropsychological deficits including mental rigidity and inflexibility, learning impairments, and reduced visuospatial skills and traits reminiscent of autism (Treasure and Willmott 2020).

Brain Volume Reduction

One of the most striking somatic consequences of AN is a severe brain volume loss that appears to affect all brain structures (Seitz et al. 2016; Seitz et al. 2018). Its cause remains largely elusive; however, astrocyte reduction might play an important role in its pathophysiology. This "pseudoatrophia cerebri" is so extensive, it can be detected with the bare eye when reading MRI scans of the brain (see Fig. 1) and encompasses 3.7% brain volume loss in gray matter and 2.2% in white matter in adult patients and 7.6% in gray and 3.2% in white matter in adolescents. The extra space created intracranially by the retracting brain is reactively filled by CSF that is increased by 12.8% and 22.8%, respectively (for a recent meta-analysis, see Seitz et al. 2018).

Brain volume loss was functionally related to drive for thinness (Joos et al. 2010) and dietary restraint (McCormick et al. 2008), reductions in visuospatial skills (Castro-Fornieles et al. 2010), perceptual organization, and reasoning skills (McCormick et al. 2008). Brain volume reduction was also associated with this prognosis: Our own study found white matter volume reduction and cerebellar pseudoatrophy to be predictive of reduced body weight at 1- and 2.5-year follow-up – a frequently used proxy for clinical outcome – explaining an additional 20% of variance above known predictors like body mass index (BMI) at admission and illness duration (Seitz et al. 2015). McCormick et al. (2008) found incomplete brain

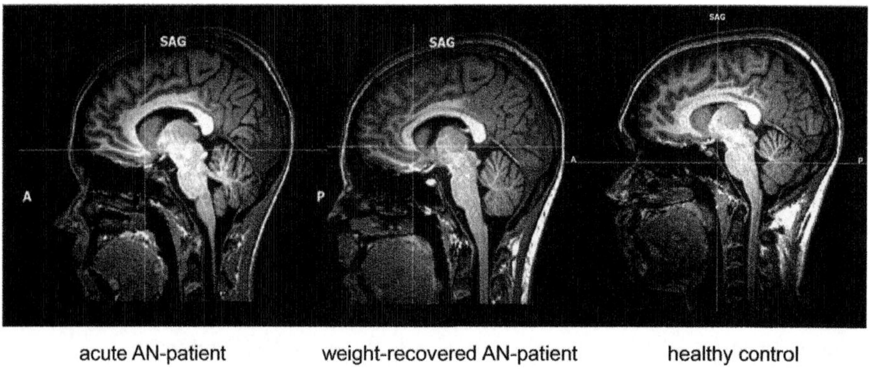

Fig. 1 Cerebral MRI images of a patient with acute AN (left), the same patient after short-term weight recovery (middle), and a healthy control (right). Note the brain volume reduction visible to the bare eye, evidenced by reduced gyri, while the sulci and ventricles are enlarged. (From Seitz et al. 2016 JNT with permission)

volume normalization upon short-term weight recovery to be predictive of relapse at 1 year (McCormick et al. 2008). Hope comes from data of short- and long-term recovered patients with about on average 50% of the brain volume reduction being reinstated after short-term weight recovery already (after about 4–6 months of treatment). Following long-term recovery (>1.5 years with normal weight), only about 0.5% of gray and 1% of white matter reductions remained in the meta-analysis, which did not reach statistical significance anymore, so it remains unclear whether some scarring and residual brain volume reduction remains or recovery is complete (Seitz et al. 2018). Importantly, this only applies for recovered, weight-restored patients – those chronically ill and underweight remain with a shrunken brain and all of its consequences.

Potentially Underlying Pathophysiology of Brain Volume Loss and the Role of Astrocytes

A systematic analysis of potential factors influencing brain volume reduction showed that absolute low BMI was most predictive, followed by BMI loss (Seitz et al. 2015). Two more studies with more chronically ill patients could also show the effect of extensive illness duration on gray matter volume reduction (Boghi et al. 2011; Fonville et al. 2014). Also regarding its reversibility upon weight gain, brain volume reduction thus appears to be a closely linked state marker for starvation. A potential confounder could be the hydration status which is known to influence also brain volumes. However, multiple studies have analyzed dehydration parameters in urine and blood serum and all do not support an influence of general (de-)hydration on brain volume reduction in AN (King et al. 2018; King et al. 2015; Vogel et al. 2016). It is interesting that gray and white matter reductions are much more pronounced in adolescents compared to adults; the still developing brain of adolescents appears to be more vulnerable. In the adolescent brain, normally, multiple developments take place in parallel that might be of importance for this process. For one, increased synapse formation in the early years is followed by pruning of unused and thus unnecessary synapses and entire neurons. Normal gray matter development includes a peak in volume representing this shift, which is region specific and follows a set course from primary sensory and motoric regions to more complex regions and ending in the prefrontal cortex (Shaw et al. 2008). Contrary to this, white matter volume normally increases due to ongoing myelination, especially of long-range fibers, well into the late 20s (Gibson 1991). Both normal developments are threatened by the lack in nourishment due to AN, especially, as new white matter fibers and those still under development have been proven to be most susceptible to insults (Gibson 1991).

Hormonal changes mentioned above may also play a significant role in these volume alterations. Increased cortisol levels, for example, are known to lead to brain atrophy (Chen et al. 2020) and have indeed been associated with brain volume reduction in AN, together with lack of trophic thyroid hormones (Castro-Fornieles et al. 2010; Chui et al. 2008; Nogal et al. 2008). Also reduced leptin and brain-derived neurotrophic growth factor (BDNF) could have an effect on brain volume changes (Holtkamp et al. 2006). Trophic gonadal hormones appear to play an important role in brain volume changes during healthy development (Neufang et al. 2009) and in AN. They have been shown to be associated with sulcal width, a proxy for brain volume reduction (Nogal et al. 2008), while our own group could show follicle stimulating hormone (FSH) to be correlated with brain volume increase following short-term weight recovery (Mainz et al. 2012). Chui et al. (2008) showed persisting amenorrhea and ensuing lack of estrogen to be the most important contributor to still decrease cognitive function in otherwise recovered patients with AN (Chui et al. 2008).

The cellular underpinnings of these volume changes remained largely unclear until recently, mostly due to the lack of postmortem analyses of deceased patients with AN. Two historic papers report a total of three analyses with signs of neuronal degeneration and altered dendritic spine morphology and ramification patterns in gray matter (Martin 1958; Neumarker 1997). Recently, animal models have been used to study the underlying cellular changes associated with this important, yet until then, elusive phenomenon of these significant brain volume reductions. The most common animal model used is the so-called activity-based anorexia (ABA) model in rodents (Frintrop et al. 2018; Routtenberg and Kuznesof 1967). Reduced food availability is coupled with running wheel access and induces a seemingly contra-intuitive increase in running wheel activity in susceptible animals, leading to further increased weight loss and even death. Food-seeking behavior by rodents is named as a possible explanation for this increased physical activity (Kas and Adan 2011), behaviorally matching the hyperactivity often seen in patients with AN. Most somatic changes experienced by patients can be studied using this animal model including reduced food intake, weight reduction, drop in body temperature, hyperactivity, amenorrhea, hypoleptinemia, and hyperghrelinemia (Frintrop et al. 2018). Our own analysis could for the first time reproduce the brain volume loss seen in patients with AN in both gray and white matter, further validating the model for the study of brain alterations. We could even show (almost) complete volume recovery after weight rehabilitation – only white matter volumes did not recover fully, potentially hinting at some enduring and longer-lasting effects (Frintrop et al. 2017; Frintrop et al. 2019). Cellular analyses of neurons by the group of Aoki et al. (2014) revealed increased GABA receptors in hippocampal neurons, especially in spines that were possibly associated with increased tonic inhibition and anxiety, further increasing hyperactivity and weight loss (Aoki et al. 2014). When investigating neuronal cell death or shrinkage as underlying mechanisms for brain volume reduction, our own analysis, however, showed a normal neuronal cell count and normal general neuronal cell sizes in ABA animals (Frintrop et al. 2017). This fits well with clinical data as extensive neuronal cell death as cause of the remarkable volume reductions would appear unlikely, as most of the volume is regained upon weight recovery – however, neurons are not known to regrow once dead.

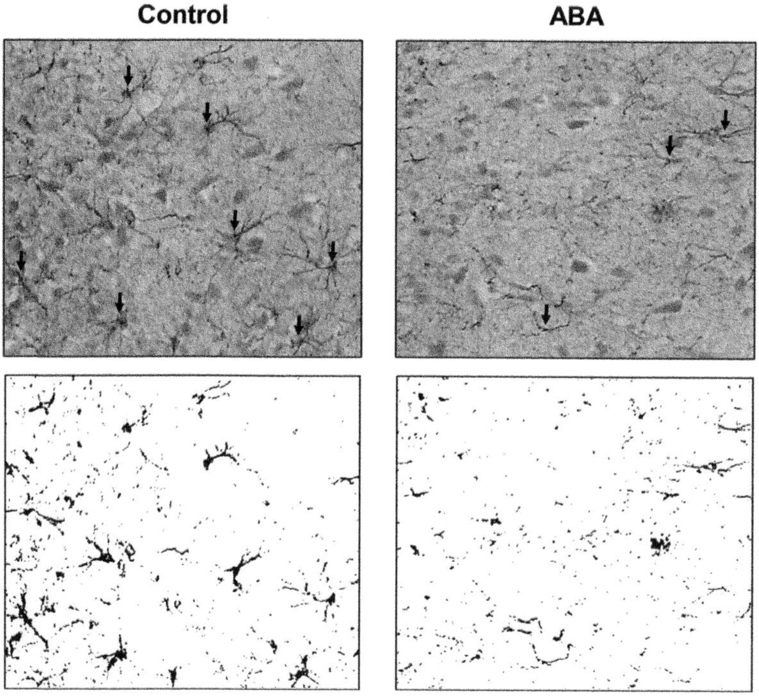

Fig. 2 Astrocyte counts in the rat cortex. Control animals (left); ABA animals after 3 weeks of semi-starvation and running wheel access (right). Upper panel GFAP immunostaining, lower panel, surface measurement using densitometry. Note the reduction of astrocyte numbers of about 50% and the even more pronounced reduction in surface area occupied by astrocytes

Conversely, significant reductions in cell numbers were found in $GFAP^{+}$ astrocytes by our group and others (Banasr et al. 2011; Frintrop et al. 2017, 2018; Reyes-Haro et al. 2015) (see Fig. 2).

Reyes-Haro et al. (2016) used a different animal model (Reyes-Haro et al. 2016). They used the effect of dehydration that leads to reduced hunger, diminished feeding, and weight loss and found slightly reduced numbers of astrocytes in the hippocampus and corpus callosum (Reyes-Haro et al. 2015; Reyes-Haro et al. 2016). Our own group could greatly expand these findings by using a chronic starvation model with a total of 3 weeks of (controlled) reduced feeding leading to a roughly 50% reduction in astrocyte cell count and even up to 90% reduction in astrocyte cell surface indicating fewer and smaller cell bodies. Our findings were corroborated with a second measurement approach by an over 50% reduction in mRNA expression of GFAP. By assessing brain volume changes and astrocytes in the same sample, we were furthermore able to uncover a direct link between decreased astrocyte cell count and volume reduction for the first time (Frintrop et al. 2017). Interestingly, these findings were not found in an acute model following only 1 week of starvation, underlining the importance of prolonged and cumulative starvation effects on the brain and its cells. Furthermore, neither neurons nor oligodendrocytes showed similar reductions, making this cell count and size reduction specific to astrocytes (Frintrop et al. 2018). In a separate study, we could replicate these findings and additionally show a complete recovery of astrocyte cell count, cell volume, and GFAP mRNA expression upon 2 weeks of refeeding, emphasizing the general reversibility of these findings, well in line with the largely reversibility of brain volume reductions (Frintrop et al. 2019). However, more research is needed with even longer periods of starvation in the animal model to study the equivalent of long and enduring AN (>3 years) with even more pronounced neuroanatomical and neuropsychological consequences in patients.

The Origin of Astrocyte Loss in AN

As astrocytes are subject to an active turnover of neogenesis and apoptosis, two potential mechanisms explaining astrocyte loss come into mind: reduced neogenesis or increased apoptosis. Here, Barbarich-Marsteller et al. (2013) showed glia cell proliferation indeed to be reduced in the dentate gyrus, dorsal hippocampus, and corpus callosum (but not in known regions with neurogenesis like the subgranular zone of the dentate gyrus) using Ki67 and BrdU staining in ABA animals (Barbarich-Marsteller et al. 2013). We could extend these findings and simultaneously analyze Ki67 as proliferation marker and Caspase 3 as apoptosis marker in the same sample described above. We found a significant reduction of neogenesis of about 50% while not evidencing any increase in apoptosis. This fits well with the catabolic state occurring during semi-starvation, where there is not enough energy to build the same amount of new astrocytes (Frintrop et al. 2019).

Potential Functional Consequences of Astrocyte Loss

Such a dramatic reduction in astrocyte cell count could have severe implications for astrocyte functioning. As astrocytes play an active role in constituting the blood-brain barrier, transporting nutrients and blood-based messengers into the brain, alimenting neurons, neurotransmitter reuptake, as well as in synapse formation and plasticity (Kaczor et al. 2015; Molofsky et al. 2012; Panatier and Robitaille 2012), their reduction by half can be hypothesized to interfere with these tasks. Astrocytes have been associated with complex phenomena such as mood and sleep – the first covered in detail in previous chapters of this book. In brief, astrocytes are reduced in numbers in patients with depression, and artificially reducing astrocytes in the frontal cortex leads to symptoms of depression in animal models (Fellin et al. 2012). Reduced numbers of astrocytes in AN could help explain concurrent symptoms of depression, sleep disturbances, and learning impairments often seen in patients with AN (Buhren et al. 2014). In our own lab, we could show learning impairments in the novel object recognition task in ABA animals, which was further linked to estrogen reduction (Paulukat et al. 2016). Altered synapse formation and neuronal plasticity due to reduced astrocytes could also help explain the increased rigidity of our patients and their decreased cognitive flexibility with increased attention to details. As patients with AN also typically display a lack of insight into the illness and a low motivation of change, these impaired learning processes and flexibility could help understand difficulties in the cognitive processes involved in psychotherapy of patients during starvation and partially explain why psychotherapy in acutely ill patients is unsuccessful. Following chronic starvation, long-lasting metabolic deprivation and astrocyte loss could help explain the behavioral neurodegenerative symptoms described above (Treasure et al. 2015). This hypothesis is further strengthened by volumetric studies showing further worsening brain volume loss in these patients with increasing illness duration.

Astroglia as Targets for Gonadal Steroid Hormones

By which mechanisms can astrocytes be influenced during semi-starvation in AN and which consequences could be expected on a cellular level? Detailed research in this field is still scarce; however, we here summarize the state of the literature and present first findings with special regard to the altered metabolic and endocrine situation of patients with AN including the chronic lack of gonadal steroid hormones, since changes in steroid synthesis, availability, and cellular signaling are well described in AN (Schorr and Miller 2017; Støving et al. 1999).

Historically, astrocytes are regarded as support cells for neurons. Research data of the past decades, however, have demonstrated a large variety of functions in the brain such as the regulation of endothelial cells and contribution to the blood-brain barrier (BBB), neurotransmitter metabolism and recycling, provision of nutrients and extracellular ion balance, synapse shielding, and others (Kimelberg and Nedergaard 2010). After CNS damage or chronic neuropathological processes, astrocytes may develop into a reactive phenotype, often referred to as A1 type, with changes in morphology and function. Astrogliosis and scar formation are typical characteristics of such pathologies (Sofroniew 2009). More recent research data additionally highlight that activated astrocytes are becoming part of the local brain-intrinsic inflammatory system (including microglia, nearby neurons, and endothelial cells) after damage and provide an array of pro-inflammatory cytokines and chemokines which regulate neighboring cells and infiltrated peripheral immune cells (Cekanaviciute and Buckwalter 2016; Johann and Beyer 2013; Scheld et al. 2016). Furthermore, we have shown that astrocytes are key sensors of and important for the surveillance of local oxidative stress phenomena and elimination of oxidants through the nuclear factor-erythroid 2 p45-related factor 2 (Nrf2)/antioxidant response element (ARE) system (Draheim et al. 2016). And finally, activated astrocytes are able to monitor and induce pyroptosis, i.e., an inflammatory form of lytic-programmed cell death. To execute this physiological response, they are equipped with several components of the intracellular multi-protein inflammasome complex which enables glia to respond to damaging stimuli in the brain and thereby produce temporarily large quantities of interkeukin-1β (IL1β) (Heitzer et al. 2017; Mortezaee et al. 2018; Slowik et al. 2018). These data altogether highlight the importance of astrocytes for proper brain functionality but also stress the issue that they are an integral component of the regulation of acute and chronic neuropathological processes.

The multitude of functions and roles in the CNS make astroglia a primary objective for well-balanced regulation and, in the case of brain-threatening events, therapeutic interventions.

As mentioned in the introduction, AN is generally associated with a number of profound endocrine alterations at different individual magnitudes. To date, it is not clear which of these AN-related changes are reactive, adaptive, or etiologic. Irrespectively, this part will discuss some aspects of steroid hormone-astroglia interactions and start with the role of gonadal steroids, i.e., estrogens and progestins. Intriguing observations in AN patients are that this disease shows a clear gender bias and is predominantly apparent in young women and that there occurs an early disturbance or even loss of the menstrual cycle and associated therewith a massive reduction of blood gonadal steroids (Young 2010). It was suggested by these authors and others that abnormal responses to estrogens may be due to an altered responsiveness of brain cells and functional abnormalities. This phenomenon might result from genetic polymorphisms in genes coding for classical nuclear estrogen receptors (nER) and membrane-associated ERs (mER). Such a disturbed estrogen signaling could be the reason for the insufficient feedback regulation during the estrous cycle in AN patients and thus account for amenorrhea (Ramoz et al. 2013). Similar observations although less detailed were reported for progesterone synthesis and signaling and functional aspects of the specific nuclear or membrane-bound progesterone receptors (nPRs, mPRs) (Klump et al. 2013; Singhal et al. 2014). What can be the connection between astrocytes and the above findings? First, rodent astrocytes cultured under starvation conditions, i.e., "hungry astrocytes," reveal a so-called A1 activation state with an unfolded protein response consisting of an accumulation of misfolded proteins in the endoplasmic reticulum ((Kogel et al. 2021) see also below). This pathophysiological condition reflects increased and unbalanced cellular oxidative stress and has been described to play a role in the development of metabolic diseases like obesity and diabetes (Yang et al. 2017). Second, astrocytes express receptors for appetite-promoting factors such as ghrelin (Frago and Chowen 2017) and are part of the brain inflammatory system which is implicated in the development of obesity and related disorders (Seong et al. 2019). Third, astroglial cells in different brain regions express the complete battery of nER/mER and nPRs/mPR, although there exist variations in expression levels depending on the developmental stage, aging, and gender but also in brain disorders and neuropathological events (Azcoitia et al. 2010; Brotfain et al. 2016; Habib et al. 2014; Johann and Beyer 2013; Karki et al. 2014; Pawlak et al. 2005a; Pawlak et al. 2005b).

How can estrogen signaling in astrocytes interfere with the manifestation or reinforcement of AN? It is well-known that ERs interact with insulin-like growth factor-1 (IGF-1) receptors in the CNS to regulate neural functions (Mendez et al. 2006).

This is of particular interest, since there exists a clear correlation between IGF-1 levels, the dysregulation of the growth hormone-IGF-1 axis, the psychopathology of AN, and related eating disorders (Brambilla et al. 2018; Fazeli and Klibanski 2012). Ghrelin, an important gut-brain axis peptide, promotes food intake, and estrogen upregulates ghrelin levels in different organ systems as shown by the use of estrogen-deficient animal experiments (Matsubara et al. 2004; Wang et al. 2015). Astrocytes are reported to be versatile metabolic sensors for the maintenance of CNS homeostasis and represent ghrelin targets (Marina et al. 2018). This might explain the role of astrocytes in mediating central effects of ghrelin on food intake (Frago and Chowen 2017). Hypothalamic astrocytes also express leptin receptors and thereby regulate hypothalamic neurons and neural circuits involved in food intake (Kim et al. 2014). Astroglia-specific leptin receptor activation caused an increased food intake after ghrelin administration (Kim et al. 2014). Despite the abovementioned evidence, no adequate literature information is available on the ghrelin/leptin system in astrocytes in human AN patients or in respective animal models.

These data convincingly point at a setscrew role for astrocytes in the brain-intrinsic control of eating, at least in the hypothalamus. Astroglial end feet surround the CNS vasculature and represent an important part of the BBB. Thus, astrocytes are ideally placed at the interface between the peripheral vascular system and the brain to sense metabolic changes and endocrine signals. BBB astrocytes on the other side possess finely branched extensions wrapping synapses and contacting nearby neurons. From a functional point of view, these specifically located astroglial cells express glutamate receptors, ion and water channels, and selective membrane transporters (reviewed by Marina et al. (2018)) which allow them to selectively tune synaptic transmission and neural communication. Particularly, the presence of the water channel protein aquaporin-4 allows astrocytes to rapidly respond to external signals and generate brain volume fluctuations (Noell et al. 2007) similar to those observed in the ABA rat model (Frintrop et al. 2017; Frintrop et al. 2019) and in human AN patients (Seitz et al. 2018). It is of interest that estrogens affect the expression and activation of aquaporin-4 water channels in astrocytes und pathological conditions (Habib et al. 2014). Thus, estrogen deficiency as observed in AN could diminish water efflux/influx and equilibrium in the brain and affect ion homeostasis by targeting astrocytes.

Patients with AN develop multiple endocrine abnormalities including a disturbance of the hypothalamus-pituitary-adrenal axis with particular changes in the activity of the growth hormone (GH)/IGF-I axis (Gianotti et al. 2002). Other lines of evidence suggest that neurotrophins and in particular brain-derived neurotrophic factor (BDNF) and glial cell line-derived neurotrophic factor (GDNF) alterations are implicated in the etiology of eating disorders and therefore could be used as biomarkers for AN (Dmitrzak-Weglarz et al. 2013; Ribasés et al. 2003). Hypothalamic astrocytes not only are targets for BDNF but also synthesize BDNF thereby influencing neuronal activity (Caruso et al. 2012). BDNF synthesis by astrocytes is regulated by estrogens which allows this steroid hormone to promote neural networks under physiological conditions but coincidently fulfill a neuroprotective role in the brain after injuries (Sohrabji and Lewis 2006). Similarly, astrocytes in the hypothalamus express GDNF which is under the control of estrogen signaling (Ivanova et al. 2002). Both gonadal steroid hormones dampen central and brain-intrinsic immune responses and regulate local growth factor supply mediated by astrocytes. This complex modulation requires the gamut of steroid-dependent signaling pathways (Kipp et al. 2012). The identification of molecular and cellular targets of sex steroids and the understanding of cell-cell interactions in the pathogenesis of AN will offer promise of novel therapy strategies generally during brain pathologies but particularly also for AN.

Another intriguing aspect of astrocyte function is the removal and recycling of the neurotransmitter glutamate. Therefore, astrocytes express highly specific glutamate transporters which belong to the family of membrane-bound excitatory amino acid transporters (EAAT). Astroglial glutamate-aspartate transporter (GLAST, EAAT1 as human homolog) and glutamate transporter-1 (GLT-1, EAAT2 as human homolog) are the major transporters which take up synaptic glutamate from the synaptic cleft to maintain physiological extracellular glutamic levels, thus preventing extracellular glutamate accumulation and ensuing excitotoxicity (Mahmoud et al. 2019). GLT-1 is responsible for over 90% of glutamate reuptake within the CNS, typically found in astrocytes but only marginally in neurons. The dysregulation of GLAST/GLT-1 plays a significant role in excitotoxicity and is associated with the neuropathogenesis of other psychiatric diseases (Pajarillo et al. 2019), also described in this book. There is no direct information in the literature available associating glutamate transporters and its dysfunction with AN. However, there exist experimental data and information from human studies concerning abnormal glutamate levels in AN patients. Widespread reductions in brain glutamate concentrations were measured using high-field magnetic resonance (MRS, 7 T) imaging in women with AN (Godlewska et al. 2017). Another study identified distinct hypothalamic neurochemical dysfunctions, i.e., paradoxical glutamate reductions, and associated structural variations in hypothalamic nuclei implicated in food intake and energy balance (Florent et al. 2020). Interestingly, the regulation of the glutamate transporters GLAST and GLT-1 is highly influenced and controlled by circulating estrogens particularly in astrocytes (Pawlak et al. 2005a), whereas in neurons, estrogens selectively interfere mainly with the regulation of vesicular glutamate transporters responsible for storage and release (Mela et al. 2016). When considering this information about the glutamate-astroglia-hypothalamus-estrogen connection, it becomes evident that pathophysiological disturbances of the glutamatergic system as observed in AN may strongly contribute to the maintenance of the disease state and go hand in hand with abnormal plasma estrogen levels (Hermens et al. 2020).

Finally, we would like draw particular attention to the mitochondrial compartment of astrocytes, its involvement in AN pathology, and possible regulation by gonadal steroid hormones. Alterations of hypothalamic proteins involved in energy and mitochondrial metabolism have been observed in ABA mice where mitochondrial dynamics in the hypothalamus appear to be modified with an increase of fission without modification of fusion (Nobis et al. 2018). Hypothalamic mitochondrial respiration and uncoupling protein 2 are all described downstream targets mediating the orexigenic effects of ghrelin (Lim et al. 2011). Atypical levels of ghrelin are clearly associated with metabolic conditions such as obesity and AN. With respect to the respiratory chain components, female AN patients revealed increased activities of complexes I and II in isolated platelets (Böhm et al. 2007). Further, the mitochondrial membrane potential seems to be decreased and reactive oxygen species (ROS) production elevated in leukocytes from anorexic patients pointing at an increased oxidative stress status at the level of complex I (Victor et al. 2014). There is convincing evidence that estrogens interfere with mitochondrial functioning and that the estrogen-mediated control of the mitochondrial compartment involves the cooperation of nERs/mERs and their co-activators on the coordinate regulation of genes for respiratory chain complex proteins (Chen et al. 2009). We have previously reported that estrogen influences the gene expression of respiratory chain proteins in astrocytes of the rodent cerebral cortex (Araújo et al. 2008). Apparently, this regulation occurs in a gender-specific way suggesting male/female differences in the protection against vulnerability and adaptive capacity of the mitochondria to external pathological stimuli (Arnold et al. 2008). Thus, the mitochondrial system seems to be regulated by estrogens at the level of ATP and ROS production as well as under a structural-functional viewpoint. This implies that sex steroids can directly and indirectly interfere with mitochondrial properties via non-nuclear, presumably mitochondria-intrinsic, and nuclear signaling mechanisms. This enables this cell organelle to cope with pathological processes and provide stabile local energy homeostasis and an anti-apoptotic base setting in the brain.

Taking into account that estrogen levels are massively reduced in AN similarly to AN-related animal models, physiological parameters such as glutamate homeostasis, mitochondrial performance, ROS elimination, water fluctuations, and growth factor supply by hypothalamic astrocytes are all imbalanced up to completely dysregulated. This, in consequence, may have strong negative effects on the function of neural circuits embedded in food intake and energy balance and thus contribute to either the development of AN or the perpetuation of this eating disease. Detailed research into these functional consequences of gonadal steroid suppression and astrocyte reduction in AN is thus highly promising to help further unravel underlying pathophysiology.

Inflammation in AN and the Role of Astrocytes

We already undertook the task of studying astrocyte function under semi-starvation conditions by creating a cell culture setup of chronic starvation. This work focused mostly on inflammation properties of astrocytes, as they play a prominent role in immune regulation, inflammation, and its attenuation, while a chronic low-grade state of inflammation has been found in patients with AN in large meta-analyses, which remains poorly understood (Dalton et al. 2018; Solmi et al. 2015).

Inflammation in AN

First reports of inflammatory reactions in patients suffering from AN date back in the early 1990s. It has been known at least since 1992 that the pro-inflammatory marker TNFα is expressed at mildly elevated levels in the blood of patients with AN (Schattner et al. 1992). In addition, other studies showed moderately increased expression of other pro-inflammatory cytokines such as IL6 and IL1β or associated receptors (Misra et al. 2006; Nakai et al. 2000; Ostrowska et al. 2015). Although some studies were not able to find changes in cytokine expression between patients with AN and healthy control subjects (Brambilla et al. 2001), the increased cytokine expression has been confirmed several times over the years and has even been associated with a predisposition to develop AN (Agnello et al. 2012; Kanbur et al. 2008). In order to give a better overview of the expression of pro-inflammatory markers in patients with AN, two large meta-analyses were performed. Both studies described an elevated concentration of IL6 and TNFα in the blood of patients with AN (Dalton et al. 2018; Solmi et al. 2015). Solmi et al. (2015) additionally found an increase in IL1β concentration which was confirmed by Dalton et al. (2018) only for the restrictive subtype of AN. However, a common feature shared by all these studies is the increased expression of pro-inflammatory markers, which is the reason why a low-grade inflammatory state during AN can be assumed. This assumption is corroborated by the fact that parameters of inflammation are increasingly expressed in peripheral blood mononuclear cells (PBMCs). In PBMCs of patients with AN, increased expression of COX2 as well as phosphorylation of p38 MAPK and p-ERK points toward a stress response of these cells in combination with an inflammation (Caso et al. 2020). COX2 is activated by cytokines during inflammation and injuries. It is involved in the production of prostaglandins that can promote inflammation and mediate pain (Simon 1999). p38 MAPK and p-ERK are part of the MAPK pathway which is activated by extracellular stress stimuli like ultraviolet light, heat shock, and pro-inflammatory cytokines. The MAPK pathway seems to play a major role in apoptosis and cell senescence but also in cell proliferation (Obata et al. 2000; Zou et al. 2019). The concomitant low expression of NFκB, a transcription factor involved in inflammation, and the increased level of the anti-inflammatory cortisol and glucocorticoid receptors, further indicates an altered regulation of inflammatory signaling pathway in patients with AN (Caso et al. 2020; Luz Neto et al. 2019). In view of the previously described results, it can be said that there is strong evidence of a low-grade inflammatory response in AN. This inflammatory reaction is, however, mostly reversible. It has been shown that altered cytokine concentrations normalize after body weight rehabilitation (Allende et al. 1998; Pomeroy et al. 1994; Schattner et al. 1990; Solmi et al. 2015). In a recent longitudinal study of cytokine concentrations, a decrease of IL6 levels after 12 weeks was correlated with improved eating disorder symptoms. For this reason, IL6 may be considered as a biomarker to assess severity of AN in the future (Dalton et al. 2020).

Astrocytes and Inflammation in the CNS

Astrocyte function includes a response following injury of the central nervous system (Sofroniew and Vinters 2010). Brain injuries and neurological diseases are able to shift astrocytes into a so-called reactive state which results in astrogliosis (Liddelow and Barres 2017). Reactive astrocytes undergo multiple changes at the molecular, cellular, and functional level (Sun and Jakobs 2012).

At the molecular level, translocation of the nuclear factor ƙ-lightchain-enhancer of activated B cells (NFƙB) in the nucleus initiates the activation of astrocytes. This translocation can be caused by pro-inflammatory stimuli (Brambilla et al. 2005; Dorrington and Fraser 2019). These include, for example, tumor necrosis factor α (TNFα), interleukin 1β (IL1β), and interleukin 17 (IL17) (Hyvärinen et al. 2019; Qian et al. 2007), reactive oxygen species (ROS) (Chandel et al. 2000), phagocytosed myelin (Ponath et al. 2016), and toll-like receptor (TLR) engagement (Kawai and Akira 2007). The increased expression of glial fibrillary acidic protein (GFAP) is a prominent marker for reactivity of astrocytes (Yang and Wang 2015). Other markers associated with astrocyte activation are, for example, pro-inflammatory cytokines (IL1β, IL6, IL8, TNFα, and granulocyte-macrophage colony-stimulating factor (GM-CSF)) as well as chemokines (CC-chemokine ligand 3 (CCL3), CCL5, and C-X-C-motif chemokine ligand 10 (CXCL10)) (Choi et al. 2014). During inflammation, a bidirectional communication between astrocytes and microglia influences the inflammatory process (Matejuk and Ransohoff 2020). Microglia secrete IL1α, IL1β, and TNFα to activate a transcriptional response in astrocytes leading to an increase in neurotoxic and a decrease in neurotrophic factors (Liddelow and Barres 2017), whereas astrocytes promote microglial inflammation through the production of IL6, GM-CSF, and other signaling factors (Linnerbauer et al. 2020).

On a functional level, astrocyte reactivity and astrogliosis are originally protective responses. They are designed to ward off acute stress, to minimize tissue damage, and to restore brain homeostasis (Sidoryk-Wegrzynowicz et al. 2011) These functions can be achieved by glutamate uptake, removal of free radicals, production of neurotrophins, and scar formation which restricts the damage to a limited area (Chen and Swanson 2003). However, under certain circumstances, reactive astrocytes promote neurotoxicity (Liddelow and Barres 2017). In this context, reactive astrocytes can be differentiated into at least two phenotypes (Liddelow et al. 2017; Zamanian et al. 2012). A1-like astrocytes appear pro-inflammatory and promote neurotoxicity. They are activated by microglia releasing cytokines and the following NFƙB activation (Liddelow et al. 2017). These astrocytes loose many of their main functions including their ability to promote neuronal survival leading to toxicity for neurons and oligodendrocytes, as well as showing a decreased ability to support synapse formation and function which could lead to an impaired learning behavior (Liddelow and Barres 2017). Neurotoxicity is promoted in several ways, for example, by decreasing their expression of glutamate transporters. The result is a deficient control of neurotransmitter uptake and release (Landeghem et al. 2006). Furthermore, NFƙB signaling in astrocytes during CNS inflammation leads to an excessive production of NO and brain-derived neurotrophic factor (BDNF) (Hsiao et al. 2013; Kaltschmidt and Kaltschmidt 2015). In the healthy brain, BDNF promotes neuronal survival (Kowiański et al. 2018). Excessive levels of BDNF, however, result in NO production and neuronal cell death (Colombo et al. 2012). An A2-like astrocyte, on the other hand, is neuroprotective, upregulates neurotrophic genes, and promotes neuronal survival. (Liddelow and Barres 2017).

Astrocytes and Metabolic Diseases

Besides an immune response in the brain, astrocytes are also responsible for metabolic support of neurons. Neurons have one of the highest requirements of glucose of all cells in the whole body (Mergenthaler et al. 2013). By metabolizing lactate from glycogen, astrocytes provide energy to neurons in times of an undersupply. This way, astrocytes thus ensure the survival and function of neurons (Kasischke et al. 2004; Pellerin and Magistretti 1994). For this reason, it is not surprising that the involvement of astrocytes in metabolic diseases, such as obesity and diabetes, has already been demonstrated. For example, increased astro- and microgliosis in the hypothalamus have been linked to development and maintenance of obesity (Rahman et al. 2018). A relationship between other diseases with metabolic aspects such as anorexia nervosa (AN), neuroinflammation, and associated activation of astrocytes remains unclear.

The *Hungry* Astrocyte In Vitro

In order to determine the importance of astrocytes in the event of nutrient deficiency, our group aimed to established a cell culture model. Due to chronic nutritional deficiencies over many years (Nagl et al. 2016), patients with AN often suffer from mild hypoglycemia, which is usually asymptomatic (Levenson 2011; Mattingly and Bhanji 1995). In order to simulate this situation as accurately as possible, we supplied astrocytes in vitro with a glucose concentration just below the physiological level of the brain at 2.4 ± 0.1 mM (Kogel et al. 2021). To represent the extended course of disease as it occurs during AN, in our study, astrocytes were chronically semi-starved over 15 days (Kogel et al. 2021). This differs from previous glucose deprivation studies that use acute starvation ranging from 30 min to 48 h (Hara et al. 1989; Lee et al. 2016; Pauwels et al. 1985). Treating cells with a glucose concentration of 2 mM for at least 2 weeks is expected to induce chronic undernutrition but not massive cell death. Both assumptions were previously shown by an in vitro study performed by Abe et al. (2006) and confirmed by our data. After 3 weeks of supplying astrocytes with a glucose concentration of 2 mM, their starved cells showed an empty glycogen store and an altered metabolism (Abe et al. 2006). Astrocytes are able to store small amounts of glycogen and metabolize them via glucose to lactate in case of metabolic undersupply (Falkowska et al. 2015). In this way, the energy supply of neurons can be maintained despite an acute deficiency (Kasischke et al. 2004). Therefore, an empty glycogen store in astrocytes of the CNS is an indication of prolonged starvation. Our group could demonstrate further indications of an altered supply situation after treatment of astrocytes with 2 mM glucose over 15 days. This involved a reduced expression of connexin43 (Cx43) in chronically semi-starved astrocytes in vitro (Kogel et al. 2021). The transport of glucose and other small molecules occurs within the astrocytic network via the gap junctions (Allard et al. 2014), which in astrocytes of the CNS are mainly composed of Cx43 (Nagy and Rash 2000). Reduced expression of this component impedes the supply of neurons, since rapid distribution of lactate is no longer guaranteed (Mayorquin et al. 2018). In the same in vitro study, we did not find increased cytotoxicity or impaired general metabolic activity after semi-starvation of astrocytes for 15 days (Kogel et al. 2021) although semi-starved astrocytes switch their metabolism to lactate and glucose oxidation (Abe et al. 2006). However, our semi-starved astrocytes in vitro did show an increased inflammatory reaction. In our study, astrocytes starved for 15 days expressed a change in morphology resulting in hypertrophy of soma and processes (Kogel et al. 2021) (see Fig. 3), indicating increased reactivity (Li et al. 2019). Furthermore, we could demonstrate an upregulated expression of pro-inflammatory cytokines such as IL6 in these astrocytes indicating an inflammatory response. Increased expression of A1 markers, for example, C3 and Psmb8, with concomitant decrease of A2 markers (e.g., S100a10) in the same study indicates a phenotype switch to a pro-inflammatory and harmful phenotype (Kogel et al. 2021). All these reactions point toward an inflammatory reaction, accompanied by reactivity and astrogliosis in chronically semi-starved astrocytes.

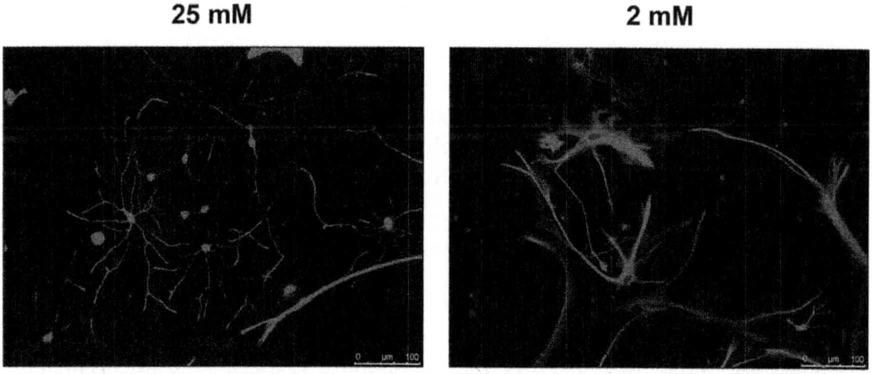

Fig. 3 Cell morphology of primary rat astrocytes cultured in glucose control medium (25 mM) (left) or after 15 days of chronic semi-starvation (2 mM) (right) indicated by an immunofluorescence staining with GFAP. Note the "activated" morphology of the semi-starved astrocytes with hypertrophy of cell bodies and reduced branching

Astrocytes in Animal Models of AN

Using the dehydration animal model of AN, Reyes-Haro et al. (2016) were able to demonstrate morphological changes in hippocampal astrocytes which also point toward an astrocyte activation. This assumption is supported by an increased expression of vimentin and nestin in the hippocampus of the same animals, indicating an elevated number of reactive astrocytes (Reyes-Haro et al. 2016). Simultaneously, increased expressions of IL6 and IL1β demonstrate an inflammatory response in the hippocampus of the same animal model (Ragu-Varman et al. 2019). This inflammatory response in the brain may be mediated by an increased level of T-lymphocyte receptors 4 (TLR4) in intestinal epithelial cells and intestinal macrophages (Belmonte et al. 2016). TLRs are able to trigger an inflammatory response (Moresco et al. 2011). Increased expression of TLR4 was found in the gut of ABA animals and associated with elevated levels of IL1β and its receptor in the hypothalamus. In addition, TLR4-deficient mice are more susceptible to the effects of the ABA model and lose weight faster than non-deficient mice (Belmonte et al. 2016). This suggests that the inflammatory response triggered by TLR4 could also in part represent a form of a protective mechanism.

In summary, AN is associated with increased inflammatory reactions and an altered function of inflammatory pathways. The extent to which this inflammation is associated with brain volume loss, found in patients with AN and decreased numbers of astrocytes observed in the ABA animal model, requires further investigation.

Gut Microbiota Alterations in AN and the Influence on Astrocytes

A rather new aspect in AN research refers to the interaction with gut microbiota and the gut-brain axis with clear implications for brain morphology including cellular and functional changes in the brain and behavioral consequences like affected learning, mood, anxiety, eating behavior, and weight gain (Seitz et al. 2019).

The Microbiome: General

The gastrointestinal tract of every human being is colonized by approximately 1–2 kg of bacterial cells collectively referred to as gut microbiota. The number of gut microbiota is at least as large as the number of eukaryotic cells in the organism (Sender and Fuchs 2016; Sgritta et al. 2019). The colonization with gut bacteria mainly takes place with birth and reaches a steady state in early childhood. Every individual hosts roughly 500 different species out of about 2000 so that a unique composition is created (Almeida et al. 2019; Arumugam et al. 2011). This composition can shift throughout life and is influenced by various factors such as birth mode, genetics, environment, and diet (Ananthakrishnan 2015; Marques et al. 2010). The colony of gut microbiota interacts with the host in a symbiotic manner by breaking down food and providing the host with energy and essential nutrients that otherwise would not be available. Moreover, the gut microbiota plays a relevant role in the development of the immune system, inflammatory processes, and regulation of appetite and body weight (Fetissov 2017; Neuman et al. 2015; Ridaura et al. 2013). Due to these manifold functions, a dysbiosis in the microbiota population, which means alterations in the species diversity and taxonomic composition, can lead to a disruption of the host's homeostasis. Initially, gastrointestinal disorders such as inflammatory bowel disease, ulcerative colitis, and allergies (e.g., celiac disease) were associated with changes in the gut microbiota (Gigante et al. 2011; Noor et al. 2010; Subramanian et al. 2006). Later, a wide range of other diseases were suggested to be affected by microbiota imbalances, among others Alzheimer's disease, multiple sclerosis, cancer, and psychiatric conditions (Cryan and Dinan 2012; Jiang et al. 2017; Mangiola et al. 2016). In the last years, evidence is growing that gut microbiota also influence mood and behavior as observed in people with mood disorders (e.g., major depression disorder and bipolar disorder) (Cryan and Dinan 2012; Huang et al. 2019; Kelly et al. 2017). It appears obvious that gut microbiota alterations might play a role in the pathophysiology of anorexia nervosa (AN) because relevant processes such as body weight and appetite regulation and metabolic and immunological processes were shown to be influenced by gut microbiota and depression represents a common comorbidity (Asano et al. 2012; Fetissov 2017; Morkl et al. 2017; Ridaura et al. 2013). However, it is often not clear whether the different bacterial composition contributes to the cause of the disease or rather represents an effect of the disorder.

In order to explain the interplay between gut microbiota and psychiatric disorders, the so-called gut-(microbiota)-brain axis needs to be considered (Rhee et al. 2009). A lot of gut bacteria produce or modulate specific metabolites and are again affected by them, e.g., short-chain fatty acids, neurotransmitters (e.g., dopamine, norepinephrine, and serotonin), and hormones (e.g., ghrelin, leptin, and glucagon-like peptide-1) in the gastrointestinal tract. Additionally, some of the mentioned effectors can execute a bidirectional signaling pathway to the brain and vice versa as part of the gut-brain axis (Clarke et al. 2014; Rhee et al. 2009). Some effector molecules are translocalized from the gut lumen into the systemic blood circulation and thereby possibly transported to the brain to transmit a signal (Ticinesi et al. 2019). Another way of gut-brain communication takes place via the vagal nerve that connects the brainstem to many organs including the gastrointestinal tract (Breit et al. 2018) and has been found to modulate gut-brain interaction (Sgritta et al. 2019).

Recently, Hata and colleagues showed that transplantation of gut bacteria populations isolated from patients with AN into germ-free mice that were born and raised under sterile conditions indeed led to reduced food intake and body weight in their offspring. Moreover, a lower food efficiency ratio defined as body weight gain per food intake was recorded in the offspring compared to controls (Hata et al. 2019).

In addition, anxiety-like behavior and compulsive behavior assessed by the open field and marble-burying tests, respectively, also common comorbidities in AN, were increased in the offspring of mice that received the fecal microbiota transplantation from patients with AN (Hata et al. 2019). These results showed for the first time a causal effect of gut microbiota on appetite and bodyweight regulation and behavioral alterations in an animal model. Similarly, changes in phenotypes after fecal microbiota transplantation can be observed with regard to other disorders such as depression and obesity (Kleiman et al. 2015; Ridaura et al. 2013).

The Microbiome: Anorexia Nervosa Animal Model

Comparable to studies in patients with AN, clear differences in the gut microbiota composition between food-restricted animals and normally fed controls were observed in studies using the activity-based anorexia (ABA) model (Breton et al. 2020; Queipo-Ortuno et al. 2013; Trinh et al. 2021).

As described before, the ABA paradigm is a translational animal model to mimic symptoms of AN using a combination of food reduction and access to a running wheel (see above). We used a chronic version of the ABA model (Frintrop et al. 2018) with separate control groups for food reduction and running wheel access and showed that food reduction significantly influenced the gut microbiota in rats, while physical activity had only marginal effects (Trinh et al. 2021).

β-diversity is a common measure in microbiome research to investigate the similarity of microbiota between samples (Lagkouvardos et al. 2017). It confirmed the difference between food-restricted animals and normally fed controls supporting that food reduction seems to be the driving factor for gut microbiota alterations rather than running-wheel activity (Trinh et al. 2021). In our study, six genera were of particular interest as they showed major shifts in relative abundance after chronic starvation. The relative abundance of the genus *Prevotella* was decreased, while the relative abundances of the genera *Odoribacter*, *Lactobacillus*, *Akkermansia*, *Bifidobacterium*, and *Ruminococcus* were increased after starvation (Trinh et al. 2021). Heterogeneous results regarding changes in single genera between patients with AN and controls or food-restricted animals and controls have been observed in previous studies that could be explained by the multifactorial genesis of AN and methodological differences in the study design. However, high levels of relative abundance of the genus *Akkermansia* were observed several times in patients with AN and in animal studies of starvation (Derrien et al. 2017; Mack et al. 2016; Remely et al. 2015; Sonoyama et al. 2009; Trinh et al. 2021). *Akkermansia* is a mucin-utilizing bacterium that belongs to the phylum *Verrucomicrobia* (Derrien et al. 2017; Mack et al. 2016). On the one hand, degradation of mucins covering the gut wall during a state of starvation is a competitive benefit compared to other microbiota that are dependent on dietary nutrients. On the other hand, however, it might increase intestinal permeability by damaging the protective mucus layer. The latter can lead to increased passage of bacterial components and products across the gut wall representing one possible pathway of the gut-brain axis (Mack et al. 2016; Seitz et al. 2019).

Regarding the gut-brain axis and altered brain morphology, we observed various associations between changes in the gut microbiota and volumetric and cellular brain parameters in food-restricted rats. Starved rats with high relative abundances of *Odoribacter* and *Bifidobacterium* showed lower levels of GFAP (glial fibrillary acidic protein) mRNA expression in white brain matter (corpus callosum) (Trinh et al. 2021), potentially contributing to the striking reduction in astrocytes described above. Furthermore, high relative abundances of *Lactobacillus* were associated with bigger brain volumes in the cerebral cortex and the corpus callosum (Trinh et al. 2021), potentially limiting the severe brain volume reduction found previously in patients with AN and in the ABA model by our group and others (Frintrop et al. 2017, 2019; Seitz et al. 2016). As described above, brain volume reductions in the ABA paradigm were linked to loss of $GFAP^+$ astrocytes (Frintrop et al. 2017, 2019). In order to prove a causal relationship between these observations, fecal microbiota transplantation transferring gut microbiota isolated from patients with AN in microbiota-free animals is needed. Additionally, supplementation with *Odoribacter* and *Bifidobacterium* is of interest to study their causal role in $GFAP^+$ astrocyte reduction. Probiotic substitution with *Lactobacillus* could represent an interesting option to study its effect on limiting brain volume loss, which would be easily transferrable to patients given successful animal experiments.

The Microbiome: Astrocytes

What are potential mechanisms that could mediate these associations of microbiota with the brain in AN? Specific research in AN is currently lacking; however, general mechanisms influencing cells in the brain include the vagal nerve, bacterial components or metabolites, immunological effects, and intestinal hormones or neurotransmitters that are able to cross the blood-brain barrier (Erny et al. 2017). From studies with germ-free animals that lack any bacterial colonization, it is clear that gut microbiota affects brain physiology. Germ-free mice, for example, show increased numbers of immature microglia with altered morphology including increased branching and longer processes compared to microglia in conventionally raised mice (Fung et al. 2017).

The activity of astrocytes seems to be influenced via microbial metabolites that activate the astrocytic aryl hydrocarbon receptor (AHR) (Fung et al. 2017; Rothhammer et al. 2016). Activation of this ligand-activated transcription factor decreases inflammation through the regulation of type 1 interferon signaling in astrocytes. Dietary tryptophan is metabolized by gut bacteria leading to the production of indole-3-aldehyde and indole-3-propionic acid that act as natural AHR agonist and thereby have a relevant effect on the inflammatory status of astrocytes and general neuroinflammation (Fung et al. 2017). As an example, *Lactobacillus reuteri*, a commonly used probiotic supplementation, is known to produce indole-3-aldehyde from dietary tryptophan (Zelante et al. 2013) and could thus influence AHR.

Furthermore, inflammatory Ly6C^{hi} monocytes have been shown to link changes in gut microbiota with cell neurogenesis in the brain and related learning capabilities (Mohle et al. 2016). The treatment of mice with antibiotics leads to loss of hippocampal neurogenesis and memory consolidation at least partially mediated by Ly6C^{hi} monocytes. Probiotic treatment with specific microbial species, in contrast, was able to restore these deficiencies in neurogenesis and learning in mice (Mohle et al. 2016).

A study by Margineanu showed that genes implicated in the astrocyte-neuron lactate shuttle (ANLS) are modulated by gut bacteria (Margineanu et al. 2020). The ANLS hypothesis states that neurons are mainly oxidative and astrocytes are mainly glycolytic cells and implies intercellular lactate transport from astrocytes to neurons for further oxidation of lactate in neurons via the citric acid cycle (Pellerin et al. 1998). By investigating mRNA expression levels of ANLS-associated genes in the hippocampus of germ-free mice, germ-free mice colonized with gut microbiota, and conventionally raised mice, a regulatory role of gut microbiota in the cellular metabolism in the brain was suggested (Hoban and Stilling 2018; Margineanu et al. 2020; Matsumoto et al. 2013). Summing up, these observations propose that gut microbiota and especially microbial metabolites may affect inflammatory and metabolic activities in the brain and especially in astrocytes. In future research, gut microbiota-targeted interventions such as dietary supplementation, pre- and probiotic treatment, and fecal microbiota transplantation might be shown to help treat AN and other psychiatric disorders.

Conclusion and Outlook

We showed that astrocyte loss could underlie the striking brain volume reduction in AN with multiple functional implications potentially including reduced learning capacity, mood changes, and sleep alterations, possibly helping to explain difficulties in psychotherapy of patients in the acutely starved state. Underlying mechanisms of this astrocyte loss include reduced neogenesis in line with the catabolic state in semi-starvation and probably involve consequences of reduced gonadal steroid hormones as well as inflammatory responses. Altered gut microbiota interactions with the brain could contribute to these numeric and functional astrocyte alterations. These results help to understand the underlying pathophysiology of AN and potentially brain-based consequences in other diseases involving cachexia like oncologic diseases and chronic inflammatory bowel diseases or of malnourished patients in developing countries and can be the basis of more astrocyte-centered research lines.

References

Abe T, Takahashi S, Suzuki N (2006) Oxidative metabolism in cultured rat astroglia: effects of reducing the glucose concentration in the culture medium and of D-aspartate or potassium stimulation. J Cerebral Blood Flow Metab 26:153–160. https://doi.org/10.1038/sj.jcbfm.9600175

Agnello E et al (2012) Tumour necrosis factor alpha and oxidative stress as maintaining factors in the evolution of anorexia nervosa. Eat Weight Disord 17:e194–e199. https://doi.org/10.1007/bf03325347

Allard C et al (2014) Hypothalamic astroglial connexins are required for brain glucose sensing-induced insulin secretion. J Cerebral Blood Flow Metab 34:339–346. https://doi.org/10.1038/jcbfm.2013.206

Allende LM et al (1998) Immunodeficiency associated with anorexia nervosa is secondary and improves after refeeding. Immunology 94:543–551. https://doi.org/10.1046/j.1365-2567.1998.00548.x

Almeida A et al (2019) A new genomic blueprint of the human gut microbiota. Nature 568:499–504. https://doi.org/10.1038/s41586-019-0965-1

Ananthakrishnan AN (2015) Epidemiology and risk factors for IBD nature reviews. Gastroenterol Hepatol 12:205–217. https://doi.org/10.1038/nrgastro.2015.34

Aoki C, Wable G, Chowdhury TG, Sabaliauskas NA, Laurino K, Barbarich-Marsteller NC (2014) α4βδ-GABAARs in the hippocampal CA1 as a biomarker for resilience to activity-based anorexia. Neuroscience 265:108–123. https://doi.org/10.1016/j.neuroscience.2014.01.011

Araújo GW, Beyer C, Arnold S (2008) Oestrogen influences on mitochondrial gene expression and respiratory chain activity in cortical and mesencephalic astrocytes. J Neuroendocrinol 20:930–941. https://doi.org/10.1111/j.1365-2826.2008.01747.x

Arnold S, de Araújo GW, Beyer C (2008) Gender-specific regulation of mitochondrial fusion and fission gene transcription and viability of cortical astrocytes by steroid hormones. J Mol Endocrinol 41:289–300. https://doi.org/10.1677/jme-08-0085

Arumugam M et al (2011) Enterotypes of the human gut microbiome. Nature 473:174–180. https://doi.org/10.1038/nature09944

Asano Y et al (2012) Critical role of gut microbiota in the production of biologically active, free catecholamines in the gut lumen of mice. Am J Physiol Gastrointest Liver Physiol 303:G1288–G1295. https://doi.org/10.1152/ajpgi.00341.2012

Azcoitia I, Santos-Galindo M, Arevalo MA, Garcia-Segura LM (2010) Role of astroglia in the neuroplastic and neuroprotective actions of estradiol. Eur J Neurosci 32:1995–2002. https://doi.org/10.1111/j.1460-9568.2010.07516.x

Banasr M, Dwyer JM, Duman RS (2011) Cell atrophy and loss in depression: reversal by antidepressant treatment. Curr Opin Cell Biol 23:730–737. https://doi.org/10.1016/j.ceb.2011.09.002

Barbarich-Marsteller NC et al (2013) Activity-based anorexia is associated with reduced hippocampal cell proliferation in adolescent female rats. Behav Brain Res 236:251–257. https://doi.org/10.1016/j.bbr.2012.08.047

Belmonte L et al (2016) A role for intestinal TLR4-driven inflammatory response during activity-based anorexia. Sci Rep 6:35813. https://doi.org/10.1038/srep35813

Boghi A, Sterpone S, Sales S, D'Agata F, Bradac GB, Zullo G, Munno D (2011) In vivo evidence of global and focal brain alterations in anorexia nervosa. Psychiatry Res 192:154–159. https://doi.org/10.1016/j.pscychresns.2010.12.008

Böhm M, Papezova H, Hansikova H, Wenchich L, Zeman J (2007) Activities of respiratory chain complexes in isolated platelets in females with anorexia nervosa. Int J Eat Disord 40:659–663. https://doi.org/10.1002/eat.20403

Brambilla F, Monti D, Franceschi C (2001) Plasma concentrations of interleukin-1-beta, interleukin-6 and tumor necrosis factor-alpha, and of their soluble receptors and receptor antagonist in anorexia nervosa. Psychiatry Res 103:107–114. https://doi.org/10.1016/s0165-1781(01)00283-9

Brambilla R et al (2005) Inhibition of astroglial nuclear factor κB reduces inflammation and improves functional recovery after spinal cord injury. J Exp Med 202:145–156. https://doi.org/10.1084/jem.20041918

Brambilla F, Santonastaso P, Caregaro L, Favaro A (2018) Growth hormone and insulin-like growth factor 1 secretions in eating disorders: correlations with psychopathological aspects of the disorders. Psychiatry Res 263:233–237. https://doi.org/10.1016/j.psychres.2017.07.049

Breit S, Kupferberg A, Rogler G, Hasler G (2018) Vagus nerve as modulator of the brain-gut axis in psychiatric and inflammatory disorders. Front Psych 9:44. https://doi.org/10.3389/fpsyt.2018.00044

Breton J et al (2020) Gut microbiota alteration in a mouse model of anorexia nervosa. Clin Nutr (Edinburgh, Scotland). https://doi.org/10.1016/j.clnu.2020.05.002

Brotfain E, Gruenbaum SE, Boyko M, Kutz R, Zlotnik A, Klein M (2016) Neuroprotection by estrogen and progesterone in traumatic brain injury and spinal cord injury. Curr Neuropharmacol 14:641–653. https://doi.org/10.2174/1570159x14666160309123554

Buhren K et al (2014) Comorbid psychiatric disorders in female adolescents with first-onset anorexia nervosa. Europ Eating Disord Rev 22:39–44. https://doi.org/10.1002/erv.2254

Caruso C, Carniglia L, Durand D, Gonzalez PV, Scimonelli TN, Lasaga M (2012) Melanocortin 4 receptor activation induces brain-derived neurotrophic factor expression in rat astrocytes through cyclic AMP-protein kinase A pathway. Mol Cell Endocrinol 348:47–54. https://doi.org/10.1016/j.mce.2011.07.036

Caso JR et al (2020) Dysfunction of inflammatory pathways in adolescent female patients with anorexia nervosa. Prog Neuro-Psychopharmacol Biol Psychiatry 96:109727. https://doi.org/10.1016/j.pnpbp.2019.109727

Castro-Fornieles J et al (2010) A cross-sectional and follow-up functional MRI study with a working memory task in adolescent anorexia nervosa. Neuropsychologia 48:4111–4116. https://doi.org/10.1016/j.neuropsychologia.2010.10.003

Cekanaviciute E, Buckwalter MS (2016) Astrocytes: integrative regulators of Neuroinflammation in stroke and other neurological diseases. Neurotherapeutics 13:685–701. https://doi.org/10.1007/s13311-016-0477-8

Chandel NS, Trzyna WC, McClintock DS, Schumacker PT (2000) Role of oxidants in NF-κB activation and TNF-α gene transcription induced by hypoxia and endotoxin. J Immunol 165:1013–1021. https://doi.org/10.4049/jimmunol.165.2.1013

Chen Y, Swanson RA (2003) Astrocytes and brain injury. J Cerebral Blood Flow Metab 23:137–149. https://doi.org/10.1097/01.Wcb.0000044631.80210.3c

Chen JQ, Cammarata PR, Baines CP, Yager JD (2009) Regulation of mitochondrial respiratory chain biogenesis by estrogens/estrogen receptors and physiological, pathological and pharmacological implications. Biochim Biophys Acta 1793:1540–1570. https://doi.org/10.1016/j.bbamcr.2009.06.001

Chen Y, Zhang J, Tan H, Li J, Yu Y (2020) Detrimental effects of hypercortisolism on brain structure and related risk factors. Sci Rep 10:12708. https://doi.org/10.1038/s41598-020-68166-0

Choi SS, Lee HJ, Lim I, Satoh J-I, Kim SU (2014) Human astrocytes: secretome profiles of cytokines and chemokines. PLoS One 9:e92325–e92325. https://doi.org/10.1371/journal.pone.0092325

Chui HT et al (2008) Cognitive function and brain structure in females with a history of adolescent-onset anorexia nervosa. Pediatrics 122:e426–e437. https://doi.org/10.1542/peds.2008-0170

Clarke G, Stilling RM, Kennedy PJ, Stanton C, Cryan JF, Dinan TG (2014) Minireview: gut microbiota: the neglected endocrine organ. Mol Endocrinol (Baltimore, MD) 28:1221–1238. https://doi.org/10.1210/me.2014-1108

Colombo E et al (2012) Stimulation of the neurotrophin receptor TrkB on astrocytes drives nitric oxide production and neurodegeneration. J Exp Med 209:521–535. https://doi.org/10.1084/jem.20110698

Cryan JF, Dinan TG (2012) Mind-altering microorganisms: the impact of the gut microbiota on brain and behaviour. Nat Rev Neurosci 13:701–712. https://doi.org/10.1038/nrn3346

Dalton B et al (2018) A meta-analysis of cytokine concentrations in eating disorders. J Psychiatr Res 103:252–264. https://doi.org/10.1016/j.jpsychires.2018.06.002

Dalton B, Leppanen J, Campbell IC, Chung R, Breen G, Schmidt U, Himmerich H (2020) A longitudinal analysis of cytokines in anorexia nervosa. Brain Behav Immun 85:88–95. https://doi.org/10.1016/j.bbi.2019.05.012

Darby A, Hay P, Mond J, Quirk F, Buttner P, Kennedy L (2009) The rising prevalence of comorbid obesity and eating disorder behaviors from 1995 to 2005. Int J Eat Disord 42:104–108. https://doi.org/10.1002/eat.20601

Derrien M, Belzer C, de Vos WM (2017) Akkermansia muciniphila and its role in regulating host functions. Microb Pathog 106:171–181. https://doi.org/10.1016/j.micpath.2016.02.005

Dmitrzak-Weglarz M et al (2013) Serum neurotrophin concentrations in polish adolescent girls with anorexia nervosa. Neuropsychobiology 67:25–32. https://doi.org/10.1159/000343500

Dorrington MG, Fraser IDC (2019) NF-κB signaling in macrophages: dynamics, crosstalk, and signal integration. Front Immunol 10. https://doi.org/10.3389/fimmu.2019.00705

Draheim T et al (2016) Activation of the astrocytic Nrf2/ARE system ameliorates the formation of demyelinating lesions in a multiple sclerosis animal model. Glia 64:2219–2230. https://doi.org/10.1002/glia.23058

Erny D, Hrabe de Angelis AL, Prinz M (2017) Communicating systems in the body: how microbiota and microglia cooperate. Immunology 150:7–15. https://doi.org/10.1111/imm.12645

Falkowska A, Gutowska I, Goschorska M, Nowacki P, Chlubek D, Baranowska-Bosiacka I (2015) Energy metabolism of the brain, including the cooperation between astrocytes and neurons, especially in the context of glycogen metabolism. Int J Mol Sci 16:25959–25981. https://doi.org/10.3390/ijms161125939

Fazeli PK, Klibanski A (2012) Neuroendocrine dysregulation and the growth hormone-IGF-1 axis in anorexia nervosa. Exp Rev Endocrinol Metab 7:223–231. https://doi.org/10.1586/eem.12.5

Fellin T, Ellenbogen JM, De Pitta M, Ben-Jacob E, Halassa MM (2012) Astrocyte regulation of sleep circuits: experimental and modeling perspectives. Front Comput Neurosci 6:65. https://doi.org/10.3389/fncom.2012.00065

Fetissov SO (2017) Role of the gut microbiota in host appetite control: bacterial growth to animal feeding behaviour. Nat Rev Endocrinol 13:11–25. https://doi.org/10.1038/nrendo.2016.150

Florent V et al (2020) Hypothalamic structural and functional imbalances in anorexia nervosa. Neuroendocrinology 110:552–562. https://doi.org/10.1159/000503147

Fonville L, Giampietro V, Williams SC, Simmons A, Tchanturia K (2014) Alterations in brain structure in adults with anorexia nervosa and the impact of illness duration. Psychol Med 44:1965–1975. https://doi.org/10.1017/s0033291713002389

Frago LM, Chowen JA (2017) Involvement of astrocytes in mediating the central effects of ghrelin. Int J Mol Sci 18.0 https://doi.org/10.3390/ijms18030536

Frintrop L et al (2017) Reduced astrocyte density underlying brain volume reduction in activity-based anorexia rats. World J Biol Psychiatry: 1–11. https://doi.org/10.1080/15622975.2016.1273552

Frintrop L et al (2018) Establishment of a chronic activity-based anorexia rat model. J Neurosci Methods 293:191–198. https://doi.org/10.1016/j.jneumeth.2017.09.018

Frintrop L, Trinh S, Liesbrock J, Leunissen C, Kempermann J, Etdoger S, Kas MJ (2019) The reduction of astrocytes and brain volume loss in anorexia nervosa-the impact of starvation and refeeding in a rodent model. Transl Psychiatry 9:159. https://doi.org/10.1038/s41398-019-0493-7

Fung TC, Olson CA, Hsiao EY (2017) Interactions between the microbiota, immune and nervous systems in health and disease. Nat Neurosci 20:145–155. https://doi.org/10.1038/nn.4476

Gianotti L, Lanfranco F, Ramunni J, Destefanis S, Ghigo E, Arvat E (2002) GH/IGF-I axis in anorexia nervosa. Eating Weight Disord 7:94–105. https://doi.org/10.1007/bf03354435

Gibson KR (1991) Brain maturation and cognitive development: comparative and cross-cultural perspectives. Brain maturation and cognitive development: comparative and cross-cultural perspectives. Aldine de Gruyter, Hawthorne

Gigante G et al (2011) Role of gut microbiota in food tolerance and allergies. Digest Dis (Basel, Switzerland) 29:540–549. https://doi.org/10.1159/000332977

Godlewska BR, Pike A, Sharpley AL, Ayton A, Park RJ, Cowen PJ, Emir UE (2017) Brain glutamate in anorexia nervosa: a magnetic resonance spectroscopy case control study at 7 Tesla. Psychopharmacology 234:421–426. https://doi.org/10.1007/s00213-016-4477-5

Gonzalez A, Kohn MR, Clarke SD (2007) Eating disorders in adolescents. Aust Fam Physician 36:614–619

Habib P, Dang J, Slowik A, Victor M, Beyer C (2014) Hypoxia-induced gene expression of aquaporin-4, cyclooxygenase-2 and hypoxia-inducible factor 1α in rat cortical astroglia is inhibited by 17β-estradiol and progesterone. Neuroendocrinology 99:156–167. https://doi.org/10.1159/000362279

Hara M, Matsuda Y, Okumura N, Hirai K, Nakagawa H (1989) Effect of glucose starvation on glucose transport in neuronal cells in primary culture from rat brain. J Neurochem 52:909–912. https://doi.org/10.1111/j.1471-4159.1989.tb02541.x

Hata T et al (2019) The gut microbiome derived from anorexia nervosa patients impairs weight gain and behavioral performance in female mice. Endocrinology 160:2441–2452. https://doi.org/10.1210/en.2019-00408

Heitzer M, Kaiser S, Kanagaratnam M, Zendedel A, Hartmann P, Beyer C, Johann S (2017) Administration of 17β-estradiol improves Motoneuron survival and Down-regulates Inflammasome activation in male SOD1(G93A) ALS mice. Mol Neurobiol 54:8429–8443. https://doi.org/10.1007/s12035-016-0322-4

Hermens DF, Simcock G, Dutton M, Bouças AP, Can AT, Lilley C, Lagopoulos J (2020) Anorexia nervosa, zinc deficiency and the glutamate system: the ketamine option. Prog Neuro-Psychopharmacol Biol Psychiatry 101:109921. https://doi.org/10.1016/j.pnpbp.2020.109921

Herpertz-Dahlmann B (2015) Adolescent eating disorders: update on definitions, symptomatology, epidemiology, and comorbidity. Child Adolesc Psychiatr Clin N Am 24:177–196. https://doi.org/10.1016/j.chc.2014.08.003

Hoban AE, Stilling RM (2018) The microbiome regulates amygdala-dependent fear recall. Mol Psychiatry 23:1134–1144. https://doi.org/10.1038/mp.2017.100

Holtkamp K, Herpertz-Dahlmann B, Hebebrand K, Mika C, Kratzsch J, Hebebrand J (2006) Physical activity and restlessness correlate with leptin levels in patients with adolescent anorexia nervosa. Biol Psychiatry 60:311–313. https://doi.org/10.1016/j.biopsych.2005.11.001

Hsiao H-Y, Chen Y-C, Chen H-M, Tu P-H, Chern Y (2013) A critical role of astrocyte-mediated nuclear factor-κB-dependent inflammation in Huntington's disease. Hum Mol Genet 22:1826–1842. https://doi.org/10.1093/hmg/ddt036

Huang TT, Lai JB, Du YL, Xu Y, Ruan LM, Hu SH (2019) Current understanding of gut microbiota in mood disorders: an update of human studies. Front Genet 10:98. https://doi.org/10.3389/fgene.2019.00098

Hyvärinen T et al (2019) Co-stimulation with IL-1β and TNF-α induces an inflammatory reactive astrocyte phenotype with neurosupportive characteristics in a human pluripotent stem cell model system. Sci Rep 9:16944. https://doi.org/10.1038/s41598-019-53414-9

Ivanova T, Karolczak M, Beyer C (2002) Estradiol stimulates GDNF expression in developing hypothalamic neurons. Endocrinology 143:3175–3178. https://doi.org/10.1210/endo.143.8.8794

Jiang C, Li G, Huang P, Liu Z, Zhao B (2017) The gut microbiota and Alzheimer's disease. J Alzheimer's Dis 58:1–15. https://doi.org/10.3233/jad-161141

Johann S, Beyer C (2013) Neuroprotection by gonadal steroid hormones in acute brain damage requires cooperation with astroglia and microglia. J Steroid Biochem Mol Biol 137:71–81. https://doi.org/10.1016/j.jsbmb.2012.11.006

Joos A et al (2010) Voxel-based morphometry in eating disorders: correlation of psychopathology with grey matter volume. Psychiatry Res 182:146–151. https://doi.org/10.1016/j.pscychresns.2010.02.004

Kaczor P, Rakus D, Mozrzymas JW (2015) Neuron-astrocyte interaction enhance GABAergic synaptic transmission in a manner dependent on key metabolic enzymes. Front Cell Neurosci 9:120. https://doi.org/10.3389/fncel.2015.00120

Kaltschmidt B, Kaltschmidt C (2015) NF-KappaB in long-term memory and structural plasticity in the adult mammalian brain front. Mol Neurosci 8:69–69. https://doi.org/10.3389/fnmol.2015.00069

Kanbur N, Mesci L, Derman O, Turul T, Cuhadaroğlu F, Kutluk T, Tezcan I (2008) Tumor necrosis factor alpha-308 gene polymorphism in patients with anorexia nervosa. Turk J Pediatr 50:219–222

Karki P, Smith K, Johnson J Jr, Lee E (2014) Astrocyte-derived growth factors and estrogen neuroprotection: role of transforming growth factor-α in estrogen-induced upregulation of glutamate transporters in astrocytes. Mol Cell Endocrinol 389:58–64. https://doi.org/10.1016/j.mce.2014.01.010

Kas MJ, Adan RA (2011) Animal models of eating disorder traits. Curr Top Behav Neurosci 6:209–227. https://doi.org/10.1007/7854_2010_84

Kasischke KA, Vishwasrao HD, Fisher PJ, Zipfel WR, Webb WW (2004) Neural activity triggers neuronal oxidative metabolism followed by astrocytic glycolysis. Science (New York, NY) 305:99–103. https://doi.org/10.1126/science.1096485

Kawai T, Akira S (2007) Signaling to NF-κB by Toll-like receptors. Trends Mol Med 13:460–469. https://doi.org/10.1016/j.molmed.2007.09.002

Kelly JR, Minuto C, Cryan JF, Clarke G, Dinan TG (2017) Cross talk: the microbiota and neurodevelopmental disorders. Front Neurosci 11:490. https://doi.org/10.3389/fnins.2017.00490

Kim JG et al (2014) Leptin signaling in astrocytes regulates hypothalamic neuronal circuits and feeding. Nat Neurosci 17:908–910. https://doi.org/10.1038/nn.3725

Kimelberg HK, Nedergaard M (2010) Functions of astrocytes and their potential as therapeutic targets. Neurotherapeutics 7:338–353. https://doi.org/10.1016/j.nurt.2010.07.006

King JA et al (2015) Global cortical thinning in acute anorexia nervosa normalizes following long-term weight restoration. Biol Psychiatry 77:624–632. https://doi.org/10.1016/j.biopsych.2014.09.005

King JA, Frank GKW, Thompson PM, Ehrlich S (2018) Structural neuroimaging of anorexia nervosa: future directions in the quest for mechanisms underlying dynamic alterations. Biol Psychiatry 83:224–234. https://doi.org/10.1016/j.biopsych.2017.08.011

Kipp M, Amor S, Krauth R, Beyer C (2012) Multiple sclerosis: neuroprotective alliance of estrogen-progesterone and gender. Front Neuroendocrinol 33:1–16. https://doi.org/10.1016/j.yfrne.2012.01.001

Kleiman SC, Watson HJ, Bulik-Sullivan EC, Huh EY, Tarantino LM, Bulik CM, Carroll IM (2015) The intestinal microbiota in acute anorexia nervosa and during Renourishment: relationship to depression, anxiety, and eating disorder psychopathology. Psychosom Med 77:969–981. https://doi.org/10.1097/psy.0000000000000247

Klump KL et al (2013) The interactive effects of estrogen and progesterone on changes in emotional eating across the menstrual cycle. J Abnorm Psychol 122:131–137. https://doi.org/10.1037/a0029524

Kogel V, Trinh S, Gasterich N, Beyer C, Seitz J (2021) Long-term glucose starvation induces inflammatory responses and phenotype switch in primary cortical rat astrocytes. J Mol Neurosci. https://doi.org/10.1007/s12031-021-01800-2

Kowiański P, Lietzau G, Czuba E, Waśkow M, Steliga A, Moryś J (2018) BDNF: a key factor with multipotent impact on brain signaling and synaptic plasticity. Cell Mol Neurobiol 38:579–593. https://doi.org/10.1007/s10571-017-0510-4

Lagkouvardos I, Fischer S, Kumar N, Clavel T (2017) Rhea: a transparent and modular R pipeline for microbial profiling based on 16S rRNA gene amplicons. PeerJ 5:e2836. https://doi.org/10.7717/peerj.2836

Landeghem FKHV, Weiss T, Oehmichen M, Deimling AV (2006) Decreased expression of glutamate transporters in astrocytes after human traumatic brain injury. J Neurotrauma 23:1518–1528. https://doi.org/10.1089/neu.2006.23.1518
Lee CY, Dallérac G, Ezan P, Anderova M, Rouach N (2016) Glucose tightly controls morphological and functional properties of astrocytes. Front Aging Neurosci 8:82. https://doi.org/10.3389/fnagi.2016.00082
Levenson JL (2011) The American Psychiatric Publishing textbook of psychosomatic medicine: psychiatric care of the medically ill. American Psychiatric Pub, Washington, DC
Li K, Li J, Zheng J, Qin S (2019) Reactive astrocytes in neurodegenerative diseases. Aging Dis 10:664–675. https://doi.org/10.14336/AD.2018.0720
Liddelow SA, Barres BA (2017) Reactive astrocytes: production, function, and therapeutic potential. Immunity 46:957–967. https://doi.org/10.1016/j.immuni.2017.06.006
Liddelow SA et al (2017) Neurotoxic reactive astrocytes are induced by activated microglia. Nature 541:481–487. https://doi.org/10.1038/nature21029
Lim CT, Kola B, Korbonits M (2011) The ghrelin/GOAT/GHS-R system and energy metabolism. Rev Endocr Metab Disord 12:173–186. https://doi.org/10.1007/s11154-011-9169-1
Linnerbauer M, Wheeler MA, Quintana FJ (2020) Astrocyte crosstalk in CNS inflammation. Neuron 108:608–622. https://doi.org/10.1016/j.neuron.2020.08.012
Luz Neto LMD, Vasconcelos FMND, Silva JED, Pinto TCC, Sougey ÉB, Ximenes RCC (2019) Differences in cortisol concentrations in adolescents with eating disorders: a systematic review. J Pediatr 95:18–26. https://doi.org/10.1016/j.jped.2018.02.007
Mack I et al (2016) Weight gain in anorexia nervosa does not ameliorate the faecal microbiota, branched chain fatty acid profiles, and gastrointestinal complaints. Sci Rep 6:26752. https://doi.org/10.1038/srep26752
Mahmoud S, Gharagozloo M, Simard C, Gris D (2019) Astrocytes maintain glutamate homeostasis in the CNS by controlling the balance between glutamate uptake and release. Cell 8. https://doi.org/10.3390/cells8020184
Mainz V, Schulte-Rüther M, Fink GR, Herpertz-Dahlmann B, Konrad K (2012) Structural brain abnormalities in adolescent anorexia nervosa before and after weight recovery and associated hormonal changes. Psychosom Med 74:574–582. https://doi.org/10.1097/PSY.0b013e31824ef10e
Mangiola F, Ianiro G, Franceschi F, Fagiuoli S, Gasbarrini G, Gasbarrini A (2016) Gut microbiota in autism and mood disorders. World J Gastroenterol 22:361–368. https://doi.org/10.3748/wjg.v22.i1.361
Margineanu MB et al (2020) Gut microbiota modulates expression of genes involved in the astrocyte-neuron lactate shuttle in the hippocampus. Europ Neuropsychopharmacol 41:152–159. https://doi.org/10.1016/j.euroneuro.2020.11.006
Marina N et al (2018) Brain metabolic sensing and metabolic signaling at the level of an astrocyte. Glia 66:1185–1199. https://doi.org/10.1002/glia.23283
Marques TM, Wall R, Ross RP, Fitzgerald GF, Ryan CA, Stanton C (2010) Programming infant gut microbiota: influence of dietary and environmental factors. Curr Opin Biotechnol 21:149–156. https://doi.org/10.1016/j.copbio.2010.03.020
Martin F (1958) Pathology of the neurological & psychiatric aspects of various manifestations of deficiency diseases with digestive & neuro-endocrine disorders. Praxis 47:197–200
Matejuk A, Ransohoff RM (2020) Crosstalk between astrocytes and microglia: an overview. Front Immunol 11:1416–1416. https://doi.org/10.3389/fimmu.2020.01416
Matsubara M, Sakata I, Wada R, Yamazaki M, Inoue K, Sakai T (2004) Estrogen modulates ghrelin expression in the female rat stomach. Peptides 25:289–297. https://doi.org/10.1016/j.peptides.2003.12.020
Matsumoto M, Kibe R, Ooga T, Aiba Y, Sawaki E, Koga Y, Benno Y (2013) Cerebral low-molecular metabolites influenced by intestinal microbiota: a pilot study. Front Syst Neurosci 7:9. https://doi.org/10.3389/fnsys.2013.00009
Mattingly D, Bhanji S (1995) Hypoglycaemia and anorexia nervosa. J R Soc Med 88:191–195

Mayorquin LC, Rodriguez AV, Sutachan J-J, Albarracín SL (2018) Connexin-mediated functional and metabolic coupling between astrocytes and neurons. Front Mol Neurosci 11:118–118. https://doi.org/10.3389/fnmol.2018.00118
McCormick LM, Keel PK, Brumm MC, Bowers W, Swayze V, Andersen A, Andreasen N (2008) Implications of starvation-induced change in right dorsal anterior cingulate volume in anorexia nervosa. Int J Eat Disord 41:602–610. https://doi.org/10.1002/eat.20549
Mela V, Vargas A, Meza C, Kachani M, Wagner EJ (2016) Modulatory influences of estradiol and other anorexigenic hormones on metabotropic, Gi/o-coupled receptor function in the hypothalamic control of energy homeostasis. J Steroid Biochem Mol Biol 160:15–26. https://doi.org/10.1016/j.jsbmb.2015.07.014
Mendez P, Wandosell F, Garcia-Segura LM (2006) Cross-talk between estrogen receptors and insulin-like growth factor-I receptor in the brain: cellular and molecular mechanisms. Front Neuroendocrinol 27:391–403. https://doi.org/10.1016/j.yfrne.2006.09.001
Mergenthaler P, Lindauer U, Dienel GA, Meisel A (2013) Sugar for the brain: the role of glucose in physiological and pathological brain function. Trends Neurosci 36:587–597. https://doi.org/10.1016/j.tins.2013.07.001
Misra M et al (2006) Uncoupling of cardiovascular risk markers in adolescent girls with anorexia nervosa. J Pediatr 149:763–769. https://doi.org/10.1016/j.jpeds.2006.08.043
Mohle L et al (2016) Ly6C(hi) monocytes provide a link between antibiotic-induced changes in gut microbiota and adult hippocampal neurogenesis. Cell Rep 15:1945–1956. https://doi.org/10.1016/j.celrep.2016.04.074
Molofsky AV et al (2012) Astrocytes and disease: a neurodevelopmental perspective. Genes Dev 26:891–907. https://doi.org/10.1101/gad.188326.112
Moresco EM, LaVine D, Beutler B (2011) Toll-like receptors. Curr Biol 21:R488–R493. https://doi.org/10.1016/j.cub.2011.05.039
Morkl S et al (2017) Gut microbiota and body composition in anorexia nervosa inpatients in comparison to athletes, overweight, obese, and normal weight controls. Int J Eat Disord 50:1421–1431. https://doi.org/10.1002/eat.22801
Mortezaee K, Khanlarkhani N, Beyer C, Zendedel A (2018) Inflammasome: its role in traumatic brain and spinal cord injury. J Cell Physiol 233:5160–5169. https://doi.org/10.1002/jcp.26287
Nagl M, Jacobi C, Paul M, Beesdo-Baum K, Höfler M, Lieb R, Wittchen H-U (2016) Prevalence, incidence, and natural course of anorexia and bulimia nervosa among adolescents and young adults. Eur Child Adolesc Psychiatry 25:903–918. https://doi.org/10.1007/s00787-015-0808-z
Nagy JI, Rash JE (2000) Connexins and gap junctions of astrocytes and oligodendrocytes in the CNS brain research. Brain Res Rev 32:29–44. https://doi.org/10.1016/s0165-0173(99)00066-1
Nakai Y, Hamagaki S, Takagi R, Taniguchi A, Kurimoto F (2000) Plasma concentrations of tumor necrosis factor-alpha (TNF-alpha) and soluble TNF receptors in patients with bulimia nervosa. Clin Endocrinol 53:383–388. https://doi.org/10.1046/j.1365-2265.2000.01091.x
Neufang S, Specht K, Hausmann M, Güntürkün O, Herpertz-Dahlmann B, Fink GR, Konrad K (2009) Sex differences and the impact of steroid hormones on the developing human brain. Cerebral Cortex (New York, NY : 1991) 19:464–473. https://doi.org/10.1093/cercor/bhn100
Neuman H, Debelius JW, Knight R, Koren O (2015) Microbial endocrinology: the interplay between the microbiota and the endocrine system. FEMS Microbiol Rev 39:509–521. https://doi.org/10.1093/femsre/fuu010
Neumarker KJ (1997) Mortality and sudden death in anorexia nervosa. Int J Eat Disord 21:205–212
Nobis S et al (2018) Alterations of proteome, mitochondrial dynamic and autophagy in the hypothalamus during activity-based anorexia. Sci Rep 8:7233. https://doi.org/10.1038/s41598-018-25548-9
Noell S, Fallier-Becker P, Beyer C, Kröger S, Mack AF, Wolburg H (2007) Effects of agrin on the expression and distribution of the water channel protein aquaporin-4 and volume regulation in cultured astrocytes. Eur J Neurosci 26:2109–2118. https://doi.org/10.1111/j.1460-9568.2007.05850.x

Nogal P, Pniewska-Siark B, Lewinski A (2008) Relation of trophic changes in the central nervous system, measured by the width of cordical sulci, to the clinical course of anorexia nervosa (II). Neuro Endocrinol Lett 29:879–883

Noor SO et al (2010) Ulcerative colitis and irritable bowel patients exhibit distinct abnormalities of the gut microbiota. BMC Gastroenterol 10:134. https://doi.org/10.1186/1471-230x-10-134

Obata T, Brown GE, Yaffe MB (2000) MAP kinase pathways activated by stress: the p38 MAPK pathway. Crit Care Med 28:N67–N77. https://doi.org/10.1097/00003246-200004001-00008

Ostrowska Z et al (2015) Selected pro-inflammatory cytokines, bone metabolism, osteoprotegerin, and receptor activator of nuclear factor-kB ligand in girls with anorexia nervosa. Endokrynol Pol 66:313–321. https://doi.org/10.5603/ep.2015.0040

Pajarillo E, Rizor A, Lee J, Aschner M, Lee E (2019) The role of astrocytic glutamate transporters GLT-1 and GLAST in neurological disorders: potential targets for neurotherapeutics. Neuropharmacology 161:107559. https://doi.org/10.1016/j.neuropharm.2019.03.002

Panatier A, Robitaille R (2012) Astrocyte, a key partner of neurons during basal synaptic transmission. Med Sci (Paris) 28:582–584. https://doi.org/10.1051/medsci/2012286009

Paulukat L et al (2016) Memory impairment is associated with the loss of regular oestrous cycle and plasma oestradiol levels in an activity-based anorexia animal model. World J Biol Psychiatry 17:274–284. https://doi.org/10.3109/15622975.2016.1173725

Pauwels PJ, Opperdoes FR, Trouet A (1985) Effects of Antimycin, glucose deprivation, and serum on cultures of neurons, astrocytes, and neuroblastoma cells. J Neurochem 44:143–148. https://doi.org/10.1111/j.1471-4159.1985.tb07123.x

Pawlak J, Brito V, Küppers E, Beyer C (2005a) Regulation of glutamate transporter GLAST and GLT-1 expression in astrocytes by estrogen brain research. Mol Brain Res 138:1–7. https://doi.org/10.1016/j.molbrainres.2004.10.043

Pawlak J, Karolczak M, Krust A, Chambon P, Beyer C (2005b) Estrogen receptor-alpha is associated with the plasma membrane of astrocytes and coupled to the MAP/Src-kinase pathway. Glia 50:270–275. https://doi.org/10.1002/glia.20162

Pellerin L, Magistretti PJ (1994) Glutamate uptake into astrocytes stimulates aerobic glycolysis: a mechanism coupling neuronal activity to glucose utilization. Proc Natl Acad Sci U S A 91:10625–10629. https://doi.org/10.1073/pnas.91.22.10625

Pellerin L et al (1998) Evidence supporting the existence of an activity-dependent astrocyte-neuron lactate shuttle. Dev Neurosci 20:291–299. https://doi.org/10.1159/000017324

Pomeroy C, Eckert E, Hu S, Eiken B, Mentink M, Crosby RD, Chao CC (1994) Role of interleukin-6 and transforming growth factor-β in anorexia nervosa. Biol Psychiatry 36:836–839. https://doi.org/10.1016/0006-3223(94)90594-0

Ponath G et al (2016) Myelin phagocytosis by astrocytes after myelin damage promotes lesion pathology. Brain 140:399–413. https://doi.org/10.1093/brain/aww298

Qian Y et al (2007) The adaptor Act1 is required for interleukin 17–dependent signaling associated with autoimmune and inflammatory disease. Nat Immunol 8:247–256. https://doi.org/10.1038/ni1439

Queipo-Ortuno MI et al (2013) Gut microbiota composition in male rat models under different nutritional status and physical activity and its association with serum leptin and ghrelin levels. PLoS One 8:e65465. https://doi.org/10.1371/journal.pone.0065465

Ragu-Varman D, Macedo-Mendoza M, Labrada-Moncada FE, Reyes-Ortega P, Morales T, Martínez-Torres A, Reyes-Haro D (2019) Anorexia increases microglial density and cytokine expression in the hippocampus of young female rats. Behav Brain Res 363:118–125. https://doi.org/10.1016/j.bbr.2019.01.042

Rahman MH, Kim M-S, Lee I-K, Yu R, Suk K (2018) Interglial crosstalk in obesity-induced hypothalamic inflammation. Front Neurosci 12:939–939. https://doi.org/10.3389/fnins.2018.00939

Ramoz N, Versini A, Gorwood P (2013) Anorexia nervosa and estrogen receptors. Vitam Horm 92:141–163. https://doi.org/10.1016/b978-0-12-410473-0.00006-4

Remely M, Hippe B, Geretschlaeger I, Stegmayer S, Hoefinger I, Haslberger A (2015) Increased gut microbiota diversity and abundance of Faecalibacterium prausnitzii and Akkermansia after fasting: a pilot study. Wien Klin Wochenschr 127:394–398. https://doi.org/10.1007/s00508-015-0755-1

Reyes-Haro D, Labrada-Moncada FE, Miledi R, Martinez-Torres A (2015) Dehydration-induced anorexia reduces astrocyte density in the rat Corpus callosum. Neural Plast 2015:474917. https://doi.org/10.1155/2015/474917

Reyes-Haro D, Labrada-Moncada FE, Varman DR, Kruger J, Morales T, Miledi R, Martinez-Torres A (2016) Anorexia reduces GFAP+ cell density in the rat Hippocampus. Neural Plast 2016:2426413. https://doi.org/10.1155/2016/2426413

Rhee SH, Pothoulakis C, Mayer EA (2009) Principles and clinical implications of the brain-gut-enteric microbiota axis Nature reviews. Gastroenterol Hepatol 6:306–314. https://doi.org/10.1038/nrgastro.2009.35

Ribasés M et al (2003) Met66 in the brain-derived neurotrophic factor (BDNF) precursor is associated with anorexia nervosa restrictive type. Mol Psychiatry 8:745–751. https://doi.org/10.1038/sj.mp.4001281

Ridaura VK et al. (2013) Gut microbiota from twins discordant for obesity modulate metabolism in mice. Science (New York, NY) 341:1241214. https://doi.org/10.1126/science.1241214

Rothhammer V, Mascanfroni ID, Bunse L (2016) Type I interferons and microbial metabolites of tryptophan modulate astrocyte activity and central nervous system inflammation via the aryl hydrocarbon receptor. Nat Med 22:586–597. https://doi.org/10.1038/nm.4106

Routtenberg A, Kuznesof AW (1967) Self-starvation of rats living in activity wheels on a restricted feeding schedule. J Comp Physiol Psychol 64:414–421

Schattner A, Tepper R, Steinbock M, Hahn T, Schoenfeld A (1990) TNF, interferon-gamma and cell-mediated cytotoxicity in anorexia nervosa; effect of refeeding. J Clin Lab Immunol 32:183–184

Schattner A, Tepper R, Steinbock M, Schoenfeld A, Vaisman N, Hahn T (1992) Cytokines in anorexia nervosa–nutritional or neuroimmunal changes? Harefuah 123:245–247. 308

Scheld M et al (2016) Neurodegeneration triggers peripheral immune cell recruitment into the forebrain. J Neurosci 36:1410–1415. https://doi.org/10.1523/jneurosci.2456-15.2016

Schorr M, Miller KK (2017) The endocrine manifestations of anorexia nervosa: mechanisms and management. Nat Rev Endocrinol 13:174–186. https://doi.org/10.1038/nrendo.2016.175

Seitz J, Walter M, Mainz V, Herpertz-Dahlmann B, Konrad K, von Polier G (2015) Brain volume reduction predicts weight development in adolescent patients with anorexia nervosa. J Psychiatr Res 68:228–237. https://doi.org/10.1016/j.jpsychires.2015.06.019

Seitz J, Herpertz-Dahlmann B, Konrad K (2016) Brain morphological changes in adolescent and adult patients with anorexia nervosa. J Neural Trans (Vienna, Austria : 1996) 123:949–959. https://doi.org/10.1007/s00702-016-1567-9

Seitz J, Konrad K, Herpertz-Dahlmann B (2018) Extend, Pathomechanism and clinical consequences of brain volume changes in anorexia nervosa. Curr Neuropharmacol 16:1164–1173. https://doi.org/10.2174/1570159x15666171109145651

Seitz J, Trinh S, Herpertz-Dahlmann B (2019) The microbiome and eating disorders. Psychiatr Clin North Am 42:93–103. https://doi.org/10.1016/j.psc.2018.10.004

Sender R, Fuchs S (2016) Revised estimates for the number of human and Bacteria cells in the body. PLoS Biol 14:e1002533. https://doi.org/10.1371/journal.pbio.1002533

Seong J, Kang JY, Sun JS, Kim KW (2019) Hypothalamic inflammation and obesity: a mechanistic review. Arch Pharm Res 42:383–392. https://doi.org/10.1007/s12272-019-01138-9

Sgritta M, Dooling SW, Buffington SA, Momin EN, Francis MB, Britton RA, Costa-Mattioli M (2019) Mechanisms underlying microbial-mediated changes in social behavior in mouse models of Autism Spectrum Disorder. Neuron 101:246–259.e246. https://doi.org/10.1016/j.neuron.2018.11.018

Shaw P et al (2008) Neurodevelopmental trajectories of the human cerebral cortex. J Neurosci 28:3586–3594. https://doi.org/10.1523/jneurosci.5309-07.2008

Sidoryk-Wegrzynowicz M, Wegrzynowicz M, Lee E, Bowman AB, Aschner M (2011) Role of astrocytes in brain function and disease. Toxicol Pathol 39:115–123. https://doi.org/10.1177/0192623310385254

Simon LS (1999) Role and regulation of cyclooxygenase-2 during inflammation. Am J Med 106:37s–42s. https://doi.org/10.1016/s0002-9343(99)00115-1

Singhal V, Misra M, Klibanski A (2014) Endocrinology of anorexia nervosa in young people: recent insights. Curr Opin Endocrinol Diabetes Obes 21:64–70. https://doi.org/10.1097/med.0000000000000026

Slowik A, Lammerding L, Hoffmann S, Beyer C (2018) Brain inflammasomes in stroke and depressive disorders: regulation by oestrogen. J Neuroendocrinol 30. https://doi.org/10.1111/jne.12482

Smink FR, van Hoeken D, Hoek HW (2012) Epidemiology of eating disorders: incidence, prevalence and mortality rates. Curr Psychiatry Rep 14:406–414. https://doi.org/10.1007/s11920-012-0282-y

Sofroniew MV (2009) Molecular dissection of reactive astrogliosis and glial scar formation. Trends Neurosci 32:638–647. https://doi.org/10.1016/j.tins.2009.08.002

Sofroniew MV, Vinters HV (2010) Astrocytes: biology and pathology. Acta Neuropathol 119:7–35. https://doi.org/10.1007/s00401-009-0619-8

Sohrabji F, Lewis DK (2006) Estrogen-BDNF interactions: implications for neurodegenerative diseases. Front Neuroendocrinol 27:404–414. https://doi.org/10.1016/j.yfrne.2006.09.003

Solmi M, Veronese N, Favaro A, Santonastaso P, Manzato E, Sergi G, Correll CU (2015) Inflammatory cytokines and anorexia nervosa: a meta-analysis of cross-sectional and longitudinal studies. Psychoneuroendocrinology 51:237–252. https://doi.org/10.1016/j.psyneuen.2014.09.031

Sonoyama K, Fujiwara R, Takemura N, Ogasawara T, Watanabe J, Ito H, Morita T (2009) Response of gut microbiota to fasting and hibernation in Syrian hamsters. Appl Environ Microbiol 75:6451–6456. https://doi.org/10.1128/aem.00692-09

Steinhausen HC, Jensen CM (2015) Time trends in lifetime incidence rates of first-time diagnosed anorexia nervosa and bulimia nervosa across 16 years in a Danish nationwide psychiatric registry study. Int J Eat Disord 48:845–850. https://doi.org/10.1002/eat.22402

Støving RK, Hangaard J, Hansen-Nord M, Hagen C (1999) A review of endocrine changes in anorexia nervosa. J Psychiatr Res 33:139–152. https://doi.org/10.1016/s0022-3956(98)00049-1

Subramanian S, Campbell BJ, Rhodes JM (2006) Bacteria in the pathogenesis of inflammatory bowel disease. Curr Opin Infect Dis 19:475–484. https://doi.org/10.1097/01.qco.0000244054.69253.f3

Sun D, Jakobs TC (2012) Structural remodeling of astrocytes in the injured CNS. Neuroscientist 18:567–588. https://doi.org/10.1177/1073858411423441

Ticinesi A, Lauretani F, Tana C, Nouvenne A, Ridolo E, Meschi T (2019) Exercise and immune system as modulators of intestinal microbiome: implications for the gut-muscle axis hypothesis. Exerc Immunol Rev 25:84–95

Treasure J, Willmott D (2020) Cognitive interpersonal model for anorexia nervosa revisited: the perpetuating factors that contribute to the development of the severe and enduring illness. J Clin Med 9. https://doi.org/10.3390/jcm9030630

Treasure J, Stein D, Maguire S (2015) Has the time come for a staging model to map the course of eating disorders from high risk to severe enduring illness? An examination of the evidence. Early Interv Psychiatry 9:173–184. https://doi.org/10.1111/eip.12170

Trinh S et al (2021) Gut microbiota and brain alterations in a translational anorexia nervosa rat model. J Psychiatr Res 133:156–165. https://doi.org/10.1016/j.jpsychires.2020.12.030

Victor VM et al (2014) Altered mitochondrial function and oxidative stress in leukocytes of anorexia nervosa patients. PLoS One 9:e106463. https://doi.org/10.1371/journal.pone.0106463

Vogel K et al (2016) White matter microstructural changes in adolescent anorexia nervosa including an exploratory longitudinal study. Neuroimage Clin 11:614–621. https://doi.org/10.1016/j.nicl.2016.04.002

Wang P, Liu C, Liu L, Zhang X, Ren B, Li B (2015) The antidepressant-like effects of estrogen-mediated ghrelin. Curr Neuropharmacol 13:524–535. https://doi.org/10.2174/1570159x13041508311120650

Yang Z, Wang KKW (2015) Glial fibrillary acidic protein: from intermediate filament assembly and gliosis to neurobiomarker. Trends Neurosci 38:364–374. https://doi.org/10.1016/j.tins.2015.04.003

Yang J et al (2017) The multiple roles of XBP1 in regulation of glucose and lipid metabolism. Curr Protein Pept Sci 18:630–635. https://doi.org/10.2174/1389203717666160627085011

Young JK (2010) Anorexia nervosa and estrogen: current status of the hypothesis. Neurosci Biobehav Rev 34:1195–1200. https://doi.org/10.1016/j.neubiorev.2010.01.015

Zamanian JL, Xu L, Foo LC, Nouri N, Zhou L, Giffard RG, Barres BA (2012) Genomic analysis of reactive astrogliosis. J Neurosci 32:6391–6410. https://doi.org/10.1523/JNEUROSCI.6221-11.2012

Zelante T et al (2013) Tryptophan catabolites from microbiota engage aryl hydrocarbon receptor and balance mucosal reactivity via interleukin-22. Immunity 39:372–385. https://doi.org/10.1016/j.immuni.2013.08.003

Zou J et al (2019) Mechanisms shaping the role of ERK1/2 in cellular senescence (review). Mol Med Rep 19:759–770. https://doi.org/10.3892/mmr.2018.9712

Part VI
Targeting Astrocytes in Therapeutic Management of Neuropsychiatric Disorders

Astroglial Serotonin Receptors as the Central Target of Classic Antidepressants

Alexei Verkhratsky, Vladimir Parpura, Caterina Scuderi, and Baoman Li

Introduction

Major depressive disorder (MDD) belongs to severe mood disorders. Global prevalence of MDD is about 4.7% with regional differences (Ferrari et al. 2013; Steel et al. 2014). The lifetime prevalence varies from 3.3% to 16% across countries (Gu et al. 2013; Otte et al. 2016). It has been estimated that MDD will be a leading cause of disability by 2030 (Mathers and Loncar 2006). MDD is associated with great individual and family suffering, disability, poor daily functioning, insomnia, and negative prognosis for comorbid medical conditions (Fröjd et al. 2008; Buysse et al. 2008; Lynch and Clarke 2006; Kupfer et al. 2012). MDD is also associated with an elevated risk of suicide, psychotic presentations (Penninx 2017; Kendler et al. 2009), and other psychiatric disorders (Cassano et al. 2004). Treating MDD usually involves pharmacological agents and/or psychotherapy. Treatment coverage of MDD has clear regional differences: the coverage of treatment of psychiatric

A. Verkhratsky
Faculty of Biology, Medicine and Health, The University of Manchester, Manchester, UK

Achucarro Center for Neuroscience, IKERBASQUE, Basque Foundation for Science, Bilbao, Spain

V. Parpura
Department of Neurobiology, The University of Alabama at Birmingham, Birmingham, AL, USA

C. Scuderi
Department of Physiology and Pharmacology "Vittorio Erspamer", SAPIENZA University of Rome, Rome, Italy

B. Li (✉)
Department of Forensic Analytical Toxicology, School of Forensic Medicine, China Medical University, Shenyang, People's Republic of China
e-mail: bmli@cmu.edu.cn

B. Li et al. (eds.), *Astrocytes in Psychiatric Disorders*, Advances in Neurobiology 26, https://doi.org/10.1007/978-3-030-77375-5_13

disorders including MDD varies between 7.0% and 20.3% across countries (Bijl et al. 2003; Demyttenaere et al. 2004). In the United States, less than 30% of adults who are positively screened for depression received any treatment (Olfson et al. 2016). In China, the discordant treatment ranges from 4% to 60% of MDD patients who ever received psychosocial therapy or pharmacotherapy (Yu et al. 2015; Liu et al. 2017; Chin et al. 2015). Overall, variations in epidemiology, clinical presentation, and treatment of MDD are significantly influenced by sociocultural and economic factors (Kleinman 2004; Compton et al. 2006; Zhong et al. 2018).

In the 1950s, imipramine and iproniazid were used, for the first time, for the pharmacotherapy of MDD, thus opening the history of antidepressant treatment (López-Muñoz and Alamo 2009). Consequently, these two agents contributed to the formulation of the first mechanistic hypothesis of depressive disorders, the monoaminergic theory. In the 1960s, the catecholamine hypothesis was further advanced, with the role of noradrenergic imbalance suggested to be the key pathological factor in the occurrence of MDD (Schildkraut 1965). Serotonin was at the focus of these hypotheses, which resulted in the invention of fluoxetine, the first selective serotonin reuptake inhibitor (SSRI). In recent decades, pathophysiological research of depression allowed pharmacological treatments to target specific neurotransmitters as well as the specific brain regions and types of MDD (Kutzer et al. 2020). For example, ketamine and esketamine are proposed to treat resistant depression (Abdallah et al. 2015; Stenovec et al. 2020), while brexanolone, a natural steroid, is the first Food and Drug Administration (FDA)-approved therapeutic medication of postpartum depression (Meltzer-Brody et al. 2018).

Despite considerable advances in studying MDD pathophysiology in recent decades, and an abundance of pharmacological agents approved for MDD treatment, a substantial number of MDD patients remains resistant to available antidepressants. A real-life sequential treatment study demonstrates that nearly 30% of MDD patients fail to achieve remission, even after multiple treatment attempts (Rush et al. 2004, 2006). This may reflect the fact that existing theories cannot fully explain pathological mechanisms of MDD; similarly, pharmacological mechanisms of action of available antidepressants are in need of further research. Overreliance on the monoaminergic hypothesis may stipulate limited success of antidepressants (Ceskova 2016; Ceskova and Silhan 2018).

Growing evidence supports the key role of astrocytes in the pathophysiology and pharmacology of MDD. Astrocytes are homeostatic cells of the central nervous system (CNS). They control multiple aspects of brain physiology at all levels of organization from molecular to organ and systemic (Verkhratsky and Butt 2013; Verkhratsky and Parpura 2015; Verkhratsky and Nedergaard 2018). An individual protoplasmic astrocyte interacts with as many as 100,000 synapses in mice and possibly up to 2,000,000 synapses in humans (Bushong et al. 2002; Oberheim et al. 2009); these astrocytic perisynaptic structures, known as synaptic cradle (Verkhratsky and Nedergaard 2014), are fundamental for maintaining neurotransmission in the CNS. Pathological changes in astrocytes, classified into astrogliosis, astrodegeneration, or atrophy with loss of function, are salient features of all neurological disorders (Pekny et al. 2016; Verkhratsky et al. 2017) including neurodevelopmental (Zeidán-Chuliá et al. 2014) and neuropsychiatric (Verkhratsky et al. 2014; Peng et al. 2015, 2016) disorders. Based on reports of decreased astroglial numbers in postmortem histopathological studies of MDD patients, the dysfunction of astrocytes was postulated to contribute to the pathophysiology of depression (Cotter et al. 2001; Rajkowska et al. 2007; Hercher et al. 2009; Verkhratsky et al. 2014).

Expression of multiple receptors and transporters makes astrocytes potential targets for specific therapy, and indeed, astrocytes respond to many antidepressants (Hertz et al. 2014; Saura et al. 1992; Zhang et al. 2010). Astrocytes possess serotonin receptor subtypes associated with the pharmacology and pathophysiology of depression (Hertz et al. 2010, 2015; Ding et al. 2013; Kirischuk et al. 1996). In this chapter, we focus on astrocytic serotonin receptors and discuss their relation to MDD.

Serotonin Receptors in the Treatments of MDD

A multitude of neurotransmitter receptors is expressed in astrocytes, including ionotropic and metabotropic glutamate receptors, purinoceptors, α/β-adrenoceptors, and serotonin receptors (Hertz et al. 2015), all associated with the intracellular excitability (Verkhratsky and Parpura 2010; Verkhratsky et al. 2020, 2021). Astrocytic serotonin receptors are of particular importance for MDD because of the widespread application of SSRIs in clinical practice. Stimulation of serotonin receptors in astrocytes activates cytosolic and nuclear signaling pathways, altering cellular functions as well as gene expression (Peng et al. 2015).

There are seven subtypes of serotonin receptors designated as 5-$HT_{1A\text{-}1F}$, 5-$HT_{2A\text{-}2C}$, 5-HT_3, 5-HT_4, 5-HT_{5A}, 5-HT_6, and 5-HT_7. Among these subtypes, 5-$HT_{1A\text{-}1F}$, 5-$HT_{2A\text{-}2C}$, 5-HT_6, and 5-HT_7 receptors were discovered in cerebral or spinal astrocytes in humans and rodents (Hirst et al. 1997, 1998; Meuser et al. 2002; Merzak et al. 1996; Li et al. 2012). These serotonin receptors can be involved in the pathophysiology of MDD through numerous cellular signaling pathways. All astrocytic serotonin receptors are G-coupled protein receptors being linked to phospholipase C and inositol-1,4,5-trisphosphate ($InsP_3$)-dependent Ca^{2+} signaling (Verkhratsky et al. 2012).

5-HT_1 Receptors

The 5-HT_{1A} and 5-HT_{1B} receptors attract particular attention because of their involvement in the pathophysiology of MDD; these receptors are also pharmacological targets in the clinical therapeutics of depression (Moret and Briley 2000; Murrough et al. 2011; Murrough and Neumeister 2011; Ruf and Bhagwagar 2009; Sari 2004; Savitz et al. 2009; Tiger et al. 2014). The 5-HT_{1A} and 5-HT_{1B} receptors display 43% amino acid sequence homology, and both are the $G_{i/o}$

protein-coupled receptors (Hoyer et al. 2002). These receptor subtypes have different cellular localizations, with 5-HT_{1A} receptors mainly expressed in somata and dendrites (Sotelo et al. 1990) and 5-HT_{1B} receptors being primarily localized in axon terminals (Boschert et al. 1994). Furthermore, 5-HT_{1A} receptor is one of the most abundant in cortical regions implicated in mood and emotion (Albert and François 2010; Albert and Lemonde 2004). The 5-HT_{1B} receptors are widely distributed throughout the brain regions, with particularly dense expression in the substantia nigra and globus pallidus (Bonaventure et al. 1997; Varnäs et al. 2011).

The 5-HT_{1A} receptors act as autoreceptors and heteroreceptors. The presynaptic 5-HT_{1A} receptors are mainly the autoreceptors localized in the cell bodies and dendrites of serotonergic neurones in dorsal raphe nuclei. These 5-HT_{1A} receptors modulate the release of serotonin and cell discharge rates. Stimulation of 5-HT_{1A} autoreceptors decreases the firing activity of central serotonergic neurones and synthesis and release of serotonin by a negative feedback mechanism while promoting the depressive-like behaviors and the resistance responses to antidepressants, thus inhibiting their activity and having opposite results (Richardson-Jones et al. 2010, 2011; Milak et al. 2018). The postsynaptic 5-HT_{1A} receptors are heteroreceptors, localized in glutamatergic and γ-aminobutyric acid (GABA)-ergic neurones; these receptors show substantial expression in the hippocampus, medial septum, prefrontal cortex, and midbrain ventral tegmental regions where they regulate the release of acetylcholine, glutamate, and dopamine (Comley et al. 2015; Borroto-Escuela et al. 2018; López-Gil et al. 2008; Jeltsch-David et al. 2010; Di Matteo et al. 2008). The 5-HT_{1A} heteroreceptors play key roles in desensitization of postsynaptic 5-HT_{1A} receptors, especially in prefrontal cortex regions (Taciak et al. 2018; Naumenko et al. 2014). Inhibition of 5-HT_{1A} presynaptic autoreceptors or stimulation of 5-HT_{1A} postsynaptic heteroreceptors may produce antidepressant effects, and hence 5-HT_{1A} receptor antagonists or agonists may act as antidepressants or antidepressants potentiators, respectively (Richardson-Jones et al. 2011; Philippe et al. 2018).

There are indeed novel antidepressants based on 5-HT_{1A} receptor agonism; these include YL-0919, brexpiprazole, vilazodone, and vortioxetine hydrobromide. Some of these 5-HT_{1A} receptor agonists also act as SSRIs and as agonists or antagonists of other serotonin subunits. For example, vortioxetine hydrobromide is the agonist of 5-HT_{1A} receptor and SSRI; it is a partial agonist of 5-HT_{1B} receptors and antagonist of 5-HT_3 and 5-HT_7 receptors (Wang et al. 2019). Compound MIN-117, an antagonist of 5-HT_{1A} autoreceptors, is considered for adjunctive therapy to improve acute antidepressant effects of SSRIs (Artigas et al. 2001; Sahli et al. 2016).

Similar to 5-HT_{1A} receptors, 5-HT_{1B} receptors are classified into autoreceptors and heteroreceptors. In postmortem tissue from the suicide MDD subjects, the level of 5-HT_{1B} receptors is decreased in the hippocampus; the mRNA of 5-HT_{1B} receptors is also lower in the frontal cortex. Conversely, mRNA of 5-HT_{1B} receptors is higher in the hypothalamus, compared with a healthy group (Anisman et al. 2008). Serotonin binding to the 5-HT_{1B} receptors in serotonergic neurones decreases the formation of cAMP and the downstream signaling responses, which results in the reduction of transmitter release (Barnes and Sharp 1999; Leenders and Sheng 2005;

Middlemiss and Hutson 1990). Furthermore, 5-HT$_{1B}$ receptors could modulate serotonin transporters (SERT), and thus 5-HT$_{1B}$ autoreceptors can regulate the serotonin release from serotonergic neurones (Hagan et al. 2012; Montañez et al. 2013). An agonist of 5-HT$_{1B}$ receptor CP-93129 decreases 5-HT release in the hippocampus in rats (Hjorth and Tao 1991). In the hippocampus and cortex of wild-type mice, an agonist of 5-HT$_{1B}$ receptor decreases the level of serotonin, while an antagonist of this receptor has the opposite effect on the serotonin level. Both effects are diminished in the transgenic mice with genetically deleted 5-HT$_{1B}$ receptors (Rutz et al. 2006). In addition, 5-HT$_{1B}$ receptors localized in non-serotonergic neurones can modulate the release of glutamate, GABA, acetylcholine, and dopamine (Ruf and Bhagwagar 2009).

Serotonin release is suppressed by 5-HT$_{1B}$ receptor antagonists SB-616234-A (Dawson et al. 2006), AZD3783 (Zhang et al. 2011), and AR-A000002 (Hudzik et al. 2003), which makes them candidates for antidepressive therapeutics. Stimulation of 5-HT$_1$ receptors may neutralize the positive effects of SSRIs on serotonin levels to produce the therapeutic latency (Blier and de Montigny 1994; Nutt 2002). Therefore, the antagonists of 5-HT$_{1B}$ receptors may act as rapid antidepressants. Available 5-HT$_{1B}$ receptor antagonists are not sufficiently effective as antidepressants but can be used as potential adjuvants (Ruf and Bhagwagar 2009). As a result, the combination of SSRIs with 5-HT$_{1B}$ receptor antagonists represents a new therapeutic scheme for MDD (Matzen et al. 2000).

5-HT$_2$ Receptors

The 5-HT$_{2A}$, 5-HT$_{2B}$, and 5-HT$_{2C}$ receptors are G$_q$ protein-coupled receptors, which activate phospholipase C (PLC) or protein kinase C and trigger rapid InsP$_3$-dependent Ca^{2+} signaling or stimulate phospholipase A2 (PLA2) thus increasing arachidonic acid release (Masson et al. 2012; Li et al. 2008). The link between 5-HT$_{2A}$ receptors and astrocytes emerged in 1999 when 5-HT$_{2A}$ receptors were found to be significantly upregulated in reactive astrocytes (Wu et al. 1999). Trazodone, a triazolopyridine derivative, could play a neuroprotective role by inhibiting 5-HT$_{2A/2C}$ receptors in human reactive astrocytes, while fluoxetine also has a similar neuroprotective effects by blocking the same receptors (Daniele et al. 2015). Antagonists of 5-HT$_{2A}$ receptors such as EMD281014, FG5893a, and M100907 suppress depressive-like behavior as reported by forced swim test, the classical rodent behavioral test for depressive phenotype (Patel et al. 2004; Albinsson et al. 1994; Marek et al. 2005). Antagonist of 5-HT$_{2A}$ receptors promotes noradrenaline release under SSRI treatment (Dremencov et al. 2007), suggesting a feedback loop between 5-HT$_{1A}$ receptors in glutamatergic neurones and 5-HT$_{2A}$ receptors in GABAergic neurones. Such interaction may enhance the antidepressant activity of SSRIs and noradrenaline reuptake inhibitor (SNRIs) YM992, which is a potent 5-HT$_{2A}$ antagonist (Szabo and Blier 2002). Some recent studies show that functional disruption of the 5-HT_{2A} receptors may be related to postpartum depression disorder or psychotic disorders (Gao et al. 2018).

Even though 5-HT_{2A} receptors are widely researched with regard to the pharmacological mechanism of antidepressant action, most research focusses on neurones (Szabo and Blier 2001a, b). However, fluoxetine does not affect the mRNA expression of 5-HT_{2A} receptors in neurones or astrocytes, isolated and sorted from transgenic mice treated with chronic unpredictable mild stress (Li et al. 2012). In contrast, astrocytic 5-HT_{2B} receptors have shown significant changes.

Fluoxetine increases mRNA expression of 5-HT_{2B} receptors decreased by the treatment with chronic stress solely in astrocytes, and not in neurones (Dong et al. 2015; Li et al. 2012). Similarly, in the Parkinson's disease model, 5-HT_{2B} receptors are downregulated in astrocytes, but not in neurones; this downregulation develops in parallel with depressive-like behaviors (Diaz et al. 2016). The 5-HT_{2B} receptor is widely distributed in mammalian brains, including the frontal cortex, septal nuclei, dorsal hypothalamus, and medial amygdala (Duxon et al. 1997; Kursar et al. 1994; Li et al. 2012; Dong et al. 2015). In the primary cultured astrocytes or in the fluorescence-activated cell sorted astrocytes from *hgfap*::EGFP transgenic mice (expressing enhanced green fluorescence protein driven by a fragment of the human glial fibrillary acidic protein promoter), 5-HT_{2B} receptors are co-expressed with 5-HT_{2A} and 5-HT_{2C} receptors (Li et al. 2012; Hertz et al. 2010). However, astrocytic 5-HT_{2B} receptor has a much higher affinity for serotonin than the other two 5-HT_2 receptors subtypes (Li et al. 2010). Stimulation of 5-HT_{2B} receptors with 1 nM serotonin triggers phosphorylation of extracellular signal-regulated kinase 1/2 (ERK1/2), while 5-HT_{2C} receptor-mediated ERK1/2 phosphorylation occurs at 100 nM, and 5-HT_{2A} receptors are activated by serotonin in concentrations above 10μM (Li et al. 2010). All SSRIs including fluoxetine act as agonists of astrocytic 5-HT_{2B} receptors suggesting the role of these receptors in antidepressive action (Zhang et al. 2010; Hertz et al. 2012; Li et al. 2009a, 2012, 2017).

Astrocytic 5-HT_{2B} receptors treated with various concentrations of fluoxetine trigger different signaling pathways to regulate gene expressions. Acute treatment with fluoxetine at concentrations lower than 1μM decreases mRNA expression of c-Fos through phosphoinositide 3 kinase(PI3K)-protein kinase B (also known as AKT) signaling pathway; at higher concentrations (more than 5μM), however, fluoxetine elevates this gene expression by mitogen-activated protein kinase (MAPK)-ERK signaling pathway (Fig. 1; Li et al. 2017). Stimulation of 5-HT_{2B} receptors by fluoxetine in astrocytes leads to the transactivation of epidermal growth factor receptors (EGFR); this in turn activates downstream MAPK/ERK or PI3K/AKT signaling cascades thus regulating expressions of genes possibly related to mood disorders. These genes include Ca^{2+}-dependent phospholipase A2 (cPLA2), subtype 2 of adenosine deaminases acting on RNA's (ADAR2), or subtype 2 of kainate receptors (GluK2) (Li et al. 2009b, 2011, 2012; Rao et al. 2006). Chronic treatment with fluoxetine also induces RNA editing of ADAR2, resulting in the loss of function of 5-HT_{2B} receptors, although expression of these receptors is increased (Peng et al. 2014). Rapid increase in $[Ca^{2+}]_i$ triggered by an agonist of 5-HT_{2B} receptors was suppressed by chronic treatment with fluoxetine (Li et al. 2011; Peng

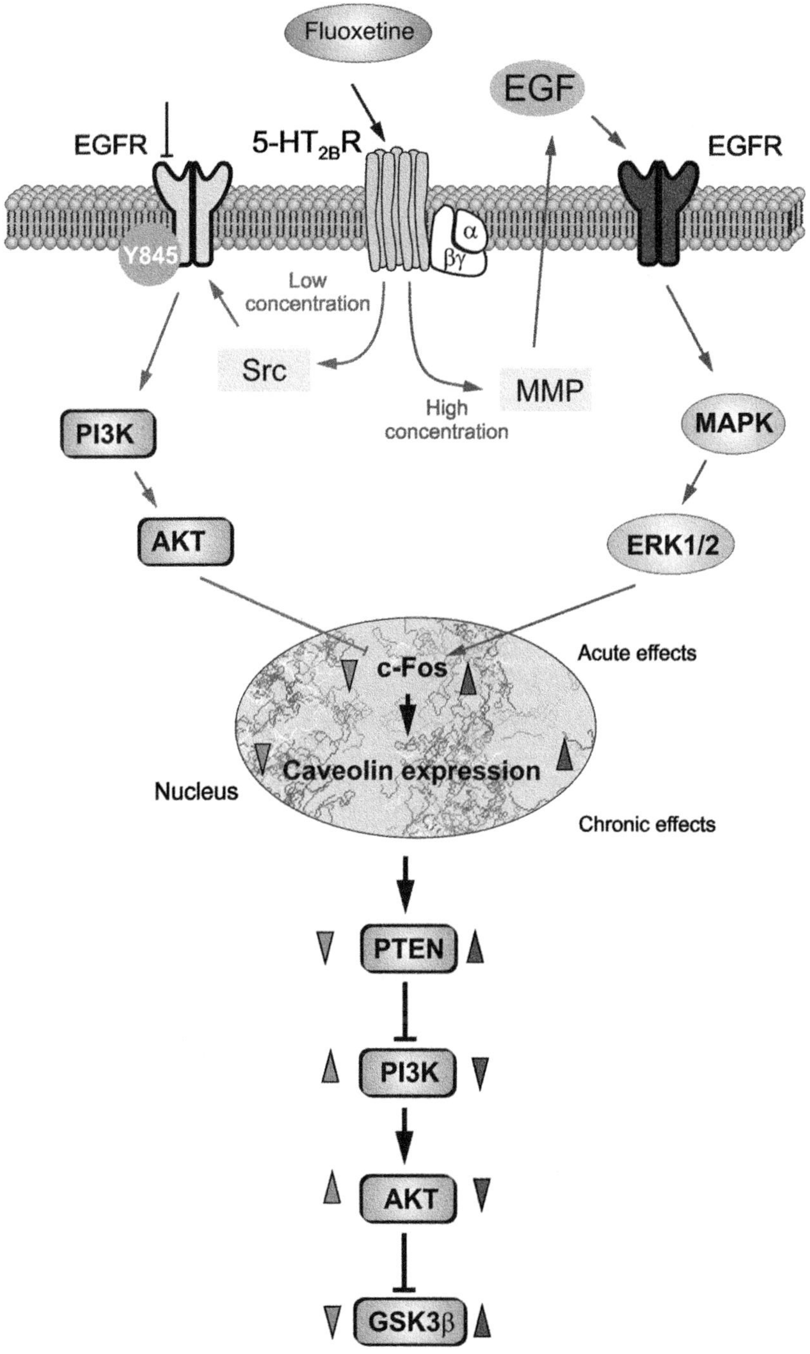

Fig. 1 Schematic illustration of biphasic concentration-dependent regulation of Cav-1 gene expression and GSK-3β activity by fluoxetine in astrocytes. Fluoxetine is an agonist to serotonin 2B receptors (5-HT_{2B}R). Acute treatment with fluoxetine at low concentrations (green arrows)

et al. 2014). Malfunction of 5-HT_{2B} receptors has been discovered in chronic stress animal models of depression and in animals with depressive-like behaviors associated with Parkinson's disease or hyperammonemia (Li et al. 2012; Zhang et al. 2015; Yue et al. 2019). This evidence indicates that 5-HT_{2B} receptors are a target for SSRIs, and it is indicative of the potential contribution of astrocytic 5-HT_{2B} receptors to the pathophysiology of MDD (Kennett et al. 1996, 1998). Agonist of 5-HT_{2B} receptors BW723C86 displays anxiolytic-like effects, whereas 5-$HT_{2C/2B}$ receptor antagonist SB200646A improves depressive-like behaviors (Kennett et al. 1995).

The 5-HT_{2C} receptors are upregulated by fluoxetine specifically in neurones sorted from *Thy1*::YFP transgenic mice, expressing yellow fluorescence protein driven by Thy1 promoter (Li et al. 2012). Antagonist of 5-HT_{2C} receptor S32006 demonstrates antidepressive effects and increases the level of dopamine and noradrenaline in the frontal cortex in rats (Dekeyne et al. 2008). An inverse agonist of 5-HT_{2C} receptor, S32212, also has antidepressant activity in behavioral tests (Dekeyne et al. 2012). Fluoxetine is reported to act as an antagonist of 5-HT_{2C} receptors (Ni and Miledi 1997), while chronic treatment with SSRIs may promote receptor desensitization (Yamauchi et al. 2004). Mianserin and several classical tricyclic antidepressants (TCAs), such as amitriptyline, imipramine, and clomipramine, all work as antagonists with moderate to high affinity at 5-HT_{2C} receptors (Jenck et al. 1993; Millan 2005). SNRIs desipramine and maprotiline as well as atypical antidepressants mirtazapine, trazodone, and nefazodone similarly act as 5-HT_{2C} antagonists (Ni and Miledi 1997).

Increased density of 5-HT_{2C} receptors was found in the frontal cortex of subjects who committed suicide (Pandey et al. 2006). This increase may reflect changes in mRNA alternative splicing and/or RNA editing (Martin et al. 2013), which has been proposed as a risk factor for suicidal attempts (Dracheva et al. 2008). Clinical studies similarly indicated increases in mRNA editing of 5-HT_{2C} receptors in the frontal cortex of depressed suicide individuals (Niswender et al. 2001; Gurevich et al. 2002; Iwamoto and Kato 2003; Schmauss 2003). A recent study demonstrated that astrocytic 5-HT_{2C} receptors are involved in the depressive- and anxious-like behaviors induced by alcohol addiction. Fluoxetine at low concentration (1μM) can alleviate these anxiety-like behaviors and motor activity induced by alcohol by

Fig. 1 (continued) stimulates Src which phosphorylates epidermal growth factor (EGF) receptors (EGFR) at Y845 residue and in turn activates phosphoinositide 3 kinase (PI3K)-AKT signaling pathway. The AKT phosphorylation by fluoxetine at low concentrations inhibits c-Fos gene expression and subsequently decreases caveolin-1 (Cav-1) gene expression (chronic effects) that in turn decreases membrane content of PTEN (phosphatase and tensin homolog), induces phosphorylation and stimulation of PI3K, and elevates glycogen synthase kinase 3 (GSK-3β) phosphorylation thus suppressing its activity. At higher concentrations, fluoxetine (red arrows) stimulates metalloproteinase (MMP) and induces shedding of growth factor which stimulates EGFR and activates mitogen-activated protein kinase (MAPK) extracellular signal-regulated kinases 1/2 (ERK1/2) signal pathway. The ERK1/2 phosphorylation by fluoxetine at high concentrations stimulates cFos gene expression and subsequently increases Cav-1 gene expression (chronic effects), which acts on PTEN/PI3K/AKT/GSK-3β in an inverse fashion. (From Li et al. 2017) suppressing the expression of ADAR2 and mRNA editing of 5-HT_{2C} receptors (Li et al. 2020).

The study in chronic social defeat animal model of depression reported increased sensitivity of 5-HT_{2C} receptors that can be reversed by chronic antidepressants (Berton et al. 2006; Rygula et al. 2005, 2006). Pretreatment with 5-HT_{2C} receptor agonist m-chlorophenylpiperazine increases behavioral presentations of defeat in hamsters (Harvey et al. 2012). In rats, however, chronic stress increases depressive-like behavior, this effect being arguably mediated through 5-HT_{2C} receptors (Kimura et al. 2008; Moreau et al. 1993). Suppression of dopamine release mediated by 5-HT_{2C} receptors has been observed in rat models of depression (Dremencov et al. 2005; Oba et al. 2013). Antagonists of 5-HT_{2C} receptors can alleviate stress-induced depressive-like behaviors in mice, while their antidepressive effects may relate to dopaminergic neurones in the ventral tegmental area (Opal et al. 2013). Chronic treatment with 5-HT_{2C} receptor agonist RO60-0175 alleviates anhedonia in stressed mice by desensitization of the receptor (Moreau et al. 1993). Depressive-like behaviors in the learned helplessness model can be improved by selective antagonists of 5-HT_{2C} receptors (Strong et al. 2009). Likewise, clomipramine, a potential antagonist of 5-HT_{2C} receptors, decreases elevated mRNA expression of 5-HT_{2C} receptors in chronic stress treated rats (Pitychoutis et al. 2012).

All these observations suggest that the dysfunctions of 5-HT_2 receptors may be a consequence of stress and can be linked to anxiety, depression, and suicide. Antidepressive effects of agonists and antagonists of 5-HT_2 receptors remain controversial and poorly understood, thus warranting more research.

Other Serotonin Receptors

The 5-HT_3 is the only ionotropic member of the serotonin receptors family. This 5HT-gated channel is assembled from five subunits, $5HT_{3A}$, $5HT_{3B}$, $5HT_{3D}$, $5HT_{3E}$, and $5HT_{3F}$, expression of which was confirmed in humans (Niesler et al. 2007). In rodents, however, only $5HT_{3A}$ and $5HT_{3B}$ subunits have been identified (Karnovsky et al. 2003). Ondansetron and zacopride, antagonists of $5HT_3$ receptors, have been used in clinical therapeutics of depression, and both reversed the depressive-like behaviors in the rat learned helplessness model (Martin et al. 1992; Thiebot and Martin 1991). Antagonists of $5HT_3$ receptors potentiate the pharmacological efficiency of clinical antidepressants (Nakagawa et al. 1998; Redrobe and Bourin 1997). Vortioxetine is a multimodal drug which binds to 5-HT_{1A}, 5-HT_{1B}, 5-HT_{1D}, and 5-HT_7 receptors and to SERT; it also acts as a potent antagonist of $5HT_3$ receptors. Vortioxetine demonstrated antidepressive effects in rodent models of depression (Guilloux et al. 2013; Mork et al. 2012; Wallace et al. 2014). Furthermore, preclinical data also show that vortioxetine as an antagonist of $5HT_3$ receptors produced antidepressant response faster than classical antidepressants (Bétry et al. 2013). In addition, vortioxetine demonstrates antidepressive performance in patients resistant to SSRIs or SNRIs (Alvarez et al. 2014). $5HT_3$ receptor antagonists may be promising therapeutic agents, although certain unresolved issues remain, including those of heterogeneity of the species used and of the inconsistent data on effective doses as antidepressants (Adrien et al. 1992; Costall et al. 1987; Johnson et al. 2002, 2006; Martin et al. 1992; Ramamoorthy et al. 2008).

$5\text{-}HT_4$ receptors are widely distributed in limbic regions, such as the amygdala and hippocampus (Hannon and Hoyer 2008; Tanaka et al. 2012). Being G_s-coupled, $5\text{-}HT_4$ receptors increase intracellular cAMP by stimulating adenylyl cyclase (Dumuis et al. 1988; Fagni et al. 1992). Decrease in $5\text{-}HT_4$ receptors was observed in the striatum of depressed patients and in mice models (Amigó et al. 2016; Madsen et al. 2014). In postmortem studies of depressed suicide victims, various binding forms of $5\text{-}HT_4$ receptors and the different levels of cAMP in distinct brain areas have been reported (Rosel et al. 2004). A selective partial agonist of $5\text{-}HT_4$ receptors, RS67333, has a rapid antidepressant effect in chronic depression models of rodents (Lucas et al. 2007).

In MDD context, studies of $5\text{-}HT_5$, $5\text{-}HT_6$, and $5\text{-}HT_7$ receptors are scarce. Administration of N-methyl-D-aspartate receptor antagonist ketamine, at a dose that leads to drug abuse, can trigger schizophrenic-like symptoms in animals (de la Salle et al. 2016). In these ketamine pretreated mice, an agonist of $5\text{-}HT_6$ receptors E-6837 as well as an antagonist of these receptors SB-271046 can both produce depressive-like behaviors (Suárez-Santiago et al. 2017). Antagonists of $5\text{-}HT_7$ receptors have faster antidepressant effects when compared to fluoxetine in the rat model of depression (Mnie-Filali et al. 2011).

To summarize, there are many uncertainties and controversies concerning MDD-related pharmacology of serotonin receptors. In general, agonists of $5\text{-}HT_{1A}$, $5\text{-}HT_{2B}$, and $5\text{-}HT_4$ receptors and antagonists of $5\text{-}HT_{1B}$, $5\text{-}HT_{2A}$, $5\text{-}HT_{2C}$, $5\text{-}HT_3$, $5\text{-}HT_6$, and $5\text{-}HT_7$ act as antidepressants. None of these drugs, however, can substantially affect clinical manifestations of depression.

Astrocytes as a Target for Classic Antidepressants

Classic antidepressants are SSRIs, SNRIs, TCAs, inhibitors of monoamine oxidase (MAO), and lithium. Major theories of MDD pathophysiology focus on neurones, with particular emphasis on monoamines (Marathe et al. 2018), aberrant signaling of brain-derived neurotrophic factor (BDNF), leptin dysregulation (Milaneschi et al. 2017), *myo*-inositol abnormalities (Yu and Greenberg 2016), and glutamate homeostasis failure (Shirayama et al. 2017). Here, we discuss the MDD from the astrocytic angle and portray astrocytes as targets for therapy.

SSRIs

SSRIs, which include fluoxetine, fluvoxamine, sertraline, paroxetine, and citalopram, are the most popular antidepressants and anxiolytics. An increase in extracellular serotonin availability following SERT inhibition has been regarded as the main and the only pharmacological mechanism of SSRIs, despite the well-known delay in clinical effect contrasting to the immediate inhibition of SERT. Recently, the contribution of astrocytes to MDD has gained increased attention. Postmortem histological analysis of the frontal cortex reported a reduced number of glia in patients suffering from MDD (Cotter et al. 2001, 2002). Decreased density of neuroglia and decrease in glia-to-neurone ratio have also been observed in the amygdala (Bowley et al. 2002; Hamidi et al. 2004). Experimental ablation of astrocytes (but not neurones) in the frontal cortex is sufficient to initiate depressive-like behavior in rats (Banasr and Duman 2008). Similarly, loss of astrocyte-derived ATP induces depressive phenotypes in rodents (Cao et al. 2013). Fluoxetine reverses the loss of hippocampal astrocytes in an animal model of depression (Czéh et al. 2006), while increasing the release of ATP from astrocytes by regulating vesicular nucleotide transporter (Kinoshita et al. 2018).

Fluoxetine acts as an agonist of astrocytic 5-HT_{2B} receptors. However, acute and chronic treatments of astrocytes with fluoxetine produce different outcomes. Acute treatment with 10μM fluoxetine triggers Ca^{2+} signals, transactivates the EGFR, and phosphorylates downstream ERK1/2 signaling cascade; ERK1/2 enters nuclei and regulates the mRNA and protein expression of *c-Fos* and *FosB*, consequently modulating the expression of several key proteins (Li et al. 2008). For instance, fluoxetine, by stimulating 5-HT_{2B} receptors, increases protein expression of cPLA2, GluK2, ADAR2, transient receptor potential canonical 1 channels, L-type $Ca_v1.2$ Ca^{2+} channels, caveolin-1, and BDNF (Li et al. 2009a, 2011, 2017, 2019; Hertz et al. 2015; Peng et al. 2018). Increased expression of ADAR2 in astrocytes following chronic treatment with fluoxetine catalyzes RNA editing, promoting loss of function of 5-HT_{2B} receptors and GluK2, which affects Ca^{2+} signaling and phosphorylation of ERK1/2 (Li et al. 2011; Hertz et al. 2012). Decreased expression of astrocytic 5-HT_{2B} receptors was also observed in chronic stress models and Parkinson's disease (Dong et al. 2015; Zhang et al. 2015). Fluoxetine can selectively upregulate the expression of 5-HT_{2B} receptors in astrocytes (Li et al. 2012). However, PI3K/AKT and MAPK/ERK signaling pathways induced by low and high concentrations of fluoxetine result in the opposite regulation of caveolin-1 and glycogen synthase kinase-3β (GSK-3β) (Fig. 1) (Li et al. 2017). Some effects of fluoxetine mediated by 5-HT_{2B} receptors may contribute to both therapeutic and adverse effects of this drug; hence, selective stimulation or inhibition of various signaling pathways triggered by fluoxetine may promote the efficacy or lessen the side effects of antidepressants.

Depressive disorders are associated with many neurological conditions including sleep disturbances, Alzheimer's, and Parkinson's diseases (Xia et al. 2018; Orgeta et al. 2017; Huang et al. 2019). Aging of astrocytes (Verkhratsky et al. 2020, 2021) may contribute to age-dependent diseases as well as to depressive phenotypes. Sleep disturbance in particular is linked to both depression and neurodegeneration. Insomnia and depression can easily combine into a vicious cycle, as decreased sleep time increases the risk for MDD, which further exacerbates sleep disorders (Roberts and Duong 2014). The glymphatic system (Iliff et al. 2012) is responsible for the clearance of brain metabolite waste via paravascular pathways (Xie et al. 2013). Astrocytic endfeet forming *glia limitans vascularis* and endfeet-polarized expression of aquaporin 4 (AQP4) support the glymphatic system. In aging, astroglial atrophy decreases glymphatic clearance facilitating extracellular accumulation of disease-related proteins, such as β-amyloid (Kress et al. 2014). Increased accumulation of β-amyloid has been reported by positron emission tomography in the right hippocampus and thalamus of human subjects having their moods affected by sleep deprivation (Shokri-Kojori et al. 2018). In animals, chronic stress increases β-amyloid$_{42}$ in the frontal cortex and hippocampus. Fluoxetine upregulates the expression of AQP4 and promotes clearance of β-amyloid$_{42}$ by improving glymphatic clearance; this also rescues depressive-like behaviors (Xia et al. 2017). Toxic load with iron also impairs the glymphatic system, thus exacerbating depressive-like phenotypes and increasing neuronal apoptosis (Fig. 2; Liang et al. 2020).

Growing evidence indicates a strong link between sleep deprivation and activation of nucleotide-binding domain and leucine-rich repeat protein-3 (NLRP3) inflammasome; this activation is mediated by astrocytic $P2X_7$ receptors in astrocytes, and it facilitates neuronal apoptosis (Xia et al. 2018; Zielinski et al. 2017). Chronic sleep deprivation selectively downregulates expression of astrocytic 5-HT_{2B} receptors, the effect of which is similarly mediated by $P2X_7$ receptors (Fig. 3; Xia et al. 2020). These changes result in depressive-like behaviors associated with the gradual elevation of extracellular ATP (Xia et al. 2020). Fluoxetine increases the phosphorylation of signal transducer and activator of transcription 3, thus suppressing astrocytic NLRP3 activation and neuronal apoptosis (Xia et al. 2018; Zielinski et al. 2017). Inhibition of astrocytic NLRP3 inflammasome by fluoxetine is also mediated by 5-HT_{2B} receptors (Li et al. 2019). In addition, fluoxetine, again acting through 5-HT_{2B} receptors, recovers BDNF levels suppressed by sleep deprivation (Li et al. 2019). Pre-treatment with the 16 kDa hormone leptin increases expression of astrocytic 5-HT_{2B} receptors, thus enhancing positive effects of fluoxetine on BDNF and depressive-like behaviors (Li et al. 2019). Of note, long-term, chronic stress may induce sleep disorders and depression associated with decreased levels of BDNF (Schmitt et al. 2016).

Pathological features and pathogenesis of sleep disturbances, depressive disorders, and neurodegenerative pathology, such as Alzheimer's disease, usually overlap, and all these may be linked to astrocytes. Morphological and functional abnormalities of astrocytes can impair BDNF signaling and the glymphatic system, with failed glymphatic clearance being a key contributor to diseases of cognition (Nedergaard and Goldman 2020), which could further worsen the deposition of brain waste and impair neuronal functions. Fluoxetine and other SSRIs may provide neuroprotection by arresting astrogliopathology through direct stimulation of 5-HT_{2B} receptors.

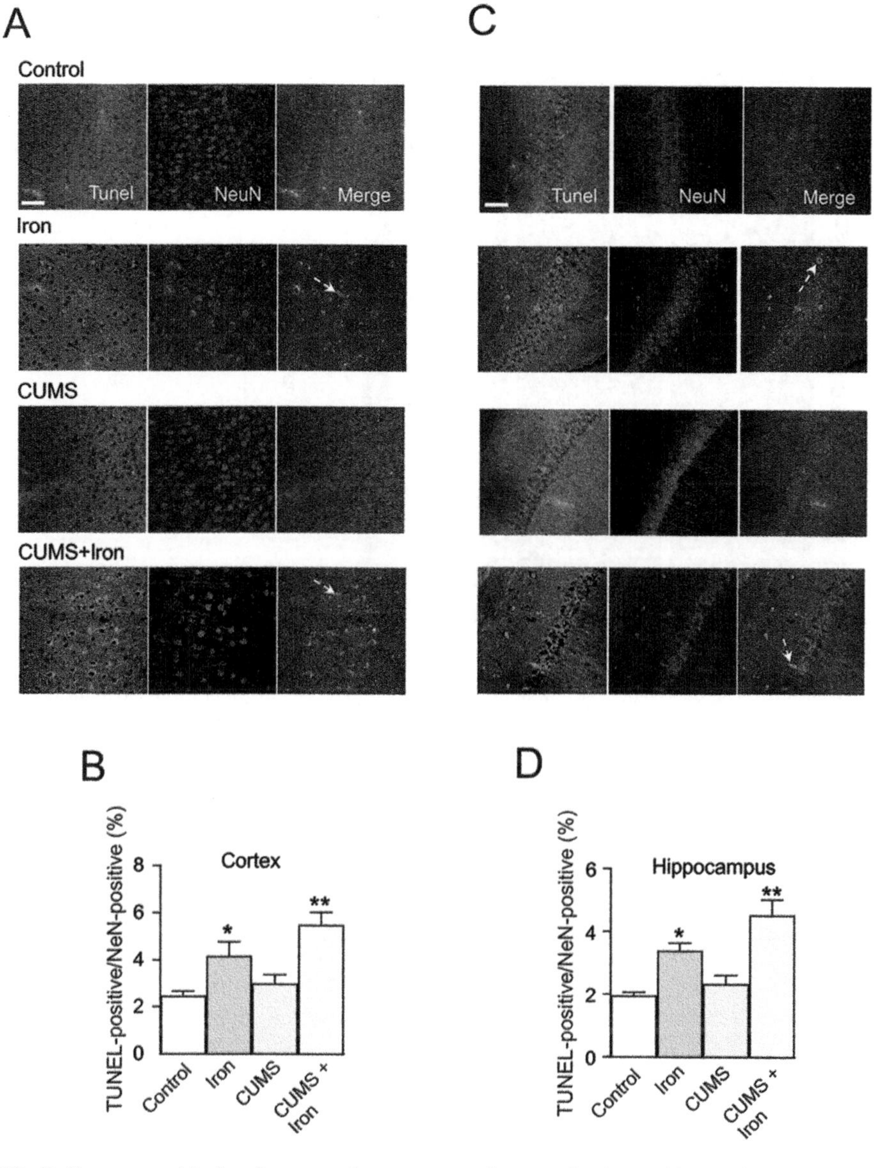

Fig. 2 Treatment with chronic stress enhances neuronal apoptosis triggered by overload with iron. Mice were pretreated without or with chronic unpredictable mild stress (CUMS) for 6 weeks; in week 6, animals were randomly separated to be injected with dextran (CUMS group) or iron dextran (CUMS+Iron group) for 6 days. Apoptosis was detected by TUNEL assay in the cortex (**a**) and hippocampus (**c**), where neuronal nuclei were stained with NeuN (red), while all the cell nuclei were labeled with DAPI (blue in merge). Percentage of cell death was determined by the ratio of TUNEL+ and NeuN+ cells, with the cortical and hippocampal regions analyzed in (**b**, **d**), respectively. Data are presented as mean ± SEM, $n = 6$. $^{*}p < 0.05$, statistically significant difference compared with control, untreated group; $^{**}p < 0.05$, statistically significant difference compared with any other group. (From Liang et al. 2020)

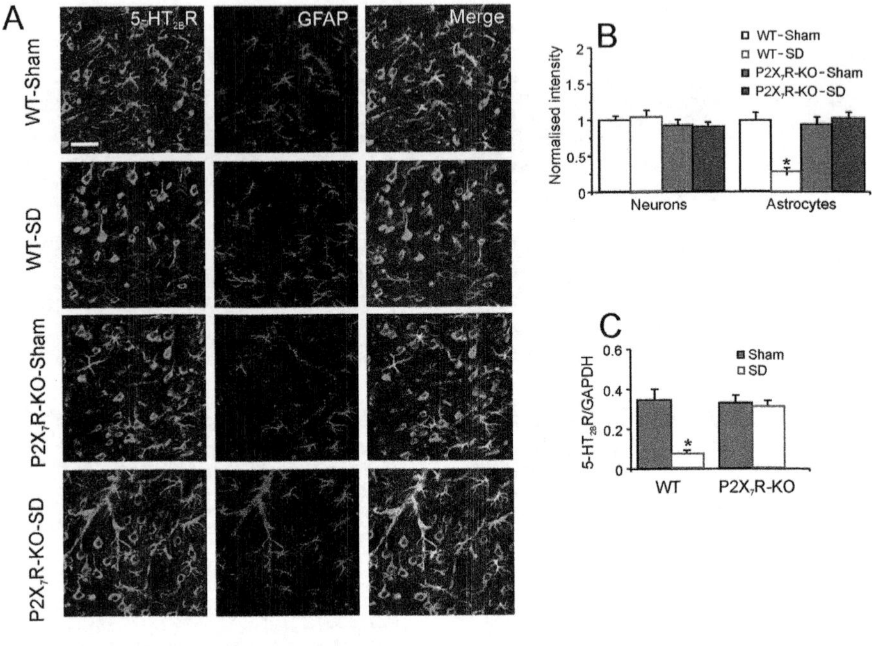

Fig. 3 $P2X_7$ receptors mediate the selective decrease of $5\text{-}HT_{2B}$ receptors in astrocytes treated with chronic sleep deprivation. (**a**) Immunolabelled $5\text{-}HT_{2B}$ receptors (green) were co-stained with glial fibrillary acidic protein (GFAP, red) and NeuN (blue) in the mice frontal cortex treated with sham (Control) or exposed to sleep deprivation for 3 weeks of wild-type (WT) and $P2X_7$ receptor knockout ($P2X_7$R-KO) mice. Scale bar, 20μm. (**b**) $5\text{-}HT_{2B}$ receptors immunolabeling intensity of neurones and astrocytes, respectively, relative to the cell-free parenchyma in the cortex, normalized to the intensity of a matching sham group. (**c**) RT-PCR analysis of $5\text{-}HT_{2B}$ receptors mRNA expression in WT or $P2X_7$R-KO mice treated with sleep deprivation for 3 weeks, expressed as the relative expression ratio of $5\text{-}HT_{2B}$ receptor/GAPDH, the latter used as loading control. Data represent mean ± SEM. *Indicates statistically significant ($p < 0.05$) difference from matching sham groups, $n = 6$. (From Xia et al. 2020)

SNRIs

Serotonin/noradrenaline reuptake inhibitors exert their effects through increasing availability of serotonin and noradrenaline in the nervous tissue, and they are commonly used to treat patients with refractory depression (Wang et al. 2014). Venlafaxine is one of the most commonly prescribed SNRI with a broad range of antidepressant effects. Accumulating evidence shows that venlafaxine may target astrocytes by affecting the metabolism of lysine, tyrosine, glutamate, methionine, ethanolamine, fructose-6-phosphate, and phosphorylethanolamine (Sun et al. 2017). Duloxetine, another SNRI widely used in the treatment of MDD (Nemeroff et al. 2002), was reported to suppress ischemia-induced astrocytic reactivity and oxidative stress (Lee et al. 2016).

TCAs and MAOIs

Astrocytic-derived neurotrophic growth factors receive attention because of their links to the mechanism of action of antidepressants (Tsybko et al. 2017). In primary cultured astrocytes, amitriptyline increases expression of BDNF, vascular endothelial growth factor (VEGF), fibroblast growth factor 2 (FGF2), and glial cell line-derived neurotrophic factor (GDNF) (Kajitani et al. 2012). Furthermore, clomipramine (TCA), fluvoxamine (SSRI), and duloxetine (SNRI) all elevate the expression of FGF2 and BDNF in astrocytes (Hisaoka-Nakashima et al. 2016; Kajitani et al. 2012). Fluoxetine and paroxetine both enhance the mRNA expression of VEGF in primary astrocytes (Allaman et al. 2011).

Mirtazapine, a tetracyclic piperazinoazepine antidepressant, has a distinct pharmacological mechanism associated with inhibition of α_2-adrenoceptors and 5-HT_2 and 5-HT_3 receptors (Benjamin and Doraiswamy 2011; Croom et al. 2009). Unlike amitriptyline and other antidepressants, mirtazapine does not affect noradrenaline or serotonin reuptake (Al-Majed et al. 2018). Moreover, mirtazapine has a faster onset time as compared with SNRIs in the treatment of acute-phase MDD (Nagao et al. 2013; Watanabe et al. 2011). In primary culture of astrocytes, mirtazapine increases the expression of GDNF and BDNF by stimulating phosphorylation of $ERK^{1/2}$ (Hisaoka-Nakashima et al. 2016).

Astrocytes in Treatments of Bipolar Disorders

Therapeutic strategy for bipolar disorders (BD) is distinct from treating MDD, due to reflecting different pathophysiology. Lithium ion (Li^+) in various salts, valproic acid (VPA), and carbamazepine (CBZ) are three classical anti-bipolar drugs. Lithium salt was first to be used for treating mania and depressive episodes of bipolar depression; importantly, lithium may also prevent relapses (Brown and Tracy 2013). Both CBZ and VPA are used for treating manic episodes and epileptic seizures (Nevitt et al. 2019; Xu et al. 2019; Post et al. 2007). These three classical anti-BD drugs have nothing in common in their chemical structures, and only a few metabolic parameters in the brain are affected by them in a similar manner (Li et al. 2007). Revealing shared effects of these drugs is important because common mechanisms of action may provide information about the pathophysiology of BP and facilitate drug development.

Astroglial cell numbers in the cortex from BD patients are reduced (Ongür et al. 1998, 2008; Rajkowska 2000). This coincides with exaggerated activity of the glutamatergic system, as demonstrated by MRI, in the brains of BD patients (Ongür et al. 2008; Michael et al. 2003; Chen et al. 2010; Eastwood and Harrison 2010), which could be due to suppressed capacity of astrocytes to maintain glutamate homeostasis. In astrocytes, riluzole can increase the expression and activity of plasmalemmal glutamate transporters in order to increase the glutamate clearance from the extracellular space (Carbone et al. 2012). Riluzole can also improve the glutamine-glutamate cycle and the glutamatergic transmission in brains of BD patients (Brennan et al. 2010) and even alleviate related disease symptoms (Zarate and Manji 2008), which is also indicative of potential astrocytic glutamatergic dysfunction.

We outline some common targets of these three classical anti-BD drugs on astrocytes. Firstly, the decreased expression of GluK2 was reported in the entorhinal cortex and the hippocampus of BD patients, most of whom had received drug therapy (Beneyto et al. 2007; Benes et al. 2001). In primary cultures of astrocytes treated with Li^+, VPA, or CBZ at clinically relevant doses the mRNA expression of GluK2 receptors decreased, whereas the expression of other subunits of GluK receptors remained unchanged (Li et al. 2009b). In vivo, after 2 weeks of intraperitoneal injection, CBZ reduced the gene expression of GluK2 which is encoded by *GRIK2* (Li et al. 2009b); of note, this gene in a genetic linkage region (6q21) is associated with BD (Schulze et al. 2004). In other studies, CBZ, VPA, or Li^+ suppresses GluK2-mediated phosphorylation of ERK1/2 in primary cultured astrocytes, while anticonvulsant topiramate is ineffective (Li et al. 2009b). Release of glutamate from astrocytes induced by ATP is inhibited by all three anti-BD drugs; this may suppress hyperactivity of the glutamatergic transmission in brains of patients with BD (Liu et al. 2015).

In addition to classical targets, Li^+ can play a key role in the metabolism of inositol. The intracellular myoinositol needs continuous re-supply from the diet (Spector and Lorenzo 1975) and de novo synthesis, which occurs in the vasculature (Wong et al. 1987). In primary culture of mouse astrocytes, chronic treatment with Li^+ at 1 mM inhibits the uptake of inositol, although this effect is absent during acute Li^+ treatment (Wolfson et al. 1998). Changes in myoinositol levels may have opposite effects depending on its concentration. For example, myoinositol at low concentration promotes the uptake of the extracellular inositol, whereas at high concentrations the uptake is inhibited. This may be related to the existence of two inositol transporters in astrocytes, the high-affinity Na^+-dependent myoinositol transporter (SMIT/SLC5A3), which accounts for most of the uptake at low (25μM) inositol concentration, and the lower-affinity H^+-dependent myoinositol transporter (HMIT/SLC2A13), which dominates the uptake at higher (50μM) inositol concentrations (Wolfson et al. 2000). Because the elevated intracellular pH (pH_i) stimulates SMIT but inhibits HMIT, the effects of anti-BD drugs on the inhibition of inositol may be caused by the intracellular alkalinization. This was further confirmed by studying the uptake of [^{3}H]myoinositol in astrocytes at various myoinositol doses or at various pH_i and corroborated by analysis of different myo-inositol transporter operation (Fu et al. 2012). Chronic treatment by any of the three anti-BD drugs increases astrocytic pH_i (Song et al. 2008, 2013). The Na^+–H^+ exchanger (NHE) exchanger (NHE) is one of the acid-controlling transporters, widely distributed in almost all cell types, including neurones and astrocytes (McAlear and Bevensee 2006). The NHE/SLC9 family has seven subtypes, NHE1-NHE7. The NHE1/SLC9A1 is expressed in cultured astrocytes from the rat hippocampus (Pizzonia et al. 1996) and from the mouse cerebral cortex (Song et al. 2008). Chronic treatment with Li^+ at 1 mM alkalinize pH_i by about 0.10, but 0.5 mM is ineffective (Song et al. 2008).

The anti-bipolar effects of lithium are often attributed to the classical hypothesis of the depletion of myoinositol induced by Li^+ (Berridge et al. 1989). According to this hypothesis, Li^+ causes the non-competitive inhibition of inositol-phosphate hydrolysis; $InsP_3$ is produced by PLC and degraded to myoinositol by inositol phosphatases, whereas diacylglycerol is converted to phosphatidate and condensed with cytidine triphosphate to form cytidine mono-phosphoryl-phosphatidate, which together with myoinositol regenerates phosphatidylinositol 4,5-bisphosphate (PIP_2) (Hallcher and Sherman 1980; Inhorn and Majerus 1987). Because Li^+ inhibits the formation of inositol from $InsP_3$, it decreases the regeneration of PIP_2 and blocks PLC pathway. The $[Ca^{2+}]_i$ response to noradrenaline in cultured astrocytes is suppressed by chronic treatments with Li^+ (Chen and Hertz 1996).

As astrocytes are the direct targets of Li^+, identification of the unique astroglial transcriptional networks is of importance. Recently, it was found that Li^+ can induce a specific astrocytic phenotype (Rivera and Butt 2019). By pharmacogenomic analyses, the roles of the extracellular matrix (ECM) regulatory enzyme lysyl oxidase (LOX) and peroxisome proliferator-activated receptor γ (PPAR-γ) were shown to modulate astrocyte morphogenesis. According to genomic analyses, LOX is the most highly regulated Li^+-responsive astroglial gene. Irreversible LOX inhibitor mimics effects of Li^+ suggesting that LOX is a major regulator of astrocyte morphology and proliferation. Treatment with Li^+ also actives PPAR-γ (Liu et al. 2011), while PPAR-γ has been reported to inhibit LOX (Segond et al. 2013). These effects may involve GSK3β-Wnt signaling, which is implicated in neuropsychiatric disorders and interacts with ECM remodeling. In addition, PPAR-γ agonists decrease TGF-β1 signaling leading to decreased fibrosis (Vallee et al. 2017), providing a further link with LOX and the actions of Li^+.

Neuroprotective effects of anti-bipolar drugs have been discovered in other neuronal diseases. Pretreatment with VPA increases expression of plasmalemmal glutamate transporter 1 (GLT-1) in astrocytes and promotes the recovery of locomotor activities in a rodent model (Johnson Jr et al. 2018). Lithium can mediate neuroprotection by inhibition of GSK3β after spinal cord injury (Li et al. 2018). Chronic treatment with Li^+ activates α1 subunit of neuronal Na^+-K^+-ATPase (Li et al. 2018). In contrast, chronic treatment with CBZ increases astrocyte specific α2 subunit containing Na^+/K^+ ATPase (Li et al. 2013).

Conclusion

In this chapter, we introduced the contribution of astrocytic serotonin receptors to the pathophysiology of MDD and to pharmacological mechanisms of classic antidepressants. Direct stimulation of astroglial $5\text{-}HT_{2B}$ receptors may be the common target of SSRIs. Activation of astrocytic serotonin receptors associates with several signaling pathways and regulates expression of numerous genes that are related to depressive-like behaviors. Recent discoveries overviewed in this chapter highlight the key roles of astrocytes in the pathogenesis and treatment of mood disorders.

Acknowledgments BL's work is supported by the National Natural Science Foundation of China (grant number 8187185), LiaoNing Revitalization Talents Program (grant number XLYC1807137), the Scientific Research Foundation for Returned Scholars of Education Ministry of China (grant number 20151098), LiaoNing Thousands Talents Program (grant number 202078), and "ChunHui"

Program of Education Ministry of China (grant number 2020703). CS's work is supported by a grant from the Italian Ministry of Education, University and Research (2015KP7T2Y_002) and a grant from the Sapienza University of Rome (RM11916B7A8D0225). VP's work is supported by a grant from the National Institute of General Medical Sciences of the National Institutes of Health (R01GM123971). VP is an Honorary Professor at the University of Rijeka, Croatia.

References

Abdallah CG, Sanacora G, Duman RS, Krystal JH (2015) Ketamine and rapid-acting antidepressants: a window into a new neurobiology for mood disorder therapeutics. Annu Rev Med 66:509–523

Adrien J, Tissier MH, Lanfumey L, Haj-Dahmane S, Jolas T, Franc B, Hamon M (1992) Central action of 5-HT3 receptor ligands in the regulation of sleep-wakefulness and raphe neuronal activity in the rat. Neuropharmacology 31:519–529

Albert PR, François BL (2010) Modifying 5-HT1A receptor gene expression as a new target for antidepressant therapy. Front Neurosci 4:35

Albert PR, Lemonde S (2004) 5-HT1A receptors, gene repression, and depression: guilt by association. Neuroscientist 10:575–593

Albinsson A, Björk A, Svartengren J, Klint T, Andersson G (1994) Preclinical pharmacology of FG5893: a potential anxiolytic drug with high affinity for both 5-HT1A and 5-HT2A receptors. Eur J Pharmacol 261:285–294

Allaman I, Fiumelli H, Magistretti PJ, Martin JL (2011) Fluoxetine regulates the expression of neurotrophic/growth factors and glucose metabolism in astrocytes. Psychopharmacology 216:75–84

Al-Majed A, Bakheit AH, Alharbi RM, Abdel Aziz HA (2018) Mirtazapine. Profiles Drug Subst Excip Relat Methodol 43:209–254

Alvarez E, Perez V, Artigas F (2014) Pharmacology and clinical potential of vortioxetine in the treatment of major depressive disorder. Neuropsychiatr Dis Treat 10:1297–1307

Amigó J, Díaz A, Pilar-Cuéllar F, Vidal R, Martín A, Compan V, Pazos A, Castro E (2016) The absence of 5-HT4 receptors modulates depression- and anxiety-like responses and influences the response of fluoxetine in olfactory bulbectomised mice: adaptive changes in hippocampal neuroplasticity markers and 5-HT1A autoreceptor. Neuropharmacology 111:47–58

Anisman H, Du L, Palkovits M, Faludi G, Kovacs GG, Szontagh-Kishazi P, Merali Z, Poulter MO (2008) Serotonin receptor subtype and p11 mRNA expression in stress-relevant brain regions of suicide and control subjects. J Psychiatry Neurosci 33:131–141

Artigas F, Celada P, Laruelle M, Adell A (2001) How does pindolol improve antidepressant action? Trends Pharmacol Sci 22:224–228

Banasr M, Duman RS (2008) Glial loss in the prefrontal cortex is sufficient to induce depressive-like behaviors. Biol Psychiatry 64:863–870

Barnes NM, Sharp T (1999) A review of central 5-HT receptors and their function. Neuropharmacology 38:1083–1152

Benes FM, Todtenkopf MS, Kostoulakos P (2001) GluR5,6,7 subunit immunoreactivity on apical pyramidal cell dendrites in hippocampus of schizophrenics and manic depressives. Hippocampus 11:482–491

Beneyto M, Kristiansen LV, Oni-Orisan A, McCullumsmith RE, Meador-Woodruff JH (2007) Abnormal glutamate receptor expression in the medial temporal lobe in schizophrenia and mood disorders. Neuropsychopharmacology 32:1888–1902

Benjamin S, Doraiswamy PM (2011) Review of the use of mirtazapine in the treatment of depression. Expert Opin Pharmacother 12:1623–1632

Berridge MJ, Downes CP, Hanley MR (1989) Neural and developmental actions of lithium: a unifying hypothesis. Cell 59:411–419

Berton O, McClung CA, Dileone RJ, Krishnan V, Renthal W, Russo SJ, Graham D, Tsankova NM, Bolanos CA, Rios M, Monteggia LM, Self DW, Nestler EJ (2006) Essential role of BDNF in the mesolimbic dopamine pathway in social defeat stress. Science 311:864–868

Bétry C, Pehrson AL, Etiévant A, Ebert B, Sánchez C, Haddjeri N (2013) The rapid recovery of 5-HT cell firing induced by the antidepressant vortioxetine involves 5-HT_3 receptor antagonism. Int J Neuropsychopharmacol 16:1115–1127

Bijl RV, de Graaf R, Hiripi E, Kessler RC, Kohn R, Offord DR, Ustun TB, Vicente B, Vollebergh WA, Walters EE, Wittchen HU (2003) The prevalence of treated and untreated mental disorders in five countries. Health Aff (Millwood) 22:122–133

Blier P, de Montigny C (1994) Current advances and trends in the treatment of depression. Trends Pharmacol Sci 15:220–226

Bonaventure P, Schotte A, Cras P, Leysen JE (1997) Autoradiographic mapping of 5-HT1B- and 5-HT1D receptors in human brain using [3H]alniditan, a new radioligand. Recept Channels 5:225–230

Borroto-Escuela DO, Narváez M, Ambrogini P, Ferraro L, Brito I, Romero-Fernandez W, Andrade-Talavera Y, Flores-Burgess A, Millon C, Gago B, Narvaez JA, Odagaki Y, Palkovits M, Diaz-Cabiale Z, Fuxe K (2018) Receptor-receptor interactions in multiple 5-HT1A heteroreceptor complexes in raphe-hippocampal 5-HT transmission and their relevance for depression and its treatment. Molecules 23:1341

Boschert U, Amara DA, Segu L, Hen R (1994) The mouse 5-hydroxytryptamine1B receptor is localized predominantly on axon terminals. Neuroscience 58:167–182

Bowley MP, Drevets WC, Ongür D, Price JL (2002) Low glial numbers in the amygdala in major depressive disorder. Biol Psychiatry 52:404–412

Brennan BP, Hudson JI, Jensen JE, McCarthy J, Roberts JL, Prescot AP, Cohen BM, Pope HG Jr, Renshaw PF, Ongür D (2010) Rapid enhancement of glutamatergic neurotransmission in bipolar depression following treatment with riluzole. Neuropsychopharmacology 35:834–846

Brown KM, Tracy DK (2013) Lithium: the pharmacodynamic actions of the amazing ion. Ther Adv Psychopharmacol 3:163–176

Bushong EA, Martone ME, Jones YZ, Ellisman MH (2002) Protoplasmic astrocytes in CA1 stratum radiatum occupy separate anatomical domains. J Neurosci 22:183–192

Buysse DJ, Angst J, Gamma A, Ajdacic V, Eich D, Rössler W (2008) Prevalence, course, and comorbidity of insomnia and depression in young adults. Sleep 31:473–480

Cao X, Li LP, Wang Q, Wu Q, Hu HH, Zhang M, Fang YY, Zhang J, Li SJ, Xiong WC, Yan HC, Gao YB, Liu JH, Li XW, Sun LR, Zeng YN, Zhu XH, Gao TM (2013) Astrocyte-derived ATP modulates depressive-like behaviors. Nat Med 19:773–777

Carbone M, Duty S, Rattray M (2012) Riluzole elevates GLT-1 activity and levels in striatal astrocytes. Neurochem Int 60:31–38

Cassano GB, Rucci P, Frank E, Fagiolini A, Dell'Osso L, Shear MK, Kupfer DJ (2004) The mood spectrum in unipolar and bipolar disorder: arguments for a unitary approach. Am J Psychiatry 161:1264–1269

Ceskova E (2016) Current pharmacotherapy of depression – focused on multimodal/multifunctional antidepressants. Expert Opin Pharmacother 17:1835–1837

Ceskova E, Silhan P (2018) Novel treatment options in depression and psychosis. Neuropsychiatr Dis Treat 14:741–747

Chen Y, Hertz L (1996) Inhibition of noradrenaline stimulated increase in [Ca2+]iin cultured astrocytes by chronic treatment with a therapeutically relevant lithium concentration. Brain Res 711:245–248

Chen CH, Suckling J, Ooi C, Jacob R, Lupson V, Bullmore ET, Lennox BR (2010) A longitudinal fMRI study of the manic and euthymic states of bipolar disorder. Bipolar Disord 12:344–347

Chin WY, Chan KT, Lam CL, Lam TP, Wan EY (2015) Help-seeking intentions and subsequent 12-month mental health service use in Chinese primary care patients with depressive symptoms. BMJ Open 5:e006730

Comley RA, van der Aart J, Gulyás B, Garnier M, Iavarone L, Halldin C, Rabiner EA (2015) In vivo occupancy of the 5-HT1A receptor by a novel pan 5-HT1(A/B/D) receptor antagonist, GSK588045, using positron emission tomography. Neuropharmacology 92:44–48

Compton WM, Conway KP, Stinson FS, Grant BF (2006) Changes in the prevalence of major depression and comorbid substance use disorders in the United States between 1991–1992 and 2001–2002. Am J Psychiatry 163:2141–2147

Costall B, Domeney AM, Gerrard PA, Kelly ME, Naylor RJ, Tyers MB (1987) Effects of the 5-HT3 receptor antagonists GR38032F, CS 215-930 and BRL 43694 in tests for anxiolytic activity. Br J Pharmacol 91:195P

Cotter D, Mackay D, Landau S, Kerwin R, Everall I (2001) Reduced glial cell density and neuronal size in the anterior cingulate cortex in major depressive disorder. Arch Gen Psychiatry 58:545–553

Cotter D, Mackay D, Chana G, Beasley C, Landau S, Everall IP (2002) Reduced neuronal size and glial cell density in area 9 of the dorsolateral prefrontal cortex in subjects with major depressive disorder. Cereb Cortex 12:386–394

Croom KF, Perry CM, Plosker GL (2009) Mirtazapine: a review of its use in major depression and other psychiatric disorders. CNS Drugs 23:427–452

Czéh B, Simon M, Schmelting B, Hiemke C, Fuchs E (2006) Astroglial plasticity in the hippocampus is affected by chronic psychosocial stress and concomitant fluoxetine treatment. Neuropsychopharmacology 31:1616–1626

Daniele S, Zappelli E, Martini C (2015) Trazodone regulates neurotrophic/growth factors, mitogen-activated protein kinases and lactate release in human primary astrocytes. J Neuroinflammation 12:225

Dawson LA, Hughes ZA, Starr KR, Storey JD, Bettelini L, Bacchi F, Dawson LA, Hughes ZA, Starr KR, Storey JD, Bettelini L, Bacchi F, Arban R, Poffe A, Melotto S, Hagan JJ, Price GW (2006) Characterisation of the selective 5-HT1B receptor antagonist SB-616234-A (1-[6-(cis-3,5-dimethylpiperazin-1-yl)-2,3-dihydro-5-methoxyindol-1-yl]-1-[2′-methyl-4′-(5-methyl-1,2,4-oxadiazol-3-yl)biphenyl-4-yl]methanone hydrochloride): in vivo neurochemical and behavioural evidence of anxiolytic/antidepressant activity. Neuropharmacology 50:975–983

de la Salle S, Choueiry J, Shah D, Bowers H, McIntosh J, Ilivitsky V, Knott V (2016) Effects of ketamine on resting-state EEG activity and their relationship to perceptual/dissociative symptoms in healthy humans. Front Pharmacol 7:348

Dekeyne A, Mannoury la Cour C, Gobert A, Brocco M, Lejeune F, Serres F, Sharp T, Daszuta A, Soumier A, Papp M, Rivet JM, Flik G, Cremers TI, Muller O, Lavielle G, Millan MJ (2008) S32006, a novel 5-HT2C receptor antagonist displaying broad-based antidepressant and anxiolytic properties in rodent models. Psychopharmacology 199:549–568

Dekeyne A, Brocco M, Loiseau F, Gobert A, Rivet JM, Di Cara B, Cremers TI, Flik G, Fone KC, Watson DJ, Papp M, Sharp T, Serres F, Cespuglio R, Olivier B, Chan JS, Lavielle G, Millan MJ (2012) S32212, a novel serotonin type 2C receptor inverse agonist/α2-adrenoceptor antagonist and potential antidepressant: II. A behavioral, neurochemical, and electrophysiological characterization. J Pharmacol Exp Ther 340:765–780

Demyttenaere K, Bruffaerts R, Posada-Villa J, Gasquet I, Kovess V, Lepine JP, Angermeyer MC, Bernert S, de Girolamo G, Morosini P, Polidori G, Kikkawa T, Kawakami N, Ono Y, Takeshima T, Uda H, Karam EG, Fayyad JA, Karam AN, Mneimneh ZN, Medina-Mora ME, Borges G, Lara C, de Graaf R, Ormel J, Gureje O, Shen Y, Huang Y, Zhang M, Alonso J, Haro JM, Vilagut G, Bromet EJ, Gluzman S, Webb C, Kessler RC, Merikangas KR, Anthony JC, Von Korff MR, Wang PS, Brugha TS, Aguilar-Gaxiola S, Lee S, Heeringa S, Pennell BE, Zaslavsky AM, Ustun TB, Chatterji S, WHO World Mental Health Survey Consortium (2004) Prevalence, severity, and unmet need for treatment of mental disorders in the World Health Organization World Mental Health Surveys. JAMA 291:2581–2590

Diaz SL, Narboux-Nême N, Boutourlinsky K, Doly S, Maroteaux L (2016) Mice lacking the serotonin 5-HT2B receptor as an animal model of resistance to selective serotonin reuptake inhibitors antidepressants. Eur Neuropsychopharmacol 26:265–279

Di Matteo V, Di Giovanni G, Pierucci M, Esposito E (2008) Serotonin control of central dopaminergic function: focus on in vivo microdialysis studies. Prog Brain Res 172:7–44

Ding F, O'Donnell J, Thrane AS, Zeppenfeld D, Kang H, Xie L, Wang F, Nedergaard M (2013) α1-adrenergic receptors mediate coordinated Ca2+ signaling of cortical astrocytes in awake, behaving mice. Cell Calcium 54:387–394

Dong L, Li B, Verkhratsky A, Peng L (2015) Cell type-specific in vivo expression of genes encoding signalling molecules in the brain in response to chronic mild stress and chronic treatment with fluoxetine. Psychopharmacology 232:2827–2835

Dracheva S, Chin B, Haroutunian V (2008) Altered serotonin2C receptor RNA splicing in suicide: association with editing. Neuroreport 19:379–382

Dremencov E, Newman ME, Kinor N, Blatman-Jan G, Schindler CJ, Overstreet DH, Yadid G (2005) Hyperfunctionality of serotonin-2C receptor-mediated inhibition of accumbal dopamine release in an animal model of depression is reversed by antidepressant treatment. Neuropharmacology 48:34–42

Dremencov E, El Mansari M, Blier P (2007) Noradrenergic augmentation of escitalopram response by risperidone: electrophysiologic studies in the rat brain. Biol Psychiatry 61:671–678

Dumuis A, Bouhelal R, Sebben M, Cory R, Bockaert J (1988) A nonclassical 5-hydroxytryptamine receptor positively coupled with adenylate cyclase in the central nervous system. Mol Pharmacol 34:880–887

Duxon MS, Flanigan TP, Reavley AC, Baxter GS, Blackburn TP, Fone KC (1997) Evidence for expression of the 5-hydroxytryptamine-2B receptor protein in the rat central nervous system. Neuroscience 76:323–329. Editing. Neuroscientist. 9:237–42.

Eastwood SL, Harrison PJ (2010) Markers of glutamate synaptic transmission and plasticity are increased in the anterior cingulate cortex in bipolar disorder. Biol Psychiatry 67:1010–1016

Fagni L, Dumuis A, Sebben M, Bockaert J (1992) The 5-HT4 receptor subtype inhibits K+ current in colliculi neurones via activation of a cyclic AMP-dependent protein kinase. Br J Pharmacol 105:973–979

Ferrari AJ, Somerville AJ, Baxter AJ, Norman R, Patten SB, Vos T, Whiteford HA (2013) Global variation in the prevalence and incidence of major depressive disorder: a systematic review of the epidemiological literature. Psychol Med 43:471–481

Fröjd SA, Nissinen ES, Pelkonen MU, Marttunen MJ, Koivisto AM, Kaltiala-Heino R (2008) Depression and school performance in middle adolescent boys and girls. J Adolesc 31:485–498

Fu H, Li B, Hertz L, Peng L (2012) Contributions in astrocytes of SMIT1/2 and HMIT to myo-inositol uptake at different concentrations and pH. Neurochem Int 61:187–194

Gao J, Wu R, Davis C, Li M (2018) Activation of 5-HT2A receptor disrupts rat maternal behavior. Neuropharmacology 128:96–105

Gu L, Xie J, Long J, Chen Q, Chen Q, Pan R, Yan Y, Wu G, Liang B, Tan J, Xie X, Wei B, Su L (2013) Epidemiology of major depressive disorder in mainland China: a systematic review. PLoS One 8:e65356

Guilloux JP, Mendez-David I, Pehrson A, Guiard BP, Repérant C, Orvoën S, Gardier AM, Hen R, Ebert B, Miller S, Sanchez C, David DJ (2013) Antidepressant and anxiolytic potential of the multimodal antidepressant vortioxetine (Lu AA21004) assessed by behavioural and neurogenesis outcomes in mice. Neuropharmacology 73:147–159

Gurevich I, Tamir H, Arango V, Dwork AJ, Mann JJ, Schmauss C (2002) Altered editing of serotonin 2C receptor pre-mRNA in the prefrontal cortex of depressed suicide victims. Neuron 34:349–356

Hagan CE, McDevitt RA, Liu Y, Furay AR, Neumaier JF (2012) 5-HT(1B) autoreceptor regulation of serotonin transporter activity in synaptosomes. Synapse 66:1024–1034

Hallcher LM, Sherman WR (1980) The effects of lithium ion and other agents on the activity of myo-inositol-1-phosphatase from bovine brain. J Biol Chem 255:10896–10901

Hamidi M, Drevets WC, Price JL (2004) Glial reduction in amygdala in major depressive disorder is due to oligodendrocytes. Biol Psychiatry 55:563–569

Hannon J, Hoyer D (2008) Molecular biology of 5-HT receptors. Behav Brain Res 195:198–213

Harvey ML, Swallows CL, Cooper MA (2012) A double dissociation in the effects of 5-HT2A and 5-HT2C receptors on the acquisition and expression of conditioned defeat in Syrian hamsters. Behav Neurosci 126:530–537

Hercher C, Turecki G, Mechawar N (2009) Through the looking glass: examining neuroanatomical evidence for cellular alterations in major depression. J Psychiatr Res 43:947–961

Hertz L, Lovatt D, Goldman SA, Nedergaard M (2010) Adrenoceptors in brain: cellular gene expression and effects on astrocytic metabolism and [Ca(2+)]i. Neurochem Int 57:411–420

Hertz L, Li B, Song D, Ren J, Dong L, Chen Y, Peng L (2012) Astrocytes as a 5-HT2B-mediated SERT-independent SSRI target, slowly altering depression-associated genes and function. Curr Signal Transduction Ther 7:65–80

Hertz L, Song D, Li B, Du T, Xu J, Gu L, Chen Y, Peng L (2014) Signal transduction in astrocytes during chronic or acute treatment with drugs (SSRIs, antibipolar drugs, GABA-ergic drugs, and benzodiazepines) ameliorating mood disorders. J Signal Transduct 2014:593934

Hertz L, Rothman DL, Li B, Peng L (2015) Chronic SSRI stimulation of astrocytic 5-HT2B receptors change multiple gene expressions/editings and metabolism of glutamate, glucose and glycogen: a potential paradigm shift. Front Behav Neurosci 9:25

Hirst WD, Price GW, Rattray M, Wilkin GP (1997) Identification of 5-hydroxytryptamine receptors positively coupled to adenylyl cyclase in rat cultured astrocytes. Br J Pharmacol 120:509–515

Hirst WD, Cheung NY, Rattray M, Price GW, Wilkin GP (1998) Cultured astrocytes express messenger RNA for multiple serotonin receptor subtypes, without functional coupling of 5-HT1 receptor subtypes to adenylyl cyclase. Brain Res Mol Brain Res 61:90–99

Hisaoka-Nakashima K, Kajitani N, Kaneko M, Shigetou T, Kasai M, Matsumoto C, Yokoe T, Azuma H, Takebayashi M, Morioka N, Nakata Y (2016) Amitriptyline induces brain-derived neurotrophic factor (BDNF) mRNA expression through ERK-dependent modulation of multiple BDNF mRNA variants in primary cultured rat cortical astrocytes and microglia. Brain Res 1634:57–1667

Hjorth S, Tao R (1991) The putative 5-HT1B receptor agonist CP-93,129 suppresses rat hippocampal 5-HT release in vivo: comparison with RU 24969. Eur J Pharmacol 209:249–252

Hoyer D, Hannon JP, Martin GR (2002) Molecular, pharmacological and functional diversity of 5-HT receptors. Pharmacol Biochem Behav 71:533–554

Huang MH, Cheng CM, Huang KL, Hsu JW, Bai YM, Su TP, Li CT, Tsai SJ, Lin WC, Chen TJ, Chen MH (2019) Bipolar disorder and risk of Parkinson disease: a nationwide longitudinal study. Neurology 92:e2735–e2742

Hudzik TJ, Yanek M, Porrey T, Evenden J, Paronis C, Mastrangelo M, Ryan C, Ross S, Stenfors C (2003) Behavioral pharmacology of AR-A000002, a novel, selective 5-hydroxytryptamine(1B) antagonist. J Pharmacol Exp Ther 304:1072–1084

Iliff JJ, Wang M, Liao Y, Plogg BA, Peng W, Gundersen GA, Benveniste H, Vates GE, Deane R, Goldman SA, Nagelhus EA, Nedergaard M (2012) A paravascular pathway facilitates CSF flow through the brain parenchyma and the clearance of interstitial solutes, including amyloid β. Sci Transl Med 4:147ra111

Inhorn RC, Majerus PW (1987) Inositol polyphosphate 1-phosphatase from calf brain. Purification and inhibition by Li^{+}, Ca^{2+}, and Mn^{2+}. J Biol Chem 262:15946–15952

Iwamoto K, Kato T (2003) RNA editing of serotonin2C receptor in human postmortem brains of major mental disorders. Neurosci Lett 346:169–172

Jeltsch-David H, Koenig J, Cassel JC (2010) Modulation of cholinergic functions by serotonin and possible implications in memory: general data and focus on 5-HT(1A) receptors of the medial septum. Behav Brain Res 195:86–97

Jenck F, Moreau JL, Mutel V, Martin JR, Haefely WE (1993) Evidence for a role of 5-HT1C receptors in the antiserotonergic properties of some antidepressant drugs. Eur J Pharmacol 231:223–229

Johnson J Jr, Pajarillo EAB, Taka E, Reams R, Son DS, Aschner M, Lee E (2018) Valproate and sodium butyrate attenuate manganese-decreased locomotor activity and astrocytic glutamate transporters expression in mice. Neurotoxicology 64:230–239

Johnson BA, Roache JD, Ait-Daoud N, Zanca NA, Velazquez M (2002) Ondansetron reduces the craving of biologically predisposed alcoholics. Psychopharmacology 160:408–413

Johnson BA, Roache JD, Ait-Daoud N, Javors MA, Harrison JM, Elkashef A, Mojsiak J, Li SH, Bloch DA (2006) A preliminary randomized, double-blind, placebo-controlled study of the safety and efficacy of ondansetron in the treatment of cocaine dependence. Drug Alcohol Depend 84:256–263

Kajitani N, Hisaoka-Nakashima K, Morioka N, Okada-Tsuchioka M, Kaneko M, Kasai M, Shibasaki C, Nakata Y, Takebayashi M (2012) Antidepressant acts on astrocytes leading to an increase in the expression of neurotrophic/growth factors: differential regulation of FGF-2 by noradrenaline. PLoS One 7:e51197

Karnovsky AM, Gotow LF, McKinley DD, Piechan JL, Ruble CL, Mills CJ, Schellin KA, Slightom JL, Fitzgerald LR, Benjamin CW, Roberds SL (2003) A cluster of novel serotonin receptor 3-like genes on human chromosome 3. Gene 319:137–148

Kendler KS, Gardner CO, Fiske A, Gatz M (2009) Major depression and coronary artery disease in the Swedish twin registry: phenotypic, genetic, and environmental sources of comorbidity. Arch Gen Psychiatry 66:857–863

Kennett GA, Bailey F, Piper DC, Blackburn TP (1995) Effect of SB 200646A, a 5-HT2C/5-HT2B receptor antagonist, in two conflict models of anxiety. Psychopharmacology 118:178–182

Kennett GA, Bright F, Trail B, Baxter GS, Blackburn TP (1996) Effects of the 5-HT2B receptor agonist, BW 723C86, on three rat models of anxiety. Br J Pharmacol 117:1443–1448

Kennett GA, Trail B, Bright F (1998) Anxiolytic-like actions of BW 723C86 in the rat Vogel conflict test are 5-HT2B receptor mediated. Neuropharmacology 37:1603–1610

Kimura Y, Naitou Y, Wanibuchi F, Yamaguchi T (2008) 5-HT(2C) receptor activation is a common mechanism on proerectile effects of apomorphine, oxytocin and melanotan-II in rats. Eur J Pharmacol 589:157–162

Kinoshita M, Hirayama Y, Fujishita K, Shibata K, Shinozaki Y, Shigetomi E, Takeda A, Le HPN, Hayashi H, Hiasa M, Moriyama Y, Ikenaka K, Tanaka KF, Koizumi S (2018) Anti-depressant fluoxetine reveals its therapeutic effect via astrocytes. EBioMedicine 32:72–83

Kirischuk S, Tuschick S, Verkhratsky A, Kettenmann H (1996) Calcium signalling in mouse Bergmann glial cells mediated by alpha1-adrenoreceptors and H1 histamine receptors. Eur J Neurosci 8:1198–1208

Kleinman A (2004) Culture and depression. N Engl J Med 351:951–953

Kress BT, Iliff JJ, Xia M, Wang M, Wei HS, Zeppenfeld D, Xie L, Kang H, Xu Q, Liew JA, Plog BA, Ding F, Deane R, Nedergaard M (2014) Impairment of paravascular clearance pathways in the aging brain. Ann Neurol 76:845–861

Kupfer DJ, Frank E, Phillips ML (2012) Major depressive disorder: new clinical, neurobiological, and treatment perspectives. Lancet 379:1045–1055

Kursar JD, Nelson DL, Wainscott DB, Baez M (1994) Molecular cloning, functional expression, and mRNA tissue distribution of the human 5-hydroxytryptamine2B receptor. Mol Pharmacol 46:227–234

Kutzer T, Dick M, Scudamore T, Wiener M, Schwartz T (2020) Antidepressant efficacy and side effect burden: an updated guide for clinicians. Drugs Context 9:2020-2-2

Lee TK, Park JH, Ahn JH, Shin MC, Cho JH, Bae EJ, Kim YM, Won MH, Lee CH (2016) Pretreated duloxetine protects hippocampal CA1 pyramidal neurons from ischemia-reperfusion injury through decreases of glial activation and oxidative stress. J Neurol Sci 370:229–236

Leenders AG, Sheng ZH (2005) Modulation of neurotransmitter release by the second messenger-activated protein kinases: implications for presynaptic plasticity. Pharmacol Ther 105:69–84

Li B, Gu L, Zhang H, Huang J, Chen Y, Hertz L, Peng L (2007) Up-regulation of cPLA(2) gene expression in astrocytes by all three conventional anti-bipolar drugs is drug-specific and enzyme-specific. Psychopharmacology 194:333–345

Li B, Zhang S, Zhang H, Nu W, Cai L, Hertz L, Peng L (2008) Fluoxetine-mediated 5-HT2B receptor stimulation in astrocytes causes EGF receptor transactivation and ERK phosphorylation. Psychopharmacology 201:443–458

Li B, Zhang S, Li M, Hertz L, Peng L (2009a) Chronic treatment of astrocytes with therapeutically relevant fluoxetine concentrations enhances cPLA2 expression secondary to 5-HT2B-induced, transactivation-mediated ERK1/2 phosphorylation. Psychopharmacology (Berlin) 207:1–12

Li B, Zhang S, Li M, Zhang H, Hertz L, Peng L (2009b) Down-regulation of GluK2 kainate receptor expression by chronic treatment with mood-stabilizing anti-convulsants or lithium in cultured astrocytes and brain, but not in neurons. Neuropharmacology 57:375–385

Li B, Zhang S, Li M, Hertz L, Peng L (2010) Serotonin increases ERK1/2 phosphorylation in astrocytes by stimulation of 5-HT2B and 5-HT2C receptors. Neurochem Int 57:432–439

Li B, Zhang S, Zhang H, Hertz L, Peng L (2011) Fluoxetine affects GluK2 editing, glutamate-evoked Ca(2+) influx and extracellular signal-regulated kinase phosphorylation in mouse astrocytes. J Psychiatry Neurosci 36:322–338

Li B, Dong L, Wang B, Cai L, Jiang N, Peng L (2012) Cell type-specific gene expression and editing responses to chronic fluoxetine treatment in the in vivo mouse brain and their relevance for stress-induced anhedonia. Neurochem Res 37:2480–2495

Li B, Hertz L, Peng L (2013) Cell-specific mRNA alterations in Na+, K+-ATPase α and β isoforms and FXYD in mice treated chronically with carbamazepine, an anti-bipolar drug. Neurochem Res 38:834–841

Li B, Jia S, Yue T, Yang L, Huang C, Verkhratsky A, Peng L (2017) Biphasic regulation of caveolin-1 gene expression by fluoxetine in astrocytes: opposite effects of PI3K/AKT and MAPK/ERK signaling pathways on c-fos. Front Cell Neurosci 11:335

Li B, Ren J, Yang L, Li X, Sun G, Xia M (2018) Lithium inhibits GSK3β activity via two different signaling pathways in neurons after spinal cord injury. Neurochem Res 43:848–856

Li X, Liang S, Li Z, Li S, Xia M, Verkhratsky A, Li B (2019) Leptin increases expression of 5-HT2B receptors in astrocytes thus enhancing action of fluoxetine on the depressive behavior induced by sleep deprivation. Front Psych 9:734

Li Z, Lu Y, Liang S, Li S, Chen B, Zhang M, Xia M, Guan D, Verkhratsky A, Li B (2020) Fluoxetine improves behavioural deficits induced by chronic alcohol treatment by alleviating RNA editing of 5-HT2C receptors. Neurochem Int 134:104689

Liang S, Lu Y, Li Z, Li S, Chen B, Zhang M, Chen B, Ji M, Gong W, Xia M, Verkhratsky A, Wu X, Li B (2020) Iron aggravates the depressive phenotype of stressed mice by compromising the glymphatic system. Neurosci Bull 36:1542–1546

Liu KJ, Lee YL, Yang YY, Shih NY, Ho CC, Wu YC, Huang TS, Huang MC, Liu HC, Shen WW, Leu SJ (2011) Modulation of the development of human monocyte-derived dendritic cells by lithium chloride. J Cell Physiol 226:424–433

Liu Z, Song D, Yan E, Verkhratsky A, Peng L (2015) Chronic treatment with anti-bipolar drugs suppresses glutamate release from astroglial cultures. Amino Acids 47:1045–1051

Liu XJ, Zhou Y, Dong L, Guo Q, Chen WC, Psychiatryet DO (2017) A survey of major depression among elderly population in Wuhan (in Chinese). Chin Ment Health J 11:851–856

López-Gil X, Artigas F, Adell A (2008) Unraveling monoamine receptors involved in the action of typical and atypical antipsychotics on glutamatergic and serotonergic transmission in prefrontal cortex. Curr Pharm Des 16:502–515

López-Muñoz F, Alamo C (2009) Monoaminergic neurotransmission: the history of the discovery of antidepressants from 1950s until today. Curr Pharm Des 15:1563–1586

Lucas G, Rymar VV, Du J, Mnie-Filali O, Bisgaard C, Manta S, Lambas-Senas L, Wiborg O, Haddjeri N, Piñeyro G, Sadikot AF, Debonnel G (2007) Serotonin(4) (5-HT(4)) receptor agonists are putative antidepressants with a rapid onset of action. Neuron 55:712–725

Lynch FL, Clarke GN (2006) Estimating the economic burden of depression in children and adolescents. Am J Prev Med 31:S143–S151

Madsen K, Torstensen E, Holst KK, Haahr ME, Knorr U, Frokjaer VG, Brandt-Larsen M, Iversen P, Fisher PM, Knudsen GM (2014) Familial risk for major depression is associated with lower striatal 5-HT_4 receptor binding. Int J Neuropsychopharmacol 18:pyu034

Marathe SV, D'almeida PL, Virmani G, Bathini P, Alberi L (2018) Effects of monoamines and antidepressants on astrocyte physiology: implications for monoamine hypothesis of depression. J Exp Neurosci 12:1179069518789149

Marek GJ, Martin-Ruiz R, Abo A, Artigas F (2005) The selective 5-HT2A receptor antagonist M100907 enhances antidepressant-like behavioral effects of the SSRI fluoxetine. Neuropsychopharmacology 30:2205–2215

Martin KF, Hannon S, Phillips I, Heal DJ (1992) Opposing roles for 5-HT1B and 5-HT3 receptors in the control of 5-HT release in rat hippocampus in vivo. Br J Pharmacol 106:139–142

Martin CB, Ramond F, Farrington DT, Aguiar AS Jr, Chevarin C, Berthiau AS, Caussanel S, Lanfumey L, Herrick-Davis K, Hamon M, Madjar JJ, Mongeau R (2013) RNA splicing and editing modulation of 5-HT(2C) receptor function: relevance to anxiety and aggression in VGV mice. Mol Psychiatry 18:656–665

Masson J, Emerit MB, Hamon M, Darmon M (2012) Serotonergic signaling: multiple effectors and pleiotropic effects. WIREs Membr Transport Signaling 1:685–713

Mathers CD, Loncar D (2006) Projections of global mortality and burden of disease from 2002 to 2030. PLoS Med 3:e442

Matzen L, van Amsterdam C, Rautenberg W, Greiner HE, Harting J, Seyfried CA, Böttcher H (2000) 5-HT reuptake inhibitors with 5-HT(1B/1D) antagonistic activity: a new approach toward efficient antidepressants. J Med Chem 43:1149–1157

McAlear SD, Bevensee MO (2006) A cysteine-scanning mutagenesis study of transmembrane domain 8 of the electrogenic sodium/bicarbonate cotransporter NBCe1. J Biol Chem 281:32417–32427

Meltzer-Brody S, Colquhoun H, Riesenberg R, Epperson CN, Deligiannidis KM, Rubinow DR, Li H, Sankoh AJ, Clemson C, Schacterle A, Jonas J, Kanes S (2018) Brexanolone injection in post-partum depression: two multicentre, double-blind, randomised, placebo-controlled, phase 3 trials. Lancet 392:1058–1070

Merzak A, Koochekpour S, Fillion MP, Fillion G, Pilkington GJ (1996) Expression of serotonin receptors in human fetal astrocytes and glioma cell lines: a possible role in glioma cell proliferation and migration. Brain Res Mol Brain Res 41:1–7

Meuser T, Pietruck C, Gabriel A, Xie GX, Lim KJ, Pierce Palmer P (2002) 5-HT7 receptors are involved in mediating 5-HT-induced activation of rat primary afferent neurons. Life Sci 71:2279–2289

Michael N, Erfurth A, Ohrmann P, Gössling M, Arolt V, Heindel W, Pfleiderer B (2003) Acute mania is accompanied by elevated glutamate/glutamine levels within the left dorsolateral prefrontal cortex. Psychopharmacology 168:344–346

Middlemiss DN, Hutson PH (1990) The 5-HT1B receptors. Ann N Y Acad Sci 600:132–147; discussion 347–48.

Milak MS, Pantazatos S, Rashid R, Zanderigo F, DeLorenzo C, Hesselgrave N, Ogden RT, Oquendo MA, Mulhern ST, Miller JM, Burke AK, Parsey RV, Mann JJ (2018) Higher 5-HT1A autoreceptor binding as an endophenotype for major depressive disorder identified in high risk offspring – a pilot study. Psychiatry Res Neuroimaging 276:15–23

Milaneschi Y, Lamers F, Bot M, Drent ML, Penninx BW (2017) Leptin dysregulation is specifically associated with major depression with atypical features: evidence for a mechanism connecting obesity and depression. Biol Psychiatry 81(9):807–814

Millan MJ (2005) Serotonin 5-HT2C receptors as a target for the treatment of depressive and anxious states: focus on novel therapeutic strategies. Therapie 60:441–460

Mnie-Filali O, Faure C, Lambás-Señas L, El Mansari M, Belblidia H, Gondard E, Etiévant A, Scarna H, Didier A, Berod A, Blier P, Haddjeri N (2011) Pharmacological blockade of 5-HT7 receptors as a putative fast acting antidepressant strategy. Neuropsychopharmacology 36:1275–1288

Montañez S, Munn JL, Owens WA, Horton RE, Daws LC (2013) 5-HT1B receptor modulation of the serotonin transporter in vivo: studies using KO mice. Neurochem Int 73:127–131

Moreau JL, Jenck F, Martin JR, Perrin S, Haefely WE (1993) Effects of repeated mild stress and two antidepressant treatments on the behavioral response to 5HT1C receptor activation in rats. Psychopharmacology 110:140–144

Moret C, Briley M (2000) The possible role of 5-HT(1B/D) receptors in psychiatric disorders and their potential as a target for therapy. Eur J Pharmacol 404:1–12

Mork A, Pehrson A, Brennum LT, Nielsen SM, Zhong H, Lassen AB, Miller S, Westrich L, Boyle NJ, Sánchez C, Fischer CW, Liebenberg N, Wegener G, Bundgaard C, Hogg S, Bang-Andersen B, Stensbøl TB (2012) Pharmacological effects of Lu AA21004: a novel multimodal compound for the treatment of major depressive disorder. J Pharmacol Exp Ther 340:666–675

Murrough JW, Neumeister A (2011) The serotonin 1B receptor: a new target for depression therapeutics? Biol Psychiatry 69:714–715

Murrough JW, Henry S, Hu J, Gallezot JD, Planeta-Wilson B, Neumaier JF, Neumeister A (2011) Reduced ventral striatal/ventral pallidal serotonin1B receptor binding potential in major depressive disorder. Psychopharmacology 213:547–553

Nagao K, Kishi T, Moriwaki M, Fujita K, Hirano S, Yamanouchi Y, Funahashi T, Iwata N (2013) Comparative clinical profile of mirtazapine and duloxetine in practical clinical settings in Japan: a 4-week open-label, parallel-group study of major depressive disorder. Neuropsychiatr Dis Treat 9:781–786

Nakagawa Y, Ishima T, Takashima T (1998) The 5-HT3 receptor agonist attenuates the action of antidepressants in the forced swim test in rats. Brain Res 786:189–193

Naumenko VS, Popova NK, Lacivita E, Leopoldo M, Ponimaskin EG (2014) Interplay between serotonin 5-HT1A and 5-HT7 receptors in depressive disorders. CNS Neurosci Ther 20:582–590

Nedergaard M, Goldman SA (2020) Glymphatic failure as a final common pathway to dementia. Science 370:50–56

Nemeroff CB, Schatzberg AF, Goldstein DJ, Detke MJ, Mallinckrodt C, Lu Y, Tran PV (2002) Duloxetine for the treatment of major depressive disorder. Psychopharmacol Bull 36:106–132

Nevitt SJ, Sudell M, Tudur Smith C, Marson AG (2019) Topiramate versus carbamazepine monotherapy for epilepsy: an individual participant data review. Cochrane Database Syst Rev 6:CD012065

Ni YG, Miledi R (1997) Blockage of 5HT2C serotonin receptors by fluoxetine (Prozac). Proc Natl Acad Sci U S A 94:2036–2040

Niesler B, Walstab J, Combrink S, Möller D, Kapeller J, Rietdorf J, Bönisch H, Göthert M, Rappold G, Brüss M (2007) Characterization of the novel human serotonin receptor subunits 5-HT3C,5-HT3D, and 5-HT3E. Mol Pharmacol 72:8–17

Niswender CM, Herrick-Davis K, Dilley GE, Meltzer HY, Overholser JC, Stockmeier CA, Emeson RB, Sanders-Bush E (2001) RNA editing of the human serotonin 5-HT2C receptor. Alterations in suicide and implications for serotonergic pharmacotherapy. Neuropsychopharmacology 24:478–491

Nutt DJ (2002) The neuropharmacology of serotonin and noradrenaline in depression. Int Clin Psychopharmacol 17(Suppl 1):S1–S12

Oba A, Nakagawasai O, Onogi H, Nemoto W, Yaoita F, Arai Y, Tan-No K, Tadano T (2013) Chronic fluvoxamine treatment changes 5-HT(2A/2C) receptor-mediated behavior in olfactory bulbectomized mice. Life Sci 92:119–124

Oberheim NA, Takano T, Han X, He W, Lin JH, Wang F, Xu Q, Wyatt JD, Pilcher W, Ojemann JG, Ransom BR, Goldman SA, Nedergaard M (2009) Uniquely hominid features of adult human astrocytes. J Neurosci 29:3276–3287

Olfson M, Blanco C, Marcus SC (2016) Treatment of adult depression in the United States. JAMA Intern Med 176:1482–1491

Ongür D, Drevets WC, Price JL (1998) Glial reduction in the subgenual prefrontal cortex in mood disorders. Proc Natl Acad Sci U S A 95:13290–13295

Ongür D, Jensen JE, Prescot AP, Stork C, Lundy M, Cohen BM, Renshaw PF (2008) Abnormal glutamatergic neurotransmission and neuronal-glial interactions in acute mania. Biol Psychiatry 64:718–726

Opal MD, Klenotich SC, Morais M, Bessa J, Winkle J, Doukas D, Kay LJ, Sousa N, Dulawa SM (2013) Serotonin 2C receptor antagonists induce fast-onset antidepressant effects. Mol Psychiatry 19:1106–1114

Orgeta V, Tabet N, Nilforooshan R, Howard R (2017) Efficacy of antidepressants for depression in Alzheimer's disease: systematic review and meta-analysis. J Alzheimers Dis 58:725–733

Otte C, Gold SM, Penninx BW, Pariante CM, Etkin A, Fava M, Mohr DC, Schatzberg AF (2016) Major depressive disorder. Nat Rev Dis Primers 2:16065

Pandey GN, Dwivedi Y, Ren X, Rizavi HS, Faludi G, Sarosi A, Palkovits M (2006) Regional distribution and relative abundance of serotonin(2c) receptors in human brain: effect of suicide. Neurochem Res 31:167–176

Patel JG, Bartoszyk GD, Edwards E, Ashby CR Jr (2004) The highly selective 5-hydroxytryptamine (5-HT)2A receptor antagonist, EMD 281014, significantly increases swimming and decreases immobility in male congenital learned helpless rats in the forced swim test. Synapse 52:73–75

Pekny M, Pekna M, Messing A, Steinhäuser C, Lee JM, Parpura V, Hol EM, Sofroniew MV, Verkhratsky A (2016) Astrocytes: a central element in neurological diseases. Acta Neuropathol 131:323–345

Peng L, Gu L, Li B, Hertz L (2014) Fluoxetine and all other SSRIs are 5-HT_{2B} agonists – importance for their therapeutic effects. Curr Neuropharmacol 12:365–379

Peng L, Verkhratsky A, Gu L, Li B (2015) Targeting astrocytes in major depression. Expert Rev Neurother 15:1299–1306

Peng L, Li B, Verkhratsky A (2016) Targeting astrocytes in bipolar disorder. Expert Rev Neurother 16:649–657

Peng L, Song D, Li B, Verkhratsky A (2018) Astroglial 5-HT2B receptor in mood disorders. Expert Rev Neurother 18:435–442

Penninx BW (2017) Depression and cardiovascular disease: epidemiological evidence on their linking mechanisms. Neurosci Biobehav Rev 74:277–286

Philippe TJ, Vahid-Ansari F, Donaldson ZR, Le François B, Zahrai A, Turcotte-Cardin V, Daigle M, James J, Hen R, Merali Z, Albert PR (2018) Loss of MeCP2 in adult 5-HT neurons induces 5-HT1A autoreceptors, with opposite sex-dependent anxiety and depression phenotypes. Sci Rep 8:5788

Pitychoutis PM, Dalla C, Sideris AC, Tsonis PA, Papadopoulou-Daifoti Z (2012) 5-HT(1A), 5-HT(2A), and 5-HT(2C) receptor mRNA modulation by antidepressant treatment in the chronic mild stress model of depression: sex differences exposed. Neuroscience 210:152–167

Pizzonia JH, Ransom BR, Pappas CA (1996) Characterization of Na+/H+ exchange activity in cultured rat hippocampal astrocytes. J Neurosci Res 44:191–198

Post RM, Ketter TA, Uhde T, Ballenger JC (2007) Thirty years of clinical experience with carbamazepine in the treatment of bipolar illness: principles and practice. CNS Drugs 21:47–71

Rajkowska G (2000) Postmortem studies in mood disorders indicate altered numbers of neurons and glial cells. Biol Psychiatry 48:766–777

Rajkowska G, O'Dwyer G, Teleki Z, Stockmeier CA, Miguel-Hidalgo JJ (2007) GABAergic neurons immunoreactive for calcium binding proteins are reduced in the prefrontal cortex in major depression. Neuropsychopharmacology 32:471–482

Ramamoorthy R, Radhakrishnan M, Borah M (2008) Antidepressant-like effects of serotonin type-3 antagonist, ondansetron: an investigation in behaviour-based rodent models. Behav Pharmacol 19:29–40

Rao JS, Ertley RN, Lee HJ, Rapoport SI, Bazinet RP (2006) Chronic fluoxetine upregulates activity, protein and mRNA levels of cytosolic phospholipase A2 in rat frontal cortex. Pharmacogenomics J 6:413–420

Redrobe JP, Bourin M (1997) Partial role of 5-HT2 and 5-HT3 receptors in the activity of antidepressants in the mouse forced swimming test. Eur J Pharmacol 325:129–135

Richardson-Jones JW, Craige CP, Guiard BP, Stephen A, Metzger KL, Kung HF, Gardier AM, Dranovsky A, David DJ, Beck SG, Hen R, Leonardo ED (2010) 5-HT1A autoreceptor levels determine vulnerability to stress and response to antidepressants. Neuron 65:40–52

Richardson-Jones JW, Craige CP, Nguyen TH, Kung HF, Gardier AM, Dranovsky A, David DJ, Guiard BP, Beck SG, Hen R, Leonardo ED (2011) Serotonin-1A autoreceptors are necessary and sufficient for the normal formation of circuits underlying innate anxiety. J Neurosci 31:6008–6018

Rivera AD, Butt AM (2019) Astrocytes are direct cellular targets of lithium treatment: novel roles for lysyl oxidase and peroxisome-proliferator activated receptor-γ as astroglial targets of lithium. Transl Psychiatry 9:211
Roberts RE, Duong HT (2014) The prospective association between sleep deprivation and depression among adolescents. Sleep 37:239–244
Rosel P, Arranz B, Urretavizcaya M, Oros M, San L, Navarro MA (2004) Altered 5-HT2A and 5-HT4 postsynaptic receptors and their intracellular signalling systems IP3 and cAMP in brains from depressed violent suicide victims. Neuropsychobiology 49:189–195
Ruf BM, Bhagwagar Z (2009) The 5-HT1B receptor: a novel target for the pathophysiology of depression. Curr Drug Targets 10:1118–1138
Rush AJ, Fava M, Wisniewski SR, Lavori PW, Trivedi MH, Sackeim HA, Thase ME, Nierenberg AA, Quitkin FM, Kashner TM, Kupfer DJ, Rosenbaum JF, Alpert J, Stewart JW, McGrath PJ, Biggs MM, Shores-Wilson K, Lebowitz BD, Ritz L, Niederehe G, STAR*D Investigators Group (2004) Sequenced treatment alternatives to relieve depression (STAR*D): rationale and design. Control Clin Trials 25:119–142
Rush AJ, Trivedi MH, Wisniewski SR, Nierenberg AA, Stewart JW, Warden D, Niederehe G, Thase ME, Lavori PW, Lebowitz BD, McGrath PJ, Rosenbaum JF, Sackeim HA, Kupfer DJ, Luther J, Fava M (2006) Acute and longer-term outcomes in depressed outpatients requiring one or several treatment steps: a STAR*D report. Am J Psychiatry 163:1905–1917
Rutz S, Riegert C, Rothmaier AK, Buhot MC, Cassel JC, Jackisch R (2006) Presynaptic serotonergic modulation of 5-HT and acetylcholine release in the hippocampus and the cortex of 5-HT1B-receptor knockout mice. Brain Res Bull 70:81–93
Rygula R, Abumaria N, Flügge G, Fuchs E, Rüther E, Havemann-Reinecke U (2005) Anhedonia and motivational defificits in rats: impact of chronic social stress. Behav Brain Res 162:127–134
Rygula R, Abumaria N, Domenici E, Hiemke C, Fuchs E (2006) Effects of fluoxetine on behavioral deficits evoked by chronic social stress in rats. Behav Brain Res 174:188–192
Sahli ZT, Banerjee P, Tarazi FI (2016) The preclinical and clinical effects of vilazodone for the treatment of major depressive disorder. Expert Opin Drug Discov 11:515–523
Sari Y (2004) Serotonin1B receptors: from protein to physiological function and behavior. Neurosci Biobehav Rev 28:565–582
Saura J, Kettler R, Da Prada M, Richards JG (1992) Quantitative enzyme radioautography with 3H-Ro 41-1049 and 3H-Ro 19-6327 in vitro: localization and abundance of MAO-A and MAO-B in rat CNS, peripheral organs, and human brain. J Neurosci 12:1977–1999
Savitz J, Lucki I, Drevets WC (2009) 5-HT(1A) receptor function in major depressive disorder. Prog Neurobiol 88:17–31
Schildkraut JJ (1965) The catecholamine hypothesis of affective disorders: a review of supporting evidence. Am J Psychiatry 122:509–522
Schmauss C (2003) Serotonin2C receptors: suicide, serotonin, and runaway RNA. Neuroscientist 9:237–242
Schmitt K, Holsboer-Trachsler E, Eckert A (2016) BDNF in sleep, insomnia, and sleep deprivation. Ann Med 48:42–51
Schulze TG, Buervenich S, Badner JA, Steele CJ, Detera-Wadleigh SD, Dick D, Foroud T, Cox NJ, MacKinnon DF, Potash JB, Berrettini WH, Byerley W, Coryell W, DePaulo JR Jr, Gershon ES, Kelsoe JR, McInnis MG, Murphy DL, Reich T, Scheftner W, Nurnberger JI Jr, McMahon FJ (2004) Loci on chromosomes 6q and 6p interact to increase susceptibility to bipolar affective disorder in the national institute of mental health genetics initiative pedigrees. Biol Psychiatry 56:18–23
Segond N, Degrelle SA, Berndt S, Clouqueur E, Rouault C, Saubamea B, Dessen P, Fong KS, Csiszar K, Badet J, Evain-Brion D, Fournier T (2013) Transcriptome analysis of PPARγ target genes reveals the involvement of lysyl oxidase in human placental cytotrophoblast invasion. PLoS One 8:e79413

Shirayama Y, Takahashi M, Osone F, Hara A, Okubo T (2017) Myo-inositol, glutamate, and glutamine in the prefrontal cortex, hippocampus, and amygdala in major depression. Biol Psychiatry Cogn Neurosci Neuroimaging 2:196–204

Shokri-Kojori E, Wang GJ, Wiers CE, Demiral SB, Guo M, Kim SW, Lindgren E, Ramirez V, Zehra A, Freeman C, Miller G, Manza P, Srivastava T, De Santi S, Tomasi D, Benveniste H, Volkow ND (2018) β-Amyloid accumulation in the human brain after one night of sleep deprivation. Proc Natl Acad Sci U S A 115:4483–4488

Song D, Du T, Li B, Cai L, Gu L, Li H, Chen Y, Hertz L, Peng L (2008) Astrocytic alkalinization by therapeutically relevant lithium concentrations: implications for myo-inositol depletion. Psychopharmacology 200:187–195

Song D, Man Y, Li B, Xu J, Hertz L, Peng L (2013) Comparison between drug-induced and K^+-induced changes in molar acid extrusion fluxes (JH^+) and in energy consumption rates in astrocytes. Neurochem Res 38:2364–2374

Sotelo C, Cholley B, El Mestikawy S, Gozlan H, Hamon M (1990) Direct immunohistochemical evidence of the existence of 5-HT1A autoreceptors on serotoninergic neurons in the midbrain raphe nuclei. Eur J Neurosci 2:1144–1154

Spector R, Lorenzo AV (1975) The origin of myo-inositol in brain, cerebrospinal fluid and choroid plexus. J Neurochem 25:353–354

Steel Z, Marnane C, Iranpour C, Chey T, Jackson JW, Patel V, Silove D (2014) The global prevalence of common mental disorders: a systematic review and meta-analysis 1980–2013. Int J Epidemiol 43:476–493

Stenovec M, Li B, Verkhratsky A, Zorec R (2020) Astrocytes in rapid ketamine antidepressant action. Neuropharmacology 173:108158

Strong PV, Greenwood BN, Fleshner M (2009) The effects of the selective 5-HT(2C) receptor antagonist SB 242084 on learned helplessness in male Fischer 344 rats. Psychopharmacology 203:665–675

Suárez-Santiago JE, Briones-Aranda A, Espinosa-Raya J, Picazo O (2017) Agonist E-6837 and antagonist SB-271046 of 5-HT6 receptors both reverse the depressive-like effect induced in mice by subchronic ketamine administration. Behav Pharmacol 28:582–585

Sun L, Fang L, Lian B, Xia JJ, Zhou CJ, Wang L, Mao Q, Wang XF, Gong X, Liang ZH, Bai SJ, Liao L, Wu Y, Xie P (2017) Biochemical effects of venlafaxine on astrocytes as revealed by 1H NMR-based metabolic profiling. Mol BioSyst 13:338–349

Szabo ST, Blier P (2001a) Response of the norepinephrine system to antidepressant drugs. CNS Spectr 6:679–684

Szabo ST, Blier P (2001b) Serotonin (1A) receptor ligands act on norepinephrine neuron firing through excitatory amino acid and GABA(A) receptors: a microiontophoretic study in the rat locus coeruleus. Synapse 42:203–212

Szabo ST, Blier P (2002) Effects of serotonin (5-hydroxytryptamine, 5-HT) reuptake inhibition plus 5–HT2A receptor antagonism on the fifiring activity of norepinephrine neurons. J Pharmacol Exp Ther 302:983–991

Taciak PP, Lysenko N, Mazurek AP (2018) Drugs which influence serotonin transporter and serotonergic receptors: pharmacological and clinical properties in the treatment of depression. Pharmacol Rep 70:37–46

Tanaka KF, Samuels BA, Hen R (2012) Serotonin receptor expression along the dorsal-ventral axis of mouse hippocampus. Philos Trans R Soc Lond Ser B Biol Sci 367:2395–2401

Thiebot MH, Martin P (1991) Effects of benzodiazepines, 5-HT1A agonists and 5-HT3 antagonists in animal models sensitive to antidepressant drugs. In: Rodgers RJ, Cooper SJ (eds) 5-HT1A agonists, 5-HT3 antagonists and benzodiazepines: their comparative behavioural pharmacology. Wiley, Chichester, pp 159–194

Tiger M, Rück C, Forsberg A, Varrone A, Lindefors N, Halldin C, Farde L, Lundberg J (2014) Reduced 5-HT(1B) receptor binding in the dorsal brain stem after cognitive behavioural therapy of major depressive disorder. Psychiatry Res 223:164–170

Tsybko AS, Ilchibaeva TV, Popova NK (2017) Role of glial cell line-derived neurotrophic factor in the pathogenesis and treatment of mood disorders. Rev Neurosci 28:219–233

Vallee A, Lecarpentier Y, Guillevin R, Vallee JN (2017) Interactions between TGFbeta1, canonical WNT/beta-catenin pathway and PPAR gamma in radiationinduced fibrosis. Oncotarget 8:90579–90604

Varnäs K, Nyberg S, Halldin C, Varrone A, Takano A, Karlsson P, Andersson J, McCarthy D, Smith M, Pierson ME, Söderström J, Farde L (2011) Quantitative analysis of [11C]AZ10419369 binding to 5-HT1B receptors in human brain. J Cereb Blood Flow Metab 31:113–123

Verkhratsky A, Butt AM (2013) Glial physiology and pathophysiology. Wiley-Blackwell, Chichester

Verkhratsky A, Nedergaard M (2014) Astroglial cradle in the life of the synapse. Philos Trans R Soc Lond Ser B Biol Sci 369:20130595

Verkhratsky A, Nedergaard M (2018) Physiology of astroglia. Physiol Rev 98:239–389

Verkhratsky A, Parpura V (2010) Recent advances in (patho)physiology of astroglia. Acta Pharmacol Sin 31:1044–1054

Verkhratsky A, Parpura V (2015) Physiology of astroglia: channels, receptors, transporters, ion signaling and gliotransmission. In: Verkhratsky A, Parpura V (eds) Colloquium series on neuroglia in biology and medicine: from physiology to disease. Morgan & Claypool Publishers, San Rafael, p 172

Verkhratsky A, Rodríguez JJ, Parpura V (2012) Calcium signalling in astroglia. Mol Cell Endocrinol 353:45–56

Verkhratsky A, Rodriguez JJ, Steardo L (2014) Astrogliopathology: a central element of neuropsychiatric diseases? Neuroscientist 20:576–588

Verkhratsky A, Zorec R, Parpura V (2017) Stratification of astrocytes in healthy and diseased brain. Brain Pathol 27:629–644

Verkhratsky A, Semyanov A, Zorec R (2020) Physiology of astroglial excitability. Function 1:zqaa016

Verkhratsky A, Augusto-Oliveira M, Pivoriunas A, Popov A, Brazhe A, Semyanov A (2021) Astroglial asthenia and loss of function, rather than reactivity, contribute to the ageing of the brain. Pflugers Arch 473:753–774

Wallace A, Pehrson AL, Sánchez C, Morilak DA (2014) Vortioxetine restores reversal learning impaired by 5-HT depletion or chronic intermittent cold stress in rats. Int J Neuropsychopharmacol 17:1695–1706

Wang J, Qiao J, Zhang Y, Wang H, Zhu S, Zhang H, Hartle K, Guo H, Guo W, He J, Kong J, Huang Q, Li XM (2014) Desvenlafaxine prevents white matter injury and improves the decreased phosphorylation of the rate-limiting enzyme of cholesterol synthesis in a chronic mouse model of depression. J Neurochem 131:229–238

Wang L, Zhang Y, Du X, Ding T, Gong W, Liu F (2019) Review of antidepressants in clinic and active ingredients of traditional Chinese medicine targeting 5-HT1A receptors. Biomed Pharmacother 120:109408

Watanabe N, Omori IM, Nakagawa A, Cipriani A, Barbui C, Churchill R, Furukawa TA (2011) Mirtazapine versus other antidepressive agents for depression. Cochrane Database Syst Rev 7:CD006528

Wolfson M, Hertz E, Belmaker RH, Hertz L (1998) Chronic treatment with lithium and pretreatment with excess inositol reduce inositol pool size in astrocytes by different mechanisms. Brain Res 787:34–40

Wolfson M, Bersudsky Y, Zinger E, Simkin M, Belmaker RH, Hertz L (2000) Chronic treatment of human astrocytoma cells with lithium, carbamazepine or valproic acid decreases inositol uptake at high inositol concentrations but increases it at low inositol concentrations. Brain Res 855:158–161

Wong YH, Kalmbach SJ, Hartman BK, Sherman WR (1987) Immunohistochemical staining and enzyme activity measurements show myo-inositol-1-phosphate synthase to be localized in the vasculature of brain. J Neurochem 48:1434–1442

Wu C, Singh SK, Dias P et al (1999) Activated astrocytes display increased 5-HT2a receptor expression in pathological states. Exp Neurol 158:529–533
Xia M, Yang L, Sun G, Qi S, Li B (2017) Mechanism of depression as a risk factor in the development of Alzheimer's disease: the function of AQP4 and the glymphatic system. Psychopharmacology (Berlin) 234:365–379
Xia M, Li X, Yang L, Ren J, Sun G, Qi S, Verkhratsky A, Li B (2018) The ameliorative effect of fluoxetine on neuroinflammation induced by sleep deprivation. J Neurochem 146:63–75
Xia M, Li Z, Li S, Liang S, Li X, Chen B, Zhang M, Dong C, Verkhratsky A, Guan D, Li B (2020) Sleep deprivation selectively Down-regulates astrocytic 5-HT2B receptors and triggers depressive-like behaviors via stimulating P2X7 receptors in mice. Neurosci Bull 36:1259–1270
Xie L, Kang H, Xu Q, Chen MJ, Liao Y, Thiyagarajan M, O'Donnell J, Christensen DJ, Nicholson C, Iliff JJ, Takano T, Deane R, Nedergaard M (2013) Sleep drives metabolite clearance from the adult brain. Science 342:373–377
Xu Z, Jing X, Li G, Sun J, Guo H, Hu Y, Sun F, Wen X, Chen F, Wang T, Lu XP (2019) Valproate decreases vitamin D levels in pediatric patients with epilepsy. Seizure 71:60–65
Yamauchi M, Tatebayashi T, Nagase K, Kojima M, Imanishi T (2004) Chronic treatment with fluvoxamine desensitizes 5-HT2C receptor-mediated hypolocomotion in rats. Pharmacol Biochem Behav 78:683–689
Yu W, Greenberg ML (2016) Inositol depletion, GSK3 inhibition and bipolar disorder. Future Neurol 11:135–148
Yu C, Lyu J, Chen Y, Guo Y, Paul S, Bian Z, Zhou H, Tan Y, Chen J, Chen Z, Li L (2015) Epidemiology of major depressive episodes among Chinese adults aged 30-79 years: data from the China Kadoorie Biobank. Zhonghua Liu Xing Bing Xue Za Zhi 36:52–56
Yue T, Li B, Gu L, Huang J, Verkhratsky A, Peng L (2019) Ammonium induced dysfunction of 5-HT2B receptor in astrocytes. Neurochem Int 129:104479
Zarate CA, Manji HK (2008) Riluzole in psychiatry: a systematic review of the literature. Expert Opin Drug Metab Toxicol 4:1223–1234
Zeidán-Chuliá F, Salmina AB, Malinovskaya NA, Noda M, Verkhratsky A, Moreira JC (2014) The glial perspective of autism spectrum disorders. Neurosci Biobehav Rev 38:160–172
Zhang S, Li B, Lovatt D, Xu J, Song D, Goldman SA, Nedergaard M, Hertz L, Peng L (2010) 5-HT2B receptors are expressed on astrocytes from brain and in culture and are a chronic target for all five conventional 'serotonin-specific reuptake inhibitors'. Neuron Glia Biol 6:113–125
Zhang M, Zhou D, Wang Y, Maier DL, Widzowski DV, Sobotka-Briner CD, Brockel BJ, Potts WM, Shenvi AB, Bernstein PR, Pierson ME (2011) Preclinical pharmacology and pharmacokinetics of AZD3783, a selective 5-hydroxytryptamine 1B receptor antagonist. J Pharmacol Exp Ther 339:567–578
Zhang X, Song D, Gu L, Ren Y, Verkhratsky A, Peng L (2015) Decrease of gene expression of astrocytic 5-HT2B receptors parallels development of depressive phenotype in a mouse model of Parkinson's disease. Front Cell Neurosci 9:388
Zhong XM, Dong M, Wang F, Zhang Q, Ungvari GS, Ng CH, Chiu HFK, Si TM, Sim K, Avasthi A, Grover S, Chong MY, Chee KY, Kanba S, Lee MS, Yang SY, Udomratn P, Kallivayalil RA, Tanra AJ, Maramis MM, Shen WW, Sartorius N, Mahendran R, Tan CH, Shinfuku N, Xiang YT (2018) Physical comorbidities in older adults receiving antidepressants in Asia. Psychogeriatrics 18:351–356
Zielinski MR, Gerashchenko D, Karpova SA, Konanki V, McCarley RW, Sutterwala FS, Strecker RE, Basheer R (2017) The NLRP3 inflammasome modulates sleep and NREM sleep delta power induced by spontaneous wakefulness, sleep deprivation and lipopolysaccharide. Brain Behav Immun 62:137–150

Ketamine Action on Astrocytes Provides New Insights into Rapid Antidepressant Mechanisms

Matjaž Stenovec, Baoman Li, Alexei Verkhratsky, and Robert Zorec

Astrocytes and Mood Disorders

Astrocytes are morphologically and functionally heterogeneous glial cells that principally provide homeostasis and defence of the central nervous system (CNS), the latter being associated with the evolutionary conserved program of reactive astrogliosis (Verkhratsky and Nedergaard 2018; Verkhratsky et al. 2017). During development, they (in the form of radial glia) guide migrating neurones towards neocortical destinations and instruct neurones to form synapses (Ullian et al. 2001). Throughout life, they promote survival of neurones (Seri et al. 2001) and signal back to them by the regulated release of gliosignalling molecules (Verkhratsky et al. 2016; Zorec et al. 2016) and factors that regulate synaptic connectivity and strength

M. Stenovec · R. Zorec (✉)
Celica BIOMEDICAL, Ljubljana, Slovenia

Laboratory of Neuroendocrinology-Molecular Cell Physiology, Institute of Pathophysiology, Faculty of Medicine, University of Ljubljana, Ljubljana, Slovenia
e-mail: matjaz.stenovec@mf.uni-lj.si; robert.zorec@mf.uni-lj.si

B. Li
Practical Teaching Centre, School of Forensic Medicine, China Medical University, Shenyang, China

Department of Poison Analysis, School of Forensic Medicine, China Medical University, Shenyang, China
e-mail: bmli@cmu.edu.cn

A. Verkhratsky
Celica BIOMEDICAL, Ljubljana, Slovenia

Faculty of Biology, Medicine and Health, The University of Manchester, Manchester, UK

Achucarro Center for Neuroscience, IKERBASQUE, Bilbao, Spain
e-mail: alexej.verkhratsky@manchester.ac.uk

B. Li et al. (eds.), *Astrocytes in Psychiatric Disorders*, Advances in Neurobiology 26, https://doi.org/10.1007/978-3-030-77375-5_14

involved in learning and memory formation (Clarke and Barres 2013; Zorec et al. 2015). Being key providers of metabolites and growth factors, and acting as stabilizers of neuronal environment, impairment of astrocytes in pathology could destabilise operation of neural circuits in brain areas implicated in mood regulation. Unsurprisingly, numerous studies already highlighted the association between psychiatric disorders such as schizophrenia, bipolar disorder or major depressive disorder (MDD) and astrocytes (Dietz et al. 2020; Peng et al. 2016; Rajkowska and Stockmeier 2013; Verkhratsky and Nedergaard 2014). The abnormalities in the amount of glutamate at synapses controlled by astrocytes have also been linked to depression, anxiety and schizophrenia (Holden 2003). Besides glutamate, astrocytes recycle serotonin, dopamine and other monoamines implicated in psychiatric disorders.

Unlike in many other neurological conditions including neurotrauma or neurodegenerative disorders, astroglial atrophy and degeneration without signs of reactivity are characteristic in psychiatric disorders and MDD in particular (Rajkowska and Stockmeier 2013; Verkhratsky et al. 2014). In postmortem samples from depressed individuals, packing density or number of Nissl-stained populations of astrocytes is reduced (Bowley et al. 2002; Cotter et al. 2001, 2002; Gittins and Harrison 2011; Ongur et al. 1998; Rajkowska et al. 1999). These changes were observed in various brain regions including orbitofrontal cortex (Rajkowska et al. 1999), dorsolateral prefrontal cortex (Cotter et al. 2002; Rajkowska et al. 1999), anterior cingulate cortex (Cotter et al. 2001; Gittins and Harrison 2011), subgenual cortex (Ongur et al. 1998) and amygdala (Bowley et al. 2002). In white matter of postmortem human brain tissue and in rats subjected to chronic stress, the density of astrocytes and glial fibrillary acidic protein (GFAP)-positive area fraction were substantially reduced when compared with healthy controls (Rajkowska and Stockmeier 2013). Following exposure to different types of stress, GFAP expression and density of astrocytes were also reduced in grey matter of rodents exhibiting depressive behaviour (Braun et al. 2009; Czeh et al. 2006). In animal models of attention-deficit disorder and MDD, downregulation of classical astroglial markers such as glutamine synthetase, plasmalemmal glutamate transporters, aquaporin 4 and astroglia-specific connexins was reported (Barley et al. 2009; Bernard et al. 2011; Sequeira et al. 2009). Pharmacological inhibition of astroglial glutamate transporters (Bechtholt-Gompf et al. 2010) and gap junction connectivity (Sun et al. 2012) triggered anhedonia indicative of depression.

Impact of Antidepressants on Astrocytes

Increased awareness of astrocyte importance in regular brain function has provided fresh momentum to an old idea that astrocytes play a causative role in the pathogenesis of psychiatric disorders and can be utilized as cellular targets for antipsychotic drugs. Classic antidepressants (such as Li^+, valproic acid or fluoxetine) indeed affect astrocyte signalling cascades and modify expression of a variety of receptors and transporters responsible for CNS homeostasis and support of synaptic transmission (Czeh and Di Benedetto 2013; Dong et al. 2015; Liu et al. 2015; Peng et al. 2018; Rivera and Butt 2019). Astroglial serotonin 5-HT_{2B} receptors interact with antidepressants belonging to serotonin specific reuptake inhibitors (SSRI) that include fluoxetine, citalopram, paroxetine, sertraline and fluvoxamine (Zhang et al. 2010). Fluoxetine and other SSRIs acting on astroglial 5-HT_{2B} receptors trigger Ca^{2+} signalling (Schipke et al. 2011) and transactivation of EGF receptor linked to MAPK/ERK and PI3K/AKT signalling cascades that modulate multiple homeostatic pathways including expression of glutamate transporters, glucose metabolism, astroglial secretion and activity of Na^+-H^+ exchanger (Hertz et al. 2012, 2015; Liu et al. 2015; Peng et al. 2018; Ren et al. 2015). Besides SSRIs, other antidepressants also affect astrocyte function. Metabolic inhibition of astrocytes by intra-hippocampal injection of fluoroacetate prevents antidepressant action of imipramine (Iwata et al. 2011). Li^+, the oldest classical antidepressant, acting through astrocyte lysyl oxidase, encoded by LOX gene, profoundly affects astrocyte morphogenesis and proliferation (Rivera and Butt 2019). Electroconvulsive therapy, regarded as the last resort for the treatment of MDD (Li et al. 2020), upregulates GFAP expression that is generally suppressed in depressive disorders (Fujiki and Steward 1997; Kragh et al. 1993). Transcranial direct current stimulation triggers cortex-wide global elevation of cytosolic Ca^{2+} in astrocytes that is mediated by $\alpha 1$ adrenoreceptor activation and astroglia-specific $InsP_3$ receptors type 2 (Monai et al. 2016).

Ketamine: The Fastest Acting Antidepressant

Ketamine, (S)-(+) and (R)-(–)-2-(2-chlorophenyl)-2-(methylamino) cyclohexanone, is an arylcycloalkylamine structurally related to phencyclidine that is occasionally used as an anaesthetic in human and veterinary medicine. It also exerts analgesia and dysphoria and is sympathomimetic. Ketamine is quickly metabolized to norketamine and nordehydroketamine that appear in venous blood 10 and 30 min after administration. (S)-(+) and (R)-(–) enantiomers have similar pharmacokinetic profiles (Domino 2010). Since first clinical studies demonstrated that a single subanaesthetic dose of ketamine evokes rapid (within 2 h) and lasting (up to 2 weeks) antidepressant effect, ketamine generated substantial interest in psychiatry (Berman et al. 2000). The speed of antidepressant effect is in striking contrast to classic SSRI antidepressants that target monoamine system and typically require weeks to produce the therapeutic effect.

Ketamine binds to the phencyclidine-binding site of the NMDA receptor and blocks the ion flux through the receptor pore. (R)-(–) ketamine has an approximately fourfold smaller affinity for NMDA receptor than (S)-(+) ketamine and shows smaller negative psychotomimetic side effects when compared to (S)-(+) ketamine (Domino 2010); its antidepressant effect is seemingly more potent than of (S)-(+) ketamine (Yang et al. 2015). The antidepressant effect of ketamine outlasts the lifetime of the drug in the organism (Mion and Villevieille 2013) and indicates that the molecular mechanism of ketamine effect extends beyond mere NMDA receptor antagonism. As other NMDA receptor antagonists do not exhibit antidepressant capabilities (Newport et al. 2015; Zanos et al. 2016), ketamine inevitably affects additional molecular targets besides NMDAR. Ketamine increases synaptogenesis, dendritic spine density, expression of α-amino-3-hydroxy-5-methyl-4-isoxazolepropionic acid (AMPA) receptor and arborisation of astrocytes (Ardalan et al. 2017; Li et al. 2010; Maeng et al. 2008). The latter control membrane transport of glutamate and glucose, glial-to-neurone gap junction communication, cell volume and blood flow (Verkhratsky and Nedergaard 2018), and ketamine affects various aspects of astrocyte physiology (Lasic et al. 2016; Mitterauer 2012; Stenovec et al. 2016; Thrane et al. 2012; Wray et al. 2019).

Ketamine Increases Astrocyte cAMP and Alters Single Vesicle Interactions with the Plasmalemma

In protoplasmic astrocytes, cAMP plays a role in the establishment of arborised cell structure (Koyama et al. 1993; Schiweck et al. 2018; Won and Oh 2000) that creates territorial domain of an individual astrocyte (Bushong et al. 2002). Within this domain, astrocyte contacts up to 100,000 synapses (in rodents) through fine, submicrometre-sized processes (Halassa et al. 2007; Ogata and Kosaka 2002; Witcher et al. 2007). These processes and astrocyte-derived factors promote synaptogenesis (Pfrieger 2010) and influence synaptic activity (Fiacco et al. 2009; Nimmerjahn 2009; Perea et al. 2009; Santello and Volterra 2009; Theodosis et al. 2008). In the absence of G-protein-coupled receptor activation, ketamine induces a sustained increase in intracellular cAMP concentration ($[cAMP]_i$) in astrocytes (Fig. 1) (Lasic et al. 2016; Wray et al. 2019). This increase in $[cAMP]_i$ may attenuate delivery of vesicle-laden channels to the plasmalemma (Potokar et al. 2013; Stenovec et al. 2016, 2020) and modify luminal cargo release or uptake of extracellular molecules through altered fusion-pore structure (Lasic et al. 2016; Stenovec et al. 2016). In C6 glioma cells, ketamine facilitates cAMP signalling in a NMDA receptor-independent manner by affecting the translocation of $G_{\alpha s}$ from lipid rafts to non-raft membrane microdomains allowing increased functional coupling of $G_{\alpha s}$ and adenylyl cyclase. Ketamine subsequently increases phosphorylation of cAMP-response element-binding protein (CREB), which, in turn, increases the expression of BDNF (Wray et al. 2019) that has been widely implicated in the pathophysiology of MDD (Autry

Fig. 1 (continued) signal increase ($k = 2.2 \pm 0.6\%$/min). Ordinate on the right displays the values of $[cAMP]_i$ estimated from equation: $[cAMP]_i = EC_{50} \times ((R - R_{min})/(R_{max} - R))^{1/n}$ (Borner et al. 2011). (**c**) Mean amplitude (±SE; ΔFRET; left panel) and initial rate of the FRET signal change (ΔFRET/Δtime; right panel) in controls (white bars) and in astrocytes treated with ketamine (KM, black bars). Changes in the FRET signal are expressed as percentages of the initial values. Numbers above the error bars depict the number of cells analysed. Mann-Whitney U test: $*P < 0.05$, $**P < 0.01$. (Reproduced with permission from Lasic and colleagues (2016))

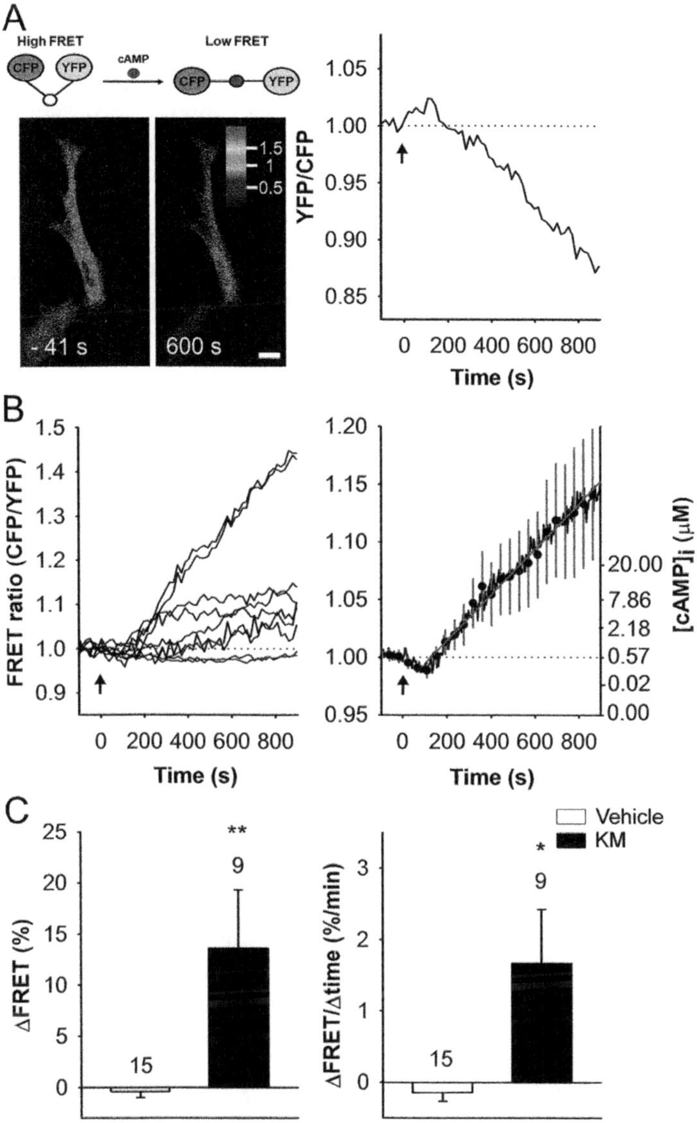

Fig. 1 Ketamine increases [cAMP]$_i$ in cultured rat astrocytes. (**a**) Schematic representation of FRET-based Epac1-camps nanosensor function (top) and pseudo-coloured FRET images of an astrocyte expressing the FRET-based nanosensor Epac1-camps before (−41 s) and after (+600 s) application of 25μM ketamine (bottom) and the corresponding normalized time-resolved trace of the Epac1-camps FRET signal (YFP/CFP; right). The pseudo-coloured scale indicates the YFP/CFP ratio values. Scale bar, 20μm. (**b**) Time-resolved (left panel) and the mean time-resolved (±SE; right panel) traces of the Epac1-camps FRET signal in astrocytes (n = 9) stimulated with ketamine at t = 0 s (black arrows). Data are expressed as the inverted FRET signal (CFP/YFP). Ketamine-evoked increase in the FRET signal reflects an increase in [cAMP]$_i$. The initial rate of change in [cAMP]$_i$ (red line; right panel) was determined by fitting the regression line to the FRET

and Monteggia 2012; Duman et al. 2016; Jiang and Salton 2013). Besides neurones, astrocytes also synthesize BDNF and are targeted by antidepressants (Autry and Monteggia 2012; Bjorkholm and Monteggia 2016; Hisaoka-Nakashima et al. 2016). In primary cortical astrocytic, but not in neuronal cultures, SSRI antidepressants fluoxetine and paroxetine, and the tricyclic antidepressants imipramine and amitriptyline upregulate the BDNF mRNA levels in a monoamine-independent manner (Allaman et al. 2011; Hisaoka-Nakashima et al. 2016; Kittel-Schneider et al. 2012; Musazzi et al. 2014; Takano et al. 2012), possibly through a direct effect on the extracellular signal-regulated kinase- and p38 mitogen-activated protein kinase-dependent signalling pathway. At single vesicle level, ketamine suppresses exocytotic release of BDNF from cultured rat astrocytes (Stenovec et al. 2016) and attenuates ATP-evoked increases in $[Ca^{2+}]_i$ in cultured cortical (Stenovec et al. 2016) and neocortical (Thrane et al. 2012) astrocytes in mice. Reduced vesicular release of BDNF and the diminished frequency of reversible and irreversible exocytotic events (Stenovec et al. 2016) can be explained by attenuation of ATP-evoked calcium signalling in ketamine-treated astrocytes. Ketamine may diminish stimulus-evoked Ca^{2+} signalling by modifying Ca^{2+} entry through TRP channels (Abdelhamid et al. 2014; Bahnasi et al. 2008), which contribute to the replenishment of the endoplasmic reticulum store (capacitive function) (Malarkey et al. 2008) and to the plateau phase of the Ca^{2+} transient (Verkhratsky et al. 2012). Moreover, ketamine disrupts synchronization of astrocytic slow inward currents presumably mediated by the extrasynaptic GluN1/GluN2B receptor activation and inhibits glutamate transmission from astrocytes to neurones (Zhang et al. 2019).

In clinically relevant concentrations, ketamine modulates interaction of single vesicles with the astrocyte plasmalemma by stabilizing a narrow fusion pore (Lasic et al. 2016). Low ketamine concentration of 0.25–2.5μM (Tassonyi et al. 2002) causes vesicles to lapse into a state of repetitive fusion pore opening and closing (Lasic et al. 2016). Flickering pore activity may hinder cargo discharge (Kreft et al. 2018) or uptake and/or vesicle recycling affecting retention of membrane receptors, transporters and ion channels at the astrocyte surface (Zorec et al. 2016). The mechanism(s) underlying flickering pore activity is/are still debated. As ketamine permeates membranes at physiological pH (Keiser et al. 2018) and accumulates inside acidified vesicles as a protonated weak base (Lester et al. 2015), it may directly affect the membranous pore (Jain et al. 1985) by electrostatically altering the anisotropicity of the pore (Kabaso et al. 2012). As ketamine principally inhibits endocytotic vesicle retrieval (Lasic et al. 2016), it could obstruct the uptake of BDNF into astrocytes (Bergami et al. 2008; Vignoli et al. 2016) and favour increased level of extracellular BDNF that is likelier to signal to nearby synapses and support long-term potentiation via sustained TrkB activation leading to enhancement of synaptic strength between neurones (Autry and Monteggia 2012; Duman et al. 2016; Jiang and Salton 2013).

Ketamine Induces Cholesterol Redistribution in the Astrocyte Plasmalemma

Synaptic transmission between neurones enables seamless operation of the central nervous system. Synaptic connectivity and synaptic function are regulated by perisynaptic astrocyte processes (Verkhratsky and Nedergaard 2014, 2018). Unless supported by astrocytes, cultured retinal ganglion neurones form few synapses that are functionally immature (Nagler et al. 2001; Pfrieger and Barres 1997; Ullian et al. 2001). Upon addition of astrocytes, the total number of synapses on neurones increases by sevenfold (Pfrieger and Barres 1997). As synapses formed in the presence of astrocytes are quickly lost after their removal, astrocyte-derived signals are also required to stabilize synapses (Ullian et al. 2001). One of the key synapse-promoting signals released by astrocytes to neurones is cholesterol (Mauch et al. 2001). In the brain, cholesterol derives almost entirely from in situ synthesis by brain cells (Dietschy and Turley 2001). As the appearance of most synapses in the developing brain temporally and spatially coincides with the development of astrocytes, synapse formation very likely depends on astrocyte-derived cholesterol (Ullian et al. 2001).

On the other side, postmortem studies in psychiatric patients (suffering from schizophrenia, bipolar disorder and MDD) revealed suspicious loss of astrocytes in brain regions important for mood, motivation and cognition (Bowley et al. 2002; Cotter et al. 2001, 2002; Gittins and Harrison 2011; Ongur et al. 1998; Rajkowska et al. 1999). The loss in glia cell numbers in animal models of depression and MDD patients or suicide victims (Rajkowska and Stockmeier 2013; Sanacora and Banasr 2013) is suggestive for hindered supply of astrocyte cholesterol to neurones that require cholesterol to build up axons, dendrites and synapses (Goritz et al. 2005; Pfenninger 2009) and to maintain functional synapses (Theodosis et al. 2008). Presynaptic vesicles with particularly high cholesterol content (40 mol%) and postsynaptic spines account for large amounts of neuronal membranes (Takamori et al. 2006). Neurones produce sufficient amount of cholesterol to survive and grow, yet they require astrocyte-derived cholesterol to form new synapse in abundance and to mature these synapses (Mauch et al. 2001). In ganglion neurones, the number of synapses and their activity (spontaneous and evoked) increases within 3 days after the addition of glia-conditioned medium or cholesterol (5μg/ml) to neuronal culture medium (Goritz et al. 2005).

Earlier in vitro studies indicated that soluble and contact-dependent astrocyte signals stimulate the formation of synapses in neurones from the hippocampus (Hama et al. 2004), retina (Mauch et al. 2001; Nagler et al. 2001; Ullian et al. 2001) and spinal cord (Peng et al. 2003), whereas a recent study (Lasic et al. 2019) revealed that ketamine instigates rapid redistribution of cholesterol in the plasmalemma of cortical astrocytes. Visible redistribution of cholesterol-enriched lipid rafts labelled with a fluorescent cholesterol-specific membrane-binding domain (D4) (Liu et al. 2017; Ohno-Iwashita et al. 1990) was observed already after 30 min ketamine treatment and indicated increase in D4-positive plasmalemmal domain density (Fig. 2).

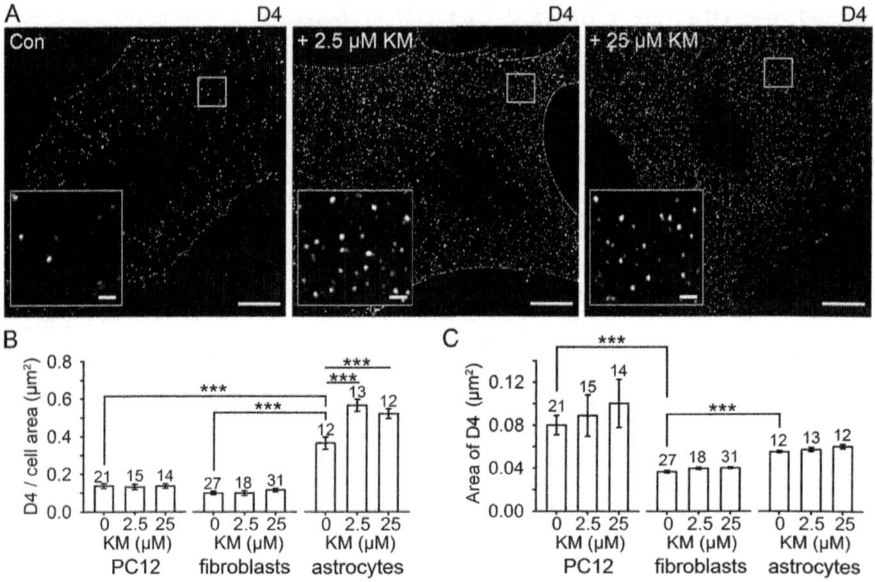

Fig. 2 Ketamine (KM) increases the surface density of cholesterol-rich plasmalemmal domains in astrocytes. (**a**) SIM images displaying mCherry-D4-labelling in non-treated controls (Con) and ketamine-treated astrocytes (2.5μM and 25μM KM, respectively). Insets show mCherry-D4 labelled domains at a higher magnification. Scale bar (inset), 10μm (1μm). (**b**) The density of the membrane cholesterol domains (the number of cholesterol domains (D4) normalised to the cell surface area) significantly increased in astrocytes after treatment with 2.5μM (0.57 ± 0.03 D4/μm²) and 25μM KM (0.52 ± 0.03 D4/μm²) when compared with controls (0.37 ± 0.03 D4/μm²) (***$P < 0.001$, Holm-Sidak one-way ANOVA). Ketamine did not affect the density of cholesterol-rich domains in the PC12 cell line and fibroblasts. The density of the D4-positive domains was higher in astrocytes (0.37 ± 0.03 D4/μm²) compared with PC12 cells (0.14 ± 0.01 D4/μm²) and fibroblasts (0.10 ± 0.01 D4/μm²) (***$P < 0.001$, Kruskal-Wallis test). (**c**) The average area of cholesterol-rich domains in vehicle-treated controls and in ketamine-treated cells (2.5 and 25μM) did not differ in the PC12 cell line, fibroblasts and astrocytes, but it differed between different cell types. The area of D4-positive domains was significantly higher in the PC12 cell line (0.080 ± 0.009μm²) and astrocytes (0.056 ± 0.001μm²) than in fibroblasts (0.037 ± 0.001μm²) (***$P < 0.001$, Kruskal-Wallis test). The data in the graphs are reported as mean ± SEM. Numbers above the bars represent the number of cells analysed. (Reproduced with permission from Lasic and colleagues (2019))

As astrocytes promote maturation of synapses (Mazzanti and Haydon 2003; Nagler et al. 2001; Ullian et al. 2001) and maintain their structure and function (Slezak and Pfrieger 2003; Ullian et al. 2001), increased plasmalemmal cholesterol in ketamine treated astrocytes may transiently boost cholesterol flux to neurones and counteract the pathophysiological reduction in dendritic spine number and function of neurones in animal model of depression (Radley et al. 2006) or reduction in the number of synapses in MDD subjects (Kang et al. 2012). How exactly ketamine increases the density of plasmalemmal cholesterol-enriched domains is still unclear. As an overall increase in astrocyte cholesterol production is unlikely (Kritchevsky et al.

1976; Saranteas et al. 2005), an alternative explanation is needed. In astrocytes, subanaesthetic ketamine dose evokes flickering activity of narrow fusion pore that hinders vesicle retrieval (Lasic et al. 2016). Prolonged flickering activity may contribute to an increased density of plasmalemmal cholesterol, as cholesterol-enriched domains are less effectively internalized via vesicle endocytosis. Cholesterol enrichment in the outer leaflet of the astrocyte plasmalemma (Lasic et al. 2019) may thus result in an enhanced flux of cholesterol to neurons, where it is required for changes in synaptic plasticity (Goritz et al. 2005), leading to sustained strengthening of excitatory synapses necessary for improvement of depressive phenotype (Zanos and Gould 2018). Such proposal, however, requires further experimental study.

Ketamine Attenuates Mobility of Vesicles Delivering Kir4.1

As our lives are shaped by numerous opposing forces, our brain is equipped with the similarly opposing systems guiding our actions and influencing our mood. Neurones in the mesolimbic system promote reward-seeking behaviour, whereas neurones in the lateral habenula (LHb) suppress this behaviour (Howe and Kenny 2018). In LHb neurones, a distinctive pattern of burst firing is causally associated with depressive phenotype and reduced extracellular concentration of K^+ ($[K^+]_o$) (Cui et al. 2018). One of the major conduits for the movement of K^+ between the extracellular space and astrocytes is an inward-rectifier K^+ channel termed $K_{ir}4.1$ (Breslin et al. 2018; Nwaobi et al. 2016). In rat models of depression, this channel was found to be upregulated at the transcript, protein and functional levels (Cui et al. 2018). When overexpressed, $K_{ir}4.1$ hyperpolarises membrane potential and causes burst firing of LHb neurones, which results in depressive behaviour. Decrease in $[K^+]_o$ thus mimics the depressive phenotype. Conversely, disruption of $K_{ir}4.1$ function by viral-mediated knockdown or expression of dominant negative construct depolarised membrane potential and causes tonic firing of LHb neurones. As depression-associated burst firing in neurones is favoured by overexpressed astrocyte $K_{ir}4.1$ lowering $[K^+]_o$ (Yang et al. 2018), modulating $K_{ir}4.1$ activity in the brain areas involved in the regulation of mood and motivation might be a way to treat psychiatric disorders. Despite recent studies demonstrated that ketamine infusion into LHb blocks burst firing and causes rapid antidepressant effects in depression-prone rats, none of them addressed the possibility that ketamine also alters trafficking and/or

Fig. 3 (continued) (top; Con) or treated with 2.5μM (middle) or 25μM (bottom) ketamine (KM). The mask images (white, right) display co-localized pixels. Scale bars, 20μm. (**b**) Ketamine treatment diminishes directional mobility of $K_{ir}4.1$-EGFP vesicles. (**c**) Co-localization (mean ± SE; %) of FM4–64 versus Kir4.1-EGFP fluorescence indicates diminished localization of $K_{ir}4.1$-EGFP at plasmalemma in astrocytes treated with 2.5 or 25μM ketamine (KM). The numbers above the top and bottom of the bars indicate the number of cells (**b**, **c**) and vesicles (**b**) analysed, respectively. $*P < 0.05$ (ANOVA on ranks followed by Dunn's test; **b**, **c**). (Reproduced with permission from Stenovec and colleagues (2020))

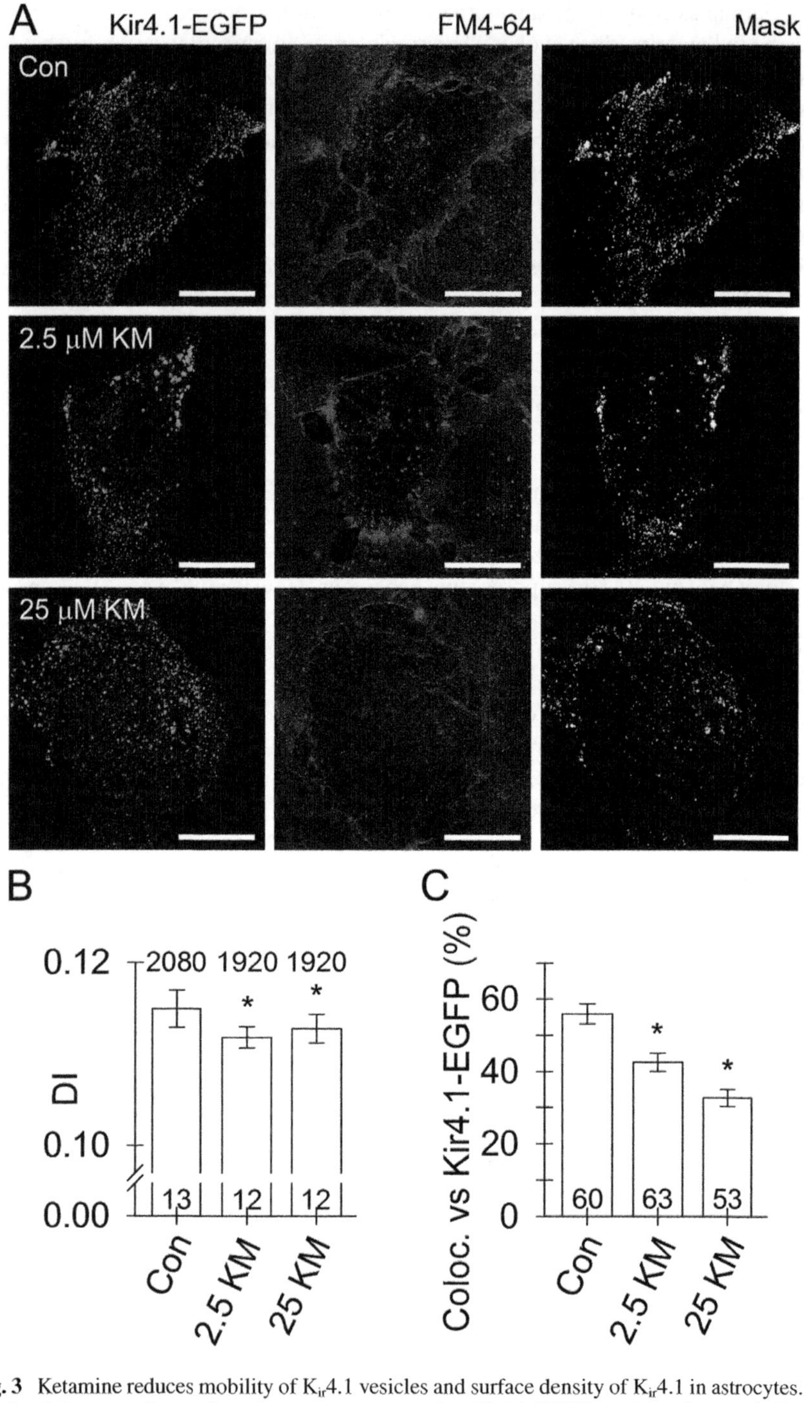

Fig. 3 Ketamine reduces mobility of $K_{ir}4.1$ vesicles and surface density of $K_{ir}4.1$ in astrocytes. (**a**) Confocal images of transfected astrocytes expressing $K_{ir}4.1$-EGFP vesicles (green, left) and labelled with the membrane styryl dye FM4–64 (red, middle) that were not treated with ketamine

surface density of $K_{ir}4.1$ or even inhibits this channel. Notably, $K_{ir}4.1$ is reversibly inhibited by antidepressants of the tricyclic and SSRI class, such as nortriptyline and fluoxetine (Furutani et al. 2009; Ohno et al. 2007; Su et al. 2007). Whether ketamine also inhibits $K_{ir}4.1$ function is currently unknown. Ketamine, however, reduces cytoplasmic mobility of vesicles carrying the inward-rectifier potassium channel ($K_{ir}4.1$) and diminishes $K_{ir}4.1$ density at astrocyte surface (Fig. 3; Stenovec et al. 2020), which may raise extracellular K^+ concentration and alleviate depressive behaviour in animal models of depression (Cui et al. 2018) and possibly also in humans (Berman et al. 2000).

Seemingly diverse but not mutually exclusive mechanisms may act synergistically to evoke changes in synaptic plasticity leading to sustained strengthening of excitatory synapses necessary for antidepressant effect of ketamine. Studies in astrocytes may provide alternative ground for the development of novel astrocyte-targeted therapeutic drugs suitable for the treatment of psychiatric disorders.

Acknowledgements The authors acknowledge financial support from the Slovenian Research Agency (research core funding P3-310) and projects J3 6790, J3 6789, J3 7605 and J3 9266.

References

Abdelhamid RE, Kovács KJ, Nunez MG, Larson AA (2014) Depressive behavior in the forced swim test can be induced by TRPV1 receptor activity and is dependent on NMDA receptors. Pharmacol Res 79:21–27

Allaman I, Fiumelli H, Magistretti PJ, Martin JL (2011) Fluoxetine regulates the expression of neurotrophic/growth factors and glucose metabolism in astrocytes. Psychopharmacology 216:75–84

Ardalan M, Rafati AH, Nyengaard JR, Wegener G (2017) Rapid antidepressant effect of ketamine correlates with astroglial plasticity in the hippocampus. Br J Pharmacol 174:483–492

Autry AE, Monteggia LM (2012) Brain-derived neurotrophic factor and neuropsychiatric disorders. Pharmacol Rev 64:238–258

Bahnasi YM, Wright HM, Milligan CJ, Dedman AM, Zeng F, Hopkins PM, Bateson AN, Beech DJ (2008) Modulation of TRPC5 cation channels by halothane, chloroform and propofol. Br J Pharmacol 153:1505–1512

Barley K, Dracheva S, Byne W (2009) Subcortical oligodendrocyte- and astrocyte-associated gene expression in subjects with schizophrenia, major depression and bipolar disorder. Schizophr Res 112:54–64

Bechtholt-Gompf AJ, Walther HV, Adams MA, Carlezon WA Jr, Ongur D, Cohen BM (2010) Blockade of astrocytic glutamate uptake in rats induces signs of anhedonia and impaired spatial memory. Neuropsychopharmacology 35:2049–2059

Bergami M, Santi S, Formaggio E, Cagnoli C, Verderio C, Blum R, Berninger B, Matteoli M, Canossa M (2008) Uptake and recycling of pro-BDNF for transmitter-induced secretion by cortical astrocytes. J Cell Biol 183:213–221

Berman RM, Cappiello A, Anand A, Oren DA, Heninger GR, Charney DS, Krystal JH (2000) Antidepressant effects of ketamine in depressed patients. Biol Psychiatry 47:351–354

Bernard R, Kerman IA, Thompson RC, Jones EG, Bunney WE, Barchas JD, Schatzberg AF, Myers RM, Akil H, Watson SJ (2011) Altered expression of glutamate signaling, growth factor, and glia genes in the locus coeruleus of patients with major depression. Mol Psychiatry 16:634–646

Bjorkholm C, Monteggia LM (2016) BDNF – a key transducer of antidepressant effects. Neuropharmacology 102:72–79

Borner S, Schwede F, Schlipp A, Berisha F, Calebiro D, Lohse MJ, Nikolaev VO (2011) FRET measurements of intracellular cAMP concentrations and cAMP analog permeability in intact cells. Nat Protoc 6:427–438

Bowley MP, Drevets WC, Ongur D, Price JL (2002) Low glial numbers in the amygdala in major depressive disorder. Biol Psychiatry 52:404–412

Braun K, Antemano R, Helmeke C, Buchner M, Poeggel G (2009) Juvenile separation stress induces rapid region- and layer-specific changes in S100ss- and glial fibrillary acidic protein-immunoreactivity in astrocytes of the rodent medial prefrontal cortex. Neuroscience 160:629–638

Breslin K, Wade JJ, Wong-Lin K, Harkin J, Flanagan B, Van Zalinge H, Hall S, Walker M, Verkhratsky A, McDaid L (2018) Potassium and sodium microdomains in thin astroglial processes: a computational model study. PLoS Comput Biol 14:e1006151

Bushong EA, Martone ME, Jones YZ, Ellisman MH (2002) Protoplasmic astrocytes in CA1 stratum radiatum occupy separate anatomical domains. J Neurosci 22:183–192

Clarke LE, Barres BA (2013) Emerging roles of astrocytes in neural circuit development. Nat Rev Neurosci 14:311–321

Cotter D, Mackay D, Landau S, Kerwin R, Everall I (2001) Reduced glial cell density and neuronal size in the anterior cingulate cortex in major depressive disorder. Arch Gen Psychiatry 58:545–553

Cotter D, Mackay D, Chana G, Beasley C, Landau S, Everall IP (2002) Reduced neuronal size and glial cell density in area 9 of the dorsolateral prefrontal cortex in subjects with major depressive disorder. Cereb Cortex 12:386–394

Cui Y, Yang Y, Ni Z, Dong Y, Cai G, Foncelle A, Ma S, Sang K, Tang S, Li Y, Shen Y, Berry H, Wu S, Hu H (2018) Astroglial Kir4.1 in the lateral habenula drives neuronal bursts in depression. Nature 554:323–327

Czeh B, Di Benedetto B (2013) Antidepressants act directly on astrocytes: evidences and functional consequences. Eur Neuropsychopharmacol 23:171–185

Czeh B, Simon M, Schmelting B, Hiemke C, Fuchs E (2006) Astroglial plasticity in the hippocampus is affected by chronic psychosocial stress and concomitant fluoxetine treatment. Neuropsychopharmacology 31:1616–1626

Dietschy JM, Turley SD (2001) Cholesterol metabolism in the brain. Curr Opin Lipidol 12:105–112

Dietz AG, Goldman SA, Nedergaard M (2020) Glial cells in schizophrenia: a unified hypothesis. Lancet Psychiatry 7(3):272–281

Domino EF (2010) Taming the ketamine tiger. 1965. Anesthesiology 113:678–684

Dong L, Li B, Verkhratsky A, Peng L (2015) Cell type-specific in vivo expression of genes encoding signalling molecules in the brain in response to chronic mild stress and chronic treatment with fluoxetine. Psychopharmacology 232:2827–2835

Duman RS, Aghajanian GK, Sanacora G, Krystal JH (2016) Synaptic plasticity and depression: new insights from stress and rapid-acting antidepressants. Nat Med 22:238–249

Fiacco TA, Agulhon C, McCarthy KD (2009) Sorting out astrocyte physiology from pharmacology. Annu Rev Pharmacol Toxicol 49:151–174

Fujiki M, Steward O (1997) High frequency transcranial magnetic stimulation mimics the effects of ECS in upregulating astroglial gene expression in the murine CNS. Brain Res Mol Brain Res 44:301–308

Furutani K, Ohno Y, Inanobe A, Hibino H, Kurachi Y (2009) Mutational and in silico analyses for antidepressant block of astroglial inward-rectifier Kir4.1 channel. Mol Pharmacol 75:1287–1295

Gittins RA, Harrison PJ (2011) A morphometric study of glia and neurons in the anterior cingulate cortex in mood disorder. J Affect Disord 133:328–332

Goritz C, Mauch DH, Pfrieger FW (2005) Multiple mechanisms mediate cholesterol-induced synaptogenesis in a CNS neuron. Mol Cell Neurosci 29:190–201

Halassa MM, Fellin T, Takano H, Dong JH, Haydon PG (2007) Synaptic islands defined by the territory of a single astrocyte. J Neurosci 27:6473–6477
Hama H, Hara C, Yamaguchi K, Miyawaki A (2004) PKC signaling mediates global enhancement of excitatory synaptogenesis in neurons triggered by local contact with astrocytes. Neuron 41:405–415
Hertz L, Li B, Song D, Ren J, Dong L, Chen Y, Peng L (2012) Astrocytes as a 5-HT2B-mediated SERT-independent SSRI target, slowly altering depression-associated genes and function. Curr Signal Transduct Ther 7(1):65–80
Hertz L, Rothman DL, Li B, Peng L (2015) Chronic SSRI stimulation of astrocytic 5-HT2B receptors change multiple gene expressions/editings and metabolism of glutamate, glucose and glycogen: a potential paradigm shift. Front Behav Neurosci 9:25
Hisaoka-Nakashima K, Kajitani N, Kaneko M, Shigetou T, Kasai M, Matsumoto C, Yokoe T, Azuma H, Takebayashi M, Morioka N, Nakata Y (2016) Amitriptyline induces brain-derived neurotrophic factor (BDNF) mRNA expression through ERK-dependent modulation of multiple BDNF mRNA variants in primary cultured rat cortical astrocytes and microglia. Brain Res 1634:57–67
Holden C (2003) Psychiatric drugs. Excited by glutamate. Science 300:1866–1868
Howe WM, Kenny PJ (2018) Burst firing sets the stage for depression. Nature 554:304–305
Iwata M, Shirayama Y, Ishida H, Hazama GI, Nakagome K (2011) Hippocampal astrocytes are necessary for antidepressant treatment of learned helplessness rats. Hippocampus 21:877–884
Jain MK, Jahagirdar DV, Van Linde M, Roelofsen B, Eibl H (1985) Solute-induced acceleration of transbilayer movement and its implications on models of blood-brain barrier. Biochim Biophys Acta 818:356–364
Jiang C, Salton SR (2013) The role of neurotrophins in major depressive disorder. Transl Neurosci 4:46–58
Kabaso D, Calejo AI, Jorgacevski J, Kreft M, Zorec R, Iglic A (2012) Fusion pore diameter regulation by cations modulating local membrane anisotropy. ScientificWorldJournal 2012:983138
Kang HJ, Voleti B, Hajszan T, Rajkowska G, Stockmeier CA, Licznerski P, Lepack A, Majik MS, Jeong LS, Banasr M, Son H, Duman RS (2012) Decreased expression of synapse-related genes and loss of synapses in major depressive disorder. Nat Med 18:1413–1417
Keiser M, Hasan M, Oswald S (2018) Affinity of ketamine to clinically relevant transporters. Mol Pharm 15:326–331
Kittel-Schneider S, Kenis G, Schek J, van den Hove D, Prickaerts J, Lesch KP, Steinbusch H, Reif A (2012) Expression of monoamine transporters, nitric oxide synthase 3, and neurotrophin genes in antidepressant-stimulated astrocytes. Front Psych 3:33
Koyama Y, Ishibashi T, Hayata K, Baba A (1993) Endothelins modulate dibutyryl cAMP-induced stellation of cultured astrocytes. Brain Res 600:81–88
Kragh J, Bolwig TG, Woldbye DP, Jorgensen OS (1993) Electroconvulsive shock and lidocaine-induced seizures in the rat activate astrocytes as measured by glial fibrillary acidic protein. Biol Psychiatry 33:794–800
Kreft M, Jorgacevski J, Stenovec M, Zorec R (2018) Angstrom-size exocytotic fusion pore: implications for pituitary hormone secretion. Mol Cell Endocrinol 463:65–71
Kritchevsky D, Tepper SA, Davidson LM, Story JA (1976) Influence of ketamine, phenylcyclidine, and phenobarbital on cholesterol metabolism in rats. Proc Soc Exp Biol Med 151:445–447
Lasic E, Rituper B, Jorgacevski J, Kreft M, Stenovec M, Zorec R (2016) Subanesthetic doses of ketamine stabilize the fusion pore in a narrow flickering state in astrocytes. J Neurochem 138:909–917
Lasic E, Lisjak M, Horvat A, Bozic M, Sakanovic A, Anderluh G, Verkhratsky A, Vardjan N, Jorgacevski J, Stenovec M, Zorec R (2019) Astrocyte specific remodeling of plasmalemmal cholesterol composition by ketamine indicates a new mechanism of antidepressant action. Sci Rep 9:10957
Lester HA, Lavis LD, Dougherty DA (2015) Ketamine inside neurons? Am J Psychiatry 172:1064–1066

Li N, Lee B, Liu RJ, Banasr M, Dwyer JM, Iwata M, Li XY, Aghajanian G, Duman RS (2010) mTOR-dependent synapse formation underlies the rapid antidepressant effects of NMDA antagonists. Science 329:959–964

Li M, Yao X, Sun L, Zhao L, Xu W, Zhao H, Zhao F, Zou X, Cheng Z, Li B, Yang W, Cui R (2020) Effects of electroconvulsive therapy on depression and its potential mechanism. Front Psychol 11:80

Liu Z, Song D, Yan E, Verkhratsky A, Peng L (2015) Chronic treatment with anti-bipolar drugs suppresses glutamate release from astroglial cultures. Amino Acids 47:1045–1051

Liu SL, Sheng R, Jung JH, Wang L, Stec E, O'Connor MJ, Song S, Bikkavilli RK, Winn RA, Lee D, Baek K, Ueda K, Levitan I, Kim KP, Cho W (2017) Orthogonal lipid sensors identify transbilayer asymmetry of plasma membrane cholesterol. Nat Chem Biol 13:268–274

Maeng S, Zarate CA Jr, Du J, Schloesser RJ, McCammon J, Chen G, Manji HK (2008) Cellular mechanisms underlying the antidepressant effects of ketamine: role of alpha-amino-3-hydroxy-5-methylisoxazole-4-propionic acid receptors. Biol Psychiatry 63:349–352

Malarkey EB, Ni Y, Parpura V (2008) Ca2+ entry through TRPC1 channels contributes to intracellular Ca2+ dynamics and consequent glutamate release from rat astrocytes. Glia 56:821–835

Mauch DH, Nagler K, Schumacher S, Goritz C, Muller EC, Otto A, Pfrieger FW (2001) CNS synaptogenesis promoted by glia-derived cholesterol. Science 294:1354–1357

Mazzanti M, Haydon PG (2003) Astrocytes selectively enhance N-type calcium current in hippocampal neurons. Glia 41:128–136

Mion G, Villevieille T (2013) Ketamine pharmacology: an update (pharmacodynamics and molecular aspects, recent findings). CNS Neurosci Ther 19:370–380

Mitterauer BJ (2012) Ketamine may block NMDA receptors in astrocytes causing a rapid antidepressant effect. Front Synaptic Neurosci 4:8

Monai H, Ohkura M, Tanaka M, Oe Y, Konno A, Hirai H, Mikoshiba K, Itohara S, Nakai J, Iwai Y, Hirase H (2016) Calcium imaging reveals glial involvement in transcranial direct current stimulation-induced plasticity in mouse brain. Nat Commun 7:11100

Musazzi L, Rimland JM, Ieraci A, Racagni G, Domenici E, Popoli M (2014) Pharmacological characterization of BDNF promoters I, II and IV reveals that serotonin and norepinephrine input is sufficient for transcription activation. Int J Neuropsychopharmacol 17:779–791

Nagler K, Mauch DH, Pfrieger FW (2001) Glia-derived signals induce synapse formation in neurones of the rat central nervous system. J Physiol 533:665–679

Newport DJ, Carpenter LL, McDonald WM, Potash JB, Tohen M, Nemeroff CB, APA Council of Research Task Force on Novel Biomarkers and Treatments (2015) Ketamine and other NMDA antagonists: early clinical trials and possible mechanisms in depression. Am J Psychiatry 172:950–966

Nimmerjahn A (2009) Astrocytes going live: advances and challenges. J Physiol 587:1639–1647

Nwaobi SE, Cuddapah VA, Patterson KC, Randolph AC, Olsen ML (2016) The role of glial-specific Kir4.1 in normal and pathological states of the CNS. Acta Neuropathol 132:1–21

Ogata K, Kosaka T (2002) Structural and quantitative analysis of astrocytes in the mouse hippocampus. Neuroscience 113:221–233

Ohno Y, Hibino H, Lossin C, Inanobe A, Kurachi Y (2007) Inhibition of astroglial Kir4.1 channels by selective serotonin reuptake inhibitors. Brain Res 1178:44–51

Ohno-Iwashita Y, Iwamoto M, Ando S, Mitsui K, Iwashita S (1990) A modified theta-toxin produced by limited proteolysis and methylation: a probe for the functional study of membrane cholesterol. Biochim Biophys Acta 1023:441–448

Ongur D, Drevets WC, Price JL (1998) Glial reduction in the subgenual prefrontal cortex in mood disorders. Proc Natl Acad Sci U S A 95:13290–13295

Peng HB, Yang JF, Dai Z, Lee CW, Hung HW, Feng ZH, Ko CP (2003) Differential effects of neurotrophins and schwann cell-derived signals on neuronal survival/growth and synaptogenesis. J Neurosci 23:5050–5060

Peng L, Li B, Verkhratsky A (2016) Targeting astrocytes in bipolar disorder. Expert Rev Neurother 16:649–657

Peng L, Song D, Li B, Verkhratsky A (2018) Astroglial 5-HT2B receptor in mood disorders. Expert Rev Neurother 18(5):435–442
Perea G, Navarrete M, Araque A (2009) Tripartite synapses: astrocytes process and control synaptic information. Trends Neurosci 32:421–431
Pfenninger KH (2009) Plasma membrane expansion: a neuron's Herculean task. Nat Rev Neurosci 10:251–261
Pfrieger FW (2010) Role of glial cells in the formation and maintenance of synapses. Brain Res Rev 63:39–46
Pfrieger FW, Barres BA (1997) Synaptic efficacy enhanced by glial cells in vitro. Science 277:1684–1687
Potokar M, Stenovec M, Jorgacevski J, Holen T, Kreft M, Ottersen OP, Zorec R (2013) Regulation of AQP4 surface expression via vesicle mobility in astrocytes. Glia 61:917–928
Radley JJ, Rocher AB, Miller M, Janssen WG, Liston C, Hof PR, McEwen BS, Morrison JH (2006) Repeated stress induces dendritic spine loss in the rat medial prefrontal cortex. Cereb Cortex 16:313–320
Rajkowska G, Stockmeier CA (2013) Astrocyte pathology in major depressive disorder: insights from human postmortem brain tissue. Curr Drug Targets 14:1225–1236
Rajkowska G, Miguel-Hidalgo JJ, Wei J, Dilley G, Pittman SD, Meltzer HY, Overholser JC, Roth BL, Stockmeier CA (1999) Morphometric evidence for neuronal and glial prefrontal cell pathology in major depression. Biol Psychiatry 45:1085–1098
Ren J, Song D, Bai Q, Verkhratsky A, Peng L (2015) Fluoxetine induces alkalinization of astroglial cytosol through stimulation of sodium-hydrogen exchanger 1: dissection of intracellular signaling pathways. Front Cell Neurosci 9:61
Rivera AD, Butt AM (2019) Astrocytes are direct cellular targets of lithium treatment: novel roles for lysyl oxidase and peroxisome-proliferator activated receptor-gamma as astroglial targets of lithium. Transl Psychiatry 9:211
Sanacora G, Banasr M (2013) From pathophysiology to novel antidepressant drugs: glial contributions to the pathology and treatment of mood disorders. Biol Psychiatry 73:1172–1179
Santello M, Volterra A (2009) Synaptic modulation by astrocytes via Ca2+–dependent glutamate release. Neuroscience 158:253–259
Saranteas T, Zotos N, Lolis E, Stranomiti J, Mourouzis C, Chantzi C, Tesseromatis C (2005) Mechanisms of ketamine action on lipid metabolism in rats. Eur J Anaesthesiol 22:222–226
Schipke CG, Heuser I, Peters O (2011) Antidepressants act on glial cells: SSRIs and serotonin elicit astrocyte calcium signaling in the mouse prefrontal cortex. J Psychiatr Res 45:242–248
Schiweck J, Eickholt BJ, Murk K (2018) Important shapeshifter: mechanisms allowing astrocytes to respond to the changing nervous system during development, injury and disease. Front Cell Neurosci 12:261
Sequeira A, Mamdani F, Ernst C, Vawter MP, Bunney WE, Lebel V, Rehal S, Klempan T, Gratton A, Benkelfat C, Rouleau GA, Mechawar N, Turecki G (2009) Global brain gene expression analysis links glutamatergic and GABAergic alterations to suicide and major depression. PLoS One 4:e6585
Seri B, Garcia-Verdugo JM, McEwen BS, Alvarez-Buylla A (2001) Astrocytes give rise to new neurons in the adult mammalian hippocampus. J Neurosci 21:7153–7160
Slezak M, Pfrieger FW (2003) New roles for astrocytes: regulation of CNS synaptogenesis. Trends Neurosci 26:531–535
Stenovec M, Lasic E, Bozic M, Bobnar ST, Stout RF Jr, Grubisic V, Parpura V, Zorec R (2016) Ketamine inhibits ATP-evoked exocytotic release of brain-derived neurotrophic factor from vesicles in cultured rat astrocytes. Mol Neurobiol 53:6882–6896
Stenovec M, Bozic M, Pirnat S, Zorec R (2020) Astroglial mechanisms of ketamine action include reduced mobility of Kir4.1-carrying vesicles. Neurochem Res 45(1):109–121
Su S, Ohno Y, Lossin C, Hibino H, Inanobe A, Kurachi Y (2007) Inhibition of astroglial inwardly rectifying Kir4.1 channels by a tricyclic antidepressant, nortriptyline. J Pharmacol Exp Ther 320:573–580

Sun JD, Liu Y, Yuan YH, Li J, Chen NH (2012) Gap junction dysfunction in the prefrontal cortex induces depressive-like behaviors in rats. Neuropsychopharmacology 37:1305–1320

Takamori S, Holt M, Stenius K, Lemke EA, Gronborg M, Riedel D, Urlaub H, Schenck S, Brugger B, Ringler P, Muller SA, Rammner B, Grater F, Hub JS, De Groot BL, Mieskes G, Moriyama Y, Klingauf J, Grubmuller H, Heuser J, Wieland F, Jahn R (2006) Molecular anatomy of a trafficking organelle. Cell 127:831–846

Takano K, Yamasaki H, Kawabe K, Moriyama M, Nakamura Y (2012) Imipramine induces brain-derived neurotrophic factor mRNA expression in cultured astrocytes. J Pharmacol Sci 120:176–186

Tassonyi E, Charpantier E, Muller D, Dumont L, Bertrand D (2002) The role of nicotinic acetylcholine receptors in the mechanisms of anesthesia. Brain Res Bull 57:133–150

Theodosis DT, Poulain DA, Oliet SH (2008) Activity-dependent structural and functional plasticity of astrocyte-neuron interactions. Physiol Rev 88:983–1008

Thrane AS, Rangroo Thrane V, Zeppenfeld D, Lou N, Xu Q, Nagelhus EA, Nedergaard M (2012) General anesthesia selectively disrupts astrocyte calcium signaling in the awake mouse cortex. Proc Natl Acad Sci U S A 109:18974–18979

Ullian EM, Sapperstein SK, Christopherson KS, Barres BA (2001) Control of synapse number by glia. Science 291:657–661

Verkhratsky A, Nedergaard M (2014) Astroglial cradle in the life of the synapse. Philos Trans R Soc Lond Ser B Biol Sci 369:20130595

Verkhratsky A, Nedergaard M (2018) Physiology of astroglia. Physiol Rev 98:239–389

Verkhratsky A, Rodriguez JJ, Parpura V (2012) Calcium signalling in astroglia. Mol Cell Endocrinol 353:45–56

Verkhratsky A, Rodriguez JJ, Steardo L (2014) Astrogliopathology: a central element of neuropsychiatric diseases? Neuroscientist 20:576–588

Verkhratsky A, Matteoli M, Parpura V, Mothet JP, Zorec R (2016) Astrocytes as secretory cells of the central nervous system: idiosyncrasies of vesicular secretion. EMBO J 35:239–257

Verkhratsky A, Zorec R, Parpura V (2017) Stratification of astrocytes in healthy and diseased brain. Brain Pathol 27:629–644

Vignoli B, Battistini G, Melani R, Blum R, Santi S, Berardi N, Canossa M (2016) Peri-synaptic glia recycles brain-derived neurotrophic factor for LTP stabilization and memory retention. Neuron 92:873–887

Witcher MR, Kirov SA, Harris KM (2007) Plasticity of perisynaptic astroglia during synaptogenesis in the mature rat hippocampus. Glia 55:13–23

Won CL, Oh YS (2000) cAMP-induced stellation in primary astrocyte cultures with regional heterogeneity. Brain Res 887:250–258

Wray NH, Schappi JM, Singh H, Senese NB, Rasenick MM (2019) NMDAR-independent, cAMP-dependent antidepressant actions of ketamine. Mol Psychiatry 24(12):1833–1843

Yang C, Shirayama Y, Zhang JC, Ren Q, Yao W, Ma M, Dong C, Hashimoto K (2015) R-ketamine: a rapid-onset and sustained antidepressant without psychotomimetic side effects. Transl Psychiatry 5:e632

Yang Y, Cui Y, Sang K, Dong Y, Ni Z, Ma S, Hu H (2018) Ketamine blocks bursting in the lateral habenula to rapidly relieve depression. Nature 554:317–322

Zanos P, Gould TD (2018) Mechanisms of ketamine action as an antidepressant. Mol Psychiatry 23:801–811

Zanos P, Moaddel R, Morris PJ, Georgiou P, Fischell J, Elmer GI, Alkondon M, Yuan P, Pribut HJ, Singh NS, Dossou KS, Fang Y, Huang XP, Mayo CL, Wainer IW, Albuquerque EX, Thompson SM, Thomas CJ, Zarate CA Jr, Gould TD (2016) NMDAR inhibition-independent antidepressant actions of ketamine metabolites. Nature 533:481–486

Zhang S, Li B, Lovatt D, Xu J, Song D, Goldman SA, Nedergaard M, Hertz L, Peng L (2010) 5-HT2B receptors are expressed on astrocytes from brain and in culture and are a chronic target for all five conventional "serotonin-specific reuptake inhibitors". Neuron Glia Biol 6:113–125

Zhang Y, Wu S, Xie L, Yu S, Zhang L, Liu C, Zhou W, Yu T (2019) Ketamine within clinically effective range inhibits glutamate transmission from astrocytes to neurons and disrupts synchronization of astrocytic SICs. Front Cell Neurosci 13:240

Zorec R, Horvat A, Vardjan N, Verkhratsky A (2015) Memory formation shaped by astroglia. Front Integr Neurosci 9:56

Zorec R, Verkhratsky A, Rodriguez JJ, Parpura V (2016) Astrocytic vesicles and gliotransmitters: slowness of vesicular release and synaptobrevin2-laden vesicle nanoarchitecture. Neuroscience 323:67–75

Epilogue

Neuropsychiatric diseases have been almost universally considered from the neuron-centric point of view, with neurons being the central cellular element of pathological process. However, brain homeostasis lies at the fulcrum of healthy brain function, and its compromise invariably results in disease. Astrocytes control homeostasis of the nervous system at organ (establishment and maintenance of the blood-brain barrier), tissue (creating micro-architecture of the nervous tissue), cell (providing for neuron- and glio-genesis), and molecular (by controlling ion and neurotransmitter homeostasis) levels. These glial cells are also central for regulating nervous system microcirculation and neuronal energy support. Astrocytes are also capable of gliotransmission, i.e., releasing transmitters to communicate among themselves (homocellular signaling) and with other cell types (heterocellular signaling) residing in the nervous system. Astrocytic excitation underlying cell-cell signaling relies on excitability of these glial cells based on variations of intracellular calcium (and other) ion levels rather than electrical excitability as seen in neurons. Finally, astrocytes contribute to neural defense, and every lesion to the nervous tissue initiates evolutionary conserved and universal glial reaction manifested by reactive astrogliosis.

Multiple roles of astrocytes, which determine the progression and outcome of neuropsychiatric diseases, are emerging, and it is becoming clear that astrocytes are involved in various aspects of disease initiation, progression, and resolution. In this edited volume, we aimed to integrate the body of information that has accumulated in recent years revealing the active role of astrocytes in neuropsychiatric pathology. Understanding the roles of astrocytes in pathology will provide new targets for medical intervention and aid in the development of much needed therapeutics. By no means have we claimed that astrocytes are the sole cell type underlying psychiatric disorders. That would be a simple-minded act especially when considering the multifactorial nature of psychiatric disorders. Quite contrary, we firmly support the notion that all neural cells, as well as cells of vasculature (the neurovascular unit), with all their morpho-functional and dynamic interactions and signaling cascades,

B. Li et al. (eds.), *Astrocytes in Psychiatric Disorders*, Advances in Neurobiology 26, https://doi.org/10.1007/978-3-030-77375-5

contribute to psychiatrist pathologies. Furthermore, we are also in support of an even more holistic vision in which the brain's interaction with other body systems, such as endocrine, immune, lymphatic, gastrointestinal (e.g., the gut-brain axis), and others, could contribute to etiology and pathogenesis of psychiatric disorders.

February 18, 2021
Birmingham, AL, USA
Manchester, UK
Rome, Italy
Shenyang, P.R. China

Vladimir Parpura
Alexei Verkhratsky
Caterina Scuderi
Baoman Li

Index

B. Li et al. (eds.), *Astrocytes in Psychiatric Disorders*, Advances in Neurobiology 26, https://doi.org/10.1007/978-3-030-77375-5

R

S

Printed by Printforce, the Netherlands